CLIMATIC CHANGE: IMPLICATIONS FOR THE HYDROLOGICAL CYCLE AND FOR WATER MANAGEMENT

ADVANCES IN GLOBAL CHANGE RESEARCH

VOLUME 10

The titles in this series are listed at the end of this volume.

CLIMATIC CHANGE: IMPLICATIONS FOR THE HYDROLOGICAL CYCLE AND FOR WATER MANAGEMENT

Edited by

Martin Beniston

Department of Geosciences,
University of Fribourg, Switzerland

KLUWER ACADEMIC PUBLISHERS
DORDRECHT / BOSTON / LONDON

A C.I.P. Catalogue record for this book is available from the Library of Congress.

ISBN 1-4020-0444-3

Published by Kluwer Academic Publishers,
P.O. Box 17, 3300 AA Dordrecht, The Netherlands.

Sold and distributed in North, Central and South America
by Kluwer Academic Publishers,
101 Philip Drive, Norwell, MA 02061, U.S.A.

In all other countries, sold and distributed
by Kluwer Academic Publishers,
P.O. Box 322, 3300 AH Dordrecht, The Netherlands.

Printed on acid-free paper

Printed in the Netherlands.

Table of contents

List of contributors

ANDREINI, Marc: Center for Development Research, University of Bonn, Germany

BALIN, Ioan: Swiss Federal Institute of Technology, Lausanne, Switzerland

BENISTON, Martin: Department of Geosciences, University of Fribourg, Switzerland

BERGSTRÖM, S.: Swedish Meteorological and Hydrological Institute, Norrköping, Sweden

BOTZET, Michael: Max-Planck-Institute for Meteorology, Hamburg, Germany

BOUWER, Laurens M.: Institute for Environmental Studies, Free University, Amsterdam, The Netherlands

BRASSEUR, Olivier: Institut National Polytechnique de Grenoble, France

BRONSTERT, Axel: Institute for Geo-Ecology, University of Potsdam, and Potsdam Institute for Climate Impacts Research (PIK), Potsdam, Germany

BÜRGER, Gerd: Potsdam Institute for Climate Impacts Research (PIK), Potsdam, Germany

CALPINI, Bertrand: Swiss Federal Institute of Technology, Lausanne, Switzerland

CARLSSON, B.: Swedish Meteorological and Hydrological Institute, Norrköping, Sweden

CAYA, Daniel: University of Quebec at Montreal (UQAM), Montreal, Quebec, Canada

CHRISTENSEN, Jens Hesselbjerg: Danish Meteorological Institute, Copenhagen, Denmark

CHRISTENSEN, Ole Bøssing: Danish Meteorological Institute, Copenhagen, Denmark

CREUTIN, J.-D.: Institut National Polytechnique de Grenoble, France

FRIGON, Anne: University of Quebec at Montreal (UQAM), Montreal, Quebec, Canada

GALLEE, Hubert: Institut National Polytechnique de Grenoble, France

GALLEGO PUYOL, David: Universidad Complutense de Madrid, Spain

GARCIA HERRERA, Ricardo: Universidad Complutense de Madrid, Spain

GARDELIN, Marie: Swedish Meteorological and Hydrological Institute, Norrköping, Sweden

GIMENO PRESA, Luis: Campus as Lagoas, Universidad de Vigo, Spain

GRAHAM, L.P.: Swedish Meteorological and Hydrological Institute, Norrköping, Sweden

HASLER, Natalia: Swiss Federal Institute of Technology, Lausanne, Switzerland

HERNANDEZ MARTIN, Emiliano: Universidad Complutense de Madrid, Spain

HOFF, Holger: Potsdam Institute for Climate Impacts Research (PIK), Potsdam, Germany

IORGULESCU, I.: Swiss Federal Institute of Technology, Lausanne, Switzerland

JONES, Roger N.: CSIRO Atmospheric Research, Aspendale, Victoria, Australia

JUNG, Gerlinge: Frauenhofer Institute for Atmospheric Environmental Research, Garmisch-Partenkirchen, Germany

KRYSANOVA, Valentina: Potsdam Institute for Climate Impacts Research (PIK), Potsdam, Germany

KUNDZEWICZ, Zbigniew W.: Agricultural and Forest Environment Research Centre, Polish Academy of Sciences, Poznan, Poland

KUNSTMANN, Harald: Frauenhofer Institute for Atmospheric Environmental Research, Garmisch-Partenkirchen, Germany

KWADIJK, J. C. J.: WL/Delft Hydraulics, Delft, The Netherlands

LARCHEVEQUE, Gilles: Swiss Federal Institute of Technology, Lausanne, Switzerland

LEBEL, T.: Institut National Polytechnique de Grenoble, France

LIEBE, Jens: Center for Development Research, University of Bonn, Germany

LINDSTRÖM, G.: Swedish Meteorological and Hydrological Institute, Norrköping, Sweden

LISCHKE, Heike: Swiss Federal Institute for Forest, Snow and Landscape Research (WSL), Birmensdorf, Switzerland

LISTON, G. E.: Colorado State University, Fort Collins, Colorado, USA

LIU, Changming: Institute of Geographical Sciences and Natural Resources Research, Chinese Academy of Sciences, Beijing, China

LUDWIG, Rolf: Institute for Geography, University of Munich, Germany

LUTERBACHER, Urs: Graduate Institute of International Studies, Geneva, Switzerland

MARBAIX, Philippe: Institut National Polytechnique de Grenoble, France

MARES, Constantin: National Institute of Meteorology and Hydrology, Bucarest, Romania

MARES, Ileana: National Institute of Meteorology and Hydrology, Bucarest, Romania

MARTILLI, A.: Swiss Federal Institute of Technology, Lausanne, Switzerland

MAUSER, Wolfram: Institute for Geography, University of Munich, Germany

MAY, Wilhelm: Danish Meteorological Institute, Copenhagen, Denmark

MENZEL, Lucas: Potsdam Institute for Climate Impacts Research (PIK), Potsdam, Germany

MICHAELESCU, Mihaela: Agricultural University, Bucarest, Romania

MIDDELKOOP, Hans: Department of Physical Geography, The University of Utrecht, The Netherlands

NIEHOFF, Daniel: Potsdam Institute for Climate Impacts Research (PIK), Potsdam, Germany

PITTOCK, A. Barrie: CSIRO Atmospheric Research, Aspendale, Victoria, Australia

QUAGLIA, Philippe: Swiss Federal Institute of Technology, Lausanne, Switzerland

RIBERA RODRIGUEZ, Pedro: Campus as Lagoas, Universidad de Vigo, Spain

ROECKNER, Erich: Max-Planck-Institute for Meteorology, Hamburg, Germany

SCHLAEPFER, Rodolphe: Swiss Federal Institute of Technology, Lausanne, Switzerland

SIMEONOV, Valentin: Swiss Federal Institute of Technology, Lausanne, Switzerland

SLIVITZKY, Michel: Institut national de recherche scientifique (INRS), Quebec, Canada

TREMBLAY, Denis: Hydro-Québec, Montreal, Quebec, Canada

VAN ASSELT, M. B. A.: ICIS, Maastricht, The Netherlands

VAN DE GIESEN, Nick: Center for Development Research, University of Bonn, Germany

VAN DEN BERGH, Hubert: Swiss Federal Institute of Technology, Lausanne, Switzerland

VAN DEURSEN, W. P. A.: Carthage Consultancy, Rotterdam, The Netherlands

VELLINGA, Pier: Institute for Environmental Studies, Free University, Amsterdam, The Netherlands

VINER, David: Climatic Research Unit, University of East Anglia, Norwich, UK

VLEK, Paul L. G.: Center for Development Research, University of Bonn, Germany

VOSS, Reinhard: Max-Planck-Institute for Meteorology, Hamburg, Germany

WECHSUNG, Frank: Potsdam Institute for Climate Impacts Research (PIK), Potsdam, Germany

WIEGANDT, Ellen: Graduate Institute of International Studies, Geneva, Switzerland, and University Institute Kurt Boesch, Sion, Switzerland

ZHENG, Hongxing: Institute of Geographical Sciences and Natural Resources Research, Chinese Academy of Sciences, Beijing, China

ZIERL, Bärbel: Swiss Federal Institute for Forest, Snow and Landscape Research (WSL), Birmensdorf, Switzerland

Preface: The rationale behind this volume

MARTIN BENISTON[1], ELLEN WIEGANDT[2], AND HOLGER HOFF[3]
[1]*Department of Geosciences, University of Fribourg, Swizerland*
[2]*Graduate Institute of International Studies, Geneva, Switzerland and University Institute Kurt Boesch, Sion, Switzerland*
[3]*Potsdam Institute for Climate Impacts Research (PIK), Potsdam, Germany*

The Year-2000 edition of the Wengen Workshops on Global Change Research, which is the 6th conference in the series since 1995, was held from September 27-29 in Wengen (Bernese Alps, Switzerland). One of its primary aims was to bring together physical and social scientists to discuss some of the critical issues related to water resources in the world and, especially, how these are likely to be modified in a changing global climate in the course of the 21st Century. Over 70 participants were present throughout the three-day meeting, representing 20 different nationalities.

It is generally accepted today that a warming climate as projected by General Circulation Models will, in the course of coming decades enhance the hydrological cycle, which implies higher rates of evaporation, and a greater proportion of liquid precipitation with respect to solid precipitation. These physical mechanisms, associated with potential changes in precipitation amount and seasonality, will affect soil moisture, groundwater reserves, and the frequency of floods or droughts. Though water is present in water is present in ample quantity globally, supply is often limited regionally. The supply of water is limited and governed by the renewal processes associated with the global hydrological cycle. With the expansion of human settlements and the growth of industrial activities, water has been increasingly used in new ways, including for the assimilation and discharge of wastes. This resource has been taken for granted, and only in the past few decades has increasing water shortage and declining water quality from pollution drawn attention to the inherent fragility and scarcity of water. This has led to concerns about water availability to meet the requirements of the 21st century.

Because of increasing population and changing patterns of use, the additional demand will be accompanied by a sharp decline in water availability per capita. While a consumption of 1,000 m^3 of water per year and per capita is considered a standard for «well-being» in the industrialized world, projections of annual water availability per capita by the 2020 for North Africa are 210 m^3, 700 m^3 for Central Asia, and 1,100 m^3 for southern Asia. This trend is declining in all parts of the world, including those that are considered to have ample water resources.

Water resources will come under increasing pressure under conditions of global change. Significant changes in environmental conditions such as climate, land-use and other factors, will affect demand, supply and water quality. In regions which are currently sensitive to water stress (arid and semi-arid regions), any shortfalls in water supply will enhance competition for water use for a wide range of economic, social, and environmental applications. In the future, such competition will be sharpened as a result of larger populations, which will lead to heightened demand for irrigation and perhaps also for industrial activities, at the expense of drinking water. Disputes over water resources may well be a significant social consequence in an environment degraded by pollution and stressed by climatic change.

Against the backdrop of physical and socio-economic problems, the Workshop program was divided into 5 separate sessions, namely:

1. Sensitivity of precipitation regimes to climatic change : observations and models

2: Floods and droughts : past, present and future

3: Sensitivity of precipitation and runoff to climatic signals such as the NAO

4: Climate projections and hydrological assessments

5: Water resource management : property rights, conflicts, transboundary sharing

In addition, a full-day session was organized under the auspices of the International Geosphere-Biosphere Program (IGBP); this special session was coordinated by the IGBP-BAHC (Biological Aspects of the Hydrological Cycle) program, coordinated by the Potsdam Institute for Climate Impacts Research (PIK), Germany.

Session 1 addressed the problems and approaches allowing the use of climate information at scales typical of General Circulation Models (GCMs) at catchment or finer scales. The most commonly-adopted methods today include statistical downscaling, by which large-scale synoptic climate fields are correlated to changes in local and regional climate characteristics, and dynamical downscaling. This is a physically-based method and is thus more robust to climate change than the statistical approaches, but there is an essential need to undertake computationally-intensive multiple and multi-year simulations in order to separate noise in the system from the climate change signal. Several contributing papers focused on case studies using Regional Climate Models (RCMs) linked to hydrological models, applied to the analysis of runoff under conditions of convective activity and extreme precipitation, in regions of complex topography, or stakeholder-driven investigations such as water runoff simulations in Quebec undertaken for a major utility. Thorough analyses of GCM results for the 20th Century were reported at the Workshop, in order to illustrate the improvements in model results which have taken place in recent years, and the increasing confidence with which the models can be used for projecting climatic change in coming decades. However, there is still much room for improvement; there is also a need to address more fully the manner in which climate and impacts models (e.g., hydrological models) can be linked, in terms of consistency and the overlap between different scales, the underlying physical assumptions, and the parameterizations used.

Session 2 was devoted to the two extremes of water resources, namely floods and droughts, the focus here being to identify the climate change component in river floods. These have significant economic implications, as was shown by several scientists from Western and Central Europe. Many long time series have been studied worldwide with the aim of detection of nonstationarities, yet there is no conclusive evidence of climate-related changes in flow records, in general. Even if baseline data (with no substantial human impact) are available, detection of weak (if any) climate-related change amidst strong natural noise is very difficult. Estimation of future changes of flood risk, due to climate change, is possible in qualitative terms only. Many studies presented here point to the probable increase in precipitation and runoff in Europe, although southern and central Europe may expect extended dry periods, particularly during summer months. Droughts can occur in all regions, even those which are not commonly associated with such extremes. For example, a case study for Swiss forests was shown, illustrating the results from a forest/hydrological model pointing to regions of increasing drought stress on trees; this stress is a complex interaction between climatic, edaphic, and biological factors. In all the talks of this session, it was emphasized that there is a need for climatological data of high quality, both from the observational point of view as well as from future climate projections. In this respect, the point was raised that access to hydrological and climatological data is very often a problem, especially in Central and Eastern Europe and in the developing world, which have considerable difficulties in finding the funds for accessing data when these are not freely available. A further point of interest is the increasing public perception that extreme events are a direct consequence of global warming as, for example, the devastating 1997 floods in Germany, Poland and the

Czech Republic. This is an issue which brings to the forefront the need to communicate uncertainty and the fact that, at present, no statistically-meaningful cause-to-effect relationship can be made.

Session 3 provided an overview of the links between strong decadal-scale climatic fluctuations such as the North Atlantic Oscillation (NAO) or the El-Niño/Southern Oscillation (ENSO) on precipitation regimes and river discharges. Several case studies were presented here, such as the sensitivity of precipitation variations in the Canary Islands and Morocco (regions which are close to the centers of action of the NAO), summer moisture patterns in Romania and Eastern Europe, and river discharges in Central Europe to changes in the NAO. For example, a strong shift occurs in the maximum discharge of the Vistula River in Poland during periods when the NAO index (the surface pressure difference between the Azores High and the Iceland Low) is high (approx. 3,000 m^3/s) compared to periods when the NAO index is low (approx. 4,700 m^3/s). High NAO values are associated with strong zonal flow over the North Atlantic and storm tracks which cross Europe at more northerly latitudes than when the NAO index is low.

Session 4 addressed hydrological assessments for the future, and the requirements in terms of climate projections and other data. The problem of scale is a dominant feature of such assessments, in that it is necessary to link GCM data to hydrological models, as was mentioned in the overview presentations of Session 1. In several examples given in Session 4, advanced multiple "cascade" methods (i.e., the processes linking models at one scale to those at a finer scale) have sought to couple global to regional climate models, and on down to hydrological models at the very fine spatial scales. Further down the "cascade" sequence, one study made use of a crop model to assess the potential change in the agricultural output of the Elbe River basin in response to changing conditions of hydrological stress. All presentations in this session emphasized the uncertainties involved in applying the methodologies developed; these are related to each step in the cascade process, whereby errors transferred from one model may amplify or contaminate results of the model at finer scales. It was shown in many of the studies that changes in extremes yield the greatest uncertainties.

Session 5 contained a number of cross-cutting themes, which were also taken up in the special IGBP/IHDP/International Programs session. Much of the genuinely interdisciplinary work was presented here, as opposed to the other four sessions, essentially devoted to sectoral and technical issues. Session 5 and the special session addressed a number of important topics, in particular:

• *VULNERABILITY OF WATER RESOURCES*

Vulnerability assessments of water resources integrate multiple interacting stresses and feedbacks from climate change, climate variability and direct effects of human activities, such as changes in land and water use, changes in distribution and age-profiles of populations, changes in economic activities and settlement patterns, and changes in political environments. There is a strong human dimensions component in vulnerability assessments. Exposure to environmental stresses and reaction to such exposure are major factors, but also very important is the potential for adaptation, which is essentially region-specific. This was demonstrated for losses due to extreme flood events, where developing countries sometimes suffer severe loss of life, while losses in developed countries are mostly economic. A goal of integrated water resource management is to reduce vulnerabilities.

• *INTEGRATION*

Integrated research now has quite a number of tools at its disposal, such as coupled models, use of remote sensing data, exchange of data between different research components, etc. The full integration of human dimensions was seen to be essential to water research, e.g., for understanding the driving forces underlying change, including behavioral and institutional aspects.

• *UNCERTAINTY*

Uncertainty is an inherent component for climate simulations and downscaling techniques, as well as climate variability and runoff analyses. There exist parallel uncertainties about the evolution of economic growth and changes in preferences. These uncertainties are then introduced into various predictive and impacts models and hence into water management applications, such as design and operation of dams, adaptation of agriculture, and allocation of water for economic sectors. Similarly, projections of future water use as well as economic assessments, e.g., of damage and adaptation costs of changing water availability, have considerable uncertainty built into them.

• *COMMUNICATION*

Although research has the ultimate aim of reducing uncertainty, there is a need to communicate existing uncertainty beyond the scientific community. Communication between scientists, the general public, policy makers, water resource managers and further participation of stakeholders is essential throughout all phases of research, including from the beginning of research planning in order to define relevant research questions, and also later on in order to apply research results, in particular to water resources management. Communication should make uncertainties part of the knowledge base for decision making. A number of ways were identified for communicating uncertainties, such as scenarios and alternatives for decision makers, results of risk assessments (to be performed jointly with stakeholders) and probabilities of exceeding critical thresholds, whereby stakeholders must be included in the definition of these critical thresholds.

• *EXTREME EVENTS*

Extreme events such as droughts and floods may well become the dominating effects of climate change in water resources. Currently a trend of increasing economic losses due to extreme flood events is observed. It is not clear at this stage, whether there is an underlying trend in frequency and intensity of extreme flood events and if this may be attributed to climate change, or whether there is only an increase in exposed economic value. Hence a systematic and comprehensive database of past flood events is required. This is an area for collaboration with the private sector, in particular insurance and re-insurance companies.

• *CONFLICT*

Groundwater and surface waters are common-pool or shared resources, often shared by more than one nation. This raises the question of who owns the water, of access and of property rights. Furthermore, water is often viewed as a strategic resource (for water supply, for food production, and for energy and industrial use), and hence economic and scientific arguments may only play a minor role in decision making. Property rights are often ill-defined for common pool resources or manifest very unequal access rights. Moreover, existing agreements on property rights and water resource planning often do not take into account climate and hydrological variability / trends. There is consequently the potential for conflict over shared water resources. Presentations were made in this sense, focusing on the Tigris-Euphrates conflict between Turkey and Syria, and the Jordan River Basin resources shared between Israel and its Arab neighbors. The need to achieve cooperative solutions was stressed, in order to maximize the joint benefit of resource use ("win-win strategies"). Unequal water use can be a fundamental cause for conflict between countries or between regions within countries. These agreements should be established proactively, and not in response to emergency situations. Cost efficient solutions, that also incorporate equity issues, may increasingly involve the private sector in water supply and management.

• *REGIONAL STUDIES*

A number of integrated regional studies are now underway for various river basins in different parts of the world. Changes in land and water use are generally recognized to be of more immediate and direct relevance to water resources than changes or variability in climate. As a general principle, changes in runoff are much stronger than the underlying changes in precipitation, because of complex interactions at and beneath the surface, as well as direct human factors which interact with the natural system. Furthermore, teleconnections are increasingly found between changes in land and water use in one region, and effects upon water resources in another region, sometimes spanning several countries. The links are generated mostly through atmospheric or river transport processes.

A final plenary session addressed the future needs of the scientific community to help address the issues raised and discussed at the Wengen-2000 Workshop. Among the most pressing points, the participants identified the following:

- Better focus on the vulnerability of systems
- Need to live with uncertainty, but improve its analysis and communication. Emphasize the aspects which are known today with a reasonable degree of confidence, such as the agreement on the sign of change in many regions as projected by climate models
- Refine the problems of scaling requirements, from the global to the local scales
- Convergence of approaches between the climate/hydrological sciences and the social and economic sciences, e.g., aggregation of individual decision making in social sciences, downscaling vs upscaling, feedback mechanisms. Steps towards genuine integration.

- Inventory of existing regional studies on floods and drought for past, present and future climate
- Address the problem of signal to noise ratio in the future, i.e., climate variability vs climatic change
- It was further suggested that:
- less well studied regions should, where possible, benefit from transfer of results (example: urban-rural coupling);
- results from regional studies could be used in global assessments of water resources, including the identification of particularly vulnerable and critical regions in the future

Changes in the mean and extremes of the hydrological cycle in Europe under enhanced greenhouse gas conditions in a global time-slice experiment

WILHELM MAY[1,*], REINHARD VOSS[2] AND ERICH ROECKNER[2]
*Corresponding author
[1]Danish Meteorological Institute, Copenhagen, Denmark
[2]Max-Planck-Institute for Meteorology, Hamburg, Germany

Abstract: The response of the hydrological cycle in Europe to the increasing atmospheric concentrations of the important greenhouse gases is investigated on the basis of a time-slice experiment. The time-slice experiment has been performed with the ECHAM4 atmospheric general circulation model at a high horizontal resolution of T106 for two time-slices representing the present-day climate (1970-1999) and the future climate after an effective doubling of the carbon dioxide concentration in the atmosphere (2060-2089). The changes in the greenhouse gas concentrations have been prescribed according to the Intergovernmental Panel on Climate Change scenario IS92a. Both changes in the mean state and changes in the extremes of the hydrological cycle are considered.

The time-slice experiment predicts an overall intensification of the hydrological cycle in Europe in the future, affecting both the atmospheric and the ground components. The magnitude of the changes as well as their geographical distribution varies, however, in the course of the year. The magnitude of the changes is largest in autumn, and the regions with an increase or decrease in the respective component are shifted in the meridional direction. In winter precipitation is, for instance, increased in the entire European area except for southern Europe and the south-eastern part of the Atlantic, while it is reduced in the entire area except for northern Scandinavia and the northern part of the Atlantic in summer. The frequency as well as the intensity of heavy precipitation events are generally enhanced in the future, also in those areas, where the mean amount of precipitation is reduced, such as southern Europe. As for the river discharge, the annual mean values are increased in northern Europe but decreased in central and southern Europe, depending on the kind of changes in precipitation and evaporation in the

M. Beniston (ed.),
Climatic Change: Implications for the Hydrological Cycle and for Water Management, 1–29.

respective catchments. The extremes of the river discharge are, however, reduced in all of Europe due to either a reduction of the snow depth (northern Europe), a rather strong increase in evaporation (central Europe) or a relatively strong decrease in precipitation (southern Europe).

1. INTRODUCTION

During the last decade it has become apparent that the uncertainties in assessing the effects of global scale perturbations to the climate system are primarily due to an inadequate understanding of the hydrological cycle, that is the cycling of water in the oceans, atmosphere and biosphere (e.g. Chahine 1992). The hydrological cycle influences climate in various ways. The exchange of moisture and heat between the atmosphere and the earth's surface, for instance, affects the dynamics and the thermodynamics of the climate system. In the forms of vapour, clouds, liquid, snow and ice, as well as during phase transitions water plays opposing roles in heating and cooling the environment. About 50% of the surface cooling results, for instance, from evaporation. Besides, water vapour in the atmosphere acts as an essential greenhouse gas, and clouds control the climate by altering the earth's radiation budget.

Given these various roles and the complexity of the exchange processes, the hydrological cycle has to be treated as a whole in order to cover its full behaviour and its intricate non-linear feedback loops. Further, the hydrological cycle has to be captured fairly realistically in general circulation models (GCMs) in order to obtain not only a realistic simulation of the present-day climate but also a reliable prediction of the future climate by these models. It are in particular the non-linear feedback loops including the water vapour in general and clouds in particular that account for a large part of the uncertainties of the predictions of the future climate based on climate model simulations.

Such predictions of the future climate, which consider the effects of the anticipated increase of the atmospheric concentrations of numerous greenhouse gases, generally reveal a warming of several degrees near the ground by the end of the 21st century (e.g. Houghton et al. 1996). This warming trend is accompanied not only by a general, that is on the global scale and for the temporal mean state, intensification of the hydrological cycle, there also are indications of an increase in the probability of extremes associated with the hydrological cycle such as flooding, drought and heavy precipitation events. Recently, Zwiers and Kharin (1998) as well as Kharin and Zwiers (2000) found, for instance, an increase of precipitation extremes in most parts of the globe due to the increasing greenhouse gas concentrations. Some other model simulations have revealed an enhanced probability of droughts in the land regions of the Northern Hemisphere midlatitudes for the future (Gregory et al. 1997, Wetherald and Manabe 1999).

Regional climate in general and precipitation in particular are, however, often affected by processes on spatial scales that are considerably smaller than the scales typically resolved by the coupled climate models (e.g. Giorgi and Mearns 1991), which are on the order of 300 to 500 km. In order to obtain information about the climate on these small scales three methods, also called "regionalization" techniques are currently used, namely a) high resolution and variable resolution so-called "time-slice" simulations with atmospheric general circulation models (AGCMs), b) dynamical downscaling with regional climate models (RCMs) and c) statistical downscaling methods. The main advantage of the dynamical downscaling approach (e.g. Giorgi et al. 1992, Marinucci and Giorgi 1992, Jones at al. 1997) is that a much higher horizontal resolution, presently of the order of a few tens of km, can be used than in the coupled atmosphere ocean models (AOGCMs), since the model domain is considerably smaller. The main limitations of this approach are that the regional information is only available in a particular region of interest and, more fundamentally, that interactions between the regional and the global scales are excluded. For the time-slice approach (e.g. Bengtsson et al. 1995, Cubasch et al. 1995, Déqué et al. 1998) these arguments are reversed. Scale interactions between the regional and the global scales are included, but the horizontal resolution, currently about 100 km, is considerably coarser than for RCMs, since the entire globe is covered. One limitation of both of these methods is, however, that coupled atmosphere-ocean modes cannot develop in these atmosphere-only models. The third method, namely statistical downscaling (e.g. von Storch et al. 1993, Wilby and Wigley 1997), is quite different, since it does not require any additional modelling but statistical methods are used for empirically deriving relationships between the regional climate and the large-scale circulation. The crucial question is, however, whether these empirical relationships, which are derived for the present-day climate, are also valid for the future climate.

In our study we use simulations based on the time-slice approach (May and Roeckner 2001) in order to investigate the possible changes in the hydrological cycle caused by the anticipated increase in the atmospheric concentrations of the important greenhouse gases in the European area. We consider both the changes in the mean state and the changes in the extremes associated with the hydrological cycle such as heavy precipitation events, flooding and drought. The investigation includes not only those components of the hydrological cycle, which typically are considered in climate research such as precipitation, evaporation and atmospheric moisture, but also the soil water content and river discharge, which usually are regarded as part of the hydrological cycle in the classical surface hydrology framework. By this we are able to obtain a very comprehensive picture of the possible changes of the water cycle in Europe, which may have major environmental and socio-economic impacts on the area.

The paper is organised as follows: In Section 2 we describe the time-slice experiment, on which our study is based. Subsequently we present the changes in the different components of the hydrological cycle under the enhanced greenhouse conditions considering both the atmospheric (Section 3 and Section 4) and the ground components (Section 5 and Section 6). A summary and the conclusions follow in Section 7.

2. TIME-SLICE EXPERIMENT

The model employed in the time-slice experiment is the ECHAM4 AGCM (e.g. Roeckner et al. 1996a). The model has been developed at the Max-Planck-Institute for Meteorology for simulating the global present-day climate and a possible global change in climate due to enhanced emissions of greenhouse gases. It is based on the global forecasting system that is used at European Centre for Medium-Range Weather Forecasts, but several major changes have been made, in particular to the physical parameterisations in order to make the model suitable for climate simulations. Details on the climate statistics of ECHAM4 in its "standard" configuration at a horizontal resolution of T42 (corresponding to 64×128 grid points on a Gaussian grid) and 19 vertical levels can be found in Roeckner et al. (1996a).

In the time-slice experiment we have used ECHAM4 with a horizontal resolution of T106 (corresponding to 160×320 grid points on a Gaussian grid) and 19 vertical levels. In ECHAM4 most of the free parameters in the physical parameterisations are independent of resolution (but note they have been selected at a horizontal resolution of T42) instead of tuning the model by choosing an optimal set of parameters at the respective resolution. Only the parameterisations of a few processes, which turned out to be extremely scale dependent, have been tuned individually at different horizontal resolutions, such as the parameterisations of gravity wave drag, horizontal diffusion and formation of precipitation in stratiform clouds (e.g. Stendel and Roeckner 1998). Thus generally speaking the high resolution does not automatically reduce all the systematic model errors, since some of the physical parameterisations, which have been tuned at the lower resolution, are possibly scale dependent. Nevertheless the high resolution leads to a more realistic simulation of the present-day climate by ECHAM4 due to the inclusion of a much wider spectrum of spatial scales and hence the non-linear interactions between them and a more realistic representation of the topography (Stendel and Roeckner 1998).

We have performed two simulations with ECHAM4 for the high resolution over a period of 30 years each. These two so-called time-slices have been chosen so that one 30-year period represents the present-day climate and the other one the climate at a time when the atmospheric concentration of CO_2 has doubled. The first time-slice covers the period 1970 through 1999 and the second one the period 2060 through 2089. In each of these time-slices the lower boundary forcing, i.e., monthly mean values of the sea surface temperatures (SSTs) and of the sea-ice extent and sea-ice thickness has been prescribed as obtained from a climate change simulation performed with a coupled atmosphere-ocean model run with low horizontal resolution. Moreover, the temporal evolution of the concentrations of the important greenhouse gases has been prescribed in the same way as in that climate change simulation (see below). Further details on the experimental design of our time-slice experiment are given in May (1999), and a thorough discussion of the changes in the mean climate inferred from these two time-slices can be found therein as well as in May and Roeckner (2001).

The climate change experiment has been performed at Max-Planck-Institute for Meteorology. The coupled atmosphere-ocean model consists of the ECHAM4 AGCM with a horizontal resolution of T42 and 19 vertical levels and an extension (level 3) of the OPYC ocean model (Oberhuber 1993) including a sea-ice model. OPYC has 11 layers and a varying horizontal resolution: poleward of 36° latitude the resolution is identical to that of the low resolution AGCM, that is circa 2.8°. At low latitudes, the meridional grid spacing is gradually decreased to 0.5° at the equator in order to allow for a better representation of the equatorial wave guide in the model and, hence, the El Niño/Southern Oscillation phenomenon. For further details on the coupling technique and the performance of the model we refer to Roeckner at al. (1996b) and Bacher et al. (1998).

This coupled model has been used for a control experiment and in the meantime for three different time-dependent forcing experiments which are thoroughly described in Roeckner et al. (1999). At the time, when we decided to perform our time-slice experiment, only the greenhouse gas experiment covering the period 1860 through 2100 (referred to as "GHG" in Roeckner et al. (1999)) was available, so that we could only use this particular climate change experiment for extracting the boundary forcing for our time-slice experiment. In that climate change simulation the concentrations of various gases have been prescribed as a function of time: the greenhouse gases CO_2, CH_4, N_2O and several industrial gases such as chlorofluorocarbons (CFC-11, 12, 113, 114, 115), hydrochlorofluorocarbons (HCFC-22, 123, 141b), hydrofluorocarbons (HFC-125, 134a, 152a), carbon tetrachloride (CCl_4) and methylchloroform (CH_3CCl_3). From 1860 to 1990 the annual mean concentrations of these gases have been prescribed as observed and after 1990 according to the Intergovernmental Panel on Climate Change (IPCC) scenario IS92a (Houghton et al. 1992). For the industrial gases the IS92a-scenario has been updated to be consistent with a "Copenhagen-like" emission scenario (Houghton et al. 1996). Ozone and aerosols, on the other hand, have been prescribed in the same way as in the control run, that is as climatological distributions for the present-day conditions.

3. PRECIPITATION

3.1 Mean state

In this section we investigate the changes in the seasonal mean values of daily precipitation in the European area due to the increase in the atmospheric concentrations of the important greenhouse gases. The seasons are defined as consecutive 3-month periods including December, January and February (DJF; winter), March, April and May (MAM; spring), June, July and August (JJA; summer) as well as September, October and November (SON; autumn). As for the quality of the simulation of precipitation in the Atlantic/European area in the time-slice experiment we refer to May (2001), where the simulation of precipitation in the time-slice experiment under present-day conditions was evaluated with respect to different observational data sets. It was found that the high horizontal resolution of the model used in the time-slice experiment allowed for a realistic simulation of the amount of precipitation in the vicinity of the major mountain ranges. Further, the amount of precipitation originating from individual cyclones was in quite good agreement with observations. A problem was, however, the relatively strong underestimation of precipitation on the European continent during summer.

Figure 1 shows in addition to the changes in the seasonal mean values of daily precipitation for the different seasons (b-e) the annual mean values in the European area under present-day conditions (a). The latter panel has been included in order to give an idea of the mean distribution of precipitation. According to this, the time-slice experiment predicts in all seasons an increase in precipitation over the northern part of the Atlantic/European area and a decrease in the southern part for the future.

Further, the amount of precipitation is enhanced over the western part of the Atlantic in all seasons but summer (Fig. 1d). The response pattern shifts, however, its location in the course of the year. During winter we find, for instance, an increase in precipitation over most of the European continent and a reduction only in southern Europe and the Mediterranean area (Fig. 1b). In summer, on the other hand, the rainfall is considerably reduced in most of Europe, and only in northern Scandinavia is the amount of precipitation enhanced (Fig. 1d). In spring (Fig. 1c) and autumn (Fig. 1e) the boundary between the northern part of the region with an increase and the southern part with a decrease in precipitation is located at about 50° N. The time-slice experiment predicts the strongest changes in precipitation in autumn (Fig. 1e). In particular the increase in precipitation in the vicinity of the Norwegian west coast and over the northeastern part of the Atlantic is so pronounced that in the future the mean amount of precipitation in this area is higher in autumn than in winter. But also the reduction of precipitation over the southwestern part of the Atlantic and in the eastern Mediterranean area is relatively strong in autumn.

3.2 Daily variability

In this section we investigate, how the intensity of individual precipitation events changes in the future. Especially changes in the heavy precipitation events may lead to substantial damages and, hence, have marked environmental and socio-economic impacts.

The distributions of precipitation intensity have been examined by means of gamma distributions, which have been fitted to the time series of daily precipitation of both time-slices separately. The gamma distribution, which is suitable for positively constrained variables, provides a comprehensive description of the distribution by a set of two parameters. The probability density function is by

$$f(x) = \frac{(x/\beta)^{\alpha-1}\exp(-x/\beta)}{\beta\Gamma(\alpha)}$$

where α denotes the shape parameter, β the scale parameter and Γ the gamma function. For both the argument x and the parameters α and β only positive values are allowed, i.e., x, α, $\beta > 0$. The product $\alpha\cdot\beta$ gives the mean of the converges to a Gaussian distribution. The gamma distributions have been fitted using the maximum likelihood method (e.g. Wilks 1995). Only days with precipitation exceeding 0.01 mm/day are considered.

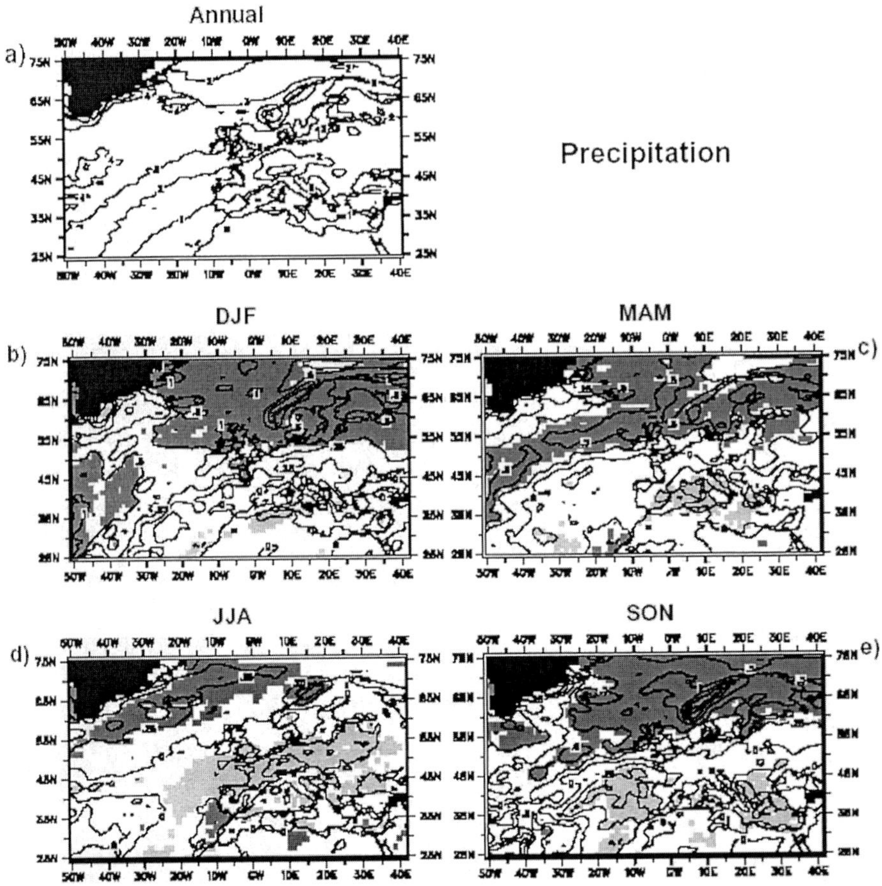

Figure 1. Annual mean daily precipitation for the simulation of the present-day climate (a) as well as the changes in the seasonal mean daily precipitation between the simulations for the future and the present-day climate for different seasons: b) winter (December, January and February), c) spring (March, April and May), d) summer (June, July and August) and e) autumn (September, October and November). Units are [mm/day]. The contour interval is 1.0 mm/day (a) and 0.5 mm/day (b-e). For the differences also the ±0.25 contour lines are given. Also the significance of the differences according to a two-sided t-test is indicated (b-e). Positive (negative) difference exceeding a significance level of 97.5% are marked by the dark (light) shading. Those grid points, where the elevation exceeds 1500 m, are blanked out.

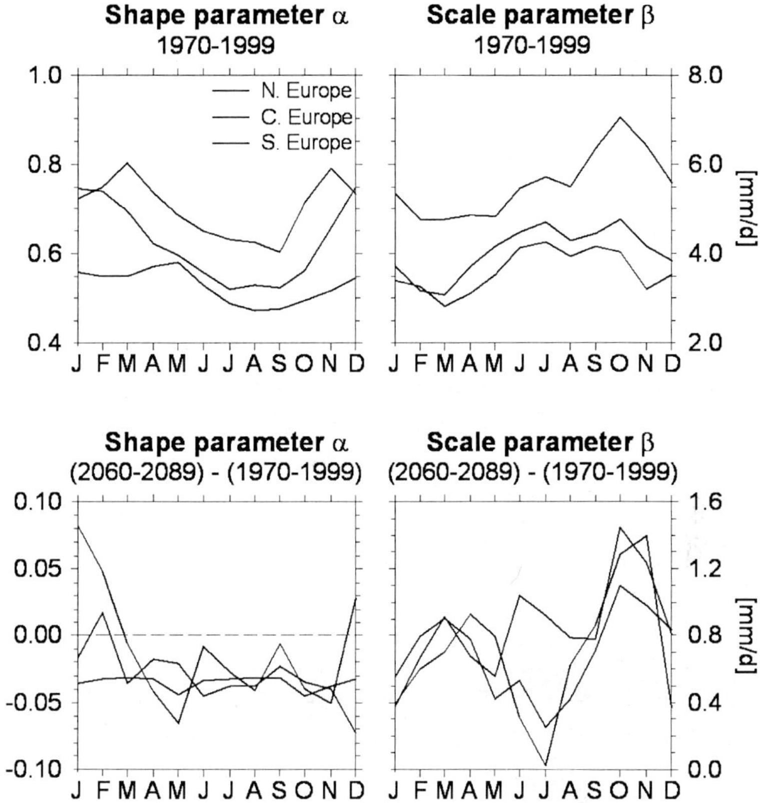

Figure 2. Shape parameter α and scale parameter β of the gamma distribution (see text) of daily precipitation for the simulation of the present-day climate (upper panels) as well as the changes between the simulations for the future and the present-day climate (lower panels) distinguishing between different areas (Table 1).

In order to depict the results for the two parameters determining the distribution of daily precipitation, we have divided Europe into three regions depending on the geographical latitude distinguishing between northern, central and southern Europe (Table 1), and we consider only those grid points that correspond to land areas. Figure 2 shows the estimates of the two parameters obtained from the time-slice for the present-day climate and the changes of these parameters associated with the changes in the daily variability of precipitation due to the increase of the greenhouse gas concentrations. Here we distinguish between the different calendar months.

Table 1. Definition of the different areas under consideration (see Section 3.2). Only those grid points that correspond to land areas are included.

Region	Latitude range	Longitude range
Northern Europe	60°N - 70°N	10°W - 45°E
Central Europe	50°N - 60°N	10°W - 45°E
Southern Europe	35°N - 50°N	10°W - 45°E

In northern Europe the highest values occur in spring and autumn, in central Europe in winter and in southern Europe in late spring. That means a) that in southern Europe the distribution is more strongly skewed to the right, i.e. the probability of heavy precipitation events is higher than in central and northern Europe (where the probability is generally smallest) and b) that the probability of heavy events is rather high in summer and reduced in the other seasons, such as in winter in central Europe. The shape parameter changes only moderately (about 5%) in the future, and the changes are generally negative. That means that the probability of heavy precipitation events is generally enhanced in the future climate. The only marked exception is found in northern Europe in winter, where the shape parameter increases, i.e. the probability of heavy events decreases. But this increase in the shape parameter is not so strong that it exceeds a value of 1, so that the character of the distribution (skewed to the right) is preserved.

The scale parameter β, on the other hand, decreases from the north to the south for the present-day climate, but the scale parameter is markedly larger in southern Europe than for the rest of the continent. The values are generally highest in summer and autumn, but in southern Europe the maximum in autumn is substantially stronger than the one in summer. That means a) that for similar values of α heavy precipitation events are more intense in southern Europe than in the other parts of the continent and b) that heavy events are rather intense in summer and autumn. As for the future climate, we find a general increase in the scale parameter, indicating an intensification of heavy precipitation events. In northern and central Europe rather strong increases occur in spring and autumn, while in southern Europe relatively strong increases are found in all seasons but winter peaking in March, July and October, respectively. In all of the three regions the strongest increases occur, however, in autumn, namely in October or November.

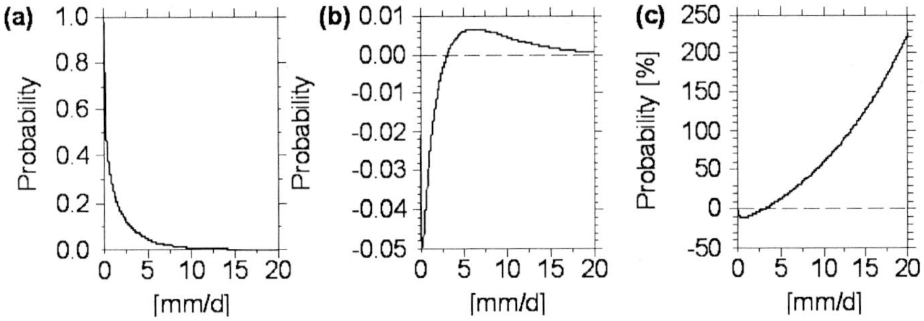

Figure 3. Probability density functions of daily precipitation for northern Europe in April with the values of $\alpha = 0.74$; $\beta = 3.1$ mm/day for the present-day climate and $\alpha = 0.69$; $\beta = 4.0$ mm/day for the future climate: a) distribution for the simulation of the present-day climate, b) change in the distribution between the simulations for the future and the present-day climate and c) change in the distribution between the simulations for the future and the present-day climate relative to the present-day climate.

The consequences of the aforementioned changes in the two parameters for the precipitation distribution are illustrated in Figure 3, where the probability density functions are shown for Northern Europe in April. This is a quite typical example, since general characteristics are also valid for the other parts of Europe and other months. As for the simulation of the present-day climate, the parameters have values of 0.74 for α and of 3.1 mm/day for β. Figure 3a reveals the common distribution of daily precipitation with light precipitation events occurring frequently and heavy precipitation events occurring rarely. In the future climate with a decrease of α by 0.05 and an increase of β by 0.9 mm/day, heavy precipitation events occur more frequently at the expense of relatively light events (Fig. 3b). Even though in this particular example the increase in the probability of heavy precipitation events exceeds 200% for events exceeding 20 mm/day in the future climate (Fig. 3.c), the probability for such heavy events is still quite small in the future. The increase in the probability of heavy precipitation events is mainly accounted for by the increase of the scale parameter, while the decrease in the shape parameter is of minor importance in this respect. This result has actually also been found in observations. Groisman et al. (1999) found, for instance, in their investigation of summer precipitation that the observed changes in the variation of daily precipitation during the 20th century could by and large be described by changes in scale parameter β.

The geographical distribution of the precipitation originating from heavy precipitation events is illustrated by the 95% percentile of daily precipitation. For the present-day climate maxima of the 95% percentile occur in the vicinity of the major mountain ranges (Fig. 4). On the Norwegian west coast, to the north of the Alps as well as on the Balkans the 95% percentile exceeds, for instance, a value of 15 mm/day. The climatic change due to the enhanced greenhouse gas concentrations leads to an increase of the 95% percentile on the entire European continent, and only over the Mediterranean Sea a reduction can be found. The strongest increase of up to 4 mm/day corresponding to 25% of the present-day values occurs in southwestern Norway.

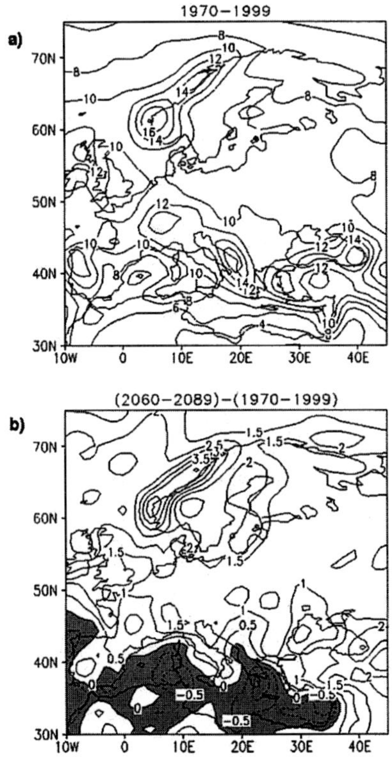

Figure 4. 95% percentile of daily precipitation for the simulation of the present-day climate (a) and b) the change in the 95% percentile of daily precipitation between the simulations for the future and the present-day climate. Units are [mm/day].

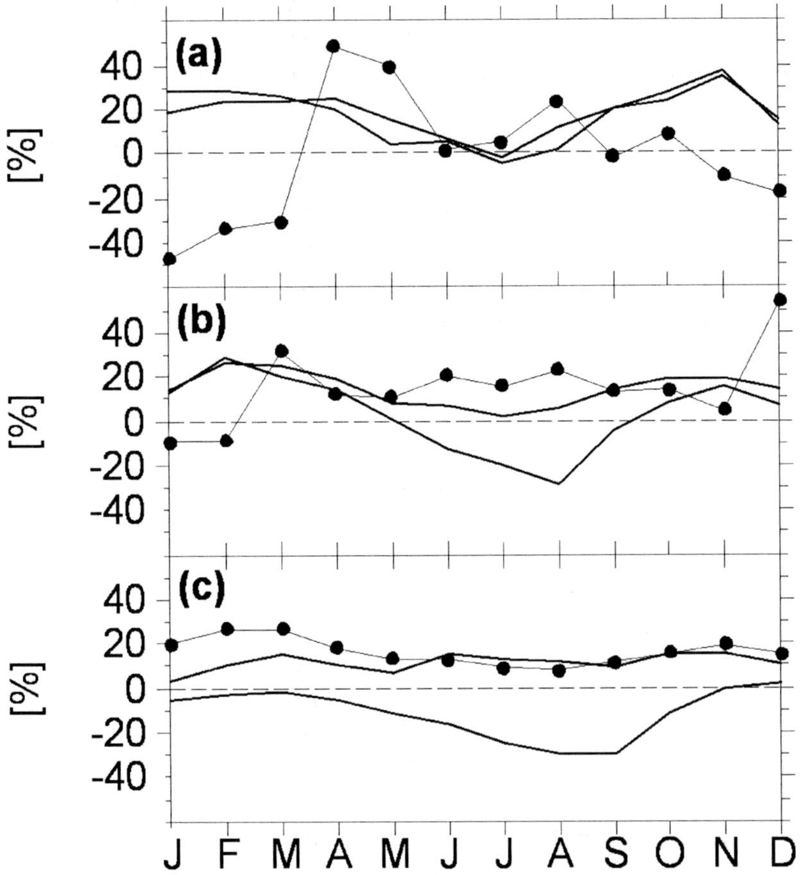

Figure 5. Changes between the simulations of the future and the present-day climate relative to the present-day climate for a) the total daily precipitation (solid line), b) the 95% percentile of daily precipitation (dashed line) and c) the number of dry days (dotted line).

When we distinguish between the different calendar months, the time-slice experiment reveals an increase in the 95% percentile and, hence, in the amount of precipitation associated with heavy precipitation events throughout the entire year (Fig. 5). In northern and central Europe the increase varies, however, between 20-30% in spring and autumn, respectively, and 0% in July, while in southern Europe the increase varies only slightly in the course of the year. In this region relative maxima occur in March, July and October in accordance with the relatively strong increase in the scale parameter β during these months (see Fig. 2). Apparently, during parts of the year the amount of precipitation originating from heavy precipitation events increases though the mean amount of precipitation decreases. In southern Europe this is the case throughout the entire year, in central Europe in summer and in northern Europe in July.

A meteorological variable connected with the change in the 95% percentile relative to the change in the mean amount of precipitation is the number of dry days, defined as days with less than 0.01 mm of precipitation. Whenever the increase in the 95% percentile is larger than in the mean precipitation, which actually may be reduced, the number of dry days is increased in the future climate and vice versa (Fig. 5). In southern Europe the number of dry days is, for instance, enhanced during the entire year, particularly in late winter, while in northern Europe the number of dry days increases during the warm season between April and October. In winter the number of dry days is actually reduced by up to more than 40%. Along with number of dry days the duration of extreme dry spells changes in the future (Voss et al. 2001). On the basis of general extreme value distributions (see Section 6) they found significant extensions of the length of dry spells by 10 to 20 days for the 10-year return values in central and southern Europe, respectively.

Other modelling studies have revealed results similar to our main findings, such as a study by McGuffie et al. (1999) based on a set of experiments with AGCMs coupled to mixed-layer oceans. In a study using a coupled climate model Gregory et al. (1997) investigated changes in dry spells in southern Europe in summer. In addition to a reduction of the mean precipitation, they found an increase in the likelihood of long dry spells. Cubasch et al. (1995) also found an extension of dry spells in the land areas of the Northern Hemisphere midlatitudes. In agreement with this, Wetherald and Manabe (1999) found indications for longer dry spells investigating the soil moisture changes in a set of coupled simulations.

4. EVAPORATION AND ATMOSPHERIC MOISTURE

In this section we investigate the changes in the seasonal mean evaporation in the European area associated with the climatic changes due to the increase in the atmospheric concentrations of the important greenhouse gases. These changes affect, in turn, the local source term of atmospheric moisture and the atmospheric moisture transport.

4.1 Evaporation

The time-slice experiment simulates relatively strong evaporation over the oceans and relatively weak evaporation over the land areas under present-day conditions (Fig. 6a). Over the oceans evaporation is strongest in the southern regions such as the southern part of the Atlantic, the Mediterranean Sea and the Black Sea and weakest to the north of Iceland, where the sea is covered with ice during the cold season (May 1999). By this the distribution of evaporation as simulated in the time-slice experiment for the present-day climate is in general agreement with observational data sets such as the NCEP/NCAR reanalyses (Trenberth and Guillemot 1998).

As for the change in evaporation, we find a general increase over the oceans in all seasons due to the general warming of the SSTs in the Atlantic and the Mediterranean Sea (May and Roeckner 2001). Rather strong increases occur over the northern Atlantic in association with a relatively strong warming of the SSTs in that area, primarily due to the retreat of the sea-ice to the north of Iceland caused by the future warming (Figs. 6b, c). In summer and autumn evaporation decreases over most of the European continent (Figs. 6d, e), while during winter and spring evaporation increases over all but southern Europe (Figs. 6b, c). Apparently the pattern of the change in evaporation over the land areas varies in a similar way in the course of the year as the changes in precipitation (see Fig. 1). This reflects the overall changes in the availability of water accounted for by the increased (winter and spring) or reduced (summer and autumn) amount of precipitation, respectively, in the future. Similar to the changes in precipitation (see Figs. 1b, e), the increase in evaporation over the northeastern part of the Atlantic and northern Europe as well as the increase over the southeastern part of the Atlantic and the decrease in southern Europe are stronger in autumn (Fig. 6e) than in winter.

4.2 Source of atmospheric moisture

The difference between evaporation and precipitation describes the source term of atmospheric moisture. When evaporation exceeds precipitation in a particular region, moisture is added to the atmosphere and when precipitation exceeds evaporation, moisture is withdrawn from the atmosphere. The geographical distribution of the difference between evaporation and precipitation reveals sources of atmospheric moisture over the oceans in the southern part of the region, i.e. over the southern part of the Atlantic, the Mediterranean Sea and the Black Sea and sinks over the northern part of the Atlantic, the Baltic Sea as well as over the land areas (Fig. 7a). This distribution is the consequence of the rather strong evaporation over the oceans in the southern part of the area and the relatively weak evaporation over the land areas (see Fig. 6a) on one hand and the relatively strong precipitation over the northern part of the area (see Fig. 1a) on the other. In particular the areas with a high amount of precipitation in the vicinity of the major mountain ranges stand out as sinks of atmospheric moisture.

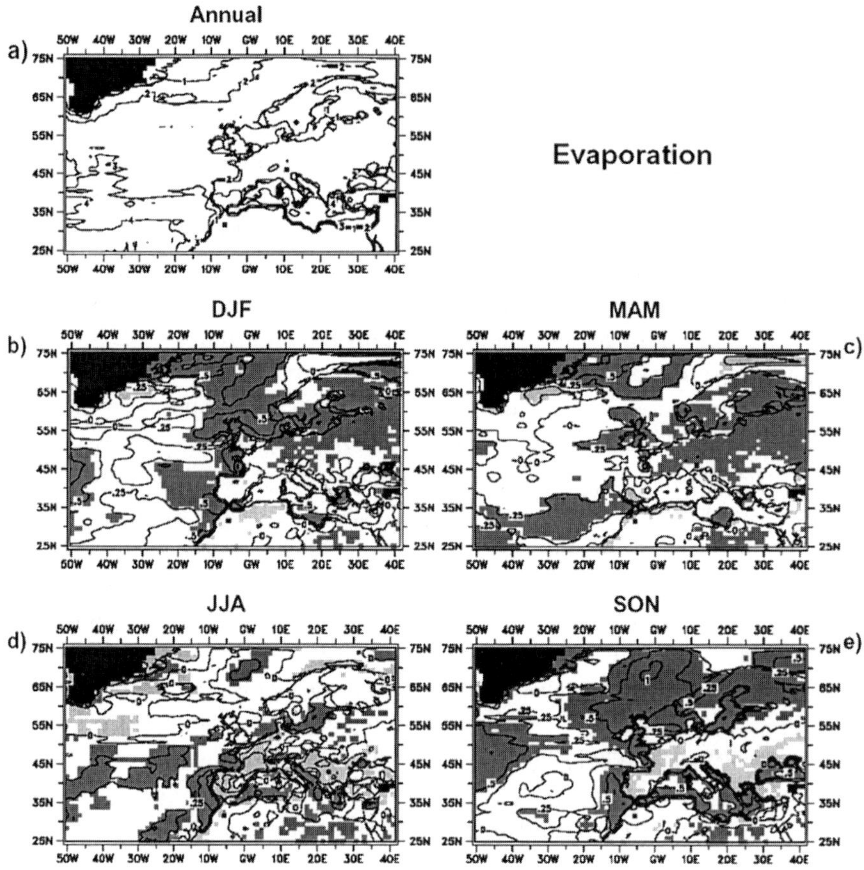

Figure 6. As Fig. 1, but for daily evaporation.

When we consider the change in the atmospheric moisture in the future climate, two effects need to be considered. Firstly, the general warming of the atmosphere due to the increase in the greenhouse gas concentrations (May and Roeckner 2001) leads to an enhancement of the specific humidity and, hence, of the atmospheric moisture content. Secondly, the changes in the difference between evaporation and precipitation associated with the climatic changes caused by the increase in the greenhouse gas concentrations may either lead to an enhancement or a reduction of the atmospheric moisture content. It is the latter effect that we investigate in this section. The changes in evaporation and precipitation, respectively, in the future lead by and large to a reduction of atmospheric moisture over the northern part of the area and to an enhancement over the southern part (Fig. 7). The variation of the change in the source term of atmospheric moisture in the course of the year is dominated by the seasonal variation of the change in precipitation, in particular in the northern part of the area (see Fig. 1). In the southern part, on the other hand, where rather strong changes in evaporation have been found (see Fig. 6), the relatively strong increase in evaporation in autumn (see Fig. 6e) leads, for instance, to a very strong source of moisture over the southeastern Atlantic and the Mediterranean Sea (Fig. 7e). Considering the rather strong increase in precipitation along the Norwegian west cost in this season (Fig. 1e) we, therefore, find the strongest changes in the source term of atmospheric moisture in autumn. In summer, on the other hand, the stronger reduction in evaporation than in precipitation in western and southern Europe is associated with a reduction in the source term of atmospheric moisture in these regions (Fig. 7d).

4.3 Atmospheric moisture transport

The atmosphere transports water vapour into those areas, where precipitation exceeds evaporation, and away from the areas with an excess in evaporation. The time-slice experiment therefore reveals the strongest atmospheric moisture fluxes over the western Atlantic in connection with the Atlantic storm track and the excessive precipitation originating from the intense cyclones in this area (Fig. 8a). The fluxes are directed eastward and reach far into the European continent, leading to an excess of precipitation in Europe. Over the southern part of the Atlantic we find, on the other hand, an anticyclonic flow pattern transporting moisture away from the area with excessive evaporation. By this the distribution of the atmospheric moisture transport as simulated in the time-slice experiment under present-day conditions is in good agreement with the estimates obtained from the NCEP/NCAR reanalyses (Trenberth and Guillemot 1998).

The changes in precipitation and evaporation associated with the climatic change are going along with changes in the atmospheric moisture transport (Figs. 8b-e). In accordance with a general reduction of the source term of atmospheric moisture in the northern part of the Atlantic/European region and an increase in the southern part (see Figs. 7b-e), the time-slice experiment reveals an intensification of the eastward moisture fluxes in the northern part of the area and a weakening in the southern part. The increase over the northeastern part of the Atlantic and northern Europe is strongest in autumn (Fig. 8e), when also the increase in precipitation in the respective area is strongest. In western and central Europe the change in the atmospheric moisture transport is rather small in summer (Fig. 8d) due to the merely slight change in the source term of atmospheric moisture in this part of Europe.

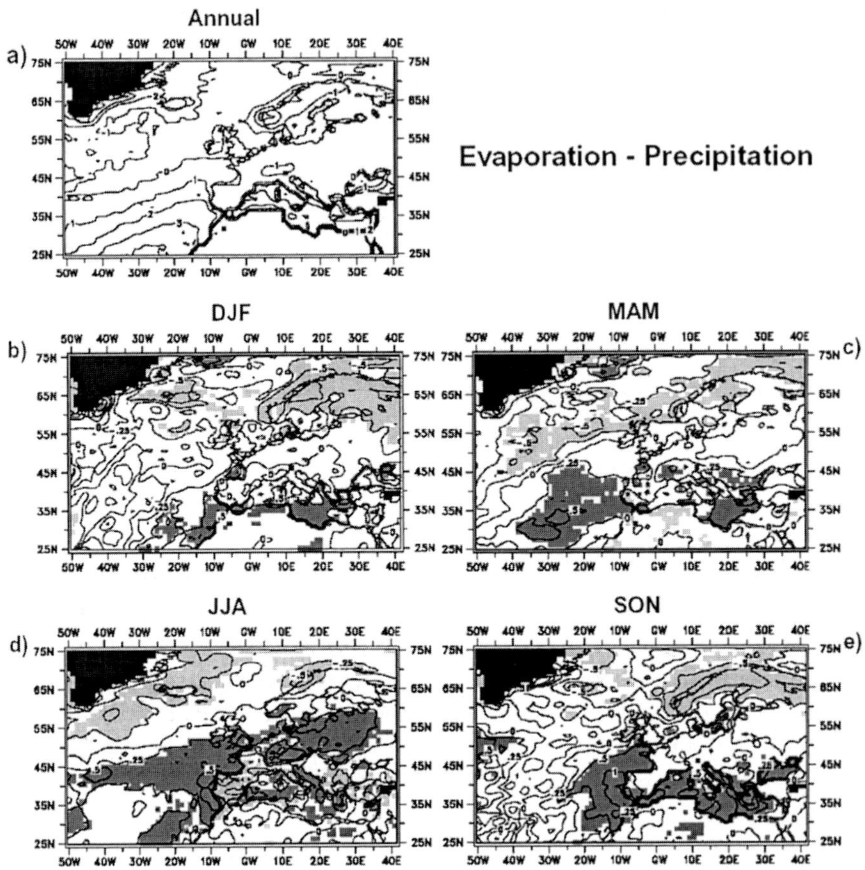

Figure 7. As Fig. 1, but for the source term of atmospheric moisture defined as the difference between evaporation and precipitation.

5. SOIL MOISTURE

For the present-day climate the time-slice experiment reveals the highest soil water content of about 400 mm in a band between about 45 and 55° N including the British Isles, central and eastern Europe (Fig. 9a). Values of about 200 mm occur in southern and northern Europe except for Norway, where the soil water content is about 100 mm due to the high alpine character of this country. The simulation of soil moisture depends crucially in the characteristics of the GCM, in particular on the parameterisations of the soil processes and of the fluxes in the atmospheric boundary layer. Therefore, an assessment of the quality of the simulated soil water content would be beyond the scope of this study. But given an underestimation of precipitation in Europe in the warm season (May 2001), the simulated soil water content certainly is too small. It is presumably the limited storage capacity of water in the ground that actually causes the unrealistic drying of the ground in the beginning of the warm season and, hence, the underestimation of precipitation.

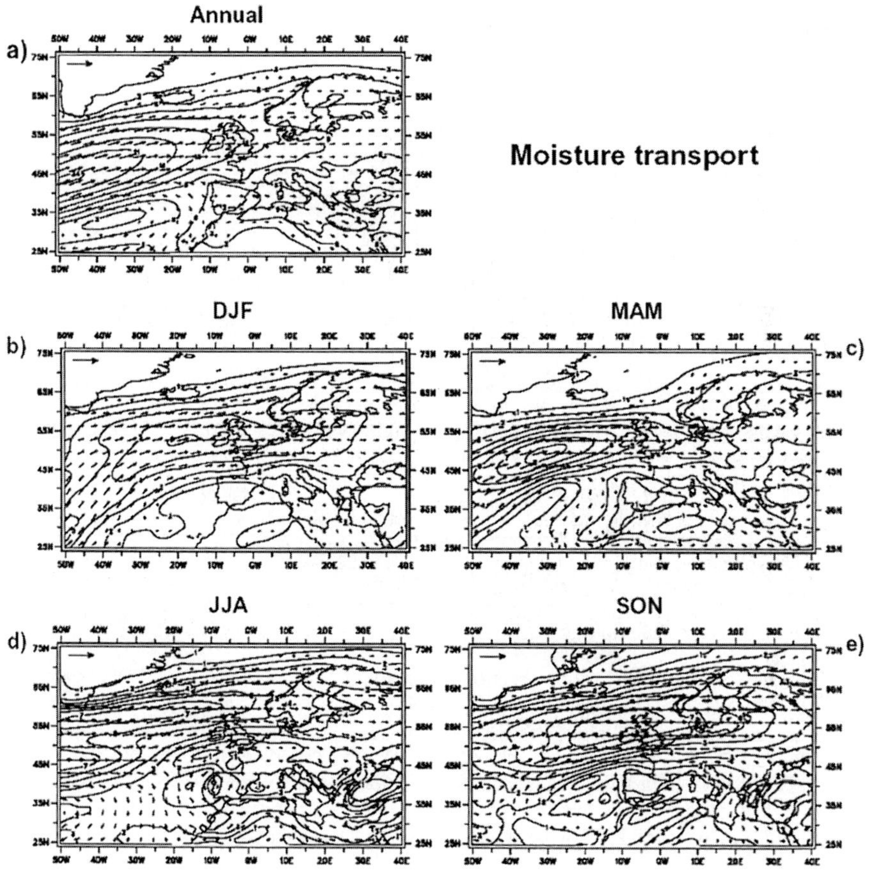

Figure 8. As Fig. 1, but for the vertically integrated atmospheric moisture transport. The arrows represent the actual transports and the contour lines their strength. Units are [10 kg/(m·s)]. The arrows in the upper left corner represent 420 kg/(m·s) (a) and 140 kg/(m·s) (b-e), respectively.

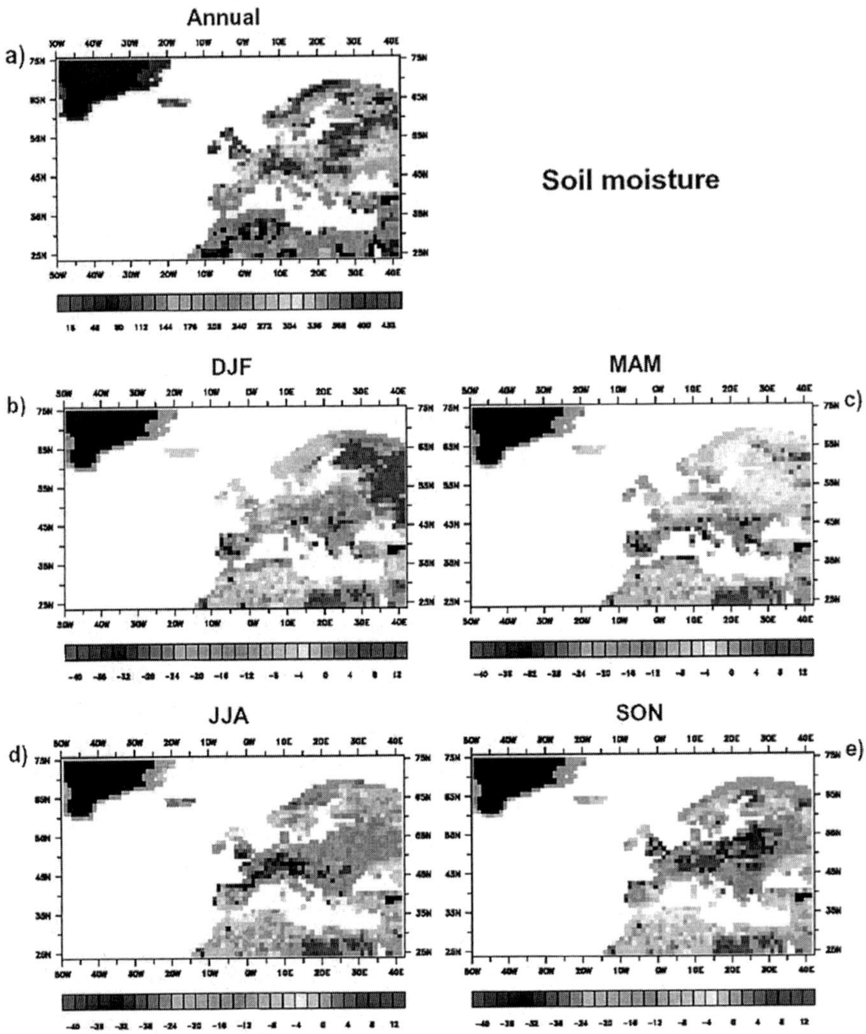

Figure 9. As Fig. 1, but for the soil moisture content. Units are [mm H2O].

In winter we find a reduction of the soil water content in all regions but eastern Europe (Fig. 9b). A comparison with the respective distribution for the source term of atmospheric moisture (see Fig. 7b) reveals that the expected relationship with negative changes in the difference between evaporation and precipitation corresponding to positive changes in the soil moisture and vice versa is not valid for northwestern Europe. Apparently the additional amount of precipitation in these regions cannot be stored in the ground, while the increase in evaporation due to the warmer temperatures leads to a loss of soil moisture. This behaviour could be due to a change in the distribution of daily precipitation with more heavy precipitation events.

In this case less water would be stored in the ground and more water would run into the rivers, as the ground is not able to store the abundant volume of water that becomes available for a very short period. Another less physical, though, explanation could be that the ground is not able to store the excessive amount of precipitation because it is already saturated under present-day conditions. Given the fact that the probability of heavy precipitation events in northern Europe increases only slightly in winter (see Fig. 2), the change in the soil moisture we have obtained may be an artifact of the model. A similar change in the soil water content in northwestern Europe is found in spring (Fig. 9c), but in this season the probability of heavy precipitation events actually is considerably increased. In western and southern Europe the soil water content is reduced in summer and autumn (Figs. 9d, e), though the change in the difference between evaporation and precipitation is negative (see Figs. 7d, e), again possibly due to the higher probability of heavy precipitation events in the future (see Fig. 2).

6. RIVER DISCHARGE

Changes in precipitation, evaporation and soil moisture, which have been described in the previous sections, affect the river discharge. Here, we focus on both the changes in the annual mean values and the changes in the annual maximum of the river discharge caused by the predicted climatic changes. Both may have considerable environmental and socio-economic impacts.

6.1 Discharge model

In order to investigate the changes in the river discharge in detail, a global hydrological discharge model has been applied (Hagemann and Dümenil 1998). The model distinguishes between the base, the overland and the river flow. The latter two components are represented by a cascade of linear reservoirs, while the base flow is modelled by a single reservoir. The discharge model is forced by daily values of surface runoff and drainage provided by the AGCM. The surface runoff is part of the overland flow and the drainage part of the base flow, while the import from adjacent areas feeds into the river flow. All three flow components contribute to the outflow of a certain grid box. Anthropogenic impacts like dams, reservoirs or irrigation are not considered in the discharge model. The performance of the hydrological discharge model is investigated in Hagemann and Dümenil (1998) for the global scale and in Hagemann and Dümenil (1999) for the Baltic Sea region. For some rivers the discharge model leads, however, to estimates of the river discharge that deviate considerably from observations.

These deviations are either due to the aforementioned anthropogenic influences, which are not considered in the AGCM, deficiencies of the AGCM or deficiencies of the discharge model itself. Nevertheless, the application of the discharge model provides valuable indications of the possible future changes in the river discharge in Europe.

6.2 Annual means

In order to illustrate the regional response of the river discharge to the increase in the greenhouse gas concentrations we have selected a number of representative rivers in Europe (Fig. 10). As for the long-term means, the discharge reflects the balance between precipitation and evaporation in the respective catchment. Hence, we can distinguish between three different categories of characteristic changes in the river discharge due to the different kinds of changes in precipitation and evaporation, respectively, in the different parts of the continent (Table 2). These three categories actually represent three latitudinal bands, as the changes in precipitation and evaporation depend on the geographical latitude (see Figs. 1 and 6).

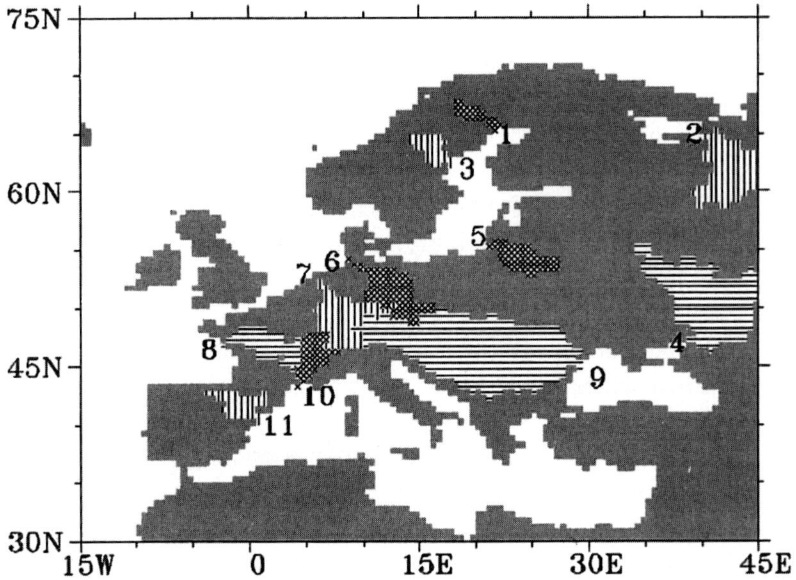

Figure 10. **Selected European river catchments (see Table 2 for the numbering).**

Table 2. Changes in the annual mean and maximum (expressed via the 10-year return values) river discharge for selected European river catchments (see Fig. 10). Stars denote a significant change for the maximum discharge (see Section 6.3).

			Annual mean		10-year return value	
			[m³/s]	[%]	[m³/s]	[%]
$\Delta P > 0$ $\Delta E > 0$ $\Delta P > \Delta E$	1	Luleaelven	115	31	-575*	-33*
	2	Northern Dvina	556	18	-6248*	-29*
	3	Angermanaelven	127	23	-270	-5
	4	Don	224	19	655	11
$\Delta E > 0$ $\Delta P < \Delta E$	5	Neman	-30	-6	-606*	-25*
	6	Elbe	-70	-9	-406	-17
	7	Rhine	-182	-11	-159	-3
$\Delta P < 0$ $\Delta E < 0$ $\Delta P < \Delta E$	8	Loire	-90	-17	-440	-20
	9	Danube	-969	-27	-3779*	-32*
	10	Rhone	-130	-15	-401	-11
	11	Ebro	-33	-25	-58	-12

As for the first category (rivers no. 1-4), both precipitation and evaporation increase in the future with the increase in precipitation exceeding the increase in evaporation in the respective catchments, which leads to an enhancement of the annual mean river discharge. Except for the Don, the rivers in this category are located in northern Europe. For the second category of rivers in central Europe (rivers no. 5-7) evaporation is increased, while the changes in precipitation are only minor, leading to a reduction of the mean discharge. Also for the third category of rivers in southern Europe (rivers no. 8-11) the mean discharge is reduced in the future. But in southern Europe both precipitation and evaporation decrease, and since the decrease in evaporation exceeds the decrease in precipitation in the respective catchments, the mean discharge is reduced.

6.3 Annual maxima

In order to investigate the impact of the changes in climate due to the increase in the concentrations of the important greenhouse gases on the maximum discharge occurring within the course of the year, general extreme value (GEV) distributions have been fitted to the maxima of the individual rivers. The maxima of the river discharge with a long return period of several years or more can then be associated with flooding. Zwiers and Kharin (1998) and Kharin and Zwiers (2000) applied this method to investigate changes in different kinds of extreme events (but not the river discharge) in simulations with AGCMs as well as with AOGCMs under increasing CO_2 concentrations. Here we have followed their strategy and apply it to the river discharge. Further details concerning the theoretical background and the fitting technique are given in the aforementioned studies as well as in Voss et al. (2001).

The parameters of the GEV distribution have been estimated by the sample L-moments (Hosking 1990), and for the estimation of the uncertainties and, hence, the significance of the changes a parametric bootstrap method has been used (Efron 1982). For a given return value the 5% and 95% percentiles of the fits are used as a measure of uncertainty, and the change is considered to be significant, if the ranges of the two distributions that are defined by these upper and lower bounds do not overlap.

In Table 2 the changes in the 10-year return values of the annual maximum river discharge are given. These are the maxima of the river discharge, which are likely to occur once within a period of 10 years. According to this, only four out of the 11 selected rivers reveal a significant change in the 10-year return values, namely the Luleaelven and Northern Dvina, the Neman and the Danube. For those rivers the return values are reduced by about 30% in the future. As for the Luleaelevn an the Northern Dvina, the maximum discharges are reduced significantly though the annual mean discharges are enhanced by 31% and 18%, respectively. For the Neman and the Danube, on the other hand, both the maximum and mean discharge are reduced. The only river, which is characterized by a non-significant, though, increase in the maximum discharge, is the Don.

The reasons for the behaviour of the Luleaelven and the Northern Dvina can be seen in Figure 11. Both the mean discharge and the annual maximum discharge are associated with the snow melt in spring. In spite of the increase of precipitation in winter, the accumulated snow mass decreases due to the warming induced by the enhanced greenhouse gas concentrations. Hence, the snow melt in spring is reduced and therewith the spring flooding becomes weaker in the warmer climate. Moreover, the peak discharge occurs earlier in the course of the year. The considerable increase in precipitation and the moderate increase in evaporation in winter lead, on the other hand, to the enhancement of the annual mean discharge. This strong impact of the changes in the snow depth on the discharge leads also to the significant reduction in the return values of the other rivers. This behaviour has also been found in other studies analysing AOGCM simulations (Arnell 1999, Arora and Boer 2001) and in an observational study for the United States (Groisman et al. 2001). A significant impact of the increased frequency of heavy precipitation events (see Section 3.2) on the maximum discharge can, however, not be found.

As for the river discharge, the comparison with observations reveals very good agreement for both the timing and the magnitude of the maximum discharge in late spring in connection with the snow melt. In summer and autumn the discharge is, however, underestimate in the time-slice experiment, reflecting the underestimation of precipitation in the warm season (May 2001).

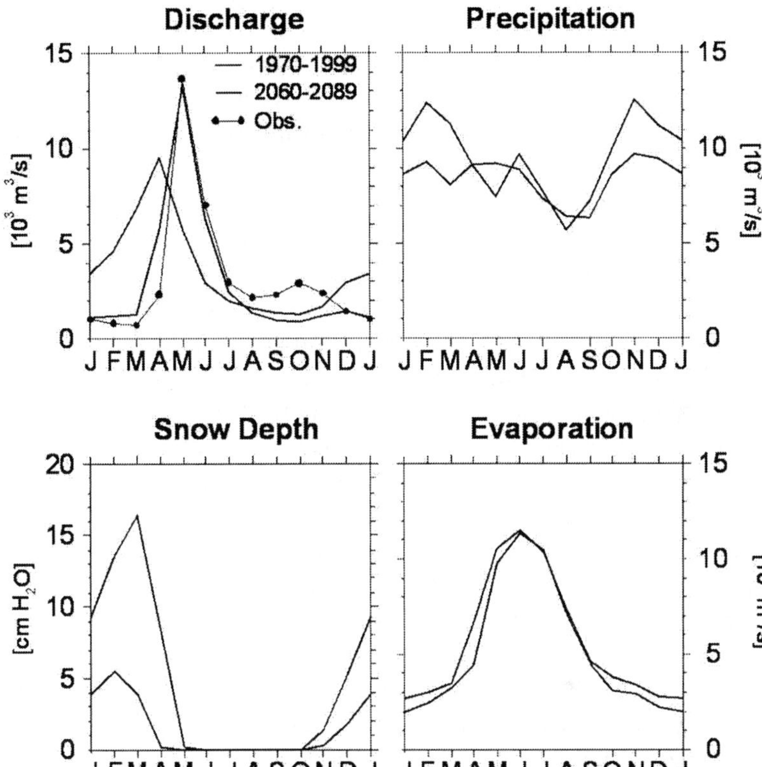

Figure 11 Monthly mean estimates of the river discharge, precipitation, snow depth and evaporation in the Northern Dvina catchment (see Fig. 10) for the simulations of the present-day (solid line) and the future climate (dashed line). For the river discharge also observations are given (thin solid line).

7. SUMMARY AND CONCLUSIONS

In this paper we have investigated the response of the hydrological cycle in Europe to the increasing atmospheric concentrations of the important greenhouse gases on the basis of a time-slice experiment. The time-slice experiment has been performed with the ECHAM4 atmospheric general circulation model at a high horizontal resolution of T106. It consists of two 30-year time-slices, one representing the present-day climate (1970-1999) and the other the future climate after an effective doubling of the carbon dioxide concentration in the atmosphere (2060-2089). We have considered both changes in the mean state and changes in the extremes of the hydrological cycle.

The time-slice experiment predicts an overall intensification of the hydrological cycle in Europe in the future. As for precipitation, we find an increase over the northern part of the Atlantic/European area and a decrease over the southern part. The response pattern shifts, however, its location in the meridional direction in the course of the year. In winter precipitation is increased in the entire European area except for southern Europe and the southeastern Atlantic, while is reduced in all of Europe except northern Scandinavia and the northern part of the Atlantic in summer. Moreover, the local changes in precipitation are most pronounced in autumn. For the daily variability of precipitation the time-slice experiment reveals a general increase in both the frequency and intensity of heavy precipitation events in most of the area except for the Mediterranean Sea and the southern part of the Atlantic. In some areas such as southern Europe, the intensification of heavy precipitation events occurs despite a reduction of the mean amount of precipitation.

As for evaporation, we find a general increase over the oceans due to the warming of the SSTs in the Atlantic and the Mediterranean Sea. Over most of the European continent evaporation is decreased in summer and autumn, but increased in winter and spring. By this the response pattern for evaporation over the land areas varies in a similar way in the course of the year as the change in precipitation, reflecting the overall changes in the availability of water associated with the respective changes in precipitation. Similar to precipitation, also for evaporation the local changes are most pronounced in autumn. The changes in precipitation associated with the climatic changes go along with changes in the source term of atmospheric moisture (defined as the difference between evaporation and precipitation) and the atmospheric moisture transport, as the atmosphere transports water vapour into those areas, where precipitation exceeds evaporation, and away from the areas with an excess in evaporation.

Also the ground components of the hydrological cycle are affected by the changes in precipitation and evaporation, such as the soil moisture content and the river discharge. Negative changes in the difference between evaporation and precipitation lead, for instance, to a surplus of water, which either is stored in the ground or transported into and by the rivers. As for the change in the mean river discharge, we can distinguish between three different categories. The rivers in the respective categories are located in different latitudinal bands, since the changes in precipitation and evaporation depend by and large on the geographical latitude. In northern Europe the time-slice experiment reveals an increase in the mean discharge due to a stronger increase in precipitation than in evaporation. In central and southern Europe we find, on the other hand, a reduction of the mean discharge. In central Europe this is caused by a rather strong increase in evaporation but in southern Europe by a relatively strong decrease in precipitation. Regarding the maxima of the river discharge or flooding, we find a reduction of up to about 30% for all but one (Don) of the catchments considered. In northern Europe the decrease is due to a reduction of the snow depth in association with the general warming trend and, hence, a reduction of the snow melt in spring. In central and southern Europe the decrease is accounted for by the aforementioned changes in precipitation and evaporation in the respective catchments.

The aforementioned changes in the hydrological cycle associated with the climatic change associated with the anticipated increase in the greenhouse gas concentrations have environmental and socio-economic impacts in the European area. In southern Europe the reduction of precipitation throughout the year will cause serious problems for the water supply in general and for agriculture in particular, which is very vulnerable due to its dependency on irrigation. The elongation of droughts in this region will worsen the living conditions, in particular in summer in combination with the warmer temperatures. In the other parts of Europe, the reduction of precipitation in the growing season is also likely to affect agriculture negatively unless different crops that require less water are cultivated. The dryness in southern Europe is, however, accompanied by an intensification of heavy precipitation events. The dryness will make the soil more vulnerable, so that the more frequent heavy precipitation events will cause serious erosion. In Norway the marked increase in heavy precipitation events may also cause erosion, but this country, which strongly depends on hydroelectric power, is likely to profit from the increase in mean precipitation. Another advantageous effect of the climatic change is the reduction of severe flooding in most of Europe and, hence, the reduction of the damages caused by these extremes of the river flow. These "advantages" of the predicted changes in the hydrological cycle are, however, unlikely to compensate for the negative effects in the other areas or for the other components of the water cycle. Therefore, we conclude on the basis of the results presented here that the future changes in the hydrological cycle will predominantly have negative environmental and socio-economic impacts in Europe.

8. ACKNOWLEDGEMENTS

This work was supported by the "Environment and Climate Programme" under contract ENV4-CT97-0640, "An enhanced resolution modelling study on anthropogenic climate change".

9. REFERENCES

Arnell NW (1999) Climate change and global water resources. Glob Env Change 9: S31-S49

Arora VK, Boer GJ (2001) The effects of simulated climate changes on the hydrology of major river basins. J Geophys Res 106: 3335-3348

Bacher A, Oberhuber JM, Roeckner E (1998) ENSO dynamics and seasonal cycle in the tropical Pacific as simulated by the ECHAM4/OPYC3 coupled general circulation model. Clim Dyn 14: 431-450

Bengtsson L, Botzet M, Esch M (1995) Hurricane-type vortices in a general circulation model. Tellus 47A: 175-196

Chahine, MT (1992) The hydrological cycle and its influence on climate. Nature 359: 373-380

Cubasch U, Waskewitz J, Hegerl G, Perlwitz J (1995) Regional climate changes as simulated in time-slice experiments. Clim Change 31: 273-304

Déqué M, Marquet P, Jones R (1998) Simulation of climate change over Europe using a global variable resolution general circulation model. Clim Dyn 14: 173-189

Efron B (1982) The Jackknife, the bootstrap and other resampling plans. SIAM: 92 pp

Giorgi F, Marinucci MR, Visconti G (1992) A 2xCO2 climate change scenario over Europe generated using a limited area model nested in a general circulation model. 2. Climate change scenario. J Geophys Res 97: 10011-10028

Giorgi F, Mearns LO (1991) Approaches to the simulation of regional climate change: A review. Rev Geophys 29: 191-216

Gregory JF, Mitchell JFB, Brady AJ (1997) Summer drought in northern midlatitudes in a time-dependant CO2 experiment. Clim Change 43: 495-511

Groisman, PY, Karl TR, Easterling DR, Knight RW, Jamason PF, Hennessy KJ, Suppiah R, Page CM, Wibig J, Fortuniak K, Razuvaev VN, Douglas A, Forland E, Zhai PM (1999) Changes in the probability of heavy precipitation: Important indicators of climate change. Clim Change 42: 243-283

Groisman PY, Knight RW, Karl TR 2001: Heavy precipitation and streamflow in the United States: Trends in the twentieth century. Bull Amer Met Soc 82: 216-246

Hagemann S, Dümenil L (1998) Development of a parameterization of lateral discharge for the global scale. Clim Dyn 14: 17-31

Hagemann S, Dümenil L (1999) Application of a global discharge model to atmospheric model simulations in the BALTEX region. Nordic Hydrology 30: 209-230

Hosking JRM (1990) L-moments: Analysis and estimation of distributions using linear combinations of order statistics. J Roy Statist Soc B52: 105-124

Houghton JT, Callandar BA, Varney SK (eds.) (1992) Climate Change 1992: The supplementary report to the IPCC scientific assessment. Cambridge, 198 pp

Houghton JT, Meira Filho LG, Callandar BA, Harris N, Kattenberg A, Maskell K (eds.) (1996) Climate Change 1995. The science of climate change. Cambridge University Press, 572 pp

Jones RG, Murphy JM, Noguer M, Keen AB (1997) Simulation of climate change over Europe using a nested regional model. II: Comparison of driving and regional model responses to a doubling of carbon dioxide. Quart J Roy Met Soc 123: 265-292

Kharin V, Zwiers F (2000) Changes in the extremes in the ensemble of transient climate simulations with a coupled atmosphere-ocean GCM. J Climate 13: 3760-3788

Marinucci MR, Giorgi F (1992) A 2xCO2 climate change scenario over Europe generated using a limited area model nested in a general circulation model. 1. Present-day seasonal climate simulation. J Geophys Res 97: 9989-10009

May W (1999) A time-slice experiment with the ECHAM4 A-GCM at high resolution: The experimental design and the assessment of climate change as compared to a greenhouse gas experiment with ECHAM4/OPYC at low resolution. DMI Scientific Report 99-2: 93 pp

May W (2001) The impact of horizontal resolution on the simulation of seasonal climate in the Atlantic/European area for present and future times. Clim Res 16: 203-223

May W, Roeckner E (2001) A time-slice experiment with the ECHAM4 AGCM at high resolution: The impact of horizontal resolution on annual mean climate change. Clim Dyn 17: 407-420

McGuffie K, Henderson-Sellers A, Holbrook N, Kothavala Z, Balachova O, Hoekstra J (1999) Assessing simulations of daily temperature and precipitation variability with global climate models for present and enhanced greenhouse climates. Int J Climatol 19: 1-26

Oberhuber JM (1993) Simulation of the Atlantic circulation with a coupled sea-ice - mixed layer - isopycnal general circulation model. Part I: Model description. J Phys Oceanogr 22: 808-829

Roeckner E, Arpe K, Bengtsson L, Christoph M, Claussen M, Dümenil L, Esch M, Giorgetta M, Schlese U, Schulzweida U (1996a) The atmospheric general circulation model ECHAM-4: Model description and simulation of present-day climate. MPI-Report No. 218: 90 pp

Roeckner E, Bengtsson L, Feichter J, Lelieveld J, Rodhe H (1999) Transient climate change simulations with a coupled atmosphere-ocean GCM including the tropospheric sulfur cycle. J Climate 12: 3004-3032

Roeckner E, Oberhuber JM, Bacher A, Christoph M, Kirchner I (1996b) ENSO variability and atmospheric response in a global coupled atmosphere-ocean GCM. Clim Dyn 12: 737-754

Stendel M, Roeckner E (1998) Impacts of horizontal resolution on simulated climate statistics in ECHAM4. MPI-Report No. 253: 57 pp

Trenberth KE, Guillemot CJ (1998) Evaluation of the atmospheric moisture and hydrological cycle in the NCEP/NCAR reanalyses. Clim Dyn 14: 213-231

von Storch H, Zorita E, Cubasch U (1993) Downscaling of global climate change estimates to regional scales: An application to Iberian rainfall in wintertime. J Climate 6: 1161-1171

Voss R, May W, Roeckner E (2001) Enhanced resolution modelling study on anthropogenic climate change: Changes in extremes of the hydrological cycle. Int J Climatol, submitted

Wetherald RT, Manabe S (1999) Detectability of summer dryness caused by greenhouse warming. Clim Change 43: 495-511

Wilby RL, Wigley TML (1997) Downscaling general circulation model output: A review of methods and limitations. Prog Phys Geography 21: 530-548

Wilks DS (1995) Statistical methods in the atmospheric sciences. International Geophysics Series Vol. 59 (Academic Press, San Diego): 464 pp

Zwiers F, Kharin V (1998) Changes in the extremes of climate simulated by CCC GCM2 under CO2 doubling. J Climate 11: 2200-2222

Investigation of the hydrologic cycle simulated by the Canadian Regional Climate Model over the Québec/Labrador territory

ANNE FRIGON[1], DANIEL CAYA[1], MICHEL SLIVITZKY[2] AND DENIS TREMBLAY[3]
[1]UQÀM (Université du Québec à Montréal), Montréal (Québec), Canada
[2]INRS (Institut national de recherche scientifique), Québec (Québec), Canada
[3]Hydro-Québec, Montréal (Québec), Canada

Abstract : This paper discusses the surface hydrology of a two-year simulation conducted over the Québec/Labrador territory with the Canadian Regional Climate Model (CRCM). The CRCM was driven by NCEP (National Center for Environmental Protection) atmospheric objective analyses and was run at a 30 km horizontal grid-point spacing. Model results are compared with available observations for the period from June 1992 to June 1994. A spatial analysis of the CRCM annual runoff shows that the model is in good agreement with basin observations although the simulated precipitation and surface evaporation are too important. The main features of seasonal runoff variation are captured by the CRCM with differences mainly attributed to the oversimplification of the surface processes in the single-layer surface scheme. We also find that spring peak runoffs are quite well synchronized with the observations but with some overestimation when a lumped hydrologic model is fed by CRCM meteorological outputs. Despite the fact that the hydrologic model was not recalibrated with the CRCM data, this first experience shows the potential use of the CRCM to feed a hydrologic model. In summary, even though our analysis shows the usefulness of the CRCM simulated surface hydrology information, results from this experience indicate the need to refine the surface scheme to represent more accurately surface hydrology processes.

M. Beniston (ed.),
Climatic Change: Implications for the Hydrological Cycle and for Water Management, 31–55.

1. INTRODUCTION

According to the latest general circulation model (GCM) simulations, global climate change is expected to be important, with globally averaged surface temperature projected to increase by 1.4 to 5.8°C over the period 1990 to 2100 (IPCC 2001). Associated to this warming, a partial change in wintertime precipitation from solid to liquid form could have serious implications for water resources management. However, the spatial resolution of present-day GCMs is inadequate to accurately represent many of the regional features of climate (IPCC 1995) and consequently for hydrological applications. For this reason, regional climate models (RCM) have been developed in the last 15 years to reach horizontal grid-point spacing of some ten kilometers. Such models, therefore, represent a more adequate tool for decision-makers in the long-term management of water resources.

In this study, we use results from a CRCM simulation driven by atmospheric analyses to evaluate the model's surface hydrology through comparison with available observations. Basin runoff is used as a validation dataset along with precipitation and temperature observations. A set of surface evaporation data is also used to help understand the CRCM's hydrological cycle. The two-year simulation presented in this paper served as a first experiment to evaluate CRCM's surface hydrology in a semi-quantitative way.

The paper is organized as follows: section 2 briefly describes the numerical model and its experimental setup. Section 3 compares hydrologic outputs of the CRCM with observed runoff at the annual scale for a number of basins and looks into its spatial distribution. It then presents an analysis of the monthly surface water budget for the Churchill Falls basin (Newfoundland's Labrador), which contains an enhanced network of surface stations that provide an exceptional observational dataset. Section 4 compares runoff output from a hydrologic model when fed alternately with CRCM meteorological information and with the usual surface weather station observations. Finally, discussion and conclusions are presented in section 5.

2. EXPERIMENTAL SETUP

This section describes the regional climate model and its configuration for the experiment, and presents the basins of interest for this study. It is important to emphasize that the CRCM was driven by atmospheric analyses so the model could be evaluated in a historical sequence of meteorological events.

2.1 CRCM description

The model used for this study is the CRCM, a limited-area regional climate model developed at UQÀM. A detailed description can be found in Caya and Laprise (1999) and in Caya (1996). Results from its first climate simulations are presented in Laprise et al. (1998). The CRCM dynamic kernel was developed by André Robert and his colleagues of the Cooperative Centre for Research in Mesometeorology (CCRM) (Tanguay et al. 1990; Laprise et al. 1997; Bergeron et al. 1994). This kernel solves the fully elastic non-hydrostatic Euler equations with an efficient semi-implicit and semi-Lagrangian numerical algorithm. This numerical formulation allows the model to use substantially larger time steps than with conventional Eulerian methods while the use of the complete Euler equations makes it adequate for all spatial scales. The model's horizontal grid is uniform in polar stereographic projection and its vertical resolution is variable with a Gal-Chen scaled-height terrain-following coordinate (Gal-Chen and Sommerville 1975). It uses an Arakawa C-type staggered grid for the location of its atmospheric variables.

CRCM's sub-grid scale physical parameterization uses the Canadian GCMii (General Circulation Model Version II; McFarlane et al. 1992) package, which was adapted to the regional model's grid and projection. The surface energy budget equation considers heat storage in the soil, net solar flux absorbed by the ground, net terrestrial flux to the ground, terrestrial flux emitted by the ground, sensible and latent heat fluxes, as well as latent heat associated with the melting (or freezing) of frozen (or liquid) soil moisture and of snow. The one-layer surface scheme consists of prognostic equations for liquid and frozen ground water contents, for surface soil temperature and for snow amount. The ground temperature is calculated using the force-restore method with a deep soil temperature defined as the mean value of surface temperature over the last 24-hour period (Giguere et al. 2000). Complete melting (or freezing) of the unique layer must be completed before the surface soil temperature can go above (or below) the 0°C value. If the temperature of the lowest atmospheric level is at the freezing point or below, then precipitation is assumed to fall in solid form. The soil moisture regime is represented through the single-layer scheme with a water holding capacity that varies over each grid point according to vegetation and soil (i.e. porosity) characteristics. Surface moisture fluxes are represented using a drag coefficient formulation that is function of the surface-layer bulk Richardson number (stability parameter). A potential evaporation efficiency that varies with vegetation type represents the effects of a vegetative canopy. This efficiency increases linearly with soil water content and reaches its maximum (i.e. unity) at a specific water content value that depends on vegetation type. For example, in the Churchill Falls basin area, with vegetation defined as evergreen needleleaf tree and short grass and forbs, the potential evaporation efficiency increases linearly to reach its maximum when water content attains a value of 0.415. On the surface, the soil layer gains water only by collection of rainwater and snow cover melt, and loses water only by evaporation; subsurface drainage is nonexistent. Finally, surface runoff is generated instantly when total soil moisture exceeds the water holding capacity, returning the water surplus to the ocean without river routing or transformation for groundwater or lake storage.

As resolution increases, the parameterization of some subgrid-scale physical processes needs to be reformulated to represent more adequately the newly resolved scales (IPCC 1995). For this reason, two important scale-dependent processes in the GCMii physical parameterization package were adapted to the CRCM enhanced resolution. First, changes were made to the convection scheme in order to correct some deficiencies in summer that were observed with the moist convective adjustment scheme present in the first version of the model. Two mesoscale convective schemes are now available in the CRCM along with a large-scale condensation process for stratiform precipitation formation. The Kain-Fritsch (Kain and Fritsch 1990; Paquin and Caya 2000) and the Bechtold-Kain-Fritsch (Bechtold et al. 2001) mesoscale convective schemes were implemented in the CRCM. Second, the GCMii cloud onset function, triggered by relative humidity as a function of height, was adapted to the CRCM's finer resolution.

The CRCM lateral boundary conditions are provided through a one-way nesting method inspired by Davies (1976) and refined by Yakimiw and Robert (1990). That is, the regional model receives information from its driving atmospheric data but does not influence the driving data in return. At the lateral boundaries exactly, the CRCM is driven by time dependent vertical profiles from the driving data's winds, air temperature, humidity and pressure. Then, over a given number of grid points from the edges of the lateral boundaries (called the nesting zone), some of the regional model's variables are gradually blended with those of the driving data. Throughout the rest of the grid (free domain), the regional model variables are not affected by the driving data.

2.2 Experimental configuration

For this experiment, the CRCM simulations were computed on a 101 X 101 points grid with a horizontal grid-point spacing of 30 km and

18 vertical levels, ranging from the surface to the model top at 29 km. The model was run with the Bechtold-Kain-Fritsch (Bechtold et al. 2001) mesoscale convective scheme along with a large-scale condensation process for stratiform precipitation formation. Figure 1 presents the domain of the CRCM experiment along with the ground cover and topography fields. The regional domain covers at total $9X10^6$ km^2 and it encompasses the province of Québec, the Maritime provinces (New Brunswick, Nova Scotia and the Prince Edward Island) and the Island of Newfoundland, located on its southeastern edge. The CRCM maximum elevation of 654 meters is found near the Québec/Labrador border. Results are analyzed on a 82 X 82 points grid, delimited by the large black square on Figure 1, in order to keep only the free domain. This CRCM simulation uses a 10-minute time step starting on the 1st March 1992 and ending on the 1st July 1994. The first three months, used to spin-up the model and allowing the different fields to adjust, were not retained for analysis. This represents a relatively short simulation from a climatological point of view but is long enough to evaluate the model's ability to simulate useful surface hydrology information for water resource managers. It took more than 2 months to run the entire 28 month CRCM simulation on our local server, explaining our choice for a shorter first experiment before attempting to launch longer simulations that are even more demanding on computer resources and data storage. This period was chosen because of the availability and reliability of surface observational datasets.

In this simulation, the driving data used are atmospheric objective observational analyses from the National Center for Environmental Protection (NCEP). These analyses are available every 12 hours with an initial horizontal grid-point spacing of 2.5° X 2.5° (approximately 275 km) and 12 vertical levels. These data were interpolated onto the CRCM grid and vertical levels and were linearly interpolated in time to the CRCM 10-minute time step. The NCEP horizontal winds were used over a nine-point nesting zone. Monthly climatological values of sea surface temperature and sea-ice cover, linearly interpolated in time to the CRCM 10-minute time step, were used for ocean grid points. Geophysical fields such as soil water content and snow cover were initialized using climatology from the Canadian GCM.

Throughout the simulation, all CRCM output fields were archived at 6-hour intervals. In order to obtain CRCM daily accumulated values, we added four consecutive 6-hour archives of cumulative runoff output (or precipitation or evaporation). In the case of screen temperature, the absolute maximum (or minimum) from four consecutive 6-hour archives of maximum (or minimum) temperature were used to obtain the maximum (or minimum) daily screen temperatures. As for the CRCM snow cover, a mean daily value was computed from four consecutive 6-hour archives.

Throughout the simulation, all CRCM output fields were archived at 6-hour intervals. In order to obtain CRCM daily accumulated values, we added four consecutive 6-hour archives of cumulative runoff output (or precipitation or evaporation). In the case of screen temperature, the absolute maximum (or minimum) from four consecutive 6-hour archives of maximum (or minimum) temperature were used to obtain the maximum (or minimum) daily screen temperatures. As for the CRCM snow cover, a mean daily value was computed from four consecutive 6-hour archives.

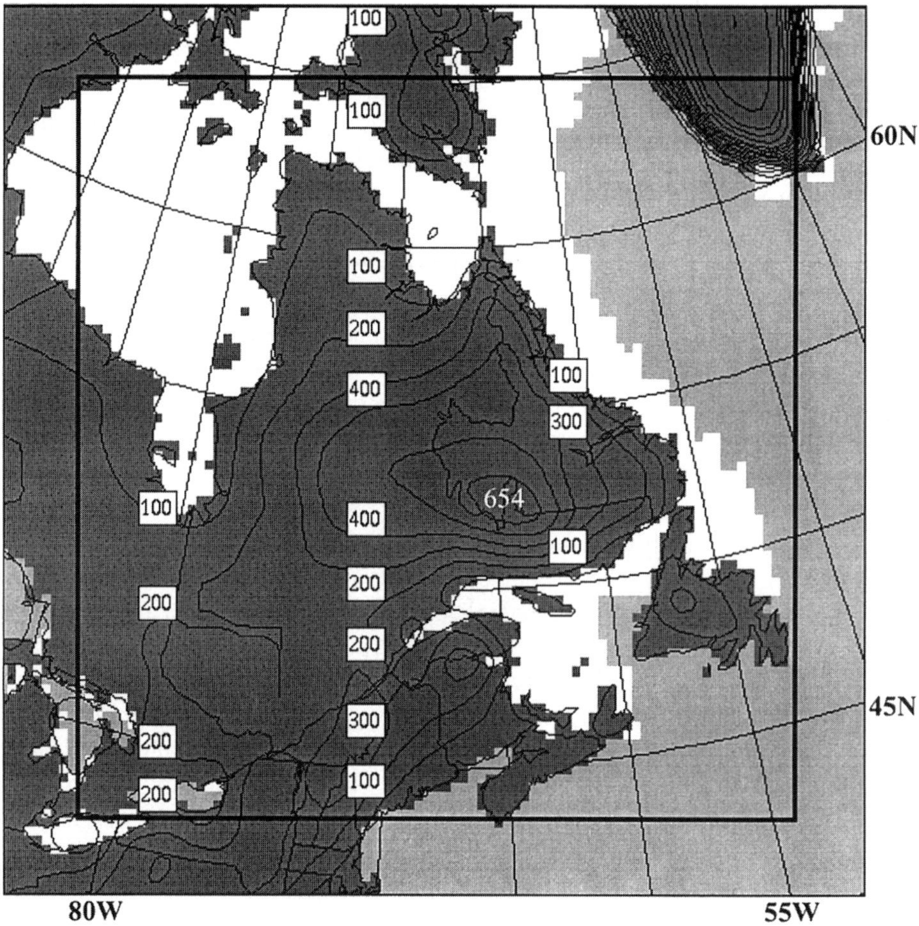

Figure 1: CRCM computational domain, land/sea/sea-ice mask and topographic height for the entire 101 X 101 grid-point domain with the inner 82 X 82 grid point free zone defined within the black square. CRCM horizontal grid-point spacing is of 30 km. Ground cover types shown are land (dark gray), open water (light gray) and January sea-ice (white). Topographic contour interval is of 100 m and the lower right corner of each label box identifies the concerned contour.

2.3 Basins of interest

Table 1 : Basins for which the CRCM surface runoff is computed. The table presents the drainage area and the number of CRCM 30 X 30 km^2 grid points within each basin. Reference to Figure 1 helps locate the basins.

Basin name	Drainage area (km^2)	Number of CRCM 30X30 km^2 grid points located within the basin
Baleine	29 800	32
Bell	22 200	28
Caniapiscau	37 870	51
Churchill	69 300	82
George	24 200	29
Manic-5	29 240	41
Rupert	40 900	53
Waswanipi	31 900	43

The CRCM outputs were analyzed over 8 basins of interest, mainly chosen because of their relatively large size. Figure 2 locates these basins and Table 1 presents their characteristics such as the drainage area and the number of CRCM grid points located within each one. All basins are located between 48°N and 58°N and 62°W to 78°W. The black star on Figure 2 identifies the Wabush surface weather station, located on the western edge of the Churchill Falls basin. This station's temperature and precipitation data were used to investigate the CRCM's behavior in the Churchill Falls basin.

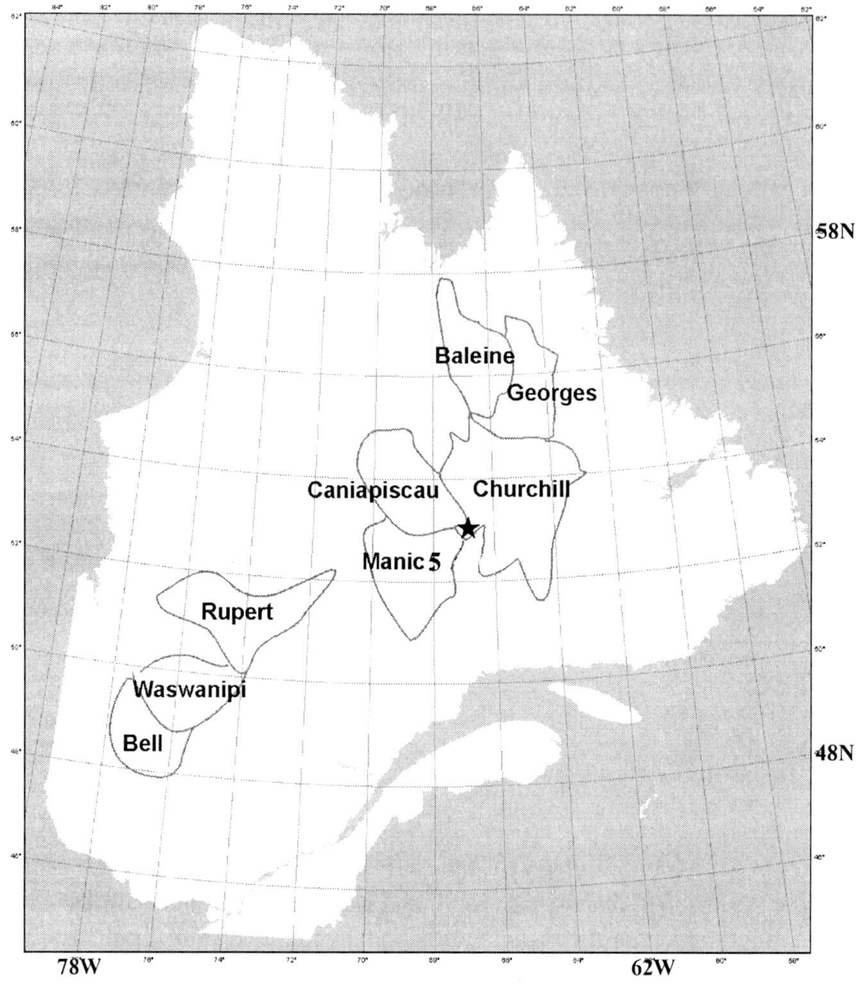

Figure 2 : Location of the 8 basins of interest. The basin's characteristics such as the drainage area and the number of CRCM grid points located within each basin are presented in Table 1. The black star on the western edge of the Churchill Falls basin identifies the location of the Wabush surface weather station.

3. RESULTS

In this section, we evaluate the model's ability to simulate the observed runoff, precipitation and surface evaporation. In the first subsection, we define the observed runoff, which serves as basis for the evaluation. Secondly, we present an annual mean basin analysis that also looks at the spatial structure of the CRCM simulated runoff. We finally focus on the Churchill Falls basin for an analysis of its monthly surface water budget. This basin was chosen because of the presence of an enhanced network of surface stations that provide an exceptional observational dataset for that region.

3.1 Observed runoff

Basin runoff was used as a validation dataset to evaluate the CRCM. Observational data for streamflow were obtained from different sources. For the Caniapiscau and Manic-5 basins, daily streamflow data was provided by Hydro-Québec while data for the Churchill Falls basin came from the Churchill Falls Labrador Corporation (CFLCo). For the other basins, the daily mean flows measured by the Hydrometric Service of the Québec Department of Environment were accessible through the published HYDAT CD-ROM (1998). For comparison with model runoff, all streamflow data was converted to mm (as defined by Chow 1964) by dividing by the area drained by a given streamflow gauge.

3.2 Annual basin hydrology

We compare annual mean hydrologic outputs from the CRCM with observed runoffs for a number of basins within the domain. The CRCM outputs were analyzed over the 8 basins of interest (Fig. 2 and Table 1) while the observed annual runoffs were computed for the same 8 basins and for 13 additional basins for reference. We chose to examine an annual period that ends when the subsurface and lake storage reach a minimum value. For the central area of the domain, this period is usually reached by the end of March, just before the beginning of the spring snow cover melt. We focused our attention on the hydrologic year included between the 1st April 1993 and the 31st March 1994. Figure 3 presents the observed runoff for the 1993-1994 hydrologic year for a total of 21 basins as well as the CRCM simulated runoff results for the 8 basins of interest. The CRCM basin values represent the average obtained from the grid points located within each basin. In summary, the CRCM simulated annual runoff overestimates the observations by about 15-20% for the Churchill Falls (14%), Manic-5 (18%) and Caniapiscau (21%) basins and underestimates the observations by about 15% for the Baleine (13%), George (16%) and Rupert (13%) basins. As for the southwestern basins of Waswanipi and Bell, the simulated annual runoffs underestimate the observations by 31%.

It is interesting to note that the hydrologic year 1993-1994 is the driest in the Churchill Falls basin record history since 1950. In fact, going through the

1969-1996 common record period for the 8 basins of interest, we find that the 1993-1994 runoff is in a dry period for the northern (Baleine, George) and central (Manic-5, Caniapiscau, Churchill Falls) basins, while the southwestern basins (Rupert, Waswanipi, Bell) are close to normal.

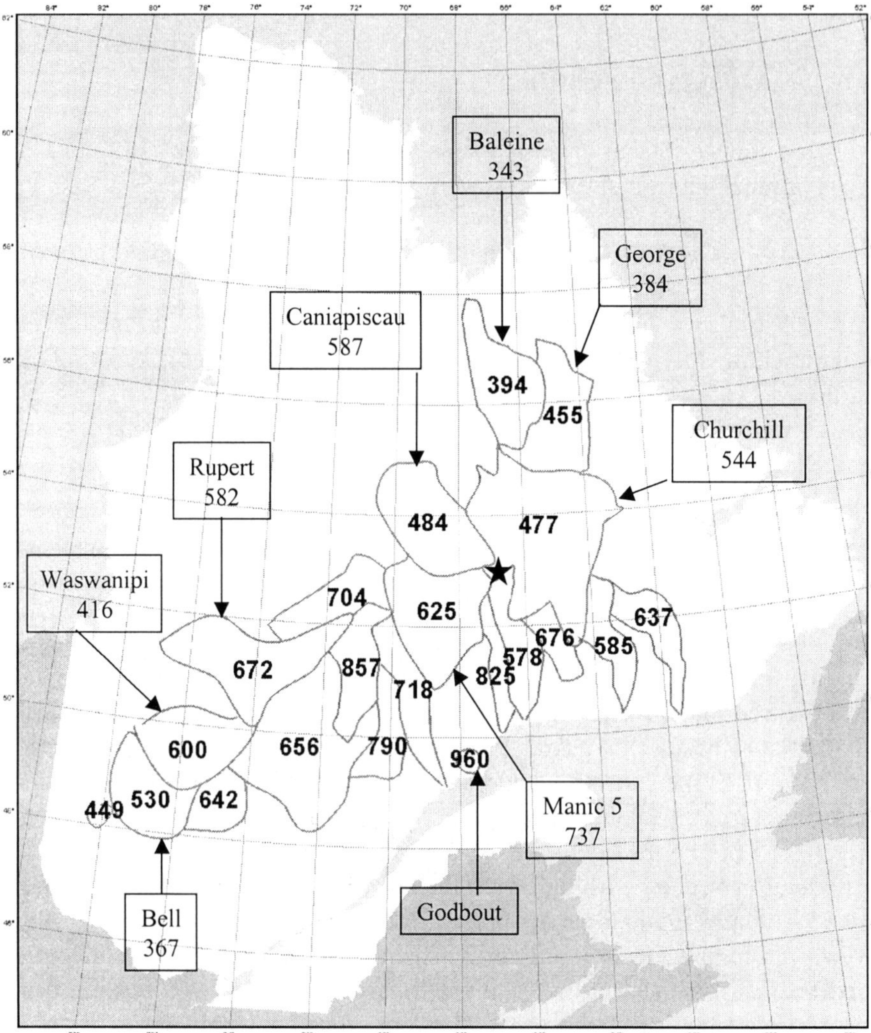

Figure 3 : Total annual runoff for the period 1st April 1993 to 31st March 1994. The observed values are presented within the basin contours and the CRCM simulated results are shown in the black squares. The CRCM basin values represent the average obtained from the grid points located within the basins. The black star on the western edge of the Churchill Falls basin identifies the location of the Wabush surface weather station. The values are shown in millimeters for each basin. The runoff observational data were obtained from different sources.

To get a better picture of the spatial distribution of the 1993-1994 runoff, Figure 4 presents the CRCM simulated results over the free domain. The high values simulated in the central part of the domain seem to agree in general with observed basin values (Fig. 3). The CRCM maximum annual runoff values of 870 mm and 715 mm correspond approximately to some high values observed close-by. A high annual runoff of 960 mm was observed that year on the small basin of Godbout (area: 1570 km^2). From the center of the domain, the CRCM simulated runoff decreases when one moves to the northeast and to the southwest, as seen with the observed basin values (Fig. 3). However, the CRCM values decrease more rapidly than the observations towards the Waswanipi and Bell basins. We are currently investigating the model's behavior in these southwesternmost areas.

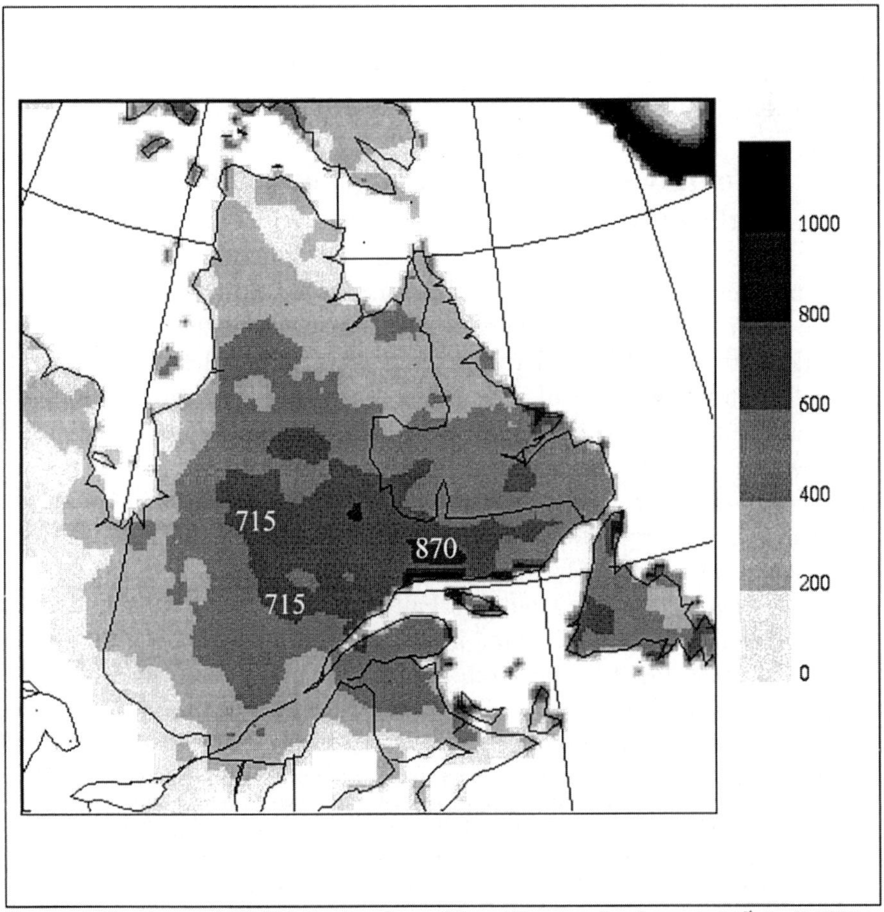

Figure 4 : Total annual CRCM simulated runoff in millimeters for the period 1st April 1993 to 31st March 1994. Contour interval is 200 mm. This Figure covers the free domain of 82 X 82 grid points.

As opposed to annual runoff and precipitation, the annual evaporation pattern (not shown) does not show pronounced spatial variability. This was also noted by Giorgi et al. (1994) for a simulation over the United States with RegCM, the Regional Climate Model developed at the National Center for Atmospheric Research. Since there are no evaporation observations available to verify this, we turned to estimated surface evaporation from Willmott and Matsuura (2000a). This monthly time series dataset was computed on a 0.5° X 0.5° grid using monthly precipitation and air temperature records on the same grid (Willmott and Matsuura 2000b), along with liquid and solid surface water budget equations with a fixed soil moisture capacity of 150 mm (Willmott et al. 1985; Willmott 1977). Annual surface evaporation from Willmott and Matsuura (2000a) shows the weak spatial variability we observed within the CRCM domain, which is mainly linked to the north-south temperature gradient. For our 1993-1994 hydrologic year, a general comparison over the simulation domain indicates that the CRCM total annual surface evaporation is greater than the values of Willmott and Matsuura (2000a) for that same year by about 35%.

3.3 Monthly surface water budget over the Churchill Falls basin

In this section, we look into monthly time series to examine the CRCM's surface water budget for the Churchill Falls basin. While the climate station network north of 50°N is relatively scarce, the Churchill Falls basin is of particular interest because of its enhanced network of 17 surface stations, operated by the CFLCo (Churchill Falls and Labrador Corporation) since 1973, which report monthly precipitation and snow cover. The Churchill Falls basin values were obtained from the average of the 17 CFLCo station observations, referred to as CFLCo-mean. Screen temperature and precipitation observations from the Wabush surface weather station, located on the western edge of the Churchill Falls basin (Fig. 2), were also used to investigate the CRCM's behavior in the basin. Finally, the CRCM basin values represent the average from the 82 grid-points located within the Churchill Falls basin.

Figure 5 presents monthly precipitation from June 1992 to June 1994 obtained from CFLCo-mean observations, from the Wabush surface weather station and from the CRCM simulated values over the Churchill Falls basin. Even though the simulated values seem to agree with the observations for winter months (January, February, March), the CRCM generally produces too much precipitation especially from April to October. For the total 25-month period from June 1992 to June 1994, CRCM monthly precipitation overestimates the CFLCo-mean observations by about 55% while the

Wabush station observations show less precipitation than the CFLCo-mean by an average of 5%.

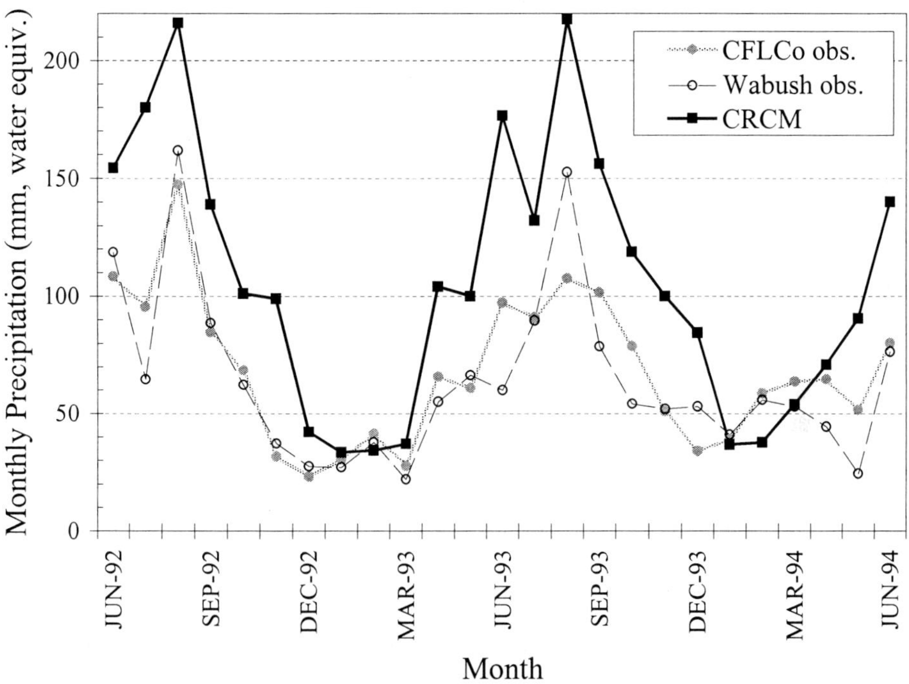

Figure 5 : Monthly precipitation from June 1992 to June 1994 obtained from the average of the 17 CFLCo stations (grey line with circles), from the Wabush surface weather station (dashed line with open circles) and from the CRCM simulated values over the Churchill Falls basin (black line with squares). The CRCM basin values represent the average obtained from the 82 grid points located within the basin. Location of the Wabush station is presented on Figure 2. All values are equivalent water contents in millimeters.

Figure 6 presents the monthly mean screen temperature difference from June 1992 to June 1994 between the CRCM simulated value on the Churchill Falls basin and the Wabush surface weather station observation. Monthly data for Wabush were obtained from the Historical Canadian Climate Database Version 2 (Vincent and Gullett 1999; Vincent 1998). Figure 6 shows that CRCM monthly mean temperatures generally follow the observations during summer, are too cold in winter and spring, and clearly too warm in the fall. It is important to mention that the CRCM temperature represents the average of 82 grid-point values located within the Churchill Falls basin, while the observation data comes from only one station located on the edge of the basin. However, this information is useful to investigate the CRCM's thermal behavior near the surface, which is discussed later in this section.

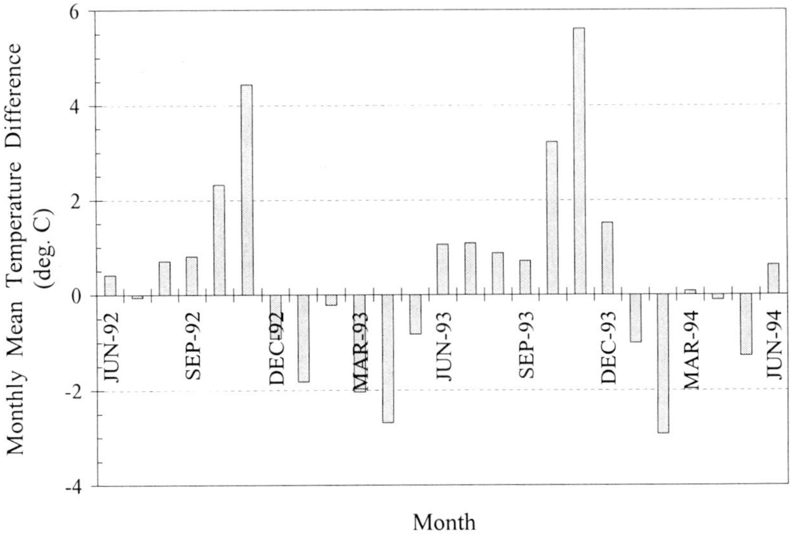

Figure 6 : . Monthly mean screen temperature difference (CRCM-OBS) from June 1992 to June 1994 between the CRCM simulated value on the Churchill Falls basin and the Wabush surface weather station observations. The CRCM basin values represent the average obtained from the 82 grid points located within the basin. Location of the Wabush station can be seen on Figure 2.

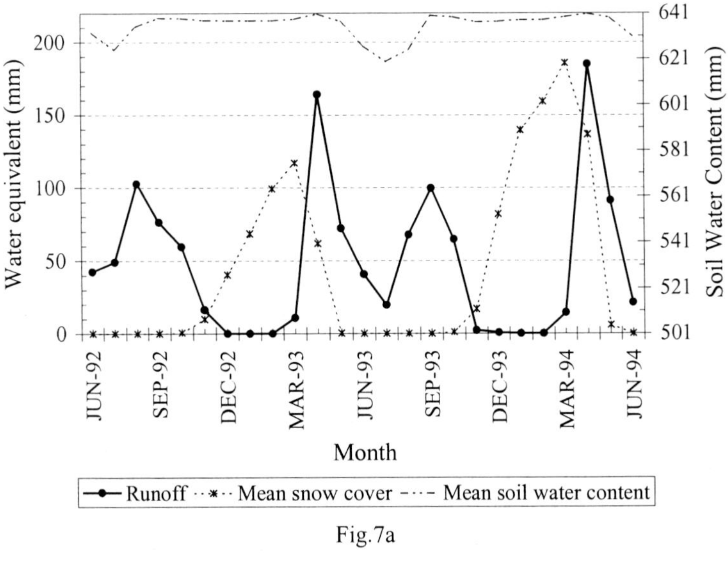

Fig.7a

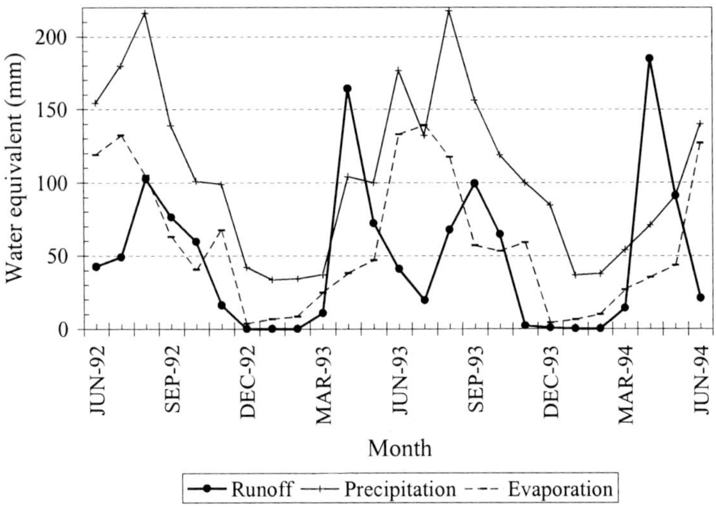

Fig.7b

Figure 7 : Monthly values simulated by the CRCM on the Churchill Falls basin from June 1992 to June 1994. Figure 7a presents monthly runoff (thick black line with circles) with mean snow cover (dashed line with stars) and mean soil water content (dashed line) while Figure 7b shows monthly runoff (thick black line with circles) along with precipitation (black line with pluses) and surface evaporation (dashed line with minuses). The CRCM basin values represent the average obtained from the 82 grid points located within the basin. For reference, the basin mean soil water holding capacity, representing soil saturation, is 641 mm. All values are equivalent water contents in millimeters.

Figures 7a and 7b show the monthly evolution over the Churchill Falls basin of CRCM's precipitation, evaporation, runoff, as well as its mean snow cover and soil water content. For reference, the basin mean soil water holding capacity, representing soil saturation, is 641 mm. It is important to note that the basin mean soil water content represents an average, not the exact saturation state of each of the 82 CRCM grid points within the basin. It is therefore possible to attain saturation at some grid points within the basin, and produce runoff, even tough the basin mean soil water content is lower than the saturation threshold of 641 mm. These simulated fields help understand the model's behavior regarding the surface water budget and the surface runoff it produces. As can be seen from Figure 7a, the CRCM builds its snow cover from November to March, producing hardly any surface runoff during that period. The model's snow cover melts in April and May. This replenishes the soil water content and causes the spring peak runoff in May. CRCM's high evaporation rates in summer (Fig. 7b) cause soil moisture to reach a minimum (Fig. 7a), decreasing its runoff during summer. Soil water recharge in the model occurs in August and September (Fig. 7a) when monthly precipitation is at its annual maximum (Fig. 7b), producing a local maximum in its surface runoff. Hence, the soil remains mostly saturated throughout winter (Fig. 7a).

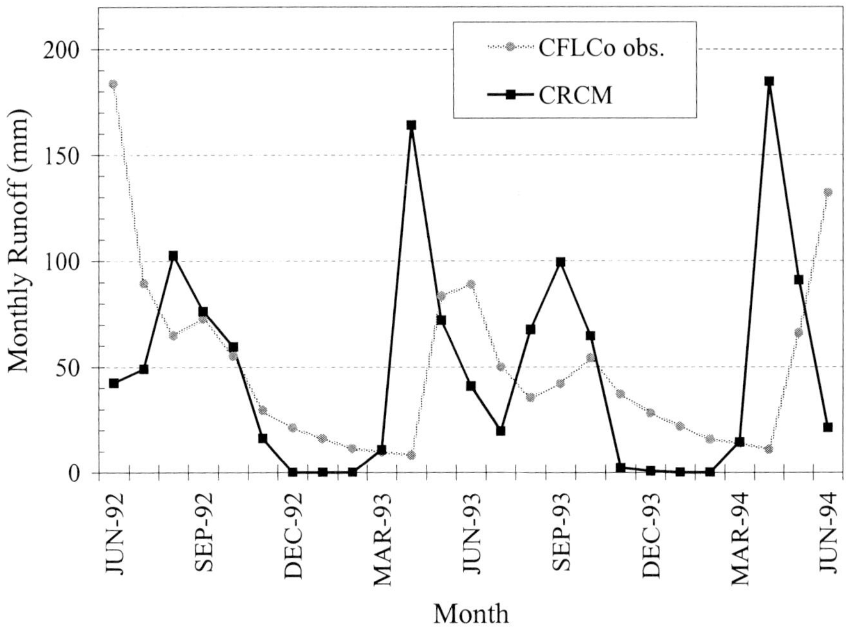

Figure 8 : Comparison of CFLCo monthly runoff observations (grey line with circles) and CRCM monthly simulated runoff (black line with squares) for the Churchill Falls basin from June 1992 to June 1994. The CRCM basin values represent the average obtained from the 82 grid points located within the basin.

Figure 8 compares monthly runoff from the observations and from the CRCM simulation over the Churchill Falls basin from June 1992 to June 1994 but the analysis focuses mainly on the 1993-1994 hydrologic year. Figure 8 shows a spring peak in runoff observations (May-June 1993) caused by the snow cover melt. Afterward, the observed runoff decreases in summer, followed by a slight maximum in fall (October 1993). The monthly runoff finally decreases throughout winter to attain its annual minimum in the spring (April 1994). The CRCM produces a stronger spring peak runoff (April 1993) about a month earlier than the observations. In spring, with its saturated single soil layer, all water from melting the CRCM snow cover produces direct surface runoff. In the real system, a large part of that water is used to replenish groundwater reserves and lake storage. In that sense, the model produces an important and immediate runoff without draining water into the subsurface aquifer. After its spring peak, the general behavior of CRCM's runoff follows that of the observations, with a decrease in summer and an increase in fall, except that the model is ahead by a month and presents more important variations. Finally, in winter (November 1993-March 1994), the CRCM doesn't produce any runoff since the surface scheme does not have a long term, deep and subsurface storage nor lake storage components that usually sustain river flow during the winter months. For the Churchill Falls basin, the subsurface component (estimated from the ratio of November to March observed runoff to the annual observations) can represent 10% to 30% of the total annual runoff.

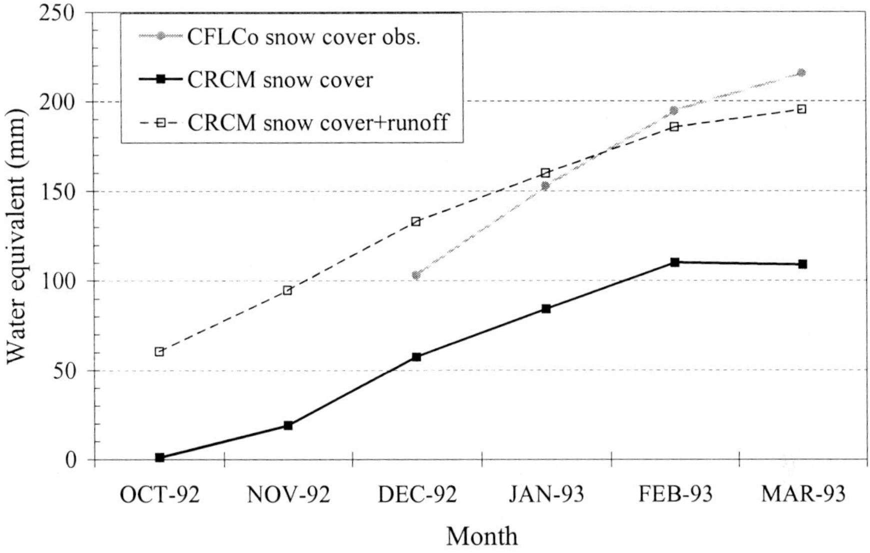

Figure 9 . Monthly evolution from October 1992 to March 1993 of the simulated CRCM snow cover (black line with squares) and snow cover plus runoff (dashed line with open squares) for the Churchill Falls basin as well as the average observed snow cover from the 17 CFLCo stations (grey line with circles). The CRCM basin values represent the average obtained from the 82 grid points located within the basin. The CRCM snow cover is taken at the end of the month and is related to the CFLCo measurements of the beginning of the next month. Note that there are no values for the CFLCo snow cover for October and November. All values are equivalent water contents in millimeters.

During spring and fall, some differences between the CRCM simulated monthly runoff and the observations are related to the thermal behavior of the CRCM single-layer surface scheme. During spring, this scheme first uses the available energy to melt the snow cover before heating the ground, producing a rapid snow cover melt (and consequently a rapid runoff) with surface temperatures maintained colder than observed around 0°C (Fig. 6). For the Churchill Falls basin, this transition period happens in the months of April and May and can be identified by the rapid snow cover melt associated with the soil saturation and the annual maximum peak of runoff (Fig. 7a). In the fall, the inverse behavior is observed because the ground layer must freeze throughout before cooling its surface below 0°C, and the CRCM is usually too warm near the ground (Fig. 6). This has an influence on precipitation, which is in the form of rain instead of snow because temperature remains near or just above 0°C. With that rain, the CRCM generates more runoff than is observed. For the Churchill Falls basin, the fall transition period stretches from October until mid-November, during which the soil is mostly saturated (Fig. 7a).

Figure 9 shows the monthly evolution from October 1992 to March 1993 of the CRCM snow cover and snow cover plus runoff for the Churchill Falls basin as well as the average observed snow cover from CFLCo-mean observations. As was just mentioned, with its fall surface temperature too warm, the CRCM produces liquid precipitation rather than snow and thus generates more surface runoff than is observed. Figure 9 illustrates this effect on the CRCM snow cover which is already less important in December (1992) than the first winter observations. Since CRCM winter precipitations compare well with the CFLCo-mean observations (Fig. 5), the model ends up with a thinner snow cover at the end of the winter. However, if the CRCM surface temperature were not as warm during October and November, precipitation would fall as snow, therefore reducing the runoff and contributing to produce an earlier and thicker CRCM snow cover. To verify this, we have added CRCM runoff to its snow cover, starting in October. Figure 9 shows that the CRCM available surface water curve (CRCM runoff plus its snow cover) reaches values close to the CFLCo-mean snow cover observations at the end of winter. The warm fall bias resulting from the single-layer surface scheme indicates that a better representation of the surface's thermal behavior in the CRCM would help in generating a snow cover closer to observations. For hydrological purposes, a good end of winter snow cover is essential as it is the major source of the spring runoff, which represents close to 50% of the annual volume.

To summarize, the CRCM simulated runoff reproduces the main features of the observed annual cycle but is a month ahead of the observations (Fig. 8). Some of the differences can be attributed to the inexistence of the subsurface flow in the CRCM. For example, in spring, with its saturated single soil layer, all water from melting the CRCM snow cover produces direct surface runoff. The model produces an important and immediate spring runoff while in the real system a large part of that water is used to replenish groundwater reserves and lake storage. In winter, the CRCM doesn't produce any runoff while observations show a slowly decreasing runoff usually sustained by the long term, deep and subsurface storage and lake storage components.

Finally, over the Churchill Falls basin, CRCM's annual basin runoff resulting from the atmospheric water flux convergence over the domain is correct but the model overestimates precipitation. This is explained by the fact that the CRCM also overestimates surface evaporation, because the model recycles too much water through precipitation and surface evaporation. This is a known deficiency of the CRCM single-layer surface scheme (Verseghy 1996), overestimating its latent heat exchanges with the atmosphere compared to its sensible heat part. For the 1993-1994 hydrologic year, on this basin, the CRCM overestimates annual CFLCo runoff observations by only 68 mm while it overestimates total annual CFLCo-mean precipitation observations by 470 mm and total annual surface evaporation data (Willmott and Matsuura 2000a) by 316 mm.

4. THE CRCM AS INPUT TO A HYDROLOGIC MODEL

In addition to a direct comparison between the observed and CRCM simulated runoff, we also compared the runoff produced by a hydrologic model when fed with the CRCM simulated meteorological information and with the usual surface weather station observations. Hydrologic models need information on the quantity of water received at a basin through precipitation events, usually taken from precipitation observations from the operational surface weather station network. However, the location of these stations and the density of the network are definitely not optimal for most basins in the province of Québec and the Labrador territory (Newfoundland). The idea here is to take advantage of the complete coverage of the high resolution meteorological information available from the CRCM. In this case, the CRCM should give better precipitation accumulation for events that are not detected by the scarce surface station network. Hence, we used the virtual meteorological output from the CRCM to feed the hydrologic model. This section presents results of a preliminary study for the Churchill Falls basin to evaluate the potential of such an approach.

The hydrologic model used is HSAMI (Bisson and Roberge 1983; Fortin 1999) and was developed by Hydro-Québec. It is a lumped model which needs as input the basin average values of daily liquid and solid precipitation and daily maximum and minimum temperatures. Using this information, HSAMI forms its own snow cover which ages and melts following a degree-day approach. Evapotranspiration is estimated from daily maximum and minimum temperatures through an empirical formula. This hydrologic model was calibrated with precipitation and temperature observations from surface weather stations interpolated to the basin using the Thiessen polygons approach. Through an iterative process using historical input-output data over a number of years, the calibration of HSAMI not only handles the subsurface and lake storage components of streamflow but also introduces the proper streamflow routing delays. In its calibration process, HSAMI ensures to respect not only the observed average annual runoff volumes but also its average distribution throughout the year and particularly during the spring (which represents close to 50% of the annual volume). This hydrologic model needs about 5 years of data to be calibrated, which is not possible with the present two-year CRCM simulation.

The hydrologic model was run once using the usual daily surface weather station observations and another time using the CRCM daily output for the same fields; these runs are respectively named HSAMI(STNS) and HSAMI(CRCM). Figure 10 presents daily runoff results from HSAMI for the Churchill Falls basin over the 25-month period of the CRCM simulation (1st June 1992 to 1st July 1994). As a reference, the observed runoff is also shown in Figure 10. With the conventional surface weather observations, HSAMI(STNS) does not reproduce very well the timing of the spring peak runoff observations in spring 1993 and 1994 but it respects the annual runoff volume and its average distribution through the year because HSAMI was calibrated with this data. Otherwise, with the CRCM information, the spring peak runoffs of HSAMI(CRCM) are quite well synchronized with the observations but are overestimated by a factor of 2. It is important here to emphasize that both simulations (STNS and CRCM) were done using HSAMI calibrated with surface weather station observations. Therefore, HSAMI should have been recalibrated to adjust itself to the CRCM data, which is of a different nature in terms of spatial and time resolution (Droegemeier et al. 2000). Even though this recalibration could not be done, CRCM's simulation being too short, the results obtained by HSAMI(CRCM) are promising. This first experience shows the potential use of the CRCM to feed the hydrologic model.

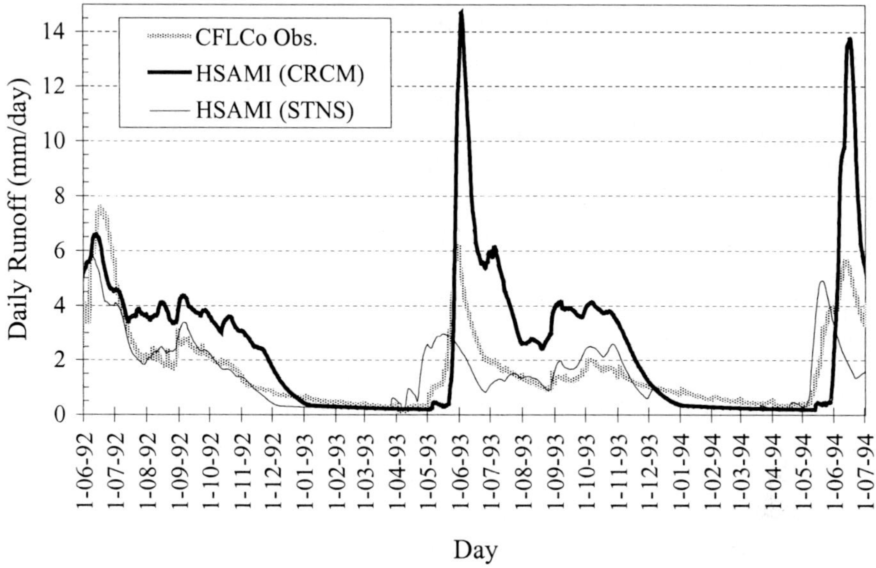

Figure 10 : Daily runoff results for the Churchill Falls basin from the HSAMI hydrologic model when fed: 1) with usual daily observations from surface weather stations (thin black line; HSAMI(STNS)) and 2) with meteorological information from the CRCM (thick black line; HSAMI(CRCM)). As inputs, HSAMI needs basin average of daily liquid and solid precipitation and daily maximum and minimum temperatures. For a reference, the CFLCo observed runoff is overlayed (thick grey line). Results are presented for the 25-month period of the CRCM simulation extending from 1st June 1992 to 1st July 1994.

5. DISCUSSION AND CONCLUSIONS

Validating surface hydrology components of a regional climate model is a complicated task. With surface observational networks in which stations can be some 400 km apart, it is difficult to estimate parameters as spatially variable as precipitation to validate regional climate models with grid point spacing that can be of only 30 km. This motivated the use of basin runoff observations as a validation dataset, benefiting from the fact that the basin acts as a space and time integrator. In that sense, observational runoff data is more similar in scale to the regional climate model output. In this work, the CRCM was driven by the NCEP atmospheric analyses so the model could be evaluated in a historical sequence of meteorological events. Results from a two-year simulation of the CRCM, from June 1992 to June 1994, served as a basis to investigate the model's ability to simulate the observed runoff, precipitation and surface evaporation over the Québec/Labrador territory.

A spatial analysis of the CRCM annual runoff showed that the model was in good agreement with basin observations. Based on the laws of physics, the regional model's atmospheric humidity budget produces an atmospheric water vapor flux convergence that generates reliable basin annual runoffs within the domain. It is interesting to note that, as was found by Giorgi et al. (1994), CRCM surface evaporation patterns present much less spatial variability than do the precipitation and runoff patterns. On a seasonal basis, the CRCM simulated runoff reproduces the main features of the observed annual cycle with differences mainly attributed to the oversimplification imposed by the single-layer surface scheme. We must point out that the CRCM surface runoff is compared to an observed runoff where both surface and subsurface components are present. The CRCM available surface water (precipitation plus surface runoff) gets close to the observations (snow cover) as winter progresses. However, the CRCM snow cover is underestimated because of the fall warm bias resulting from the single-layer surface scheme. We also found that the CRCM produces reliable precipitation in winter but that it overestimates precipitation from April to October. It seems that the CRCM recycles too much water through precipitation and surface evaporation, particularly in summer.

We also evaluated runoff simulated by the lumped hydrologic model HSAMI when fed by the CRCM meteorological information. Using basin daily precipitation and temperature data as input, this model is usually run with conventional surface weather station observations. The idea was to take advantage of the complete coverage of meteorological information available form the CRCM, which should represent more realistic basin conditions than the scarce surface weather station network. Hence, with the CRCM data as input, HSAMI produced spring peak runoffs quite well synchronized with the observations but with greater values than the observed runoff. It is rather difficult to fully evaluate the performance of HSAMI with the CRCM information as input because HSAMI was calibrated with surface weather station observations. However, this first experience shows the potential use of the CRCM to feed a hydrologic model.

There is a need to improve parameterization of land surface processes. For that reason, we have just completed implementation of CLASS (Canadian Land Surface Scheme; Verseghy 2000) in the CRCM and are currently running the first control simulations. CLASS is a new generation multi-layer surface scheme that improves representation of the effects of vegetation, snow and ground on the exchanges of heat and humidity with the atmosphere. With its three soil layers, it also considers infiltration and subsurface drainage. Our plans for future simulations include running

CRCM with CLASS to evaluate its effect on surface temperature, evaporation and on monthly runoff. We also hope to launch longer simulations over the same domain to pursue our understanding of the CRCM. We favor an approach by stages in which we conduct a relatively long simulation after implementation of each new element into the CRCM (cf. convection, surface scheme), allowing us to evaluate the effect of individual elements on the simulated climate system.

6. ACKNOWLEDGEMENTS

We are grateful to Hydro-Québec for the financial support for this project and for supplying runoff data. We are also thankful to the Churchill Falls and Labrador Corporation for supplying runoff, precipitation and snow cover data for the Churchill Falls basin. The CRCM project was supported by a Strategic grant of the National Sciences and Engineering Research Council (NSERC) of Canada, by the Meteorological Service of Canada (MSC) through its Science Subvention program and through the Canadian Climate Research Network (CCRN) operated by the Canadian Institute for Climate Studies (CICS), by Forestry Canada (Ontario Region), and by the Climate Change Action Fund (CCAF). We thank reviewers for very valuable comments which helped in improving our manuscript.

7. REFERENCES

Bechtold, P., E. Bazile, F. Guichard, P. Mascart et E. Richard, 2001 : A Mass Flux Convection Scheme for Regional and Global Models. *Quart. J. Roy. Meteorol. Soc.*, 127, 869-886.

Bergeron, G., R. Laprise et D. Caya, 1994 : Formulation of the Mesoscale Compressible Community (MC2) model. Cooperative Centre for Research in Mesometeorology (CCRM). 165 p.

Bisson, J.L., and F. Roberge, 1983 : Prévision des apports naturels : Expérience d'Hydro-Québec. Compte-rendu de l'Atelier sur la prévision du débit. Toronto, novembre 1983.

Caya, D., 1996 : Le modèle régional de climat de l'UQÀM. Thèse de doctorat, Université du Québec à Montréal. 133 p.

Caya, D., and R. Laprise, 1999: A Semi-Implicit Semi-Lagrangian Regional Climate Model: The Canadian RCM. *Mon. Wea. Rev.*, 127, 341-362.

Chow, V.T., 1964: Handbook of Applied Hydrology. McGraw Hill, pp. 14-2,14-3.

Davies, H.C., 1976: A lateral boundary formulation for multi-level prediction models. *Quart. J. Roy. Meteorol. Soc.*, 102, 405-418.

Droegemeier, K.K., J.D. Smith, S. Businger, C. Doswell III, J. Doyle, C. Duffy, E. Foufoula-Georgiou, T. Graziano, L.D.James, V. Krajewski, M. LeMone, D. Lettenmaier, C. Mass, R. Pielke Sr., P. Ray, S. Rutledge, J. Schaake, and E. Zpiser, 2000: Hydrological Aspects of Weather Prediction and Flood Warnings: Report of the Ninth Prospectus Development Team of the U.S. Weather Research Program. *Bull. Amer. Meteor. Soc.*, 81, 2665-2680.

Fortin, V., 1999 : Le modèle météo-apport HSAMI : historique, théorie et application. Rapport de recherche, Institut de recherche d'Hydro-Québec. Révision 1.4, 14 décembre 1999. 68 p.

Gal-Chen, T., and R.C. Sommerville 1975: On the use of a coordinate transformation for the solution of Navier-Stokes. *J. Comput. Phys.*, 17, 209-228.

Giorgi, P., S.W. Hostetler, and C. Shields Brodeur, 1994: Analysis of the surface hydrology in a regional climate model. *Quart. J. Roy. Meteorol. Soc.*, 120, 161-183.

Giguere, M., R. Laprise, D. Caya and S. Biner, 2000: An implicit scheme for the ground energy equation in the CRCM. In WMO Research Activities in atmospheric and oceanic modelling, Report No. 30, February 2000, H. Ritchie Ed., WMO/TD-No.987, 4.13-4.14.

HYDAT 1998: Canadian hydrological data available on CD-ROM. National Water Data Archive, Environment Canada. Available through the Environment Canada Web site at http://www.cmc.ec.gc.ca/climate/hydatintro.htm.

IPCC, 2001: Intergovernmental Panel on Climate Change Third Assessment Report from Working Group I. Climate Change 2001: The Scientific Basis. Summary for Policymakers, p. 13. Available through the IPCC Web site at http://www.ipcc.ch.

IPCC, 1995: Climate Change 1995. The Science of climate change. Contribution of Working Group I to the second assessment report of the IPCC. Houghton, Meira Filho, Callander, Harris Kattenberg and Maskell (Eds), Cambridge University Press, Cambridge, UK, 572 p.

Kain, J.S., and J.M. Fritsch 1990: A one-dimensional entraining/detraining plume model and application in convective parameterization. *J. Atmos. Sci.*, 47, 2784-2802.

Laprise, R., D. Caya, G. Bergeron, and M. Giguere, 1997: The Formulation of André Robert MC2 (Mesoscale Compressible Community) Model. *Atmos.-Ocean*, 35 (2), 195-220.

Laprise, R., D. Caya, M. Giguere, G. Bergeron, H. Cote, J.-P. Blanchet, G.J. Boer, and N.A. McFarlane, 1998: Climate and Climate Change in Western Canada as simulated by the Canadian Regional Climate Model. *Atmos.-Ocean*, 36 (2), 119-167.

McFarlane, N.A., G.J. Boer, J.-P. Blanchet, and M. Lazare 1992: The Canadian Climate Centre second generation general circulation model and its equilibrium climate. *J. Clim.*, 5, 1013-1044.

Paquin, D., and D. Caya, 2000: New convection scheme in the Canadian Regional Climate Model. In WMO Research Activities in atmospheric and oceanic modelling, Report No. 30, February 2000, H. Ritchie Ed., WMO/TD-No.987, 7.14-7.15.

Tanguay, M.A., A. Robert, and R. Laprise, 1990: A semi-implicit semi-Lagrangian fully compressible regional forecast model. *Mon. Wea. Rev.*, 118, 1970-1980.

Verseghy, D.L., 1996: Local Climates Simulated by Two Generations of Canadian GCM Land Surface Schemes. *Atmos.-Ocean*, 34 (3), 435-456.

Verseghy, D.L., 2000: The Canadian Land Surface Scheme (CLASS): Its History and Future. *Atmos.-Ocean*, 38 (1), 1-13.

Vincent, L.A., 1998: A technique for the identification of inhomogeneities in Canadian temperature series. *J. Clim.*, 11 (5), 1094-1104.

Vincent, L.A., and D.W. Gullett, 1999: Canadian Historical and Homogeneous Temperature Datasets for Climate Change Analyses - Canada. *Intern. J. Climatol.*, 19 (12), 1375-1388. Data available through the Meteorological Service of Canada Web site at http://www.cccma.bc.ec.gc.ca/hccd.

Willmott, C.J., 1977: WATBUG: A FORTRAN IV algorithm for calculating the climatic water budget. *Publ. in Climatol.*, 30, 1-55.

Willmott, C.J., C.M. Rowe, and Y. Mintz, 1985: Climatology of the terrestrial seasonal water cycle. *J. of Climatol.*, 5, 589-606.

Willmott, C.J., and K. Matsuura, 2000a: Terrestrial Water Budget Data Archive: Monthly Time Series (1950-1996). Version 1.0.1, released January 31, 2000. Data available through the University of Delaware, Center for Climatic Research Web site at http://www.climate.geog.udel.edu/~climate.

Willmott, C.J., and K. Matsuura, 2000b: Terrestrial Air Temperature and Precipitation: Monthly and Annual Time Series (1950-1996). Version 1.0.1, released January 31, 2000. Data available through the University of Delaware, Center for Climatic Research Web site at http://www.climate.geog.udel.edu/~climate.

Yakimiw, E., and A. Robert, 1990: Validation experiments for a nested grid-point regional forecast model. *Atmos.-Ocean*, 28, 466-472.

Major influences on precipitation in the Canary Islands

DAVID GALLEGO PUYOL[1], RICARDO GARCÍA HERRERA[1], EMILIANO HERNÁNDEZ MARTÍN[1], LUIS GIMENO PRESA[2], AND PEDRO RIBERA RODRÍGUEZ[1]

[1]*Dpto. Física de la Tierra II. Facultad de Ciencias Físicas Universidad Complutense de Madrid.*
[2]*Campus As Lagoas, Universidad de Vigo.*

1. INTRODUCTION

The Canary Islands constitute an isolated spot of great interest which has been poorly studied from a climatic point of view. The archipelago is located in the subtropical North Atlantic under the influence of the semipermanent high pressure system centered centered at 30ºN—40°N in the northeastern Atlantic Ocean. However, the Islands' steep orography makes them much wetter that is usual at these latitudes.

The dominant mode of winter atmospheric variability in the North Atlantic area is the North Atlantic Oscillation (NAO) (Walker and Bliss 1932). This oscillation modulates the climate of much of Europe (van Loon and Rogers 1978; Hurrell 1995; Ulbrich et al. 1999; Wallace et al. 1995) and its influence has been detected as far south as in North Africa (Lamb and Peppler 1987; Ward et al. 1999). On the other hand, previous papers have shown (Dong et al. 2000; Huang et al. 1998), that this subtropical area exhibits some kind of ENSO-related variability.

This paper focuses on a study of the possible influence of NAO and SO on the precipitation recorded in the Canary Islands. A preliminary statistical analysis provides the basis for deeper study. Then, an automatic objective

M. Beniston (ed.),
Climatic Change: Implications for the Hydrological Cycle and for Water Management, 57–73.

detection algorithm is applied for the detection of synoptic systems leading to precipitation in the Islands. Finally, the influence of NAO and ENSO on these patterns is discussed.

2. CANARY ISLANDS PRECIPITATION OVERVIEW

From the values recorded at individual stations, total precipitation has been estimated for every island through a weighting procedure whose details can be found in (García et al. 2001). These values have been used to characterize the behavior of the precipitation on monthly and higher time scales. Precipitation in the islands exhibits a marked seasonality, with a common pattern for all the islands. Fig. 1 shows the average monthly pluviograms and the geographic distribution of precipitation for every island for the entire archipelago. It is clear that the period November through March constitutes the rainy season, while during the during the other months precipitation is very scarce, even in the western and wetter islands. The precipitation is more abundant in the northern side of the islands. Fig 2 shows the evolution of standardized annual precipitation for every island and the temporal coverage. Tenerife Island time series span from 1901 to 1998 being the most reliable series. The rest of the island's series cover from mid-40s or mid 50s to 1998.

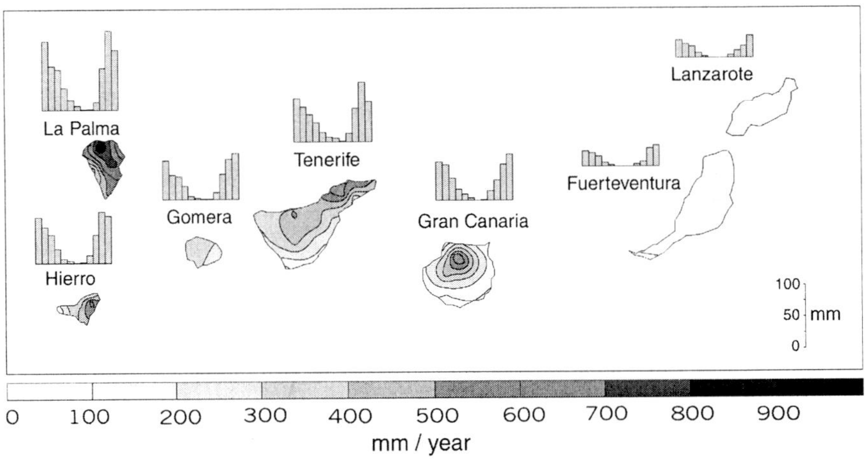

Figure 1. Precipitation regime in the Canary Islands. Shaded contours show annual precipitation (mm per hydrologic year). Average pluviograms (mm per month) are also shown.

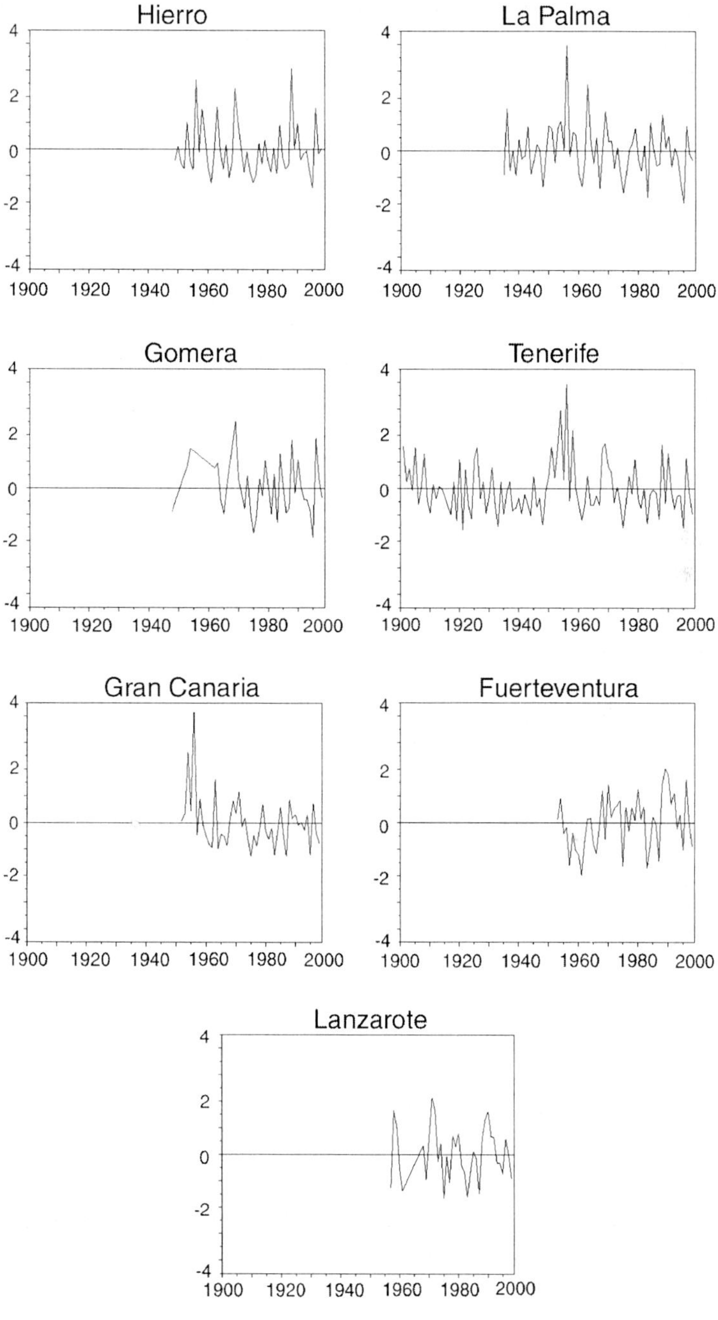

Figure 2. Normalized annual precipitation time series for the Canary Islands.

To evaluate the relationship among precipitation in the different islands, the correlation coefficients between annual values for every pair of islands have been computed, and the results shown in Table 1.

Table 1. Correlation between rainfall time series of the Canary Islands. (hydrological years from September to August).

	Hierro	La Palma	Gomera	Tenerife	G. Can.	Fuertev.
La Palma	0.84**					
Gomera	0.86**	0.84**				
Tenerife	0.69**	0.75**	0.82**			
G. Can.	0.62**	0.82**	0.65**	0.87**		
Fuertev.	0.27	0.29	0.36*	0.39**	0.37*	
Lanzarote	0.29	0.44**	0.21	0.50**	0.68**	0.70**

** 99% significant correlation
* 95% significant correlation

The four western islands behave as a cluster, and are poorly corelated with the two eastern ones, while centrally located Gran Canaria, is highly correlated with both groups of islands. This differential behavior stems mainly from the differences in orography and their relation with the atmospheric vertical structure. The Canary Islands are under the direct influence of the Azores High, and typically, a subsidence inversion is found at elevations between 700 and 1500 m, making the usual weather very stable and dry. Precipitation events can occurs when this situation is broken by any kind of disturbance (Font 1956). The five western island reach heights above 1500 m and the orography usually acts as triggering factor. On the other hand, Lanzarote and Fuerteventura are flatter, reaching elevations of 600 m at most, below of the trade wind inversion layer, and therefore, stronger disturbances are needed to register precipitation. This atmospheric structure along with the orography plays a key role into the instability caused by the presence of cold air at upper levels. The trade winds blow mainly against the north side of the islands, which can rise over their slope frequently leading to condensation and cloud development. However, the cloud growth is obstructed by the inversion layer. When cold air at upper levels reach the Canary latitude, the instability generated allow for cloud growth and the development of precipitation.

The location of the islands, in the southern edge of the Azores High makes the precipitation sensitive to the influence of the North Atlantic Oscillation (NAO). On the other hand, there are a number of recent papers which hypothesize on the influence of ENSO in the area (Dong et al. 2000; Huang et al. 1998). Therefore, the possible relationship between these major oscillations and the precipitation has been investigated. The linear correlations between precipitation recorded at individual islands and indices of the NAO and SO systems (data taken from the Climatic Research Unit)

(Hurrell 1995a; Jones et al 1997) have been computed for every island, and the results shown in tables 2 and 3, respectively.

Table 2. Correlation between annual rainfall time series for calendar years and hydrological years and winter NAO index (NAO index computed from December to March, hydrological years from September to August).

Island	Calendar year	Hydrological year
Hierro	-0.36**	-0.43**
La Palma	-0.39**	-0.49**
Gomera	-0.22	-0.51**
Tenerife	-0.13	-0.20**
Gran Canaria	-0.27*	-0.38**
Fuerteventura	0.08	-0.04
Lanzarote	-0.12	-0.14

** 99% significant correlation.
* 95% significant correlation.

Table 3. AS table 2 but for SO index.

Island	Calendar year	Hydrological year
Hierro	0.04	0.13
La Palma	0.08	0.16
Gomera	-0.10	0.06
Tenerife	0.07	0.04
Gran Canaria	0.20	0.29*
Fuerteventura	0.02	0.15
Lanzarote	0.16	0.22

* 95% significant correlation.

Table 2 gives the coefficient values for the NAO, while, table 3 is for SOI. It can be seen that NAO and SO correlations are higher when the hydrologic year (September–August) is considered. NAO influence is clear and significant for the five western islands. Gran Canaria is the only island which exhibits some significant correlations with SO.

Daily precipitation values were used to gain more insight on the nature of these influences. Daily precipitation is available for the study period at 8 stations across the islands, whose distribution can be seen in fig. 3. For these stations, the average monthly standardized values were computed and represented against NAO and SO indexes, as is shown in fig. 4. It can be seen that most of the precipitation occurs for NAO negative values, as expected. However, it is also evident that SOI positive values (La Niña prevailing conditions) are also associated with high precipitation; furthermore, the highest levels are recorded when NAO negative and SOI positive conditions occur simultaneously. NAO seems to be linked to more frequent precipitation, but SO seems to have an influence in the development of the more intense precipitation events.

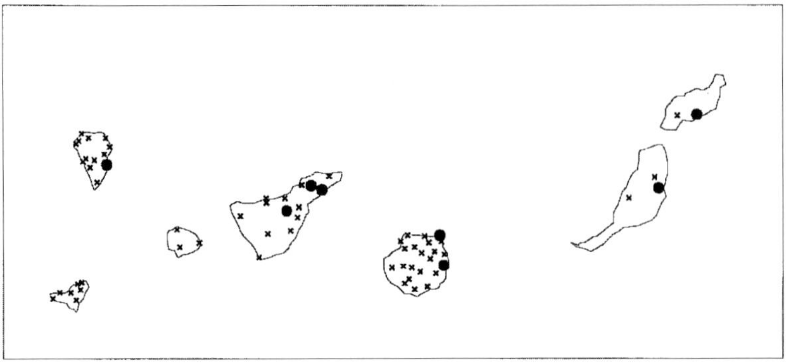

Figure 3. Location of selected Canarian rainfall station used in the study. Stations with monthly data are marked as crosses while circles indicate the stations with daily coverage

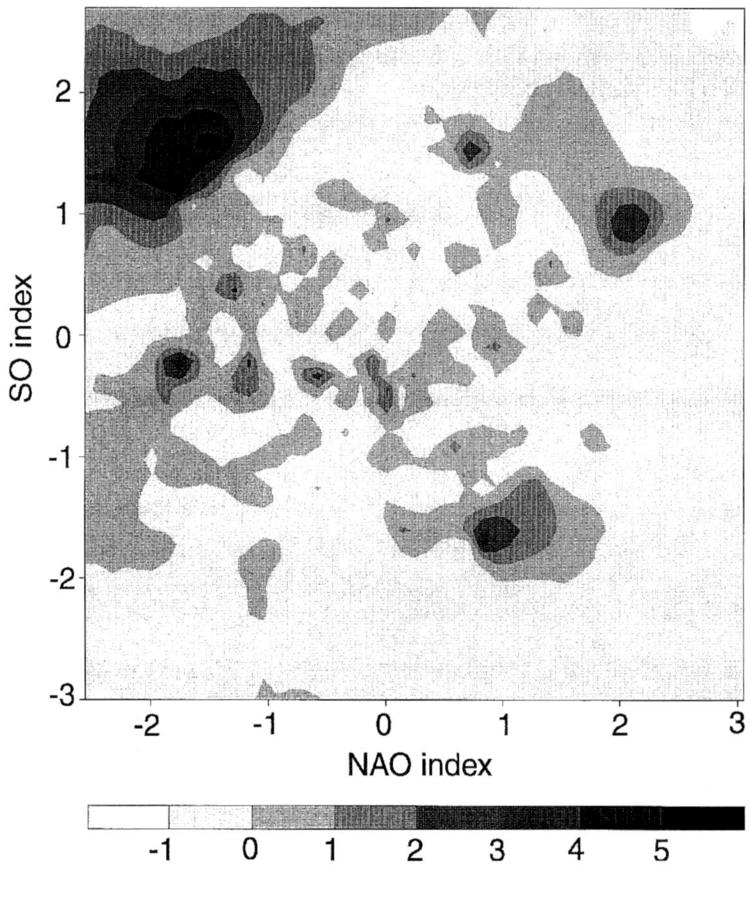

Figure 4. Average monthly standardized values (contour shades) represented against NAO (x-axis) and SO (y-axis) for the eight stations with daily coverage

3. SYNOPTIC PATTERNS ASSOCIATED TO PRECIPITATION IN THE CANARY ISLANDS

3.1 Synoptic patterns leading to precipitation in the Canary Islands

As discussed before, precipitation events on the Canary Islands are likely to occur when stable atmospheric conditions that are usually found over the archipelago, due to the presence of the Azores High, are broken by some kind of disturbance. In order to establish an objective classification criteria, these disturbed states have been grouped into three main categories:

Surface disturbances: They consist mainly of low pressure systems losing strength with height, so at the 500 hPa level they are scarcely detectable. These systems which usually form over the Atlantic to the north of the islands can reach them when the Azores High weakens or is displaced with respect to itsmean position.

High level disturbances: Most of the precipitation recorded in the islands occurs under this type of event. When cold air flows directly over the archipelago, the induced instability (which is reinforced by the orographic effect) usually produces precipitation. At 500-hPa , these disturbances can appear as troughs or as closed lows (cut-off lows). Such closed lows are often the second stage in the development of a wave disturbance.

Synoptic systems affecting the entire troposphere: Sometimes, normally in the winter season, well developed Atlantic low pressure systems can reach the Canary Islands. Their associated frontal tails often results in abundant precipitation.

3.2 Classification procedure

To get an objective classification of the disturbances throughout the study period, a set of automatic detection algorithms have been developed and applied to the geopotential data from 1955 to 1998 provided by NCEP/NCAR reanalysis (Kalnay et al 1996) for the North Atlantic area. This will allow the study of the variability of the synoptic systems and the assessment of the relative importance for each islands' precipitation of different synoptic systems. Two geopotential levels have been selected, 1000 hPa to characterise the disturbances detected near the surface and 500 hPa for those affecting upper levels.

The synoptic systems consisting of closed lows, either at surface or at 500 hPa have been automatically detected by means of an adapted version of algorithms described in previous works (Blender et al. 1997; Trigo et al.

1999), which look for local geopotential minima surrounded by enough pressure gradient to assure that they are not spurious. The algorithm has been fitted to the particular latitude of the Canaries and to the data set, reaching more than 95% of effectiveness. A new algorithm devoted to the detection of troughs at the 500 hPa level have been developed. Essentially it compares the geopotential height field over the Canary Islands with those found over a surrounding strip at the same latitudinal belt. Particular thresholds of both algorithms were chosen after a subjective analysis carried out for January, April and August of 1979, thus avoiding the effects of the seasonal changes. Fig. 5 shows a diagram of the procedure along with the thresholds employed (complete details can be found in Garcia et al 2001).

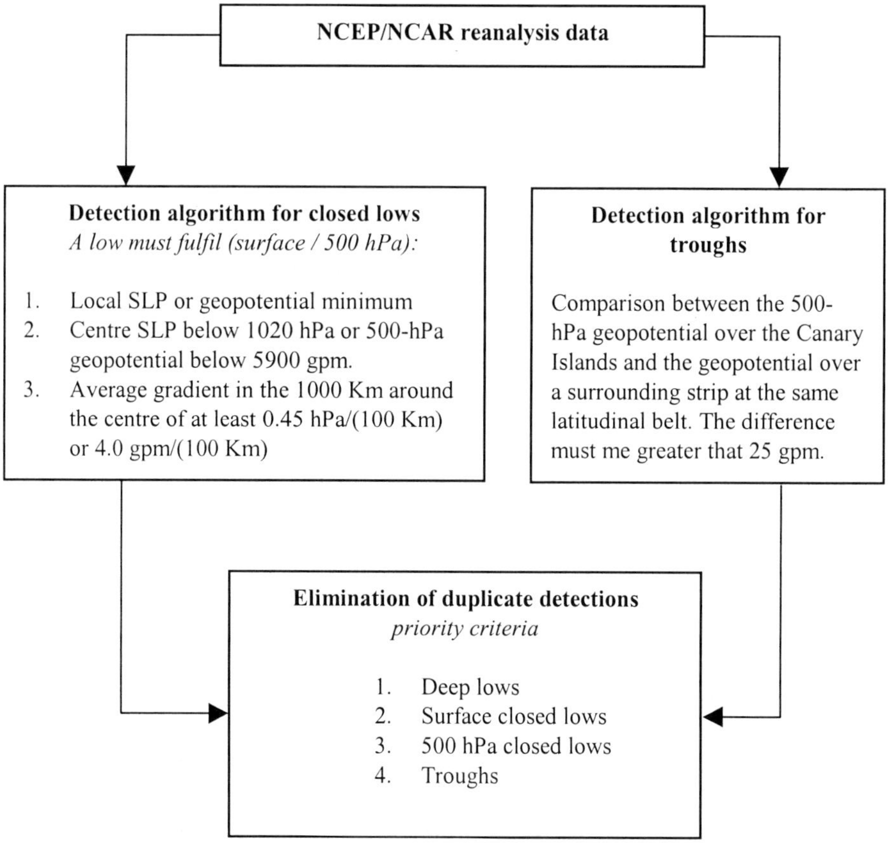

Figure 5. Synoptic system detection procedure and detection algorithm's thresholds

The principal statistics obtained for the four disturbances are shown in table 4. The more frequent synoptic situations are those found at upper levels. The 500-hPa lows occur most frequently, with almost 45 cases per year, followed by troughs. However, the more intense precipitation events occur in association with deep lows, followed by the surface lows type. As can be seen from Table 4, these disturbances are the least frequent, but they cause the most intense individual precipitation events. It is important to emphasize that the detection of a particular disturbance does not assure the occurrence of measurable rainfall. On average, about 50% of detections yield rainfall. Over 75% of total precipitation in the archipelago has been associated to one or another of the disturbances. Almost 40% is due to 500-hPa disturbances, while more that 20% is due to well-developed Atlantic lows.

Table 4. Main synoptic disturbances leading to precipitation in the Canary Islands, average detection per year (average number of events) , average precipitation per event (mm) and percentage of annual precipitation due to each disturbance.

Synoptic disturbance	Number of detections per year	Precipitation per event	% of total precipitation
Atlantic low at surface	18.1	7.2	17.0
Atlantic low at 500 hPa	44.9	4.9	21.5
Trough at 500 hPa	25.1	5.9	17.0
Deep Atlantic low	9.0	8.9	21.2

3.3 NAO influence on synoptic patterns

In section 2, the NAO influence on precipitation in the Canary Islands was established on a statistical basis. However, the correlation coefficients are not as strong as would be expected if the rainfall variability depended only on the behaviour of the Azores High, one of the poles of the North Atlantic Oscillation. A more detailed study of the rain occurrence associated with different synoptic types, in relation to the NAO is presented in this section. We wish to answer questions such as: how different is the NAO influence on the surface lows with respect to the 500-hPa lows or the troughs?, and What are the temporal characteristics of this signal? To investigate the NAO effect over the different kinds of disturbances, the five more extreme NAO years of the period from 1955 to 1998 were chosen and their respective frequency and anomaly pattern were computed.

Table 5 shows the average number of disturbances per winter depending on the NAO phase. With the notable exception of the 500-hPa lows, the disturbances are more frequent during the negative phase of the NAO, especially the surface lows, which exhibit a significant increase of more than 200%.

Table 5. Frequencies of the synoptic disturbances leading to precipitation in the Canary Islands. Frequency is expressed as number of events per winter (from December to March).

	NAO+	NAO+
Atlantic low at surface	5.6	12.4**
Atlantic low at 500 hPa	18.0	9.2*
Trough at 500 hPa	7.6	9.8
Deep Atlantic low (detected at surface and 500 hPa)	3.2	4.2

** 99% significant difference with the NAO+ value
* 95% significant difference with the NAO+ value

The amount of precipitation depends not only on the frequency but on the intensity of the disturbance. Composite anomaly maps for each disturbance type were computed. Figs. 6a and 6b show the results for the surface and 500-hPa lows, respectively.

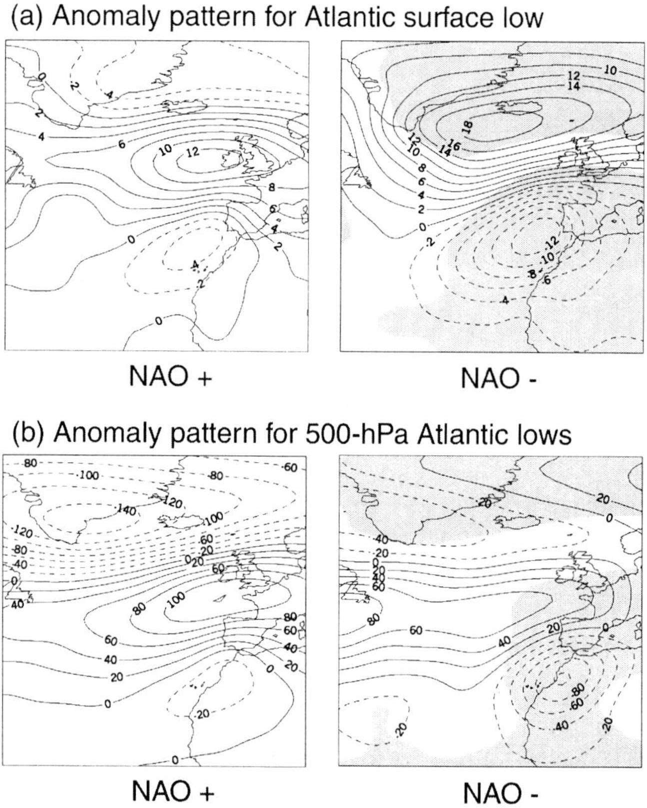

Figure 6. SLP winter anomaly patterns for Atlantic surface low (panel a) and geopotential height anomaly for 500-hPa Atlantic lows (panel b) for the five most NAO positive / negative years (left / right). Anomalies are computed respect to the 1000 hPa (a) and 500 hPa (b) geopotential winter average. Shaded areas in the NAO negative maps indicate 95% significant differences between NAO negative and NAO positive phases (1000-trials Montecarlo test).

A strong intensification of the disturbances during the NAO negative phase is indicated. Surface lows and 500-hPa lows are on average, 8 hPa and 80 gpm deeper, respectively. Similar remarks can apply to the upper air troughs while the deep Atlantic lows, show only a slight and non-significant change in the same direction (not shown). Rainfall series were split into four parts, each one representing the precipitation due to every kind of disturbance. Partial correlations between these split rainfall series and the NAO index were computed. Table 6 shows the results for Tenerife Island, which represents the behaviour of the five western islands most closely related with NAO. Two of the disturbance show significant correlation values, while the NAO signal is lost for the two remaining. As comparison, the correlation index of this island for the raw precipitation series was –0.20.

Table 6: Significant correlations between the rainfall recorded for every disturbance type (Tenerife Island). Correlations shown are 99% significant.

Disturbance	**Correlation**
Atlantic low at surface	-0.63
Atlantic low at 500 hPa	-
Trough at 500 hPa	-0.50
Deep Atlantic low (detected at surface and 500 hPa)	-

The most strongly NAO related rainfall is that due to the Atlantic lows at the surface. As was shown in table 5 and in fig. 6a, this type of event is more frequent and more intense during NAO negative phase. The same remark is valid for the precipitation attributable to the 500-hPa troughs although in this case, during NAO negative phase, the slight increment in the frequency is not significant? In contrast, deep Atlantic lows do not exhibit significant NAO relationships. Neither their frequency, nor their intensity depends on changes in the NAO, and consequently, the related precipitation does not exhibit any detectable signal. The case of 500-hPa troughs is especially interesting. In this case, the frequency and intensity of these disturbances show an opposite behaviour (both significant), the low frequencies corresponding to the deeper disturbances and vice versa. The net effect over the precipitation signal is thus cancelled, and they do not show an NAO relationship.

It must be pointed out that the NAO signal over the disturbance's characteristics is not only detected when NAO extremes are considered. Significant correlation indexes are found between the frequency and depth of the closed lows systems and the overall NAO index, when the entire period of study is considered over an annual basis (not shown).

The synoptic systems affecting the Canary Island environment have been shown to be especially sensitive both in frequency and intensity to the NAO signal. An analysis of the Atlantic storm track based on the filtered geopotential series variance in the 2.5-8 days period (Mukarami 1979; Christoph et at 1995) has been carried out in order to explore the causes and the spatial extension of the increase of synoptic activity in the Canary Islands.

The analysis of the raw geopotential variance does not reveal clearly the changes in the circulation in the region at the latitude of the Canary Islands Latitude, since the maximum variance is located in a band from 50ºN to 70ºN, which comprises the typical path of the Atlantic cyclones. However when the difference of the storm track (i.e. the difference of the geopotential variance) is computed between the NAO extremes the areas with the greatest variation are more clearly shown. Fig 7 shows the 1000 hPa storm track difference between NAO extremes for winter (December–March), January, the month for which the signal is greatest, and April, when the signal is mostly gone.

The bipolar structure indicates the southward shift of the storm track during NAO opposite phases. During NAO negative years, in winter, there is a decrease in variance of about 20% in the characteristic storm track centred at 60ºN, while at the latitude of the Canary Islands, a significant increase is seen of between 30 and 40%. January is the month with greatest differences in storm activity between extremes of the NAO, reaching increasing above 100% with respect to the NAO positive value during NAO negative phase centred exactly at the latitude of the islands. This behaviour is limited to the winter season. For example a comparison for the month of April indicates an absence of significant differences. Similar results are obtained for the 500 hPa storm track but with lesser increases.

3.4 ENSO influence on synoptic patterns

The ENSO signal over the Canary Islands precipitation is subtler than that found for the NAO (Gallego et al. 2001). The correlation between the annual precipitation series and the SOI showed slight positive correlations, which are significant for Gran Canaria Island only. In order to maximize any possible signal of ENSO on Canary Is. Precipitation, the five most intense El Niño/La Niña events for the period 1955-1998 were chosen and the changes in the patterns and frequencies of the synoptic disturbances were examined.

For the ENSO case, there are no significant changes in the frequency associated with each disturbance type, although in all cases, but for the troughs, there is a slight increase during La Niña events. The synoptic patterns corresponding to the surface level do not show significant variations. However, when the upper levels are considered, the ENSO-related changes are evident.

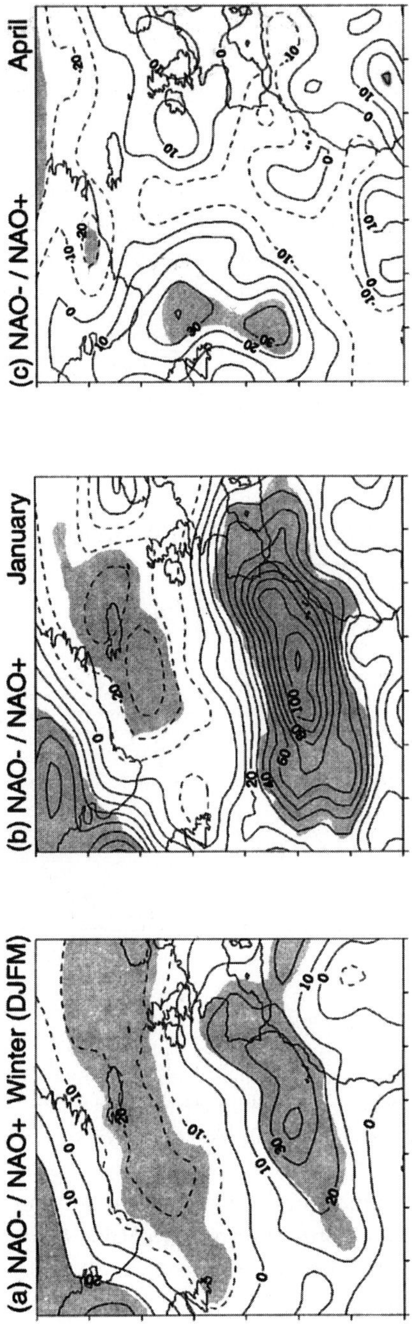

Figure 7. Comparison between the storm track based on the filtered geopotential series for NAO negative and NAO positive phases computed as 100%·[(NAO- value)-(NAO+ value)]/(NAO+ value). 99% significant differences are shaded (1000-trials Montecarlo test).

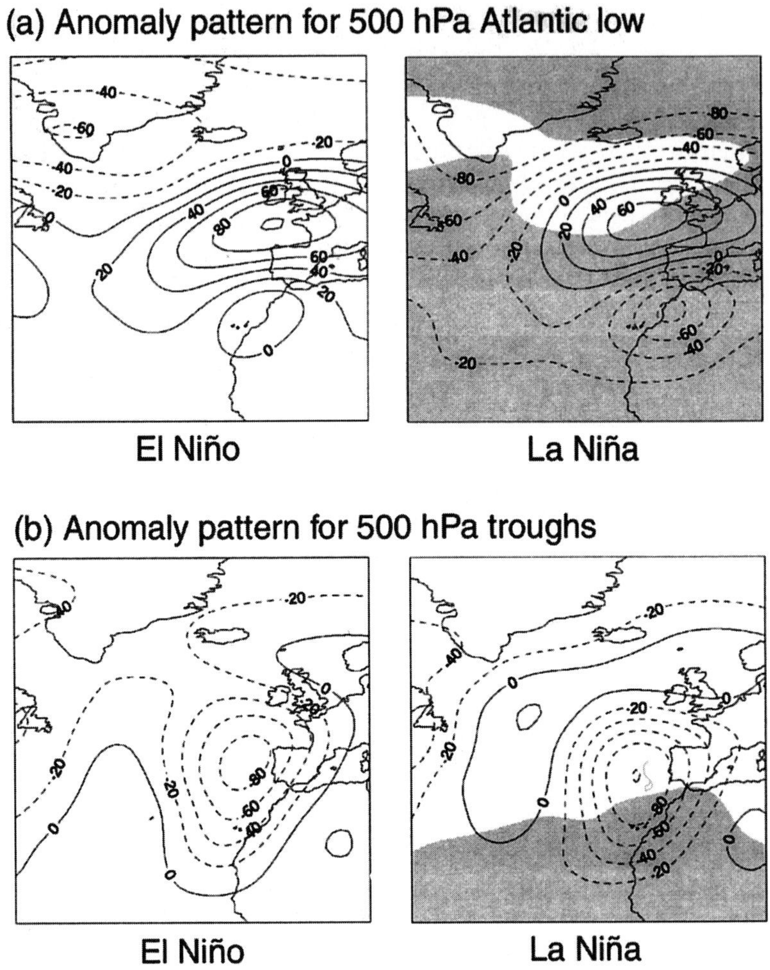

Figure 8. 500-hPa geopotential annual anomaly patterns for 500-hPa Atlantic low (panel a) and 500-hPa Troughs (panel b) for the five stronger El Niño / La Niña years (left / right). Anomalies are computed respect to the annual average 500-hPa geopotential. Shaded areas in the La Niña maps indicate 95% significant differences between El Niño and La Niña patterns (1000 trials Montecarlo test).

Figs. 8a and 8b show the anomaly pattern for the 500-hPa lows and the 500-hPa troughs. The 500-hPa low pattern structure is bipolar, with a negative anomaly causing instability over the islands, much deeper during La Niña years. Trough intensity does not increase with the ENSO phase, however their centres are displaced 10º southward, much closer to the islands. It should be noted that the appearance of a closed-low-shaped pattern for the troughs is an artefact of the averaging procedure. Even the

500–hPa anomaly pattern, corresponding to the lows detected at the surface, shows a negative centre significantly displaced southward during La Niña years. Deep Atlantic lows do not shown significant variations.

The ENSO effect on precipitation is very consistent when extreme years are considered. Table 7 shows the average rainfall for the whole archipelago and for each disturbance type. The greatest increments are found for rainfall due to high level disturbances, as expected from the synoptic analysis, although a significant signal is present for the surface Atlantic lows.

Table 7. Average precipitation intensity (mm per precipitation day) detached by disturbance type for the five more extreme El Niño and La Niña events (1955 – 1998)

Disturbance	El Niño	La Niña
Atlantic low at surface	4.9	7.9*
Atlantic low at 500 hPa	3.3	6.0**
Trough at 500 hPa	5.0	8.2*
Deep Atlantic low (detected at surface and 500 hPa)	10.3	11.3

** 99% significant difference with the El Niño value
* 95% significant difference with the El Niño value

4. CONCLUSIONS AND REMARKS

In this work, the study of the rainfall series of the Canary Islands along with the analysis of the synoptic patterns leading to precipitation has resulted in the detection of consistent NAO and ENSO signals. The implementation of objective identification methods for the detection of the main synoptic systems related to precipitation provides a qualitative description of the dynamical mechanisms involved in the observed teleconnections. The results obtained can be summarised as follows:

1. The Canary Islands are an excellent natural observatory for the atmospheric variability of the North Atlantic area. Their geographic situation, along with the orographic configuration make them very sensitive to small changes in the atmospheric circulation.
2. Four main synoptic system types have been detected that are responsible for more than 75% of the observed rainfall. These systems can be classified according to the height as low level disturbances (surface lows), upper level disturbances (500-hPa lows or troughs), or low pressure systems found through the entire troposphere (deep lows).
3. A consistent NAO signal has been detected in precipitation. During NAO negative years, precipitation is more intense and more frequent,

in agreement with results obtained in previous works for the nearby Moroccan region.

4. NAO negative years are characterised by deeper disturbances in all cases, which are more frequent as well, with the exception of the 500-hPa lows. In the region studied, the greatest sensitivity of precipitation is found for rainfall due to the surface lows and troughs, because the related disturbances are simultaneously more frequent and deeper during NAO negative phase. The 500-hPa lows exhibit an opposite behaviour, NAO negative years are characterised by much deeper anomalies than average but the frequency is reduced by a factor of two, thus cancelling an NAO signal on precipitation. Deep lows do not show NAO signal, however, the number of these systems is about 4 to 5 events per year only (winter), making the sample barely significant for the five-year study described here.
5. The increase of synoptic activity during NAO negative years is due to the southward shift of the cyclone track during the winter season, especially during January, when the precipitation signal is a maximum.
6. A subtle ENSO signal has also been found. Positive correlation coefficients between SOI and rainfall are found for all the islands, although they are only significant from Gran Canaria, indicating an increase of precipitation in La Niña years.
7. The ENSO cycle does not appear to significantly influence the frequency of the disturbances. Nonetheless, all the disturbances, especially those found at high level, are significantly deeper during La Niña years. This signal is translated into changes in the precipitation events, which are more intense for the 500-hPa disturbances. A significant ENSO signal is found even for the precipitation due to surface lows.
8. Both, NAO and ENSO signals are present during the entire period of record, but the signal becomes more significant during the extreme phase of these oscillations, but especially for the ENSO cases.

5. ACKNOWLEDGMENTS:

This study is supported by the Spanish *Comisión Interministerial de Ciencia y tecnología* under grant CLI98-0649.

6. REFERENCES

Blender, R., K. Fraedrich, and F. Lunkeit, 1997: Identification of cyclone-track regimes in the North Atlantic. *Quart. J. Roy. Meteor. Soc.*, **123**, 727-741.

Christoph, M., Ulbrich, U., and Haak, U., 1995: Faster determination of the intraseasonal variability of storm tracks using Murakami's recursive filter. *Mon. Wea. Rev.,* **123**, 578-581.

Dong, B.W., Sutton, R.T., Jewson, S.P., O'Neill, A. and Slingo, J.M., 2000: Predictable winter climate in the North Atlantic sector during the 1997-1999 ENSO cycle. *Geophys. Res. Let.*, **27**, 985-988.

Font, I. 1956: *El tiempo atmosferico en las Islas Canarias.* Servicio Meteorologico Nacional, 96 pp.

Gallego, D., García, R., Hernández E., Gimeno, L., Ribera, P. (2001): An ENSO signal in the North Atlantic subtropical area. Submitted to *Geophys. Res. Let.* (under revision)

García, R., Gallego, D., Hernández, E., Gimeno, L., Ribera, P. (2001): Influence of the North Atlantic Oscillation on the canary Islands precipitation. Submitted to *J of Climate* (under revision).

Huang, J.P., Higuchi, K. and Shabbar, A. 1998: The relationship between the North Atlantic Oscillation and El Niño-Southern Oscillation. *Geophys. Res. Let.*, **25**, 2707-2710.

Hurrell, J.W., 1995a: Decadal Trends in the North Atlantic Oscillation Regional Temperatures and Precipitation. *Science*, **269**, 676-679.

Jones, P.D., Johnson, T. and Wheeler, D., 1997: Extension to the North Atlantic Oscillation using early instrumental pressure observations from Gibraltar and South-West Iceland. *Int. J. Climatol.* **17**, 1433-1450

Kalnay E. Et al., 1996: The NCEP/NCAR 40-year reanalysis project. *Bull. Amer. Meteor. Soc.,* **77**, 437-471.

Lamb, P.J. and Peppler, R.A.: North Atlantic Oscillation: concept and an application. *Bull. Amer. Meteor. Soc.,* **68**, 1218-1225.

Murakami, M., 1979: Large-scale aspects of deep convective activity over the GATE area. *Mon. Wea. Rev.*, **107**, 994-1013.

Ward, M.N., Lamb, P.J., Portis, D.H., El Hamly, M., Sebbari, R. *Beyond El Niño: Decadal and interdecadal Climate Variability.* (Navarra, A., Ed.). Springer 374 pp.

Trigo, I.F., Davies, T.D., and Bigg, G.R., 1999: Objective climatology of cyclones in the Mediterranean region. *J. Climate*, **12**, 1685-1696.

Ulbrich, U., Christoph, M., Pinto, J.G. and Corte-Real, J., 1999: Dependence of winter precipitation over Portugal on NAO and baroclinic wave activity. *Int. J. Climatol.*, **19**, 379-390.

van Loon, H. and Rogers, J.C., 1978: The seesaw in winter temperatures between Greenland and Northern Europe. Part I: General Description. *Mon. Wea. Rev.,* **106**, 296-310.

Walker, G.T., and E.W. Bliss, 1932: World weather V. *Mem. Roy. Meteor. Soc.,* **4**, 53-84.

Wallace, J.M., Zhang, Y., and Renwick, J.A., 1995: Dynamic Contribution to hemispheric mean temperature trends. *Science*, **270**, 780-783.

High resolution simulations of precipitation over the Alps with the perspective of coupling to hydrological models

O. BRASSEUR[1], H. GALLÉE[1], J.-D. CREUTIN[1], T. LEBEL[1], AND P. MARBAIX[2]

[1] *Laboratoire d'étude des Transferts en Hydrologie et Environnement*
Institut National Polytechnique de Grenoble, Grenoble, France
[2] *Institut d'Astronomie et de Géophysique G. Lemaître*
Université catholique de Louvain, Louvain-la-Neuve, Belgium

Abstract: As the computer resources increases, regional climate simulations can be investigated at finer horizontal resolutions in order to improve the representation of surface characteristics and mesoscale circulation. This paper aims at examining the relevance of precipitation fields simulated by high-resolution regional climate models for hydrological purposes.

The MAR – Modèle Atmosphérique Régional – model was run at 20-km resolution for a four-month period over the Alps and the southeastern mountainous regions of France. The validation on the climatology of mean variables has not revealed visible drift or spurious trend. Precipitation generally reaches about 80% of the value obtained by the GPCC – Global Precipitation Climatology Center – climatology, and the spatial distribution of precipitation also compare favorably to observations. Significant sensitivity to the inclusion of cloud water fluxes at the lateral boundaries has been found for the spatial distribution of precipitation, more specifically in the representation of stratiform precipitation.

However the finest horizontal resolution used in RCM is about 20 km, which remains too coarse for the forcing of hydrological models, especially over mountainous regions. In order to improve the impact of the topographical forcing on microphysical processes, a rain disaggregation model has been applied to the precipitation simulated with MAR. Such a model has proven its usefulness for the representation of precipitation over watersheds, particularly for the situations characterized by heavy rainfall. Even if improvements are yet required, for instance considering interactive disaggregation, rain disaggregation models can be seen as a useful tool to interface hydrological and regional climate models.

M. Beniston (ed.),
Climatic Change: Implications for the Hydrological Cycle and for Water Management, 75–99.

1. INTRODUCTION

For this century, a modification of precipitation intensity is expected in response to an intensification of greenhouse effect (IPCC, 1996). Several models suggest a possibility for more extreme rainfall events in a warmer climate. Resulting changes in precipitation regime would have an impact on hydrology and, for extreme situations, on floods. This impact could be more pronounced over complex topography: Beniston and Rebetez (1996; Beniston et al., 1997; Rebetez and Beniston, 1998) have suggested that surface climate change associated with global warming presents elevation dependency for the mountain regions of the world: the warming would be more pronounced for high elevation sites than for low elevation. The consequences of this elevation dependency are important considering changes on temperature and precipitation, which may have significant implications on surface energy budgets and water resources (Giorgi et al., 1997). However reliable modeling of orographic precipitation remains elusive (e.g. Dunn, 1991), although the scientific understanding of microphysical processes has been substantially improved. In addition, accurate representation of precipitation over complex topography is of crucial importance to hydrological modeling. Progresses should be accomplished in atmospheric models in order to improve the representation of precipitation over complex topography for hydrological purposes. The present paper proposes a first step toward the elaboration of a tool to interface atmospheric and hydrological models working at very different spatial scales.

Current regional climate models (denoted RCM) consider typical horizontal resolutions between 20 and 60 km. Most of these RCM present significant orographic precipitation biases (e.g. Christensen et al., 1997). Monthly-averaged precipitation bias with a RCM at 20 km resolution at the scale of Switzerland is about 30% (Marinucci et al., 1995). the bias on precipitation can become much larger if more local scales are considered, in particular the scale of watersheds (area between 100 and 10000 km^2). The main reason explaining local fluctuations of rain is related to the representation of surface characteristics (orography, vegetation, etc) that cannot be explicitly represented in RCM, even at 20-km resolution.

Available computing resources are obviously responsible for limitations in the horizontal resolution. But even when computing resources will be available to run climate models at finer resolution, the physics of deep convection will probably limit studies to the horizontal resolution of 20 km, because no satisfying solution has been found for the representation of convection between 2- and 20-km resolutions (Molinari and Dudek, 1992). Nevertheless, such resolutions remain insufficient to accurately represent the fine scale orography. As a consequence, further developments will be necessary to improve the physics of parameterizations describing subgrid processes in function of their importance in some regions (Giorgi and Mearns, 1999).

In this framework, rain disaggregation models (denoted RDM) could progressively be considered as subgrid schemes in RCM. As the topographical forcing mostly contributes to spatially redistributing precipitation (Giorgi and Marinucci, 1996), RDM improves the realism of local precipitation since they take into account a more accurate representation of the subgrid orography.

This paper proposes to assess the relevance of precipitation simulated over the Alps with a RCM for hydrological purposes. Since a fine scale description of precipitation is required over watersheds, simulated precipitation fields have been disaggregated with the model of Sinclair (1994). The usefulness of such a model is assessed. Section 2 presents experiment setup. The next sections are devoted to the assessment of model simulations: section 3 includes the validation of mean atmospheric variables and precipitation over the simulation domain, while section 4 presents the results obtained with a RDM over four Italian watersheds.

2. EXPERIMENT SETUP

A short description of the MAR model is given in section 2.1. The choice of the simulation domain and studied period are presented in sections 2.2 and 2.3.

2.1 Model overview

The formulation of the present version of MAR is described in Gallée and Schayes (1994) and Gallée (1995, 1996). MAR is a hydrostatic primitive equation model in which the vertical coordinate is the normalized pressure,

$$\sigma = (p - p_t)/(p_s - p_t),$$

where p, p_s and p_t are the actual pressure, the surface pressure and the model top pressure respectively. The solar radiation scheme is that of Fouquart and Bonnel (1980). The longwave radiation scheme follows a wide-band formulation of the radiative transfer equation (Morcrette 1984). The heat and moisture exchanges over land are represented with the surface model of Deardorff (1978). The hydrological cycle is fully described in Gallée (1995). The MAR includes also a convective adjustment scheme derived from Fritsch and Chappell (1980) for the representation of deep convection. The boundaries are treated according a dynamic relaxation that includes a Newtonian term (Davies 1976) and a diffusion term (Davies 1983; Anthes et al. 1989).

The parameterization scheme for the surface layer is based on Businger (1973) and Duynkerke (1991) formulations. In order to well represent the vertical mixing in convective boundary layer, the Therry and Lacarrère (1982) turbulent closure including a diagnostic relation for the mixing length has been preferred to one-and-a-half order closure with prognostic equation on dissipation, or second order closures (Brasseur et al. 1998).

2.2 Simulation domain

For all experiments presented in this paper, a common three-dimensional simulation domain is used. The vertical discretization includes 33 levels, with an increased resolution close to the surface. The first model level is at 10-m height. The vertical grid size is determined with arithmetic and geometric progressions, producing a finer resolution close to the surface layer and in the boundary layer. As shown in Fig. 1, the horizontal mesh covers France and the Alps with a horizontal extent of 2240 km by 1600 km, and is centered on 55.5N, 2W. The horizontal grid spacing is 20 km. The atmospheric and surface variables are initialized and forced at the boundaries with the ECMWF (European Center for Medium-range Weather Forecast) analysis. The temporal resolution of ECMWF fields is 6 hours, the horizontal resolution is 1 degree, and the vertical discretization includes 50 levels.

2.3 Simulation period

The selected period covers August to November 1999. The choice of this period is justified by precipitation intensities generally observed during autumn over mountainous regions of South-eastern France and Northern Italy. For the regions of Cevennes, South-western Alps, Lago Maggiore, and Po valley, climatology of precipitation exhibits high precipitation activity in October (Frei and Schär, 1998). This period partly includes the main convective period over Northern Alps, which typically concerns the months from May to September.

3. RESULTS AND ANALYSIS OF THE 4-MONTH SIMULATION

This section is devoted to the assessment of the simulation results for the entire simulation domain. In section 3.1, mean variables such as mean sea level, geopotential, temperature and specific humidity are examined. The most important variable for the forcing of hydrological models, i.e. precipitation, is analyzed in section 3.2. The results of the four-month simulation presented in sections 3.1 and 3.2 have been obtained by including cloud water fluxes at the lateral boundaries. These fluxes are computed according to the formulation presented in section 3.3. Their impact on atmospheric variables and precipitation is also assessed in this sub-section.

3.1 Validation of climatology of atmospheric variables

Before examining the simulated precipitation fields, the simulated climatology of atmospheric variables is assessed in order to point out some possible bias. For all the variables presented in this sub-section, the comparison is done by analyzing the finer-resolution fields (MAR) to the grid of the coarse-resolution fields (ECMWF). In figure 1, the simulated mean distribution of mean sea level pressure (hereafter MSLP) is compared with the driving analysis from ECMWF. The error in MSLP is generally less than 1 hPa, except over the mountainous regions of Jura and the Alps, which can be explained by the topography differences between MAR and ECMWF. Note that MSLP presents a negative bias compared to ECMWF analysis.

The climatology of 850-hPa temperature (hereafter T850) is presented in figure 2. A positive bias appears clearly in nearly all regions of the domain. This bias is weak (less than 1 K) over the Atlantic Ocean and increases from West to East. Maximum values do not exceed 2 K over mountainous regions.

In figure 3, the comparison for the 500-hPa height (hereafter Z500) shows a slight trend for a positive bias. The differences between MAR and ECMWF do generally not exceed 10 m, except over the mountainous region where they can reach 25 m. This bias is consistent with the analysis of MSLP and T850, since the negative bias on MSLP and the positive bias on T850 tend to compensate and to limit the bias on Z500.

The simulation of precipitation is highly dependent on the humidity fields. Consequently, there is a particular interest to assess the specific humidity field in the boundary layer. Figure 4 presents the comparison of simulated 850-hPa specific humidity with analysis. The difference is less than 0.4 g/kg over France and the Alps, while a significant bias is found over

Spain. However this bias is probably not too concerning for the simulation of precipitation over the Alps.

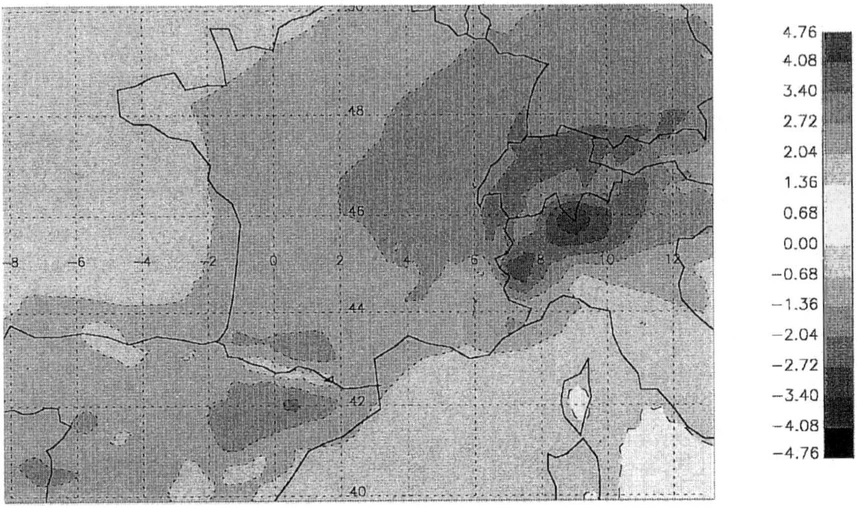

Figure 1 : Difference between simulated mean sea level pressure and driving analysis (units, hPa) for the period from August to November 1999. Positive and negative biases are denoted respectively by dashed and dotted contour lines. The represented domain ranges from 8W to 13E in longitude, and from 40N to 50N in latitude. Horizontal and vertical dotted lines represent longitude and latitude where the spacing is 2 degrees.

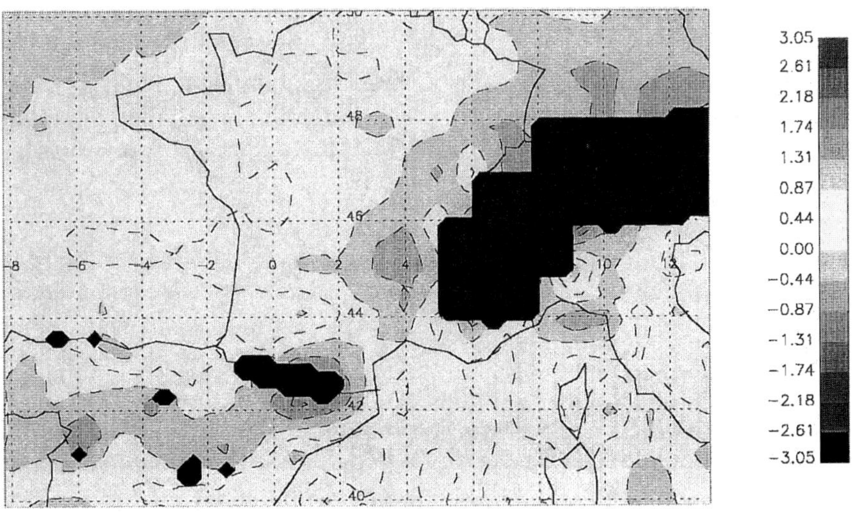

Figure 2 : As in figure 1, except that the difference between simulation and driving analysis is considered for the 850-hPa temperature (units, K). The black area on this map is due to the fact that the surface pressure is lower than 850 hPa over the mountainous regions in ECMWF or MAR models.

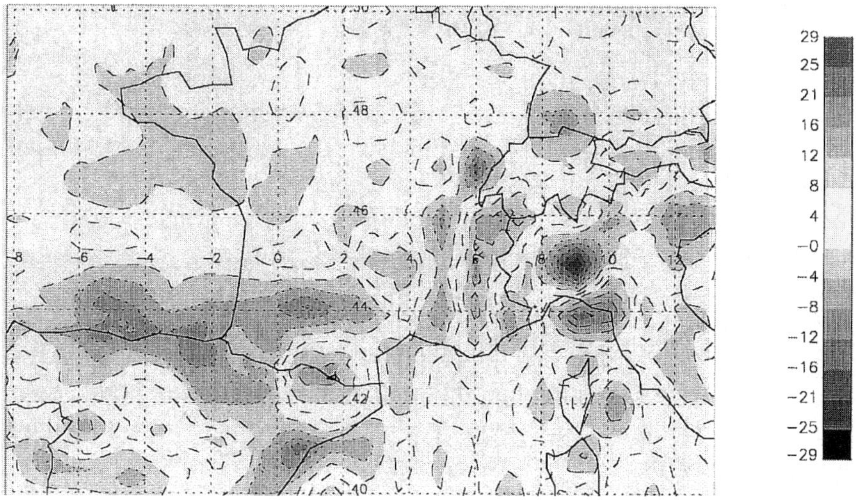

Figure 3: As in figure 1, except that the difference between simulation and driving analysis is considered for the height (units, m) of the 500-hPa surface pressure.

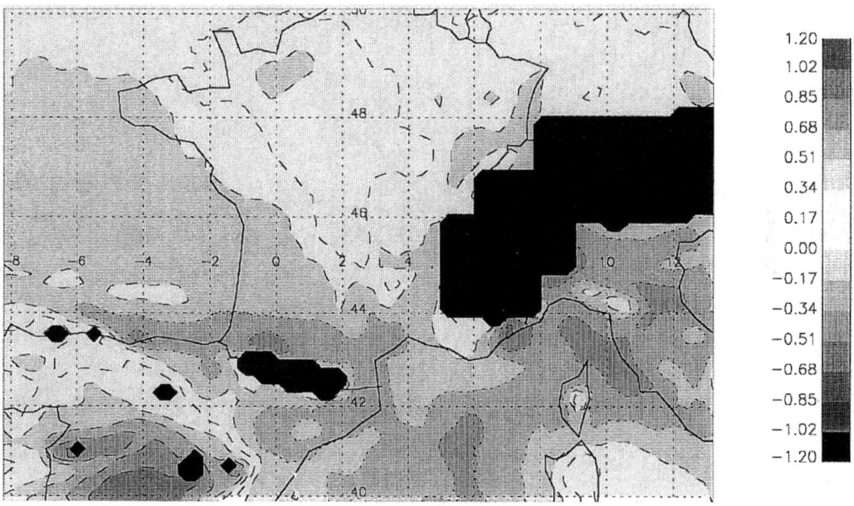

Figure 4 : As in figure 2, except that the difference between simulation and driving analysis is considered for the 850-hPa specific humidity (units, g/kg).

In conclusion, this analysis has not revealed spurious trend in the mean variables at the time scale of four months, which is an important result before assessing precipitation fields. The bias noted for MSLP, T850, Q850, and Z500 are in the range of biases found in other studies.

3.2 Precipitation analysis

The simulated precipitation has been compared to ECMWF 6-hourly precipitation and GPCC (Global Precipitation Climatology Center, 1992) monthly climatology. ECMWF precipitation is obtained from model forecast. The precipitation fields are available at 0.5-degree resolution (different from the resolution of forcing fields for lateral boundary conditions of MAR). Although they are not obtained from an analysis of observations, it can be expected that they should be close to observed precipitation over flat areas provided that the model has well assimilated the atmospheric fields. Monthly precipitation fields from GPCC are available on a 1-degree grid. As in the previous sub-section, the comparison between the different precipitation fields is done by analyzing the finer-resolution fields (MAR) to the grid of the coarse-resolution fields (ECMWF or GPCC).

3.2.1 Comparison with GPCC

Figures 5 and 6 present respectively the precipitation amounts obtained from the MAR simulation and from GPCC.

GPCC fields are not available over ocean. Over land, the amount of precipitation in a grid cell of 1 by 1 degree is obtained by aggregating observed precipitation. The accuracy of GPCC fields is related to the density of stations: as shown in figure 7, the density is particularly high over Switzerland with more than 6 stations per cell, but it is only 1 to 3 over France and Northern Italy. This weak density could particularly reduce the relevance of GPCC precipitation over the mountainous regions such as the Southern Alps and the Massif Central. In addition, observation stations in mountainous regions are often located in valley and low elevation. As mentioned in Sevruk (1975, 1989), measurements of precipitation can be underestimated by up to 30% at high elevations under strong wind conditions. Therefore, the comparison between simulation results and observations has to take into account this effect.

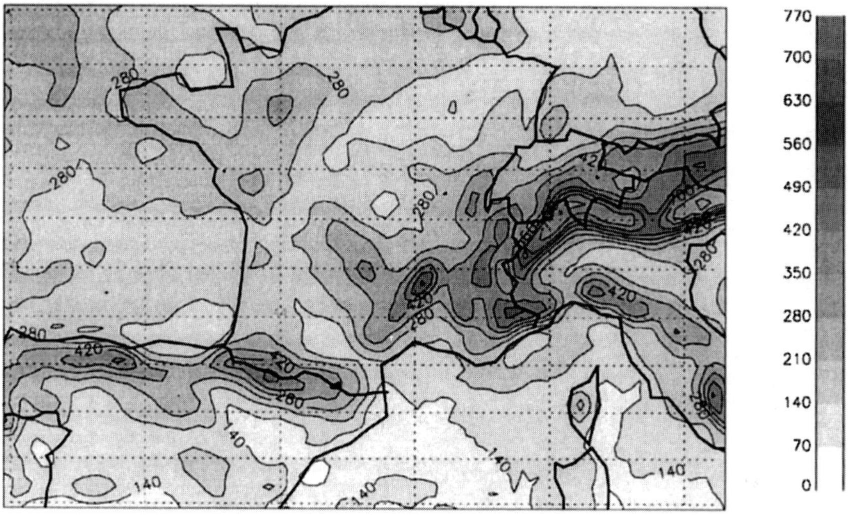

Figure 5 **: Precipitation amount (units, mm) simulated by MAR for the period from August t November 1999.**

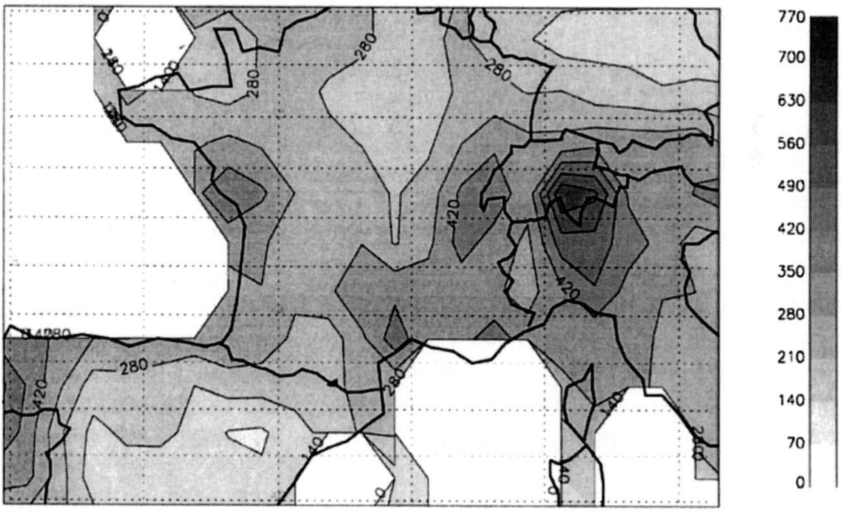

Figure 6 **: Precipitation amount (units, mm) from GPCC for the period from August to November 1999.**

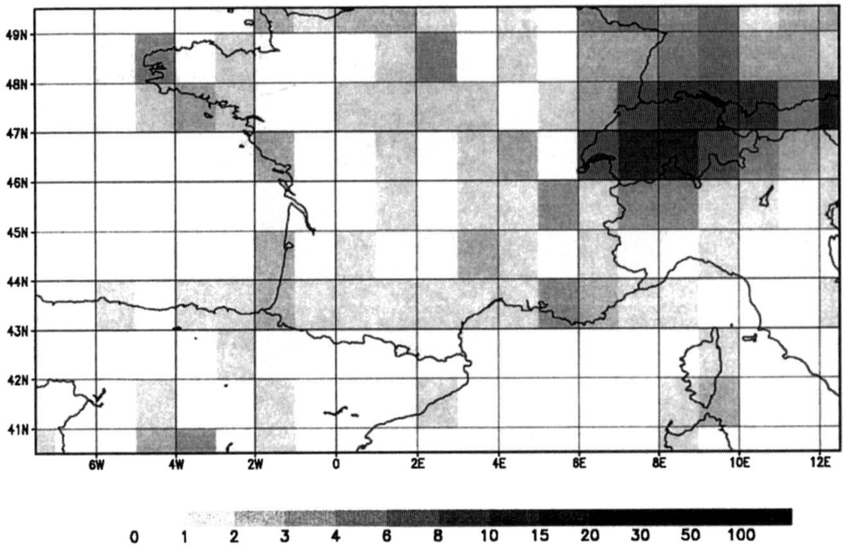

Figure 7 : Number of rain gauges available for the GPCC climatology during the period from August to November 1999.

A lot of similarities can be deduced from the comparison between MAR and GPCC precipitation fields. First, the pattern of simulated precipitation over France and Western Germany is generally close to GPCC. The amount of precipitation is underestimated by 15 to 30% over the West and the North of France. The observed maximum of rain in the West of France (Poitou – Charente) has not been reproduced by MAR. This is mainly explained by underestimated rainfall (by a factor 2) on the 19th September in the Golfe of Gascogne. However the smaller amounts of precipitation in Normandie and in the Northeast of France has been well simulated. Rain amount over the Massif Central appears also in good agreement with observations, as well as precipitation over Switzerland. Simulated precipitation over Jura seems somewhat underestimated (280 mm in MAR instead of 420 mm in GPCC). Precipitation over Southern Alps and Cevennes are probably much more difficult to interpret, because of the lack of observation stations in these mountainous regions. MAR simulates a maximum of precipitation over Pyrenees and along the northern coast of Spain, which is not present in GPCC. However it cannot be concluded that the simulated rainfall is wrong, because no or very few GPCC observation stations are available in this area.

Since the interpretation of GPCC precipitation could be difficult in grid cells where there are too few observation stations, it is interesting to assess precipitation fields as a function of the stations per cell in the GPCC grid. Table 1 shows the spatial averaged precipitation amounts computed on the GPCC grid for GPCC, MAR, and ECMWF fields. If all GPCC grid cells over land are considered (i.e. minimum number of GPCC stations per grid cell set at 0), the MAR underestimates GPCC precipitation by about 15%. But in this case, GPCC precipitation is computed with an interpolation in 24 grid cells where no observations are available. If these cells are not taken into account, i.e. setting the threshold of minimum observation stations per cell at 1, the averaged precipitation amount is not significantly modified. If this threshold is increased, the selected grid cells are progressively concentrated over only Switzerland. In this area, the bias of MAR increases from 15 to 20%. Precipitation averaged over Switzerland is within 20% of the observed values, which is in the same range as in Marinucci et al. (1995) although they considered a simulation period of one month.

Table 1: Comparison of spatial averaged precipitation amounts (units, mm) for the period from August to November 1999 and considering GPCC climatology, MAR and ECMWF simulations. The amounts are computed in function of the number of available GPCC stations. The first column indicates the minimum number of observation stations contained in a cell of the GPCC grid (1 by 1 degree). The second column gives the number of cells in the simulation domain such that each cell contains at least the minimum number of stations specified in the first column. Note that GPCC precipitation in a grid cell where there is no observation stations is computed using interpolation.

		Precipitation amount for August to November 1999				
Min. number of GPCC stations per grid cell	Number of grid cells in the simul. domain	GPCC (ref.)	MAR	MAR (diff.)	ECM-WF	ECM-WF (diff.)
0	167	317	270	-15%	472	49%
1	143	323	273	-15%	478	48%
2	91	328	279	-15%	491	50%
3	40	342	285	-17%	507	48%
4	19	355	289	-19%	504	42%
5	12	398	318	-20%	568	43%
8 (CH)	7	445	354	-20%	655	47%

3.2.2 Comparison with ECMWF

Figure 8 represent the precipitation amounts obtained from ECMWF. A quick look at this figure shows an important additional amount of precipitation in the whole domain, and particularly over the Alps, compared to GPCC climatology. From this figure, it appears that ECMWF fields tend to overestimate precipitation by a factor from 1.3 to 1.7. Overestimated ECMWF precipitation is confirmed by table 1, where it can be noted that the amounts are about 50% larger than GPCC climatology. The difference between MAR and ECMWF fields amounts up to 85% over Switzerland. This trend to overestimate precipitation has been probably attenuated with the recent changes in the ECMWF cloud physics (ECMWF, 2000).

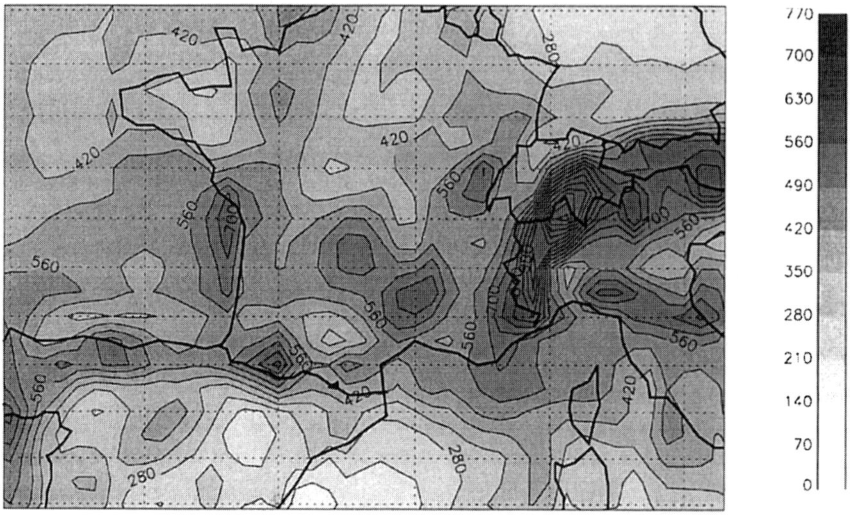

Figure 8 : Precipitation amount (units, mm) from ECMWF for the period from August to November 1999.

Despite of the systematic overestimated precipitation, the ECMWF precipitation fields can be useful to analyze some features in the spatial and temporal distributions of precipitation. In particular, it can be noted that rainfalls are simulated over Pyrenees, Northern Spain and Northern Italy by both ECMWF and MAR, while they was not present in GPCC climatology. This fact can be explained by an insufficient number of GPCC rain gauges over the Pyrenees. In addition, the climatology of "Atlas Climatique de France" (1989) tends to confirm the maximum of precipitation in this region, as simulated in MAR and ECMWF. Note also that the maximum of precipitation over the western coast of France is well marked with the ECMWF model, but it appears overestimated compared to GPCC.

The time series of ECMWF precipitation can be also considered for a comparison with MAR since the operational ECMWF model is regularly initialized with analysis and is therefore able to accurately reproduce the time evolution of rainy events. Figure 9 confirms this fact, with significant differences for the events characterized by high precipitation rates. The signal of precipitation simulated with MAR is generally in good agreement with ECMWF: the correlation coefficient between the two time series is 0.75 as indicated in table 2.

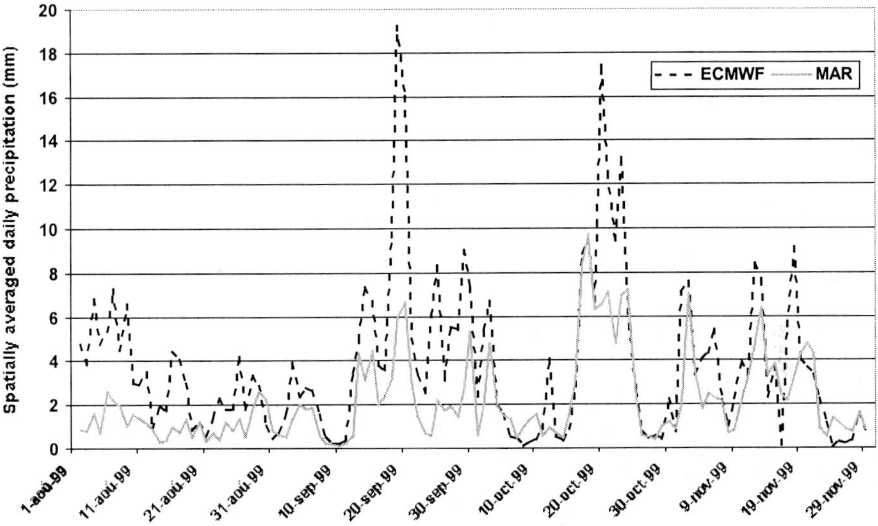

Figure 9 : Time evolution of precipitation (units, mm) spatially averaged over the simulation domain (the relaxation zone is excluded) for ECMWF (dashed black line) and MAR (solid gray line).

Since this study will mostly focus on precipitation climatology at the regional or local scale, table 2 presents the correlation of precipitation time series between ECMWF and MAR for some specific subregions. This correlation is satisfying for France (0.71) and the Massif Central (0.69). It is weaker over the Rhône valley, partly because this subregion is smaller than France or the whole simulation domain. The correlations are also presented for the Alps (South-West and East), but these results will be more extensively discussed in section 4.

Table 2 : Correlation coefficient between time series of daily precipitation from ECMWF and MAR, considering various subdomains. Each subdomain is delimited by coordinates in longitude and latitude (columns 2 to 5).

	Location				
Subdomain	Long. min	Long. max	Lat. min	Lat. max	Correlation
Simulation domain	-8	13	40	50	0.75
France	-3	6	43.5	50	0.69
Massif Central	1	5	43.4	46.5	0.71
Rhône valley	4	5.5	43.5	45.2	0.63
Alps (South-West)	5	8	43.5	47	0.58
Alps (East)	8	15	45.2	48	0.53

3.3 Impact of cloud water fluxes at the lateral boundaries

The three-dimensional fields of cloud water content are not available in ECMWF analyses. However these fields mostly influence radiation and can influence the representation of precipitation since they modify the water balance. Although the cloud water content is weak compared to the global water content, the impact of such fields on condensation processes could be important for grid points close to saturation.

In order to take into account the cloud fields that are not available in ECMWF analyses, the following formulation is proposed to assess the cloud water mixing ratio q_{cloud} :

$$q_{cloud} = q_{\max} \, e^{-\left(\frac{100-RH}{\sigma}\right)} \qquad (1)$$

where RH (units, percent) is the relative humidity obtained from ECMWF analyses, σ is a parameter fixed to 7, and q_{max} is the maximum cloud mixing ratio fixed at 0.5 g/kg, which corresponds to stratiform clouds. The evolution of q_{cloud} in function of RH is presented in figure 10. No cloud or precipitation is produced by 80% in relative humidity, which is consistent with other studies, such as Bader and Roach (1977) or Robichaud and Austin (1988).

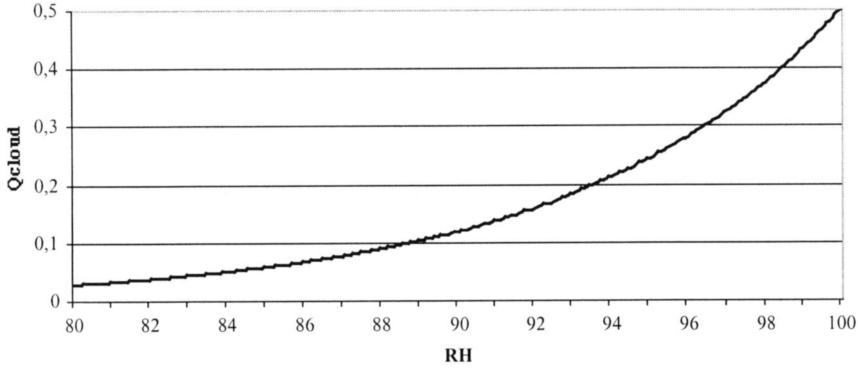

Figure 10 : Evolution the estimated cloud water content (units, g/kg) as a function of the relative humidity (in percent).

The parameter σ has been adjusted in order to ensure the consistence with the microphysics of MAR. More specifically, σ is such that the domain-averaged cloud equivalent water content diagnosed with relation (1) matches the domain-averaged cloud equivalent water content simulated by MAR running without cloud water fluxes at the lateral boundaries. This approach prevents from overestimating the cloud fluxes.

Figure 11 shows the simulated precipitation without cloud fluxes at the lateral boundaries. When compared to figure 5, it is clear that the simulation including these fluxes presents an improved representation of the precipitation field. In particular, the domain-averaged precipitation amounts to 230 and 281 mm for the simulations respectively without and with the cloud fluxes at the lateral boundaries, while observations from GPCC are about 313 mm. The RMS error is also significantly reduced (24%, from 113 to 86 mm). Another important improvement obtained with the application of relation (1) is the spatial distribution of precipitation: the spatial correlation between MAR and GPCC fields is improved from 0.66 to 0.69. In particular, the maximum of precipitation in Western France, over Jura, and over the Massif Central are better represented. This improvement will be verified with higher resolution datasets in further studies.

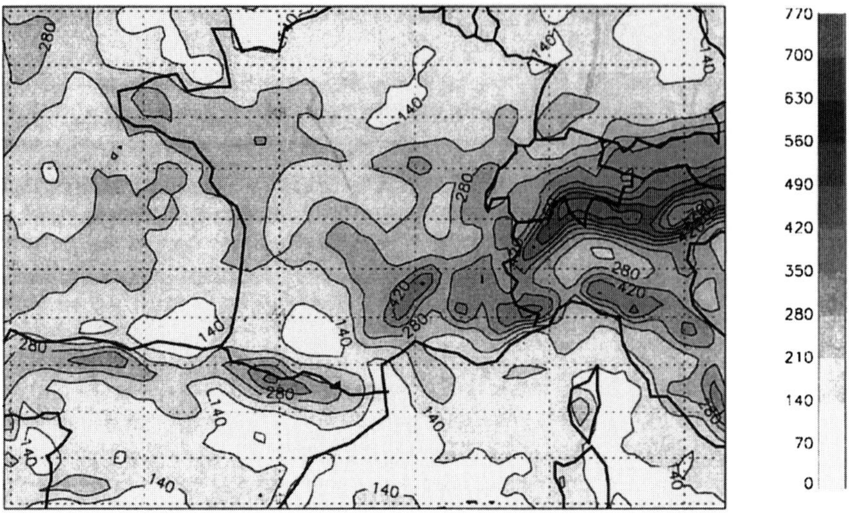

Figure 11 : Precipitation amount (units, mm) simulated by MAR without the cloud water fluxes at the lateral boundaries for the period from August to November 1999.

Table 3: Comparison between simulations without and with cloud water fluxes at the lateral boundaries. Four variables are examined: mean sea level pressure (MSLP; units, hPa), 500-hPa surface height (Z500; units, m), 850-hPa temperature (T850; units, K), and 850-hPa specific humidity (Q850; units, g/kg). The second and third columns present respectively the time-averaged values for MAR and ECMWF models, and the fourth and the fifth columns show the root mean square error and the spatial correlation between MAR and ECMWF.

	MAR	ECMWF	RMS	Correlation
Simulation without cloud fluxes at the lateral boundaries				
MSLP	1020.2	1021.5	2.6	0.90
Z500	5623.0	5619.5	29.3	0.99
T850	276.1	275.4	0.9	0.96
Q850	3.6	3.5	0.1	0.84
Simulation with cloud fluxes at the lateral boundaries				
MSLP	1020.4	1021.5	2.2	0.91
Z500	5623.0	5619.5	28.7	0.99
T850	275.9	275.4	0.5	0.98
Q850	3.7	3.5	0.1	0.84

The improvement is not only noted on precipitation, but also on mean variables such as mean sea level pressure, geopotential, and mean temperature. Table 3 presents these time-averaged variables for the entire simulation period. The inclusion of lateral cloud water fluxes has led to a small improvement of sea level pressure and 500-hPa geopotential, but its impact is particularly significant on the representation of boundary layer temperature (850 hPa) with a reduction of the RMS error nearly by a factor 2.

4. DISAGGREGATION OF SIMULATED PRECIPITATION

The representation of precipitation over complex topography is one of the most difficult problems in mountain meteorology (e.g. Alpert and Shafir, 1989; Barros and Lettenmaier, 1993; Barros and Lettenmaier, 1994). The importance of a good prediction is obvious for operational weather forecast, but it is also important for climate studies and hydrological applications.

4.1 Disaggregation model of Sinclair (1994)

The role of topography on precipitation is firstly explained by disturbances of the vertical structure of the atmosphere. Topography can act as a barrier and modify the horizontal flow, or contribute to rise air masses and to produce condensation and precipitation during their forced ascent. Sinclair (1994) proposed a diagnostic model that estimates the mesoscale variation of orographically enhanced precipitation. It is based on the same approach as in Alpert and Shafir (1989), but it includes an additional thermodynamical scheme to compute the condensation and precipitation rates.

The Sinclair's model first determines topographically induced vertical motion at the surface as $V.\nabla z$, where V is the horizontal component of wind and ∇z is the variation of topography. This vertical motion decreases with height at a specific rate. In addition, it includes the large-scale vertical motion computed by the atmospheric model. The determination of the vertical profile of vertical velocity allows the computation of the condensation rate, assuming that condensation arises by means of saturated ascent along the moist adiabat through the LCL. Precipitation is then calculated by vertically integrating the condensation rate and by considering the efficiency of precipitation as a function of low-level relative humidity.

The consistence between MAR and RDM precipitation is ensured by imposing that spatial average of precipitation computed with the Sinclair (1994) model matches the precipitation amount simulated by from MAR and averaged over the simulation domain (excepted relaxation zone at the lateral boundaries).

Note that other rain disaggregation models have been proposed. As already mentioned, the Sinclair's model shares the same approach as in Alpert and Shafir (1989). Both these models determine precipitation at small scale using diagnostic relations. Barros and Lettenmaier (1993) have proposed a prognostic scheme that simulates Lagrangian transport of moist static energy and total water, and where precipitation is the only scavenging agent of both variables. Although this model is aimed primarily at the estimation of the long-term hydroclimatology of large regions characterized by complex topography, it appears also available to represent the distribution of storm precipitation at the forecasting time-scale. Another interesting approach is the subgrid orographic precipitation scheme of Leung and Ghan (1995). The most important feature of this model is its interactivity with the atmospheric model, which contributes to improve the representation of simulated meteorological fields. This scheme includes a simple airflow model, and a thermodynamic model that treats the influence of parcel rise on temperature, water vapor, and clouds. Contrary to the above-mentioned approaches, the full physical package (i.e. radiative transfers, microphysics, exchanges with the surface, convection) of the atmospheric model is taken into account in each subgrid cells defined in function of topography classes. Let us remark that all these approaches do not consider subgrid-scale processes such as blocking, thermally forced circulations, leeside cyclogenesis, lee waves, and local effects of mesoscale.

4.2 Application of the Sinclair's model to precipitation simulated by MAR

The RDM of Sinclair (1994) has been applied on precipitation fields simulated by MAR at 20-km horizontal resolution. Several resolutions have been considered for disaggregation: 3, 5, and 8 km. From a hydrological point of view, there were three significant hydrological events during the simulation period (Benoît et al., 2000): 20 September, 26 September, and 21 October. In section 4.2.1, it is proposed to examine more specifically the event of 20 September 1999 that was marked by the most intense rainfall, while the results for the whole simulation period is presented in section 4.2.2.

4.2.1 Situation of 20 September 1999

Disaggregation of simulated precipitation is tested for the event of 20 September 1999 that was responsible for the most intense precipitation over

Northern Italy during the period from August to November 1999. The results are presented in table 4 for the watersheds of Dora, Orco, Sesia, and Toce (see figure 12 for the location). Simulated precipitation amounts without disaggregation are underestimated for all watersheds. They represent 61, 73, 43, and 65% of observed precipitation respectively over Dora, Orco, Sesia, and Toce watersheds. The use of RDM of Sinclair (1994) has significantly improved these results: simulated amounts correspond to 91, 102, 59, and 74% of observed values respectively for the four above-mentioned watersheds. The redistribution of precipitation done by the RDM has increased the rain amounts over the Italian watersheds. The better representation of surface elevation has then contributed to improve the realism of precipitation fields, despite of the relative simplicity of the RDM.

Nevertheless RDM is not able to correct all model deficiencies. For instance, let us consider the Sesia watershed where precipitation has been widely underestimated. This misrepresentation can be explained by a bad location of the maximum rainfall simulated by MAR. Considering figure 12.a, heavy precipitation is simulated at the East of Toce watershed, but not over Sesia. The RDM tends to correct the precipitation over Sesia, but its impact is limited because the precipitation amounts simulated with MAR in this area are weak. The bad location of precipitation simulated with MAR is probably explained by a smoothed representation of topography at 20-km resolution. As shown in figure 13, the Valais is poorly represented and the consequence for the precipitation during 20 September is a spreading or a shift in the precipitation pattern.

Table 4 : Comparison of precipitation amounts (units, mm) during 20 and 21 September 1999 over four Italian watersheds. MAR precipitation without disaggregation is compared with disaggregated precipitation fields at 3, 5, and 8 km, and with observations.

Watershed	Without disaggreg.	3 km disaggreg.	5 km disaggreg.	8 km disaggreg.	Observation
Dora	33.5	49.4	46.8	40.1	54.5
Orco	69.0	96.0	90.0	89.7	94.0
Sesia	69.5	95.9	92.9	97.4	162.7
Toce	100.4	113.8	110.4	111.8	154.0

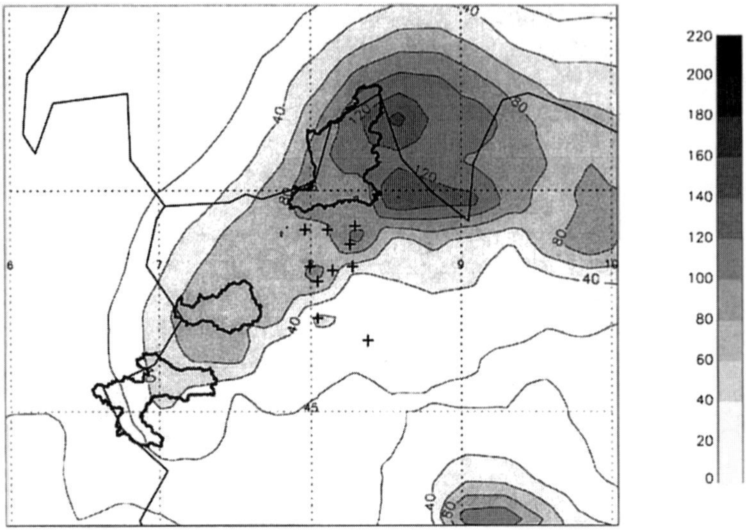

Fig.12 a

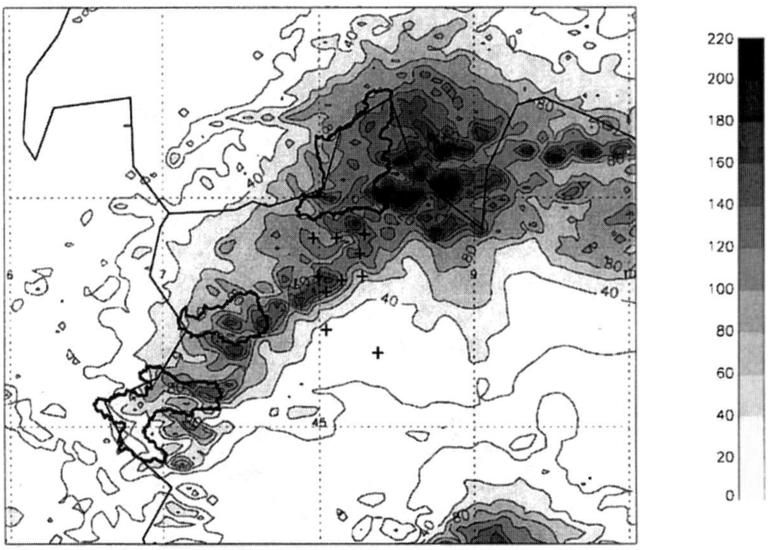

Fig. 12 b

Figure 12 : Cumulated precipitation (units, mm) during 20 and 21 September, simulated (a) by MAR without disaggregation and (b) by MAR including the RDM of Sinclair (1994). The Italian watersheds of Dora, Orco, and Toce (respectively from South to North) are delimited by orange contour lines. Observation stations in Sesia watershed are represented with (+) symbols. The represented domain ranges from 6E to 10E in longitude, and from 44.5N to 46.8N in latitude. Horizontal and vertical dotted lines represent longitude and latitude where the spacing is 1 degree.

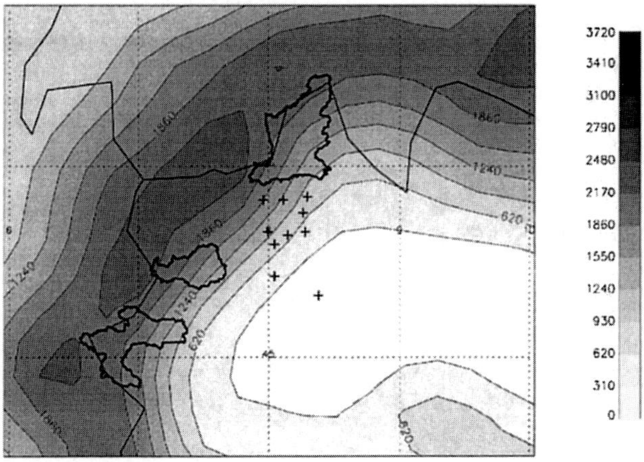

Fig.13 a

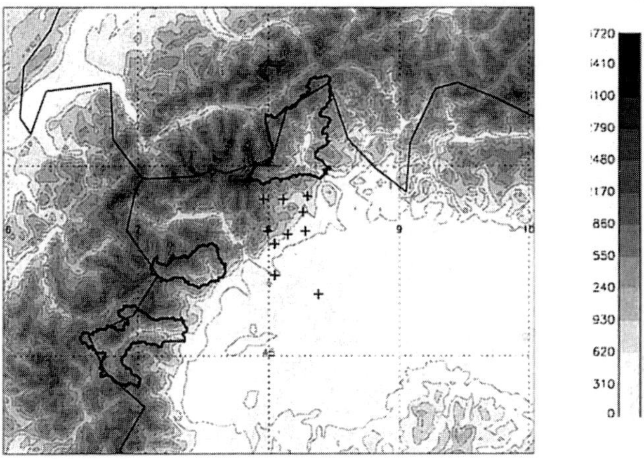

Fig.13 b

Figure 13 : (a) MAR topography (units, m) at 20 km resolution; (b) 3 km topography used with the RDM of Sinclair (1994). The represented domain is the same as in figure 12.

4.2.2 Period from August to November 1999

Results for the whole simulation period are presented in Table 5. Simulated precipitation amounts are overestimated for the four watersheds presented in the previous subsection. Nevertheless the MAR model produces more realistic precipitation than the ECMWF model. The possible reasons for the differences found have been already explained in section 3.2.2.

Analyzing the MAR simulation, it is observed that atmospheric situations responsible for precipitation overestimates over Italian watersheds are characterized by the presence of depressions over the Mediterranee that direct southerly moisture fluxes toward the Alps. In several cases, MAR has overestimated the deepening of pressure lows (most of time by 3 hPa) and consequently enhanced moisture fluxes. This fact namely explains overestimates in simulated precipitation for the situations of 3 October, 16 to 21 October, 24 October, and 3 November. As a consequence of this overestimate on simulated precipitation, applying RDM at 3-km and 8-km resolutions on these fields has increased the error on cumulated rain amounts over watersheds. This error increase could be expected, because RDM tends to enhance precipitation over high elevation sites.

Table 5 : Cumulated amounts of precipitation (units, mm) over the watersheds of Toce, Orco, Sesia, and Dora, for the period from August to November 1999. Observations are compared with disaggregations at 3 and 8 km, MAR without disaggregation ("no disag."), and with ECMWF. Correlations between time series of observations and simulated precipitation are presented in the second part of the table.

Watershed	Observation	Disag. 3 km	Disag. 8 km	No disag.	ECMWF
Cumulated precipitation over watersheds					
Toce	586	924	842	751	1221
Orco	402	885	804	694	939
Sesia	615	794	786	621	1073
Dora	243	616	521	487	856
Time correlation with observations					
Toce		0,57	0,56	0,55	0,76
Orco		0,52	0,52	0,51	0,85
Sesia		0,55	0,55	0,51	0,79
Dora		0,50	0,46	0,42	0,73

On the other hand, the correlation of time series has been improved with RDM, especially over Dora and Sesia. This means that the RDM of Sinclair (1994) has been able to physically modify precipitation fields in function of the fine scale orography. As already mentioned in the previous section, the efficiency of RDM is related to the performance of RCM. A recommendation that can be deduced from this study is the need of improving the spatial distribution of precipitation in RCM. A possible improvement is the inclusion of a sophisticated subgrid scheme in the RCM in order to allow a better representation of the topography forcing and surface characteristics. Consequently such a parameterization would produce better precipitation patterns and more accurate locations of heavy rainfall, and should improve the efficiency of RDM for computing fine grid precipitation required by hydrological models.

5. CONCLUSIONS

The MAR – Modèle Atmosphérique Régional – model was run at 20-km resolution for a four-month period over the Alps and the southeastern mountainous regions of France. The validation on the climatology of mean variables has not revealed visible drift or spurious trend. Precipitation generally reaches about 80% of the value obtained by the GPCC – Global Precipitation Climatology Center – climatology, and the spatial distribution of precipitation also compare favorably to observations. The inclusion of cloud water fluxes at the lateral boundaries has significantly improved the amount and the spatial distribution of precipitation, and the climatology of mean variables, particularly for the 850-hPa temperature.

The finest horizontal resolution currently used in RCM is about 20 km, which remains too coarse for the forcing of hydrological models, especially over mountainous regions. In order to improve the impact of the topographical forcing on microphysical processes, a RDM has been applied to the precipitation simulated with MAR. This analysis over four Italian watersheds has underlined the usefulness of RDM in the framework of climate studies, especially in the perspective of the coupling with hydrological models. Using a better representation of the surface characteristics, in particular the surface elevation, the RDM of Sinclair (1994) based on a simplified dynamics and thermodynamics significantly improve the representation of heavy rainfall. This RDM has also contributed to improve the time series of precipitation. Nevertheless it is not able to remove all deficiencies in the mesoscale model running at a coarser resolution (compared to RDM). An interesting compromise between computing cost and sophistication would be the interactive disaggregation, since a better representation of orography and surface characteristics should lead to more realistic precipitation patterns and improve the efficiency of diagnostic RDM such as Sinclair (1994) RDM. This approach will be examined in future work.

6. ACKNOWLEDGMENTS

This research is supported by the European Commission in the framework of Marie Curie Fellowship. We are grateful to the "Regione Piemonte, settore meteo-idrografico" for the daily precipitation data over the four Italian watersheds examined in this paper.

7. REFERENCES

Alpert, P., and H. Shafir, 1989 : Mesoγ-scale distribution of orographic precipitation: numerical study and comparison with precipitation derived from radar measurements. *J. Appl. Meteor.*, 28, 1105-1117.

Anthes, R. A., Kuo, Y.-H., Hsie, E.-Y., Low-Nam, S., and T. W. Bettge, 1989 : Estimation of skill and uncertainty in regional numerical models. *Quart. J. Roy. Meteor. Soc.*, 115, 763-806.

Atlas Climatique de France, 1989. Published by the "Ministère des Transports, Direction de la Météorologie", Paris.

Bader, M. J., and W. T. Roach, 1977 : Orographic rainfall in warm sectors of depressions. *Quart. J. Roy. Meteor. Soc.*, 103, 269-280.

Barros, A. P., and D. P. Lettenmaier, 1993 : Dynamic modelling of the spatial distribution of precipitation in remote mountainous areas. *Mon. Wea. Rev.*, 121, 1195-1214.

Barros, A. P., and D. P. Lettenmaier, 1994 : Dynamic modeling of orographyically induced precipitation. *Rev. of Geophys.*, 32, 265-284.

Beniston, M., and M. Rebetez, 1996 : Regional behavior of minimum temperatures in Switzerland for the period 1979-1993. *Theor. Appl. Climatology*, 53, 231-243.

Beniston, M., Diaz, H. F., and R. S. Bradley, 1997 : Climatic change at high elevation sites: An overview. *Clim. Change*, 36, 233-251.

Brasseur, O., Gallée, H., Schayes, G., Tricot, C., and K. De Ridder, 1998 : Impact of Turbulent Closures on Diurnal Temperature Evolution for Clear Sky Situations over Belgium. *Bound.-Layer Meteor.*, 87, 163-193.

Businger, J. A., 1973 : Turbulent transfer in the atmospheric surface layer. *Workshop on Micrometeorology, Amer. Meteorol. Soc.*, 67-100.

Christensen, J. H., Machenhauer, B., Jones R. G., Schär C., Ruti P. M., Castro M., and G. Visconti, 1997 : Validation of present-day regional climate simulations over Europe: LAM simulations with observed boundary conditions. *Climate Dynamics*, 13, 489-506.

Davies, H. C., 1976 : A lateral boundary formulation for multi-level prediction models. *Quart. J. Roy. Meteorol. Soc.*, 102, 405-418.

Davies, H. C., 1983 : Limitations of some common lateral boundary schemes used in regional NWP models. *Mon. Wea. Rev.*, 111, 1002-1012.

Deardorff, J. W., 1978 : Efficient prediction of ground surface temperature and moisture, with inclusion of a layer of vegetation. *J. Geophys. Research*, 83, 1889-1903.

Dunn, L. B., 1991 : Evaluation of vertical motion: Past, present, and future, *Wea. Forecast.*, 6, 65-75.

Duynkerke, P. G., 1991 : Radiation fog: a comparison of model simulation with detailed observations. *Mon. Wea. Rev.*, 119, 324-341.

ECMWF, 2000 : The IFS cycle CY21r4 made operational in October 1999. *ECMWF Newsletter* Number 87, Spring 2000.

Fouquart, Y., and B. Bonnel, 1980 : Computations of solar heating of the Earth's atmosphere: a new parameterization. *Beitr. Phys. Atmos.*, 53, 35-62.

Frei, C., and C. Schär, 1998 : A precipitation climatology of the Alps from high-resolution rain-gauge observations. *Int. J. Climatology*, 18, 873-900.

Fritsch, J. M., and C. F. Chappell, 1980 : Numerical prediction of convectively driven mesoscale pressure systems. Part I: Convective parameterization. *J. Atmos. Sci.*, 37, 1722-1733.

Gallée, H., 1995 : Simulation of the mesocyclonic activity in the Ross Sea, Antarctica. *Mon. Wea. Rev.*, 123, 2051-2069.

Gallée, H., 1996 : Mesoscale atmospheric circulations over the Southwestern Ross Sea sector, Antarctica. *J. Appl. Meteor.*, 35, 1142-1152.

Gallée, H., and G. Schayes, 1994 : Development of a three-dimensional meso-γ primitive equations model, katabatic winds simulation in the area of Terra Nova Bay, Antarctica. *Mon. Wea. Rev.*, 122, 671-685.

Giorgi, F., and M. R. Marinucci, 1996 : An investigation of the sensitivity of simulated precipitation to model resolution and its implications for climate studies. *Mon. Wea. Rev.*, 124, 148-166.

Giorgi, F., Hurrell, J. W., Marinucci, M. R., and M. Beniston, 1997 : Elevation dependency of the surface climate change signal: A model study. *J. Climate*, 10, 288-296.

Giorgi, F., and L. O. Mearns, 1999 : Regional climate modeling revisited. *J. Geophys. Res.*, 104, 6335-6352.

GPCC, 1992: The Global Precipitation Climatology Centre. Data available at the web site http://www.dwd.de/research/gpcc.

IPCC (Intergovernmental Panel on Climate Change) - Houghton, J.T., Meira Filho, L.G., Callander, B.A., Harris, N., Kattenberg, A., and Maskell, K., 1996 : Climate change 1995: The science of climate change. Cambridge University, Cambridge Press, UK, 572 pp.

Marinucci, M. R., Giorgi, F., Beniston, M., Wild, M., Tschuck, P., Ohmura, A., and A. Bernasconi, 1995 : High-resolution simulations of January and July climate over the Western Alpine region with a nested regional modeling system. *Theor. Appl. Climatology*, 51, 119-138.

Molinari, J., and Dudek, M., 1992 : Parameterization of convective precipitation in mesoscale numerical models: a critical review. *Mon. Wea. Rev.*, 120, 326-344.

Morcrette, J.-J., 1984 : Sur la paramétrisation du rayonnement dans les modèles de la circulation générale atmosphérique. *Unpublished thesis*, Université de Lille, France.

Leung, L. R., and S. J. Ghan, 1995 : A subgrid parameterization of orographic precipitation. *Theor. Appl. Climatol.*, **52**, 95-118.

Rebetez, M., and M. Beniston, 1998 : Changes in sunshine duration are correlated with changes in daily temperature range this century: An analysis of Swiss climatological data. *Geophys. Res. Letters*, 25, 3611-3613.

Robichaud, A. J., and G. L. Austin, 1988: On the modelling of warm orographic rain by seeder-feeder mechanism. *Quart. J. Roy. Meteor. Soc.*, 114, 967-988.

Sevruk, B., 1975 : Inaccuracy in precipitation measurements. A serious problem in water resources instrumentation. *Proc. Second World Congress, International Water Resources Association,* vol. III, New Delhi, 429-440.

Sevruk, B., 1989 : Reliability of precipitation gradient estimates. *Proc. XIV Int. Conf. Carpathian Meteorology,* Swiss Climatic Data Center, Sofia, Bulgaria, 402-408.

Sinclair, M. R., 1994 : A diagnostic model for estimating orographic precipitation. *J. Appl. Meteorol.*, 33, 1163-1175.

Therry, G., and P. Lacarrère, 1982 : Improving the eddy kinetic energy model for the planetary boundary-layer description. *Bound.-Layer Meteor.*, 25, 63-88.

Heavy precipitation occurrence in Scandinavia investigated with a Regional Climate Model

OLE BØSSING CHRISTENSEN[1], JENS HESSELBJERG CHRISTENSEN[1] AND MICHAEL BOTZET[2]

[1] *Danish Meteorological Institute, Copenhagen*

[2] *Max-Planck-Institute for Meteorology, Hamburg*

1. INTRODUCTION

The topography of Scandinavia varies a great deal on spatial scales which are not described properly in coarse resolution global climate models (GCMs). In order to study quantities that vary considerably throughout the Scandinavian area, it is therefore useful to apply a downscaling technique. In this work we will report on investigations of present-day as well as greenhouse-warmed climate over Scandinavia employing a regional climate model (RCM).

The application of an RCM nested into a coupled global ocean-atmosphere model enables an improved description of smaller scales, both spatially and temporally. For instance, extreme precipitation events appear to be more realistic in a high-resolution model, both due to less spatial averaging, and due to a better simulation of deep low-pressure systems. It has been shown in Christensen *et al.* (1998) –hereafter denoted CCMB – that the description of several features of the geographical distribution of precipitation over Scandinavia improves considerably with higher resolution.

Several studies have addressed simulated mean climate changes over Denmark (*e.g.*, Fenger et al., 1996; Christensen, 2000, and references therein) whereas few studies have been performed on changes in variability or extreme weather events; one example, however, is Kaas and Andersen (2000).

M. Beniston (ed.),
Climatic Change: Implications for the Hydrological Cycle and for Water Management, 101–112.

In the present study we will re-examine some results of CCMB regarding heavy precipitation and the most extreme tail of the precipitation distribution. A longer discussion of the experiments, based partly on the results reported here, may be found in Machenhauer *et al.* (1998), hereafter called MBC, and in CCMB. The temperature change in the Scandinavian integration area between control and scenario experiments is of the order of 4K over most of the area and most of the seasons, in accordance with the regional temperature change pattern extracted from the driving coupled AOGCM. As regards the temperature field there only seems to be little added value due to the dynamical downscaling.

We shall investigate the quality of the frequency of heavy precipitation as well as the effect of greenhouse warming on that. The basis of this study will be the regional model simulations of different resolution and the driving AOGCM.

2. MODEL SETUP

The regional model employed is HIRHAM4, which is based on the HIRLAM (Källén, 1996) numerical weather forecast model but using the physical parameterization schemes of the ECHAM4 global climate model (Roeckner *et al.* 1996). HIRHAM4 is documented thoroughly in Christensen *et al.* (1996).

The basis of the regional simulation is the coupled global simulation with the ECHAM4 atmospheric GCM and the OPYC3 ocean model, which was performed at the Max-Planck-Institute for Meteorology in 1996 (Roeckner *et al.* 1996, 1999; MBC). The experiment consists of a control period and a transient scenario period with greenhouse gas forcing according to the IPCC emission scenario IS92a. The experiment did not consider changes in sulfate aerosols. Because of this, as well as the late time of the scenario time window, a rather strong global warming of approximately 4 degrees K is taking place.

The global experiment has a resolution of T42, or about 300 km in the atmospheric component with a similar resolution in the ocean. Two 10-year time slices have been selected from the coupled experiment; one from the control simulation corresponding to pre-industrial time and one from the scenario, corresponding to the years 2071-2080. From these two time-slices atmospheric data and daily sea-surface temperatures (SSTs) and sea-ice concentrations were extracted every 6 hours and used as boundary conditions for a HIRHAM simulation covering Europe and the North Atlantic with a spatial resolution of about 56 km. From this medium-resolution experiment atmospheric data were again extracted every 6 hours and used as boundary conditions for a high-resolution simulation with the same model at 18 km resolution for an area covering Scandinavia and the North Sea. Fig. 1 shows the medium-resolution integration area as the entire map, and the high-resolution area inside the frame. The medium-resolution experiment was introduced in order to avoid small- and large-scale behavior to be artificially decoupled which could result from a too large discrepancy between boundary resolution and RCM resolution. This phenomenon has been recognized as important for the quality of operational limited area numerical weather prediction models.

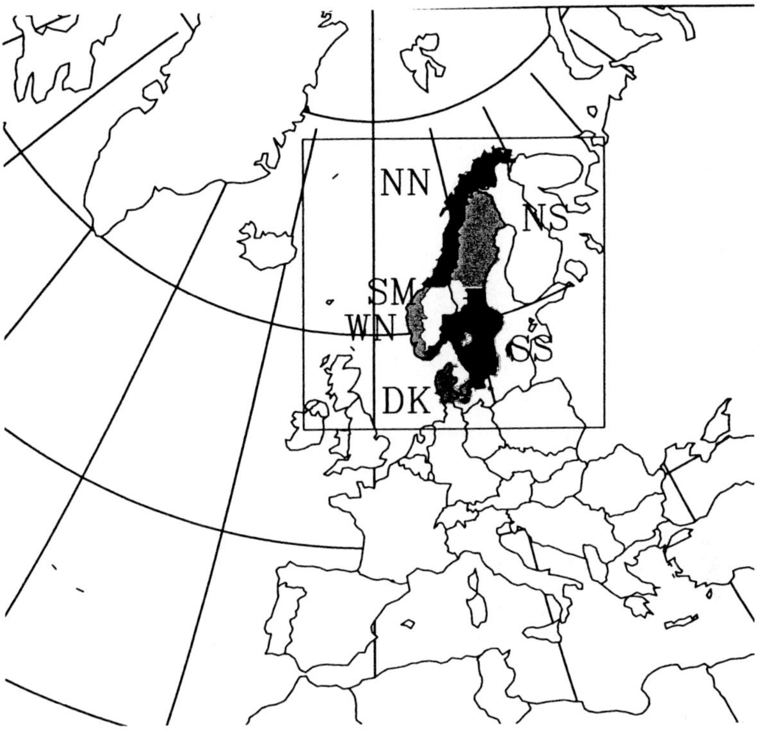

Figure 1. Integration domains for RCM simulations (from CCMB). The dark gray area labeled DK is used for the present analysis.

3. AVERAGE PRECIPITATION

The annual evolution of spatial averages of mean precipitation is shown in Fig. 2 for the Danish land area indicated in Fig. 1. The observations used are daily values from the period 1981-1999, which have been gridded in 20 km resolution (Scharling, 1999), thus approximately corresponding to the high-resolution simulation. Values for the simulations originate from all grid points within a rectangle surrounding Denmark. In general, simulated precipitation is larger than observed values. It should be noted that the observed precipitation displayed here is for a different area than the displayed simulation results, and that especially the low-resolution results are from a displaced area. The simulation results cover a rectangular sub-grid including Denmark and a small part of Sweden, whereas observations are from Denmark proper. This work is not about comparing the quality of model simulation of average precipitation, but rather about a comparison of extreme events scaled by average precipitation. Therefore, the large discrepancy between observed and simulated average precipitation is partly accounted for by a difference in areas. For a discussion of the realism of simulated average precipitation, see CCMB.

For the RCMs there is a generally positive climate change signal over most of the area and seasons. The GCM shows a very weak negative signal. Also in the rest of Scandinavia precipitation increases (not shown); especially autumn precipitation over Norway increases significantly, more than 50%. This can be attributed to an increased zonality of the weather type experienced in the scenario experiment, but also to increased fall-out from a moister atmosphere due to orographic lifting. This effect appears to be stronger in the high-resolution experiments (see MBC). Both temperature and precipitation changes have potential consequences for other fields, *e.g.*, snow cover and local runoff generation.

4. INTENSITY DISTRIBUTION OF PRECIPITATION

This section will describe the intensity distribution of precipitation, the probability density of a given day having a given amount of precipitation. The basis of this study is the simulated daily precipitation at a particular grid point and day. We sample seasons for as many years as possible with the purpose of generating the best possible statistics for the rare, very heavy precipitation events.

For a further improvement of the statistics we will collect data for several adjacent grid points. It is of course important to realize that this will not result in uncorrelated samples, as there is a considerable spatial correlation length depending on place, time of year, and weather type as well as a temporal correlation of a few days. Hence, it is difficult to estimate the statistical significance of values of the distribution function constructed in this way. Also the temporal correlation of point precipitation limits the degrees of freedom of a sample. We have truncated the plots at the probability 1 in 10,000 lacking an accurate estimate of statistical confidence.

For the low end of the intensity spectrum CCMB found that all the models have far too few dry days, *i.e.,* days with precipitation below 0.1 mm. This "drizzling problem" is common for many climate models. The problem is largest for the lowest resolutions. This is not surprising, as any averaging procedure will result in lower probabilities of total absence of precipitation.

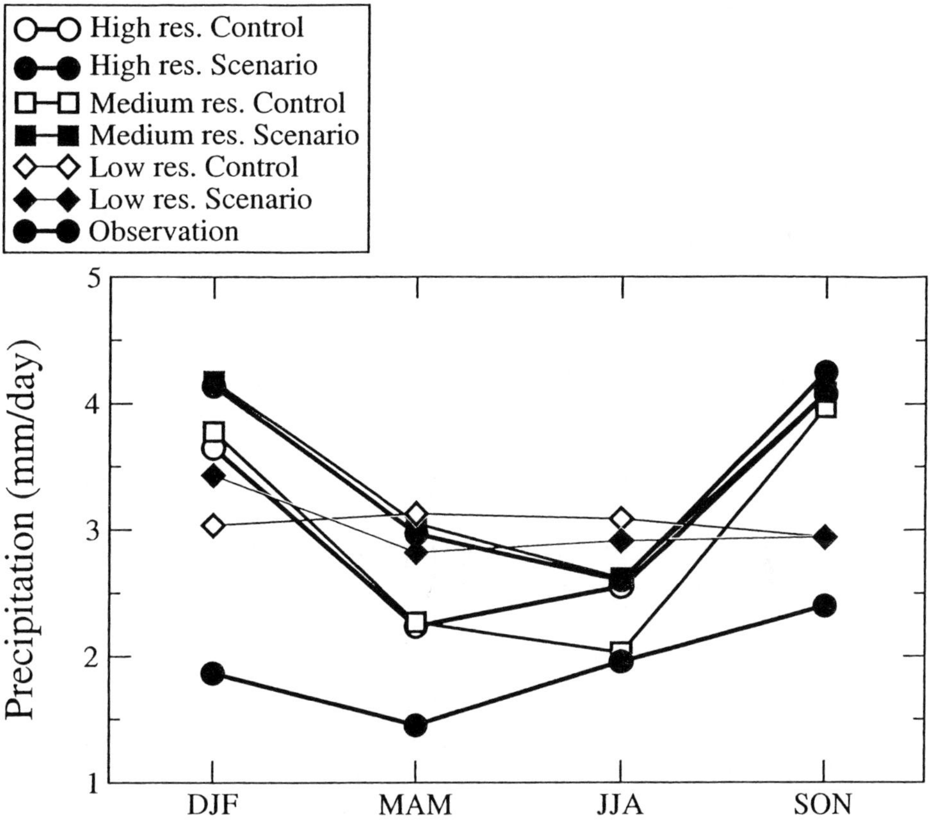

Figure 2. Precipitation over Denmark in the 4 seasons for control and scenario simulations in the different simulations. These are the values with which percentiles and their decay constants have been scaled in Figs 3-5.

Turning to the higher end of the precipitation spectrum, Fig. 3 shows observed as well as simulated percentile values for Denmark for each season. The figure contains curves for both scenario and control simulations and for low-, medium-, and high-resolution simulations as well as observations. Since the total precipitation varies from one model to another, all intensity values have been scaled with their average, resulting in dimensionless quantities.

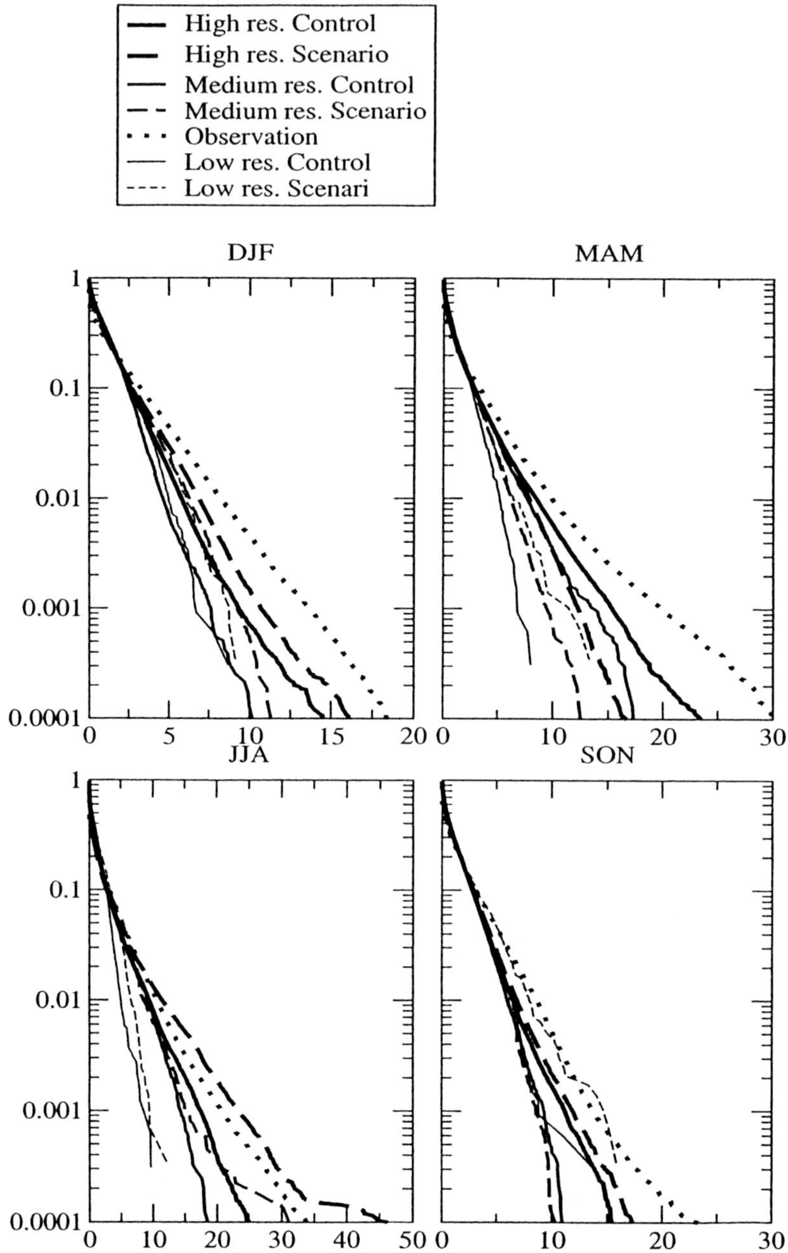

Figure 3. Percentiles of precipitation for land points over Denmark, scaled with total precipitation, i.e., a display of the probability of a day exceeding some multiple of the average. Data are shown for each of the four seasons. Dotted line: observation; thickest lines:

high resolution RCM; medium thickness: medium resolution RCM; thin lines: GCM. The solid lines are from the control experiments and the dashed lines are from scenario experiments.

The medium- and high-resolution simulations follow each other rather closely for lower precipitation intensities. However, at higher values the lower-resolution simulation falls behind, simulating fewer events. Note the differing scales on the x-axes that indicate annual variations in the probability of heavy precipitation with a maximum in summer. In general there is more heavy precipitation in the observations than in the simulations, even the high-resolution ones. This may be connected with the drizzling problem mentioned above, that days with limited amounts of precipitation occur too often. This will increase the simulated total precipitation without increasing the number of heavy-precipitation events and thus reduce the scaled quantities. Note that the vertical extent of the curves depends on the number of daily precipitation values of the calculation, which is proportional to the length of the simulation/observation period and to the squared resolution. The lowest possible probability corresponds to exactly one event in one grid point, or the probability 1/(number of grid points times number of days). The actual lowest value plotted is a bit higher, as the very tails of the curves are not shown.

The probability axis of Fig. 3 is logarithmic, so that approximately straight lines in the plots indicate an exponential relationship. Focusing on the intensive precipitation we will make use of the approximate exponential decay of the distribution functions. The curves are fitted to an exponential function $P = P_0 \exp(-b\,x)$ where P is the probability of a day showing an amount of x times the average precipitation. We will use the associated characteristic half-value $\pi_c = \ln(2)/b$ in the remainder of our analysis. An interpretation of π_c is that it measures the increase in normalized intensity that happens with half the probability. If the probability of exceeding some precipitation of x_1 times the average precipitation is P_1 then the probability of exceeding a larger value of x_2 times the average precipitation is $P_2 = P_1 \exp(-b\,(x_2 - x_1)) = P_1\, 2^{-(x2 - x1)/\pi c}$. Hence, the higher the value of π_c, the more likely the more extreme events.

4.1 Control climate vs. Observation

The lowest resolution data originating from the AOGCM show results that seem quite different from the RCM data. Obviously, any grid averaging or aggregation will act to reduce high-intensity precipitation occurrences (e.g. Durman *et al.*, 2000). In order to estimate the role of such effects, we have aggregated grid point values from both the RCM simulations as well as observations to a grid of the same coarseness as the GCM grid. Fig. 4 shows characteristic intensities for observations and simulations, both as calculated with the original spatial resolution and aggregated to the GCM resolution. Apparently the spatial averaging in itself only affects the characteristic intensity of the exponential fit slightly. The difference between resolutions therefore is concluded to be predominantly caused by the enhanced simulation of the development of precipitating weather systems and precipitation release due to flow over orography and local convection in the high-resolution models.

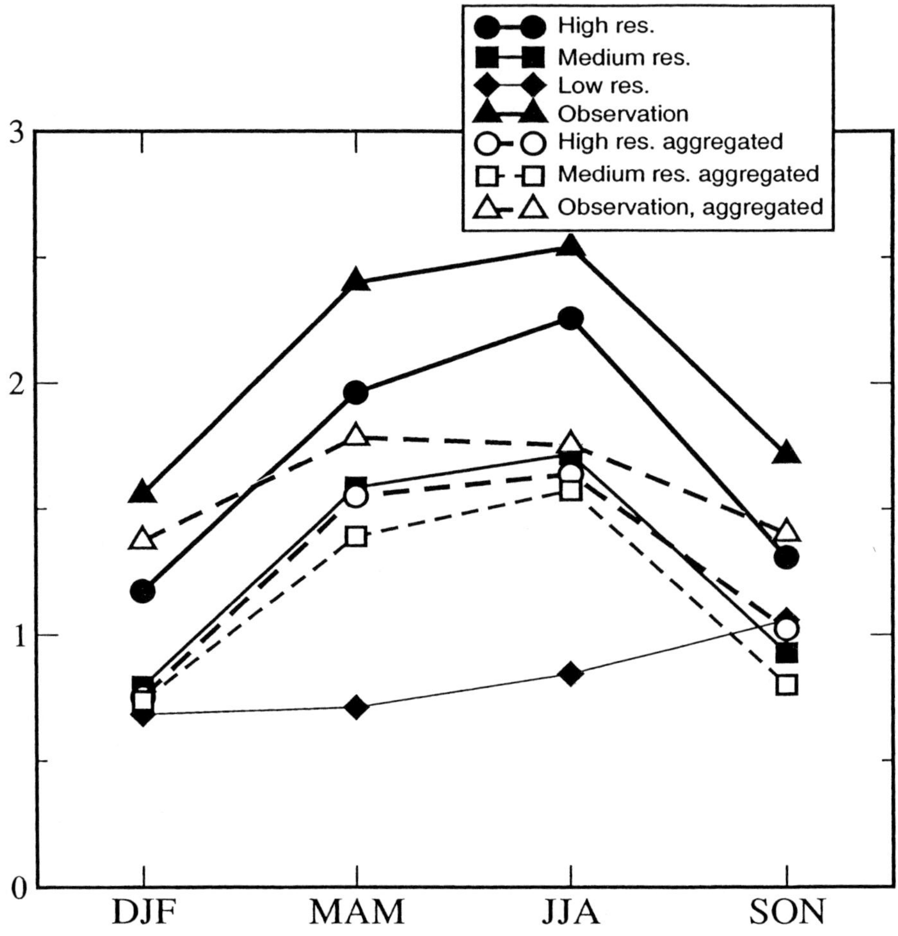

Figure 4. Exponential characteristic intensities for observed and simulated precipitation intensity spectra for land points over Denmark. Curves with full lines are original values, dashed curves are based on aggregated daily precipitation values corresponding to GCM resolution.

All RCM-based curves show a maximum in summer, as do observations. This corresponds to a high likelihood of intensive rainfall, consistent with the convectively dominated weather types of this season. This season is particularly important with the highest observed occurrence of high-intensity precipitation. In contrast, the GCM-based characteristic intensities are largest in autumn. This suggests a higher credibility of the simulated distribution function and seasonality of processes leading to precipitation events by the RCMs.

4.2 Climate change

It is found in Fig. 3 that the probability for heavy precipitation increases with the enhanced greenhouse effect approximately with the same magnitude as the change in average precipitation. The scaled curves for control and scenario simulations at various resolutions and seasons are closer than the curves of absolute values in Fig. 2.

There is a possible deviation from an exponential function in the summer time of the high-resolution scenario curve, showing an increased probability of very extreme events. It cannot be excluded, however, that this feature is due to a statistical fluctuation. Since this anomaly is due to very few, extreme events, its statistical significance is relatively low; a quantitative estimation still needs to be performed.

Note that this behavior is calculated for the Danish area and this relation may not be valid in other areas (see also Voss *et al.,* 2000). A further investigation should be performed with subsets (excluding individual years) of the entire simulation, providing insight into the significance of the right-hand end of the curves of Fig. 3. A thorough analysis, however, is beyond the scope of the present work.

Concerning the resolution dependence of climate change signal, it is interesting to inspect Fig. 5 comparing the modification of the decay constants of the exponential fit to the tail of the frequency distribution across model resolution. It is seen that the change in decay constants is most pronounced in case of the GCM (see also Durman *et al.*, 2000). Although the two RCM experiments follow each other quite closely, the direction of change is not consistent in the autumn. This again stresses the need for an assessment of the significance of the deduced decay constants.

A preliminary analysis using a non-parametric statistical method indicates that the apparent changes appear to be significant in most of the seasons in the case of RCM simulations, while this is not true for the GCM. This emphasizes the need for high resolution RCMs in climate change studies.

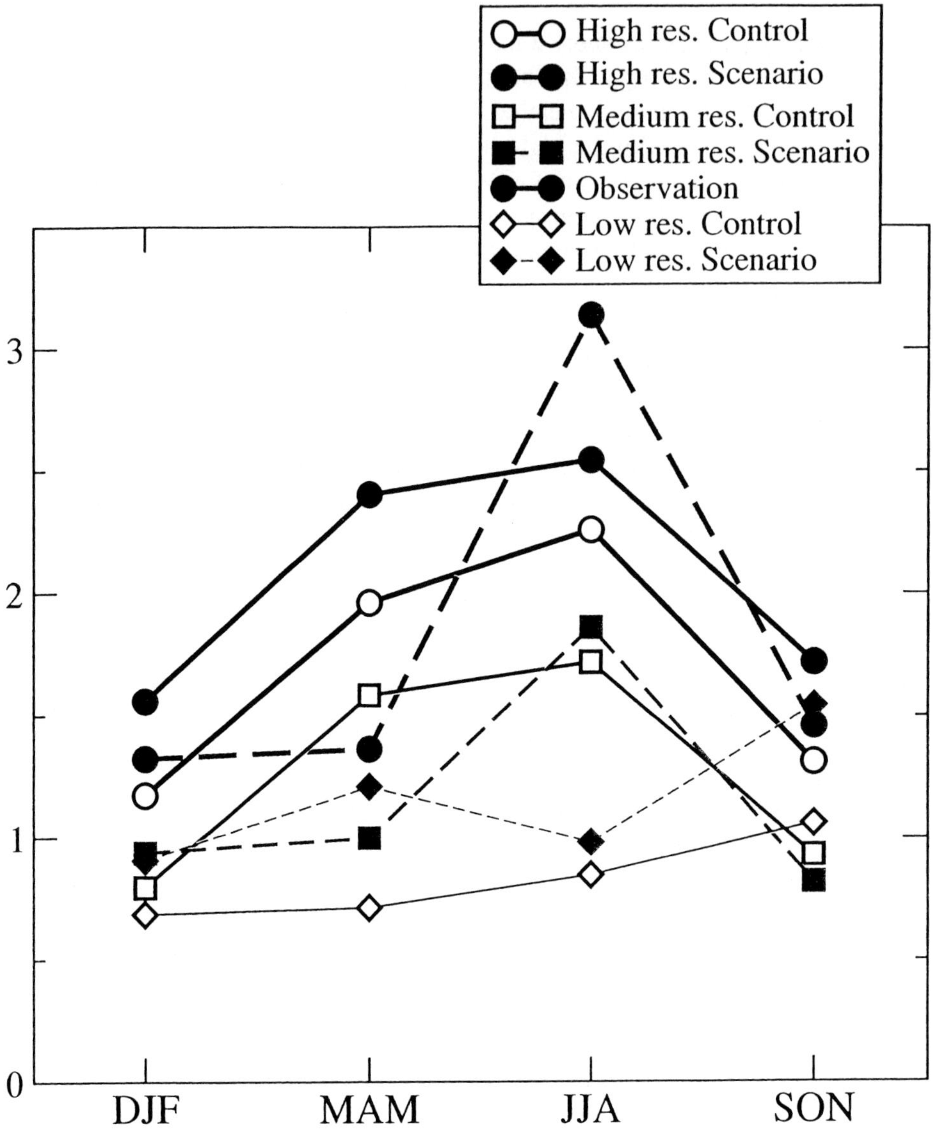

Figure 5. Exponential decay constants for simulated precipitation intensity spectra for land points over Denmark at different resolutions. Full lines/filled symbols correspond to control simulations; dashed lines/empty symbols correspond to scenario simulations.

5. SUMMARY AND CONCLUSIONS

A set of nested climate change simulations for the Nordic region and Denmark has been revisited. In the present work we have re-examined the results of CCMB and MBC with special emphasis on precipitation intensity frequencies, in particular the more extreme part of the frequency distribution. It has been demonstrated that the role of extreme precipitation events appears to be more realistically described in a high-resolution model, in terms of numerical agreement as well as seasonal variation. This is mainly due to a better simulation of deep low-pressure systems and mesoscale circulation.

In the present work a resolution effect has been identified which seems essential to the understanding of climate change effects on the extreme end of the precipitation intensity distribution. In order to analyze the role of the model resolution we have aggregated both the nested model data and observational records to the GCM grid from the driving AOGCM. It was found that, in spite of changes in absolute numbers, the seasonal behavior of decay constants does not change appreciably because of the aggregation. The RCM results show a seasonal behavior very similar to an observed data set. It is therefore concluded that the GCM has an unrealistic simulation of the dependence of heavy precipitation on climate, as manifested in seasonal variation. In contrast, the regional simulations remain close to observation in this respect. Furthermore, they agree on a conclusion that extreme precipitation generally scales with average precipitation (no significant change in decay constants were detected), but that the crucial summer season may be an exception, exhibiting an anomalous increase in heavy precipitation due to the anthropogenic greenhouse effect.

The analysis has only been performed over Denmark due to lack of daily observational data for other regions. It is, however, necessary to extend the work to other areas, for instance areas where the greenhouse effect leads to a decrease in total precipitation, in order to draw general conclusions about the relation between total precipitation and frequency of heavy-precipitation events.

6. ACKNOWLEDGMENTS

The simulations used in these analyses were carried out as part of EC Environment and Climate Research Programme activities (Contracts EV5V-CT92-0126 and EV5V-CT94-0505, Climate and Natural Hazards).

7. REFERENCES

Christensen J.H., O.B. Christensen, P. Lopez, E. van Meijgaard, and M. Botzet, 1996: The HIRHAM4 Regional Atmospheric Climate Model. *DMI Sci. Rep.* 96–4. 51pp. DMI, Copenhagen.

Christensen O.B., J.H. Christensen, B. Machenhauer and M. Botzet, 1998: Very high-resolution regional climate simulations over Scandinavia – Present climate. *J. Climate*, 11, 3204–3229.

Durman, C.F., J.M. Gregory, D.H. Hassell, R.G. Jones, 2000. The comparison of extreme European daily precipitation simulated by a global and a regional climate model for present and future climates. *Quart. J. Roy. Met. Soc.*, (Submitted).

Fenger, J., K. Halsnæs, and A.M. Jørgensen (eds), 2000. Drivhuseffekt og klimaændringer: Betydningen for Danmark set i lyset af IPCC's 1996 - rapporter. Miljø & Energiministeriet 1996 ISBN: 87-7810-692-3.

Kaas, E., and U. Andersen, 2000: Scenarios for extra-tropical storm and wave activity: Methodologies and results. In Beersma, J., M. Agnew, D. Viner, and M. Hulme (eds). Climate Scenarios for water-related and coastal impacts. *ECLAT-2 Workshop Report No. 3, KNMI, the Netherlands, 10-12 May 2000*. CRU, Norwich, UK, 144 pp.

Källén, E., (ed.), 1996: HIRLAM documentation manual, system 2.5. Swedish Meteorological and Hydrological Institute, 126 pp. Available from SMHI, S-60176 Norrköping, Sweden.

Machenhauer, B., Windelband, J., Botzet, M., Christensen, J.H., Déqué M, Jones, R., Ruti, P.M., and Visconti, G., 1998: Validation and analysis of regional present-day climate and climate change simulations over Europe. MPI Report No. 275, 80 pp., MPI, Hamburg, Germany.

Roeckner E., et al. 1996: The atmospheric general circulation model ECHAM-4: Model description and simulation of present-day climate. MPI Report No. 218, 90 pp., MPI, Hamburg, Germany.

Roeckner E., L. Bengtsson, J. Feichter, J. Lelieveld, and H. Rohde, 1999: Transient climate change simulations with a coupled atmosphere-ocean GCM including the tropospheric sulfur cycle. *J. Clim.*, 12, 3004–3032.

Scharling, M., 1999: KLIMAGRID - DANMARK Nedbør, lufttemperatur og potentiel fordampning 20*20 & 40*40 km. *DMI Tech. Rep.* 99–12. 48 pp. DMI, Copenhagen.

Voss, R., W. May, and E. Roeckner, 2000: Enhanced resolution modeling study on anthropogenic climate change: changes in extremes of the hydrological cycle. *Int. J. Climatology* (Submitted).

Testing of NAO and ENSO signals in the precipitation field in Europe

C. MARES[1], ILEANA MARES[1] AND MIHAELA MIHAILESCU[2]
[1]*National Institute of Meteorology and HydrologyBucharest, Romania*
[2]*Agricultural University, Bucharest, Romania*

Abstract: The summer Palmer Drought Severity Index (PDSI) for the 1891-1991 period is considered in five from nine homogeneous regions, as they were defined by *Briffa et al.* (1994), and in a region representative for the Romania zone (ROMZ).
In order to test the influence of NAO (North Atlantic Oscillation) or ENSO (El Nino- Southern Oscillation) on the precipitation fields or on the drought index, the correlation technique has been applied using the lags from 1 to 12 months for NAO and from 1 to 4 seasons for ENSO. Only the results with a high statistical significance level have been presented in the paper.
As we expected the NAO signal is evident in wintertime, but the most influence is not simultaneous in the same month - but with some lags.
Generally, ENSO signal is weak in the precipitation field in Europe; only for some regions the connections are significant. A negative SOI in winter and springtime favours a wetter summer in northeast part of Europe.

1. INTRODUCTION

NAO is the dominant mode of the winter climate variability in the North Atlantic region. The corresponding index varies from year to year, but also exhibits a tendency to remain in a positive or negative phase for intervals lasting several years. Over the past thirty years, NAO has steadily strengthened, rising from its low index state in the 1960s to a historic maximum in the early 1990s. This trend accounts for a significant portion of the Northern Hemisphere wintertime temperature increase over Eurasia, a major component of the recent warming.

M. Beniston (ed.),
Climatic Change: Implications for the Hydrological Cycle and for Water Management, 113–121.

Recently Garcia's et al. (2000) paper referred many publications related to NAO and its influence on the pressure field in the Atlantic-European region.

The NAO used in this paper is defined as the normalised pressure difference between Ponta Delgata (Azores) and Reykjavik (SW Iceland), according to *Jones et al*. (1997).

El Nino/Southern Oscillation (ENSO) is one of the most important phenomena occurred in the coupled interaction between ocean and atmosphere and it produces changes in the evolution of the air components at large spatial-temporal scales.

It is well known that the atmosphere and the ocean in the Pacific Ocean tropical basin have an interannual variability that affects the weather in many parts of the Earth.

Ropelewski and Halpert (1987) identified the regions of the globe that have relatively strong, consistent ENSO –related precipitation signals. The physical mechanism of the relation between ENSO and the precipitation in the southern Europe is difficult to establish, as also shown by Ropelewski and Halpert (1987). Fraederich (1994) made a review upon the latest investigations on the ENSO impact in Europe.

In order to obtain a quantification of ENSO effect we considered Southern Oscillation Index (SOI). The SOI is defined as the normalised pressure difference between Tahiti and Darwin. There are several slight variations in the SOI values calculated at various centres and in the present paper we used the SOI calculated by the method given by *Ropelewski and Jones* (1987).

The paper is structured in 2 parts: Data description (section 2) and Discussion of the results (Section 3) with two paragraphs related to NAO signal and Teleconnections between ENSO and precipitation. The NAO signal was investigated for monthly values and SOI impact by means of seasonal time series.

2. DATA

a) NAO and SOI values (1891-1997) obtained from *The IPCC Data Distribution Centre*, http: ipcc-ddc.cru.uea.ac.uk ;

b) Precipitation (31 stations) and temperature fields (33 stations) in Romania between 1950-1997;

c) The Palmer Drought Severity Index (PDSI) for the summer time over the 1891-1991 interval, as it was calculated by *Briffa et al.* (1994) defined for 6 regions:

- The northwestern Europe (NWEUR),

- The northwest Mediterranean (NWMED),
- The Danube Basin (DANUB),
- The Romania zone (ROMZ),
- The northern Fennoscandia (NFENN),
- The lower Volga (LVOLG)

These regions with coherent summer moisture variability have been objectively defined by the orthogonal rotation of the principal components of the summer PDSI.

The precipitation and temperature fields in Romania have been decomposed in extended empirical orthogonal function (EEOF) with a window of 3 months.

The advantage of EEOF against EOF consists in the fact that EEOF is a straightforward way to identify the systematic space/time variations in the analysed field (von Storch and Frankignoul, 1998). Also, the EEOF analysis is a very efficient method for detecting cyclical behaviours of time series.

The first mode from our analysis, accounted for 19% of the total variance. The spatial pattern indicates 2 principal weights situated in the northeastern and the southwestern parts of Romania and are more intense for a delay by one month, which is representative for the EEOF decomposition with a window length 3. The area from the south-western part may be associated with the Mediterranean cyclones influences and the other from north-west may indicate impact of the retrograde cyclones, which are characteristic for Romania and for neighbour zones and which generally yield very intensive precipitation in north-eastern part of Romania.

We introduced a drought index (EEOFDI), which was estimated by means of the first component EEOF for the temperature and precipitation fields for 1951-1997. This drought index has given, in the case of Romania, similar results with the Palmer Drought Severity Indices (PDSI) defined by Palmer and calculated by *Briffa et al.* (1994). The behaviours of the two indices are very similar, except for their signs that are reverse. These conclusions can be found in the study of *Mares et al.* (1996). Positive values of EEOFDI indicate a state of drought.

The power spectra for temporal EEOF1 for precipitation in Romania reveal significant quasi-periodicities of 8 and 23 months; the last of which may be associated with the quasi- biennial oscillation and this result may be used in our teleconnections with NAO and ENSO.

3. DISCUSSION OF THE RESULTS

3.1 NAO signal

3.1.1 Link between NAO and the drought index in Romania

We presented here the results of testing the correlation between NAO and the drought index (EEOFDI). In Fig. 1, the simultaneous correlation (1951-1997) between NAO and EEOFDI is presented. In Fig. 1 for the simultaneous correlation we can see that there are three months (January, February, March) with statistical significance. For 47 years (1951-1997) the correlation coefficient (r) is significant with 95% confidence level if $|r| > 0.29$ and with a 99% confidence level if $|r| > 0.38$. For the latter situation that means there is a chance $p < 0.01$ that this is the result of a random fluctuation of uncorrelated data.

The better results for correlation with different lags between EEOFDI and NAO have been obtained. These results are presented in Table 1.

Table 1: The correlation coefficients (r) between NAO and EEOFDI with different lags.

Mon1	1	1	1	2	2	3	3	9	9	10	10	10	10	12
Mon2	2	3	4	3	4	4	5	11	12	2	3	9	10	9
Lag	1	2	3	1	2	1	2	2	3	4	5	11	12	9
r	.62	.58	.42	.49	.43	.47	.44	.39	.33	.35	.30	.36	.37	.35
β (%)	99	99	99	99	99	99	99	99	95	95	95	95	95	95

Mon_1 and *Mon_2* from table 1 represent the months in which NAO and the precipitation index are considered, *Lag* denotes the difference (in months) between the two time series correlated, and β is the statistical confidence level.

From Table 1, one can see that NAO from January, February and March is a good predictor for the drought index in Romania especially for the springtime. It must be mentioned that EEOFDI from February represents the moisture state in 3 months, namely January, February and March.

3.1.2 Signal NAO in the summer Palmer index in Europe

The analysis has been carried out for 101 years (1891-1991) and only for Palmer Drought Severity Indices (PDSI) in summer time (June, July and August). A negative value of PDSI indicates a state of drought. The absolute value of the correlation coefficient must be greater than 0.19 for a 95% confidence level and greater than 0.25 for a 99% confidence level. ROMZ from Table 2 represents the Palmer index for the Romanian territory as was calculated by *Briffa et al.* (1994).

Table 2: The correlation coefficients (r) between NAO and summer Palmer Drought Severity Indices from different regions in Europe. Mon (NAO) - the month for the NAO.

Region	Mon (NAO)	r	β (%)
NWMED	2	-0.22	95
	3	**-0.31**	**99**
	4	-0.23	95
	10	0.20	95
DANUB	1	-0.24	98
	2	-0.23	95
	3	-0.22	95
ROMZ	**1**	**-0.29**	**99**
	2	0.19	95
	11	0.21	95
NFENN	1	0.19	95
LVOLG	**1**	**-0.26**	**99**
	9	**-0.26**	**99**

For the northwestern Mediterranean (NWMED) region, the best correlation has been obtained between NAO in March and PDSI in summer. For the Romanian area (ROMZ), Danube basin and lower Volga (LVOLG), NAO in January is the best predictor for the summer PDSI.

The mechanism responsible for some delay in the NAO impact on the precipitation in Europe are poorly understood, the springtime response to NAO in January may be due to the snow cover state in wintertime (Kaczmarek, 2000). For the other lags it is very difficult to explain the complex interaction between components of land-atmosphere-ocean system.

3.2 Teleconnections between ENSO and precipitation

In order to test the connections between SOI and precipitation, the seasonal values with a one to four seasons lag for the 1891-1991 period, have been used The results for the investigation by means of monthly time series are presented in Mares and Mares (1993).

By testing the connections between SOI and PDSI from 5 homogeneous areas in Europe in summertime, results with statistical significance at a 99% confidence level have been obtained only for the lower Volga, for which the SOI behaviour in wintertime can influence the moisture condition during the summer in this area (r = - 0.31). A lower SOI for instance, namely the presence of El Nino, may cause a rainy summer in a certain area (lower Volga). In the area of Romania even if this connection still exists (r = - 0.27, at the 99% significance), the most important ENSO signal exhibits in the summer season (r = - 0.35) when the warm phase of springtime ENSO favours the excess of precipitation in Romania.

These results are partially in accordance with the findings of Oldenborgh et al. (2000) who have taken into account the evolution of NINO3 index for El-Nino quantification.

This index is correlated with the precipitation in Europe for the 1857-1993 period. Fig. 3 of their paper proves the existence of 3 main nuclei, where the correlation between the NINO3 index in wintertime and the precipitation in Europe during the spring is really significant. A nucleus is located in the lower Volga area and this is the place where the above - mentioned authors' results are the same with ours; the other 2 nuclei are located in the NWEUR and NWMED regions, for which we haven't obtained any significant results yet. These discrepancies could come from the fact that we got only the PDSI in the summertime or the differences existing in the quantification of ENSO events and therefore the propagation mechanisms may be different or from the fact that the time series of data, processing having been made for, are different. Even if SOI is not a predictor as good as the NINO3 index, it seems that its behaviour during the winter and spring is very important for the humidity condition in the summer time in certain areas in eastern Europe. The explanation of the physical mechanisms, which lead to this situation is very difficult to achieve and acknowledged by common consent. Certain authors have tried to explain the propagation of the ENSO influence on precipitation in Europe referring to one-third region located in the southeast of Asia (Oldenborgh et al., 2000). As proved by the authors, in the spring, the transition to the Asian and Chinese monsoon systems makes the circulation more susceptible to perturbations. Our connection results between SOI in the springtime and PDSI during the summer should be included into these suppositions.

Another supposition is related to the Pacific influence upon Europe by the Atlantic Ocean. Thus, Rogers (1984) proved that there exist certain tendencies of association between a warm Pacific (the ENSO warm phase) and a blocking circulation over the Atlantic.

To test whether there is a connection between ENSO and SSTA within the Tropical Atlantic, we performed correlation analysis between SOI and the first 10 EOF components of SSTA analysed for the 1948-1994 period. To highlight the ENSO signal, we only considered the lag in a unique sense, i.e. we assumed that the ENSO events influence the SSTA in the Tropical Atlantic.

Some of our results are in accordance with the results of researchers like Enfield and Mayer (1997) (quoted by Dommenget and Latif, 2000) who found that the Atlantic is significantly correlated with ENSO in the 10° N - 20° N region, with a lag of 4-5 months, and Nicholson (1997) shows that the lag between the Tropical Atlantic and the Eastern Pacific is 6 months. The link between SSTA from tropical Atlantic and drought episodes in Europe we investigated in Mares et al. (2000) and we found that variability of SSTA from the tropical Atlantic in spring influences the Palmer Drought Severity Index during the summer across Europe. Therefore we can speculate that the influence of winter SOI on summer precipitation in Europe is achieved by means of SSTA behaviour from tropical Atlantic in springtime.

For the area of Romania, we had 4 stations with continuous and relatively long observations, for which we made a seasonal correlative analysis between SOI and the precipitation or the drought index for the same period (1891-1991). The results differ from one another, depending on the station location. Thus, for Bucharest-Filaret, there was obtained a single result with a 99% confidence level and this represents a positive correlation (0.25) between SOI in winter and the drought index in Bucharest in summer.

For Sulina station, placed in the Danube Delta, the single result refers to the influence of SOI during fall on the winter precipitation (-0.25).

Sibiu station placed in the centre of Romania reflects the result of Oldenborgh et al. (2000), meaning that SOI in winter may influence the precipitation in spring (with a 95% confidence level), but the best result (r= -0.31) is obtained for the pair of precipitation time series formed by SOI in summer and the precipitation in winter.

The 4th station, Timisoara, placed in the extreme west of Romania, emphases an instantaneous SOI signal in winter (with a 99% confidence level) in the precipitation time series as well as in the drought index, but also a weak connection (between a 99% and 95 % confidence interval) between SOI in winter and the drought index in the spring.

By testing the ENSO event influence quantified by SOI upon the precipitation field in Romania and the areas in the northeast of Romania, there results that the SOI behaviour in the spring is a good signal for the evolution of the humidity condition in summer time.

It is clear that further investigations are needed in order to elucidate the teleconnection mechanism between ENSO and moisture in Europe.

4. CONCLUSIONS

NAO from January, February and March influences the behaviour of the precipitation regime in Romania, especially for the springtime.

The drought index calculated by means of first EEOF of temperature and precipitation fields in Romania (EEOFDI) is more sensitive to signal NAO than the precipitation regime.

Concerning the sensitivity of the summer Palmer Drought Severity Index (PDSI) in Europe to the NAO behaviour the following significant results have been obtained: NAO in March has an impact on the PDSI in the north-western Mediterranean (NWMED); NAO in January influences the behaviour of PDSI in the Danube Basin, Romania zone and the lower Volga region. For all the cases a positive NAO determines a negative value of PDSI, namely a drier state and reverse. The drier period from the last two-decade in Romania may be attributed to the NAO positive phase.

The statistically significant results at the 99% confidence level have been obtained for Romania region for which the moisture state from summer time is sensitive to SOI anomalies in spring, namely an El-Nino event favours a wetter summer.

The lower Volga region (LVOLG) in summer is influenced by SOI state in winter. The correlation coefficient between SOI and PDSI is negative, i.e. an El Nino event favours a wetter summer in the northeastern part of Europe.

The ENSO impact on the precipitation field in Europe may be explained either by SSTA from the Indian Ocean or especially by means of SSTA behaviour from the tropical Atlantic.

For the cases where a significant correlation between a circulation index (NAO or SOI) and predictand variable was obtained, one can develop linear models to attempt in order to predict a precipitation index.

5. ACKNOWLEDGEMENTS

Authors would like to thank very much to the organisers and the Swiss National Science Foundation for the support to attend this very important Conference.

We also acknowledge the very helpful comments and suggestions of two anonymous reviewers.

6. REFERENCES

Briffa, K.R, P.D.Jones and M.Hulme, 1994: Summer moisture variability across Europe, 1892 - 1991 : An analysis based on Palmer drought severity index. *Int.J. Climatol.*,14, 475-506.

Dommenget, D. and M. Latif, 2000: Interannual to decadal variability in the tropical Atlantic. *J. Climate*, 13, 777-791.

Enfield, D.B. and D.A. Mayer, 1997: Tropical Atlantic sea surface temperature variability and its relation to El Nino-Southern oscillation. *J. Geophys. Res.,* 102, 929-945.

Fraedrich, K., 1994 : An ENSO impact on Europe ? A review. *Tellus* 46 A, 541-552.

Garcia, R., D. Gallego, E. Hernandez, A.Marcias, L. Gimeno and P. Ribera, 2000: Precipitation in the Canary Islands: reconstruction and major influences. *International Workshop on Climatic Change: Implications for the Hydrological Cycle and for water Management*, Wengen, Switzerland, September 27-29, 2000.

Jones, P.D. T. Jonsson and D. Wheeler, 1997 : Extension to the North Atlantic Oscillation using early instrumental pressure observations from Gibraltar AND South-West Iceland. *Int. J. Climatol.* 17,1433-1450.

Kaczmarek, Z., 2000 : Impact of North Atlantic Oscillation on river discharges in Europe. *International Workshop on Climatic Change : Implications for the Hydrological Cycle and for water Management*, Wengen, Switzerland, September 27-29, 2000.

Mares, C. and Ileana Mares, 1993: ENSO effects upon the low frequency components of 500 hPa geopotential field. Four International Conference on Southern Hemisphere Meteorology and Oceanography. American Meteorological Society, 29 March - 2 April 1993, Hobart, Australia, 78-79.

Mares, Ileana , C. Mares, M.Mihailescu, S. Capsuna and M. Chirovici, 1996 : Study of the extreme events in Romania in the regional climate context. : *International Symposium on Applied Agrometeorology and Agroclimatology* , 24-26 April 1996, Volos, Greece, 25-36.

Mares Ileana, C. Mares and Mihaela Mihailescu, 2000: The influence of the SSTA from tropical Atlantic on the drought episodes in Europe. 3rd European Conference on Applied Climatology, 16-20 October 2000, Pisa, Italy. CD ISBN 88-900502-0-9, CNR-IATA-Institute of Agrometeorology and Environmental Analysis of Agriculture.

Nicholson, S. 1997 : An analysis of the ENSO signal in the tropical Atlantic and Western Indian Oceans. *Int. J. Climatol.*, 17, 345-375.

Oldenborgh,G.J., G. Burgers and A. Klein Tank, 2000: On the El Nino teleconnection to spring precipitation in Europe. *Int. J. Climatol.* 20, 565-574.

Rogers, J., 1984 : The association between the Noth Atlantic Oscillation and Southern Oscillation in the Northern Hemisphere. *Mon. Wea. Rev.,*112, 1999-2015.

Ropelewski, C.F. and Jones, P.D., 1987: An extension of the Tahiti-Darwin Southern Oscillation Index. *Mon. Wea. Rev.* 115, 2161-2165.

Ropelewski, C.F. and M.S. Halpert, 1987 : Global and regional scale precipitation patterns associated with the El Nino / Southern Oscillation. *Mon. Wea. Rev*., 115, 1606-1626.

von Storch H. and C. Frankignoul, 1998: Empirical modal decomposition in coastal oceanography. In: *The Sea*, 10, edited by Kenneth H. Brink and Allan R. Robinson. John Wiley & Sons,Inc, 419-455.

Water vapor vertical profile by Raman lidar in the free troposphere from the Jungfraujoch Alpine Station

IOAN BALIN, GILLES LARCHEVÊQUE, PHILIPPE QUAGLIA, VALENTIN SIMEONOV, HUBERT VAN DEN BERGH, AND BERTRAND CALPINI
Laboratory of Air Pollution, Swiss Federal Institute of Technology – Lausanne (EPFL), Lausanne, Switzerland

Abstract: The water vapor content in the atmosphere is an important criteria for the validation of predictive results obtained from global scale atmospheric models. Due to its non-homogeneous distribution in the troposphere, both in space and time, the water vapor content in the atmosphere may still be considered today as the largest uncertainty in our understanding of the earth radiation budget. This paper presents new results obtained by Raman lidar measurements as one of the attractive method for long-term continuous observation of the water vapor content in the atmosphere. A powerful pulsed laser beam at 355 nm is emitted and the inelastic back-scatter signals (Raman shift) from nitrogen and water vapor are recorded respectively. The ratio between the water vapor Raman shifted wavelength at 408nm and the nitrogen at 387nm gives a first estimate of the relative water vapor mixing ratio with good vertical resolution. The absolute water vapor vertical profiles are retrieved using an additional *in situ* external reference value directly obtained from a local meteorological station. The Raman lidar system, operated at an altitude of 3'580 m above see level in the Swiss Alpine region at the Jungfraujoch research station, is presented, and two typical water vapor vertical profiles obtained in clear sky and in cloudy conditions are discussed and directly compared with radio sounding measurements performed by the Swiss Meteorological Station from Payerne (80 km West). A first estimate of the statistical (signal to noise) and systematic error sources is presented.

M. Beniston (ed.),
Climatic Change: Implications for the Hydrological Cycle and for Water Management, 123–138.

1. INTRODUCTION

Water vapor (H_2O) is the primary greenhouse gas and is a crucial atmospheric constituent. It induces the natural atmospheric greenhouse effect [1] and the hydrological cycle [2]. Atmospheric trace gases such as carbon dioxide (CO_2), methane (CH_4), nitrous oxide (N_2O), or chloro-fluorocarbons (CFC) play an additional role as greenhouse gases. Both, water vapor and these trace gases contribute to the fragile thermodynamics equilibrium of the atmosphere by trapping long-wave (infrared) terrestrial re-radiation.

Water vapor averages about one per cent by volume in the atmosphere, but is much variable in time and space: it comprises about 4 percent of the atmosphere by volume near the surface, but only 3-6 ppmv (parts per million by volume) above 10 to 12 km. The annual average precipitation over the globe is about 1 meter while the water-vapor column density (precipitable water) averages about 5 cm in the tropical region and less then 1 mm at the poles. Nearly 50% of the total atmospheric water is trapped into the planetary boundary layer (PBL, from 0 to 1-2 km) while less than 6 % of the water is above 5 km, and only 1 % above 12 km in the stratosphere. Despite this small amount of water vapor in the free troposphere (above 2 km), recent studies have shown that in the middle and upper troposphere the water vapor content is the main source of the OH radical on one hand, while it plays an essential role for the earth climate on the other hand [1,2,3]. In the upper troposphere and lower stratosphere, more accurate water vapor measurements are needed to cover our lack of knowledge of the water vapor vertical dynamic around the tropopause region, as well as of the nucleation processes and cirrus formations for example. In its final phase of development this lidar will make also possible the monitoring of water vapor content in the lower stratosphere where an increase of water vapor has already been acknowledged, that may partially be related to the methane oxidation. Finally the water vapor spatial distribution is also an essential input for initializing the Numerical Weather Prediction (NWP) and other Global Climate models.

Different measurement techniques are used to estimate the water vapor content in the atmosphere, with various spatial coverage (horizontal and vertical resolution) and temporal resolution, as well as detection limits and costs [3]. These methods may also be identified by their principle (sampling detector or remote sensing), by their location (balloons, satellites, aircraft or ground based method). Among these, the Raman lidar (**L**idar **D**etection **A**nd **R**anging) is a remote sensing technique with high vertical resolution (tens to hundreds of meters), good temporal resolution (typically one hour per vertical profile) and the potential for long-term continuous measurements. On the other hand this method will not bring information in terms of spatial coverage, as is the case for airborne lidar or satellite observations. Water vapor measured by Raman lidar is limited to a maximum altitude range of typically 15 km, but combined with the passive microwave techniques, a method with lower range resolution but possible data retrieval up to the mesosphere [4], these combined methods appear to give an ideal tool for 2D continuous water vertical profile throughout the atmosphere. Here we will focus on the development of a Raman lidar system at the Jungfraujoch Research Station (3'580 m ASL) in order to measure with high vertical resolution the water-vapor content in the free troposphere, present the first nighttime water-vapor profiles measurements and error sources, and give indications about the final configuration of the system.

2. THE RAMAN LIDAR METHOD

2.1 Principle

A lidar instrument is composed of a transmitting and a receiving section. The transmitter emits the pulsed laser beam in the atmosphere. This beam interacts with the atmospheric constituents upon propagation, and the back-scattered light is collected by a telescope, spectrally resolved, and recorded versus time by a detection unit. The light interacts with atmospheric gases and aerosols by a multitude of phenomena like Rayleigh (or molecular), Mie (or aerosol), Fresnel, and Raman scattering, as well as by inducing fluorescence, so as to provide the source of the lidar signal for atmospheric monitoring. A picture of the Jungfraujoch lidar system in operation at night is given in Figure 1.

The Raman lidar measurement of the water vapor is based on the spontaneous vibrational Raman scattering of an incident laser beam by atmospheric N_2 and H_2O molecules. In this experiment the third harmonic of a Nd:YAG laser at 355 nm is used. The Raman shifted wavelengths are respectively 387 nm from N_2 and 408 nm from H_2O. The water-vapor mixing ratio $H_2O(z)$ defined in units of [g/kg dry air] is calculated from these back-scattered signals assuming a constant mixing ratio for N_2 in the atmosphere [8, 9, 12-15]. It is expressed as:

$$H_2O(z) = \left[\frac{N_{H2O}(z) - b_{H2O}}{N_{N2}(z) - b_{N2}}\right] C \bullet \Gamma(z, \lambda_{H2O}, \lambda_{N2}) \tag{1}$$

where N(z) is the total number of photons detected for the H_2O and N_2 Raman signals respectively from an altitude z, and b is the background intensity due to the skylight and the detector noise.

Figure 1. The EPFL lidar in operation at night from Jungfraujoch.

The correction term $\Gamma(z,\lambda_{H20},\lambda_{N2})$ takes into account the differential extinction of the atmosphere at the water vapor (408 nm) and nitrogen (387 nm) Raman shifted wavelengths on the return path due to the total extinction coefficient α. It can be written as:

$$\Gamma(z,\lambda_{H2O},\lambda_{N2}) = \frac{\exp\left[-\int_{zo}^{z}\alpha(\lambda_{N2},r)dr\right]}{\exp\left[-\int_{zo}^{z}\alpha(\lambda_{H2O},r)dr\right]} \tag{2}$$

where:

$$\alpha = \alpha_{molec} + \alpha_{aerosol} + \alpha_{abs} \tag{3}$$

The total extinction term α is the sum of the molecular extinction (α_{molec}), the aerosol extinction ($\alpha_{aerosol}$) and the molecular absorption (α_{abs}) [9] and is wavelength dependent [7]. The altitude of the Jungraujoch lidar site (z_0) is 3'580 m ASL. This optical extinction Γ corresponds to an integrative effect over the entire range from the lidar site z_0 to the altitude of interest z.

The calibration coefficient C in Equ. 1 can be expressed as:

$$C = \frac{L_{N2}(\lambda)\sigma_{N2}(\lambda)}{L_{H2O}(\lambda)\sigma_{H2O}(\lambda)} \frac{M_{H2O}}{M_{\substack{dry\\air}}} \frac{n_{N2}}{n_{\substack{dry\\air}}} \tag{4}$$

where L is the instrument constant that takes into account the transmitter and receiver optical efficiency and the quantum efficiency of the detectors for the two channels, σ is the Raman back-scatter cross section, and M and n are the molecular mass and number density of the water vapor and nitrogen respectively. As this calibration constant C depends upon the lidar system efficiency L, it cannot be measured directly. This is why in Equ. 1, the value of the water vapor mixing ratio is only retrieved as a relative measure and an external calibration value must be added to the lidar measurements.

Let us recall here that the main difficulty in this water-vapor measurement in the middle and upper troposphere by Raman lidar relies on the rather weak Raman water vapor signal intensity: On one hand in the free troposphere, the water vapor concentration is very low and decreases by 2-3 orders of magnitude from 2 to 12 km, and on the other hand the water vapor Raman cross section (which is similar to the nitrogen Raman cross section) induces a back-scattered signal that is typically three orders of magnitude lower than the elastic back-scattered signal in clear sky or molecular condition [12].

2.2 Raman lidar setup at the Jungfraujoch Alpine Research Station

A multi-wavelength lidar system was installed in 1999 at the Jungfraujoch Research Station (Lat: 46.55°N, Long: 7.98°E, ASL 3'580m) in the astronomic dome in order to probe the atmosphere above the Swiss Alps [17]. The simplified optical layout of the lidar system is shown in Figure 2. The transmitter of the system is based on a Nd:YAG laser (Spectra Physics, Infinity) with a maximum energy of 400 mJ at 1064 nm equipped with two non-linear crystals for second (532 nm) and third (355 nm) harmonic generation. The repetition rate of the laser can be tuned continuously up to 100 Hz using a phase-conjugated mirror. Dichroïc mirrors at the laser output separate the three laser wavelengths and each beam is expanded 5 times in order to reduce the laser divergence from 0.7 to 0.14 mrad. These expanded beams are emitted to the atmosphere using 45° dielectric mirrors mounted on piezoelectric-driven stages (M).

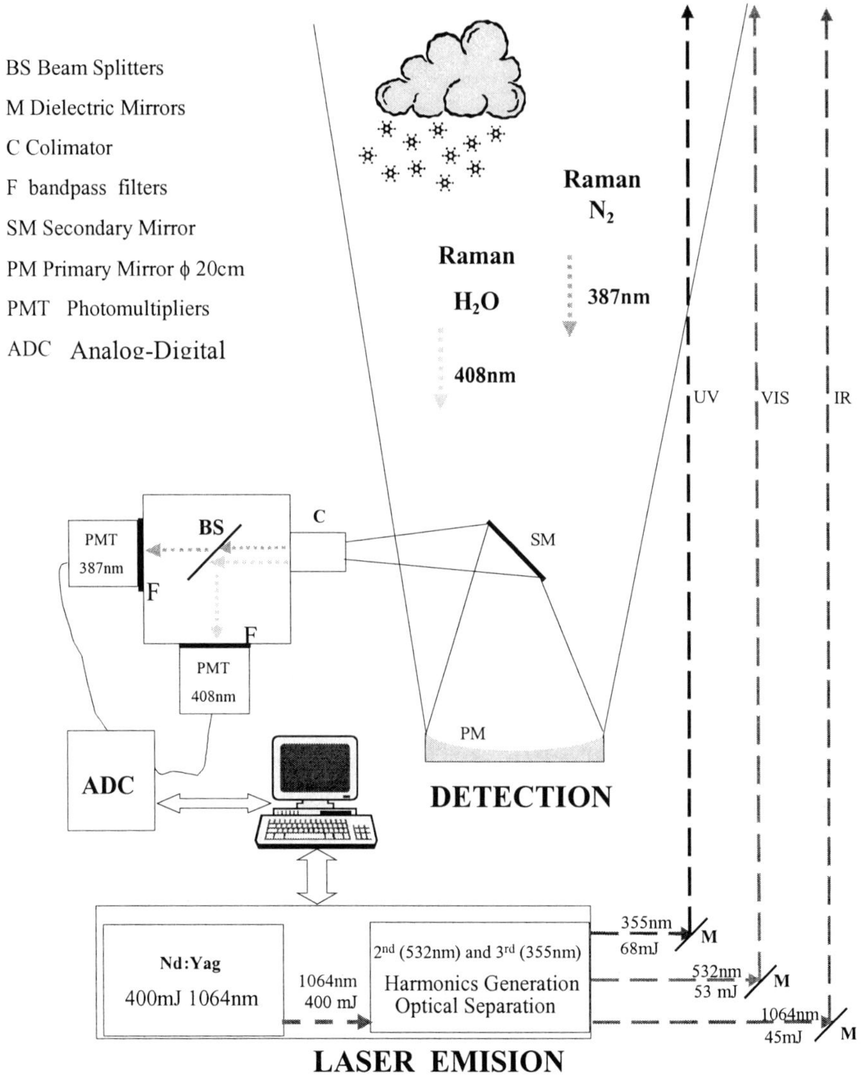

Figure 2. Raman lidar setup

The typical output energy emitted into the atmosphere is 70 mJ at 355 nm, 60 mJ at 532 nm, and 45 mJ at 1064 nm. The lidar system is working on the vertical axis. The lidar emitter is off-axis from the lidar receiver and the first data analysis can only be performed at an altitude higher than 250 m above the ground, an altitude where a full overlap of the laser beam into the telescope field of view is achieved. In its actual configuration the system is equipped with a ϕ 20 cm primary mirror diameter Newtonian telescope with a focal length of 80 cm. This small size telescope enables measurements up to 12-15 km above the lidar site. The Jungfraujoch astronomical telescope, a ϕ 76 cm primary mirror diameter Cassegrain telescope, will soon be made available for simultaneous operation and lidar detection in the upper troposphere and stratosphere.

The elastic back-scatter signals at 355 nm, 532 nm with parallel and perpendicular polarization and 1064 nm as well as the Raman shifted signals from N_2 at 387 nm and H_2O at 408 nm (pumped at 355 nm) and N_2 at 607 nm (pumped at 532 nm) are simultaneously recorded. They are used to estimate the aerosol back-scatter/extinction coefficients and the water vapor content. Since Spring 2000, the Jungfraujoch lidar is in operation on a regular basis participating in the EARLINET (A European Aerosol Research Lidar Network to Establish an Aerosol Climatology) Environment program.

Here we used only the 355 nm beam to extract the water vapor mixing ratio from the Raman back-scatter from H_2O at 408 nm and N_2 at 387 nm. The light is collected by the telescope and spectrally resolved by a set of dichroïc mirrors and filters. Two sets of interference filters at each of the signal output are used to reduce the sky background light and suppress the residual elastically back-scattered light in the Raman channels. This combination of filters acts as an equivalent narrow band filter with 0.5 nm FWHM at 408 nm and 387 nm respectively and a rejection ratio of better than 10^{-7} between 200 to 1200 nm. Two head-on photomultipliers tubes (type EMI 9829 QA) are used in photon-counting mode. The acquisition unit has a maximum counting rate of 250 MHz with a sampling rate of 20 MHz.

2.3 Measurement procedure

The lidar signals for H_2O and N_2 are acquired in files of 4’000 laser shots each with a typical repetition rate of 50 Hz or an integration time of less than 2 minutes per file. The acquisition unit is used with 3'000 channels at 50 ns per channel or an equivalent maximum altitude range of 22.5 km above the Jungfraujoch lidar site. A LabView routine allows processing the two Raman signals for retrieving the water vapor mixing ratio according to Equ. (1).

This routine also allows visualizing file by file each of the two recorded signals, calculating the signal to noise ratio and detection limit, estimating the statistical error and selecting the right temporal and spatial averaging conditions. The background value (b) of each signal is calculated as an average of the last 1’000 channels (i.e. 7'500 m),which is subtracted afterwards from the raw signal.

As the system constant C in Equ. (1) is unknown, a ground reference H_2O [g/kg] value is calculated using the relative humidity RH [%], the air pressure P_{air} [mbar] and the temperature T [°C] values that are permanently measured by Swiss Meteorological Institute (ISM) at Jungfraujoch. The ISM hygrometer used for these measurements is a Meteolabor type VTP6 (±0.15°K, 0.1% RH). This ground-based value is used as a calibration for the first H_2O Raman lidar data at an altitude of 250 m above the lidar site. This implies that the assumption of an homogeneous water vapor mixing ratio is made over this range of 250 m in the free troposphere. This assumption may in special atmospheric conditions be an additional source of error. It will be soon investigated in more details using tethered balloon measurements.

Balloon measurements launched at ~80 km West from the ISM meteorological station in Payerne are used as independent comparisons with the lidar profiles. The water vapor content from the balloon was calculated based on the temperature and the dew point measurements (VIZ hygristor detector).

Finally, as the location of this lidar system at Jungfraujoch is at an altitude of 3'580 m ASL, the system is essentially all year long in the free troposphere, thus avoiding the complex interference of the planetary boundary layer (PBL) photo-chemistry and meteorology. This will in particular help to measure with good accuracy the water vapor content in region of the atmosphere where the water content is very low, e.g. the tropopause region, or with concentrations that are essentially undetectable with a similar method if the lidar site was located at low altitude in the PBL. Furthermore as Jungfraujoch is far away from the brightly lit main urban areas, it has therefore very little "light pollution".

2.4 Error sources

The differential atmospheric transmission factor $\Gamma(z,\lambda_{H20},\lambda_{N2})$ includes the extinction contributions due to the molecular (Rayleigh), the aerosol (Mie) and the trace gas absorption. According to Whiteman et al., [9] the correction factor related to the differential atmospheric transmission at these wavelengths (N_2: 387 nm and H_2O: 408 nm) is mainly due to the Rayleigh molecular diffusion ($\alpha_{molec} \sim\lambda^{-4}$) in clear sky conditions (no aerosols) and its contribution rises to 5 % at 7 km ASL. In Figure 3 the Rayleigh extinction coefficients from N_2 and H_2O and the H_2O Rayleigh correction factor are calculated from Jungfraujoch up to the tropopause. This is done using the molecular Rayleigh differential cross section [18] and considering real atmospheric parameters for the air number density based on temperature and pressure profiles measured directly by the balloon from Payerne. The results given in Figure 3 for a temperature of 1.5 °C and a pressure of 669 mbar at

Jungfraujoch show that for these wavelengths the error increases with the altitude and reaches ~ 4 % at the tropopause.

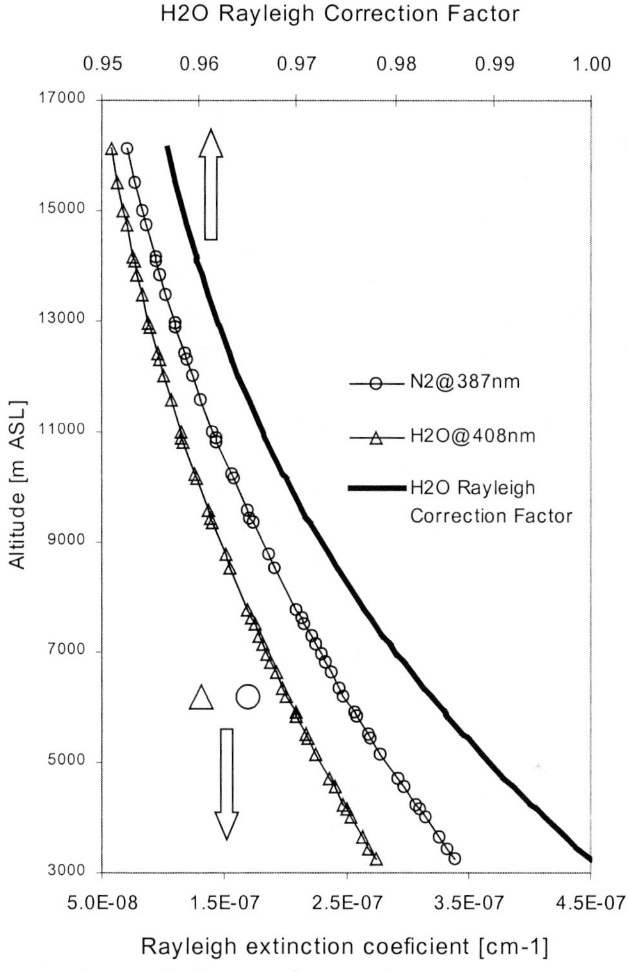

Figure 3. Evolution of the error source due to the differential Rayleigh cross section in the troposphere: the Rayleigh extinction coefficient (lower scale) for the N_2 Raman shifted wavelength at 387 nm and the water vapor at 408 nm is shown versus altitude. The Rayleigh correction factor is shown on the upper scale.

Nevertheless in the case of cloud layers or hazy conditions, the aerosol or Mie scattering must be accounted for in $\Gamma(z,\lambda_{H2O},\lambda_{N2})$. Here we used the elastic 355 nm lidar signal together with the N_2 Raman shifted signal following a data analysis proposed by Ansman et al. [6] to retrieve the aerosol extinction coefficient at 355 nm initially. The 387 nm and 408 nm aerosol extinction coefficients are retrieved following the $\alpha_{aerosol}$ proportional to λ^{-1} law. Therefore the differential aerosol extinction coefficient between the two Raman shifted wavelengths can be directly taken into account in $\Gamma(z,\lambda_{H20},\lambda_{N2})$.

Finally $\Gamma(z,\lambda_{H20},\lambda_{N2})$ is also influenced by possible interference with trace gas absorption. While this is particularly true in the planetary boundary layer where most of the primary pollution is emitted and trapped, these absorption effects tend to be less significant in the free troposphere. Furthermore the molecular absorption (α_{abs}) by ozone and other trace gases can in first approximation be neglected at the 387 and 408 nm wavelengths [6, 9, 12].

The estimation of the statistical error takes into account two error sources: the shot noise on the PMT and the background noise (estimated from the last 1'000 channels of the lidar signals or 7'500 m). The signal to noise ratio is considered as the ratio between the background subtracted signal and the average of these last 1'000 channels. The shot noise is expressed by a Poisson distribution (i.e. proportional to $\sqrt{N}$) and dominates when the signal to noise ratio is high, e.g. at low altitude, while the background noise is the principal source of the statistical error at low signal to noise ratio or high altitudes. The error source on the water vapor retrieval due to these two contributions is given in the following equation:

$$\frac{\delta_{H_2O}}{H_2O} = \frac{1}{\sqrt{N_{H_2O}}} \frac{\sqrt{1 + \frac{R}{N_{bkg}}}}{1 - R} \tag{5}$$

where N_{H2O} is the number of detected photons, N_{bkg} is the number of channels used to average the background signal (1'000 in this case) and R is the ratio of the average value of the background and the total signal [12].

For the systematic error sources, a careful control of the alignment of the lidar emitter/receiver was performed regularly. Since no misalignment was observed even over periods of some days of continuous measurements, this error source was no longer considered. Note that a systematic error would change the calibration constant C but this effect is also cancelled by considering for each H_2O profile the simultaneous external calibration value.

3. H_2O MEASUREMENTS BY RAMAN LIDAR

Since August 2000, nighttime measurements of the water vapor content were performed by Raman lidar at Jungfraujoch on a regular basis. Two typical cases of water vapor lidar profiles in clear sky conditions (A) and in presence of a cloud layer (B) will be presented. The laser parameters and the meteorological conditions during these two periods of measurements are summarized in Table 1.

Table 1: Lidar parameters and meteorological conditions

Case	Date/Time Start	Int. Time [h]	Laser 355nm	Meteorological Conditions (Jungfraujoch site)				
				T_a [°C]	P_a (mbar)	RH_a [%]	WD_a [°N]	WS_a [km/h]
A	01/08/2000 22:34	2:41	60mJ/20Hz	1.4	668.9	39.9	307.9	33
B	29/08/2000 23:26	2:28	31mJ/20Hz	-0.4	664.4	61.2	307.3	30

The water vapor vertical profile obtained by Raman lidar in clear sky (A) conditions for an integration time of 2 h 41 min with a space resolution of 75 m is shown in Figure 4a. This result is extracted from the lidar signals presented in Figure 4b where the range corrected signals RCS i.e. the raw lidar signals multiplied by the square of the distance, for four different lidar channels, are given. They correspond to the two Raman shifted signal at 408 nm (H_2O) and 387 nm (N_2) and the two elastic signals at 355 nm and 1064 nm. A water vapor profile obtained by meteorological balloon launched simultaneously from Payerne by ISM is shown for comparison in Fig. 4a. The reference value of the water vapor mixing ratio of the Raman lidar profile is directly measured at Jungfraujoch and considered as constant for the lidar calibration at the effective altitude of 3'700 m ASL.

With the use of the 20 cm Newtonian telescope the measurement range is limited to typically 8 km ASL due to insufficient signal to noise ratio at higher altitude. The statistical error bars have been calculated according to Equ. (5). The H_2O Raman lidar profile are in good agreement with the balloon profile up to about 5 km ASL. This comparison must nevertheless be considered with care since the balloon was launched from Payerne, which is located approximately 80 km west from Jungfraujoch at 490 m ASL. The rather large discrepancy between lidar and balloon data observed in the altitude range from 5-6 km is probably linked to the presence of cloud layers over Payerne and not at Jungfraujoch. Above 7 km both techniques indicates a water vapor content of less than 0.5 g/kg and the error on the lidar estimation increases due to the very low signal to noise ratio. If we integrate the water vapor column density estimated from the lidar data, a value of 4 mm of precipitable water vapor above Jungfraujoch is obtained, or about 40% of the water vapor column density measured from Payerne.

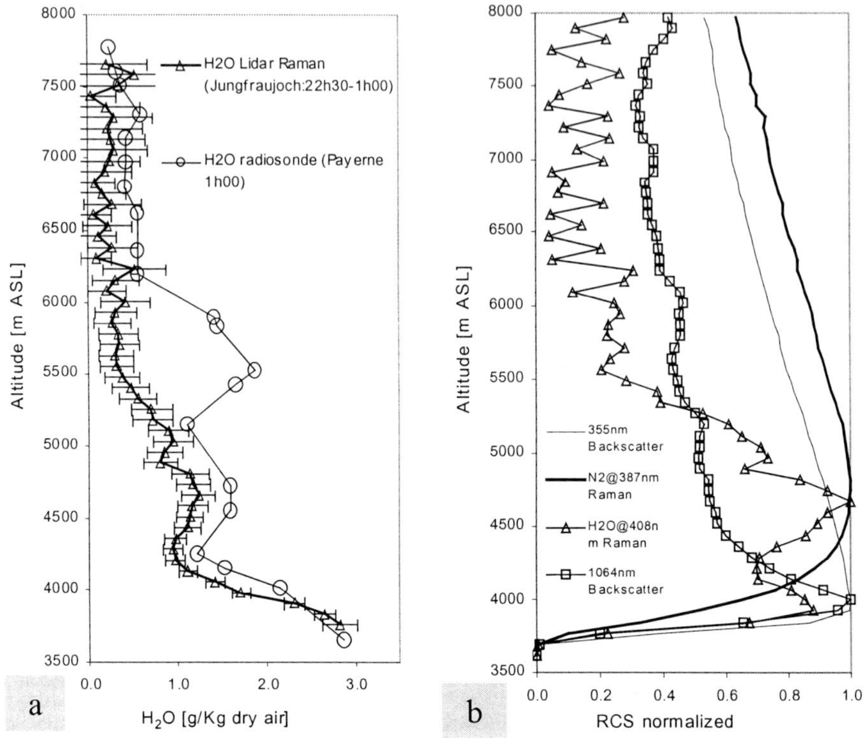

Figure 4. Vertical profiles comparison between water vapor mixing ratio measured by Raman lidar and by radiosonde (4a) in clear sky condition (case A). Range corrected signals (4b) for the two Raman wavelengths and the elastic 355 and 1064 nm.

The case (B) with a cloud layer uniformly developed over Switzerland is given in Figure 5. Fig. 5a shows a similar comparison between balloon and Raman lidar measurements as in Fig. 4a for a lidar integration time of 2 h 28 min and 75 m vertical resolution in the presence of a cloud layer around 5.5 km ASL. Here the profiles are shown only up to 6 km ASL due a strong attenuation of the laser light above the cloud. In this figure the molecular and the aerosol correction factors that are considered in $\Gamma(z,\lambda_{H2O},\lambda_{N2})$ are also shown. Up to 5 km ASL the Raman lidar water vapor profile is in reasonable agreement with the balloon measurement. The lidar profile shows an increase of the water content in the cloud despite of its low relative optical thickness. The aerosols correction which is negligible below the cloud (no aerosols) reaches high values (20 %) in the cloud. For comparison the Rayleigh molecular correction has a contribution of less than 4 % at 6 km ASL.

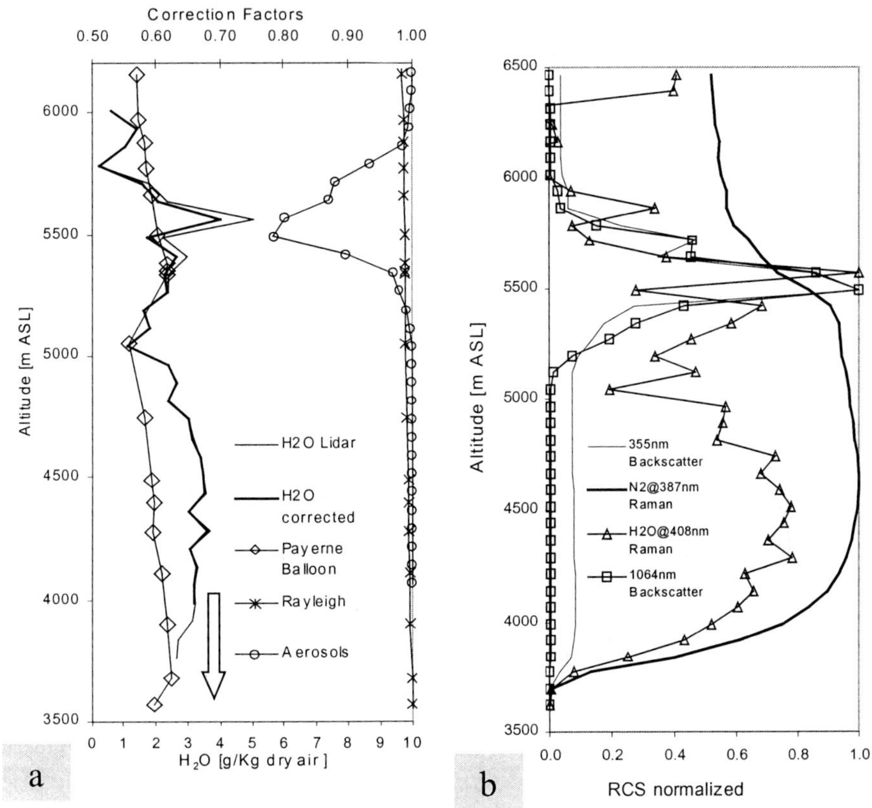

Figure 5. Vertical profiles comparison between water vapor mixing ratio (lower scale) measured by Raman lidar and by radiosonde (5a) in cloudy sky condition (case B). The correction factors (upper scale) for the water vapor content by Raman lidar are indicated both for the Rayleigh correction (less than 5%) and for the Mie scattering (up to 20 % in the cloud). Range corrected signals (5b) for the two Raman wavelengths and the elastic 355 and 1064 nm. The cloud layer at an altitude of 5,5 km ASL is clearly shown on the two elastic signals, while the Raman signals are essentially not perturbed. This is indicative of an excellent rejection ratio of the elastically back-scattered light in the Raman channels.

In Fig. 5b the two elastic lidar signals at 355 and 1064 nm are strongly perturbed by the presence of the cloud layer. This is not the case for the measurement of the N_2 Raman shifted wavelength. This is indicative of a receiver optic with excellent rejection of the elastic signal and essentially no interference on the Raman channel. In the case of the Raman water vapor channel, even if the signal is also strongly attenuated by the cloud layer, there is a distinct increase of water vapor content measured in the cloud layer. Note that this study is performed without considering multi-scattering effects, water droplets absorption, or Raman scattering from liquid water [6].

These first results demonstrate the sensitivity of the method for both clear sky and cloudy conditions. The relatively small altitude range of the measurement is due to the small active area of the telescope. From simple model calculation, the use of the astronomical telescope with an active area 15 times larger than the actual telescope will enhance our range of measurement of the water by Raman lidar over the entire troposphere.

4. CONCLUSION

At the Jungfraujoch Alpine Research Station in Switzerland, a new lidar system was operated since 1999 with continuous monitoring of the aerosol properties in the free troposphere since spring 2000. Water vapor measurements by Raman lidar have been demonstrated in this paper. These measurements are obtained in the free troposphere using an intense UV laser beam at 355 nm and detecting the range resolved Raman back-scattered signal by water molecules at 408 nm and nitrogen at 387 nm. With a calibration value measured directly at Jungfraujoch, the ratio of these two signals can be directly interpreted as a water vapor mixing ratio profile in the free troposphere. These measurements are obtained over a limited altitude range of typically 4-8 km ASL due to the initial use of a small size telescope (20 cm primary mirror diameter). The same method will soon be applied using the astronomical telescope (76 cm primary mirror diameter) and in this case water vapor profile up to the tropopause should be achieved, with a time resolution of less than one hour and an altitude range resolution of typically 100 meters. These predictions are done for nighttime measurements, while daytime Raman lidar observations are still limited by the availability of interference filters with higher daylight rejection ratio.

At the same research station, passive microwave detection of water vapor in the stratosphere and mesosphere is performed on a regular basis as well. The combination of water vapor retrieval by Raman lidar and passive microwave will provide a unique tool for the detection of the water vapor content throughout the earth atmosphere. These combined methods could ideally complement the regular (every six hours) balloon sounding performed by the Swiss Meteorological Institute from Payerne. They will therefore be an ideal data base for calibration of satellite observations.

Furthermore the combination of *in situ* aerosol characterization [19], solar radiation monitoring [20], trace gas column density by FTIR [21, 22], and range resolved observation of water vapor and ozone by passive microwave [23] and lidar, as well as temperature by rotational Raman lidar [24], will make this primary site of observation of the NDSC (Network for Detection of Stratospheric Changes) [25] a unique place for the observation of global climate change issues.

In the mid term development, the Jungfraujoch lidar site will be upgraded with an ozone DIAL [26] instrument that will give ideal insights into studies such as the characterization of fine laminar structures in the free troposphere that have been described with high water vapor content and low ozone concentration. [27].

5. ACKNOWLEDGEMENTS

The authors kindly thank Mr. P. Jeannet and Dr. A. Heimo from Swiss Meteorological Institute (Payerne station) for the radio sounding data and the use of meteorological data from Jungfraujoch, the Swiss National Foundation for the financial support and the Jungfraujoch Alpine Research Station for allowing us to use their research facilities.

6. REFERENCES

[1] Houghton, J.T., G.J. Jenkins, and Ephraim J.J. (Eds.), "Climate Change: The IPCC Scientific Assessment, Intergovernmental Panel on Climate Change", U.K. Meteorological Office, Bracknell, England, (1990).

[2] Chahine, M.T., "The hydrological cycle and its influence on climate", Nature, 359, 373, (1992).

[3] Starr, D.O. and Melfi, S.H. (Eds.), "The Role of Water-vapor in Climate: A Strategic Research Plan for the Proposed GEWEX" , Water-vapor Project (GVaP), NASA Conf. Publ., CP-3120, 50 pp., (1991).

[4] England, M.N., Ferrare, R.A., Melfi, S.H., Whiteman, D.N. and Clark, T.A. "Atmospheric water-vapor measurements: Comparison of microwave radiometry and lidar", J.Geophys. Res., 97, 899, (1992).

[5] Melfi, S.H., and Whitemann, D.N., " Observation of lower-atmospheric moisture structure and its evolution using a Raman lidar", Bulletin of the American Meteorological Society, 66(10), pp.1288-1292, (1985).

[6] Ansmann, A., Riebessel, M., Wandinger,U., et al., "Combined Raman elastic-backscatter LIDAR for vertical profiling of moisture, aerosols extinction, backscatter and lidar ratio", Applied Physics B:Photonics and Laser chemistry, vol. 55, pp. 18-28 (1992).

[7] Lazzarotto, B., Frioud, M., Larchevêque, G., Mitev, V., Quaglia, P., Simeonov, V., Thompson, A., van den Bergh, H., and Calpini B. "Ozone and water vapor measurements by Raman lidar in the planetary boundary layer: error sources and field measurements" in press, Applied Optics, (2001).

[8] Melfi, S.H., Lawrence, J.D., and McCormick, "Observation of Raman Scattering by water vapor in the atmosphere", Appl. Phys. Letters, 15(9), pp. 295-297 (1969).

[9] Whitemann, D.N., Melfi S.H., and Ferrare R. A., " Raman lidar system for the measurement of the water vapor content and aerosols in the Earth's atmosphere", Appl.Opt. 31, 3068-3082 (1992).

[10] Keckhut, P., Hauchecorne, A., and Chanin, M. L. "A critical review of the database acquired for the long-term surveillance of the middle atmosphere of the French Rayleigh lidars," J. Atm. Ocean. Techn., 10, 850--867, (1993).

[11] Platt, C. M. R., Scott, J. C., and Dilley, A. C. "Remote sensing of high clouds. Part VI: Optical properties of midlatitude and tropical cirrus," J. Atm. Sci. 44, 729--747, (1987).

[12] Sherlock, V., et al. "Implementation and validation of a Raman lidar measurement of middle and upper tropospheric water-vapor", Applied Optics, vol.38, No 27, (1999).

[13] Kwon, S.A. et al., "Vertical distribution of atmospheric particles and water-vapor densities in the free troposphere", Atmospheric Environment, vol.31, No. 10, (1997).

[14] Sherlock, V., et al. "Methodology for the independent calibration of Raman backscatter water-vapor lidar system", Applied Optics, vol.38, No 27, (1999).

[15] Hair, J.W. "A high spectral resolution Lidar at 532nm for simultaneous measurement of atmospheric state and aerosols profiles using iodine vapor filters", thesis, (1999).

[16] Lazzarotto B. , Simeonov V. , Quaglia P. , Larchevêque G. , van den Bergh H. , and Calpini B. "A Raman Differential Absorption Lidar for Ozone and Water-vapor Measurement in the Lower Troposphere"; Intern. J. of Env. An. Chem. , 74 (1-4), pp. 255-261, (1999).

[17] Quaglia, P., Balin, I., Larchevêque, G., Simeonov, V., van den Bergh, H. and. Calpini B. "A new LIDAR station at the Jungfraujoch alpine station for long term monitoring of aerosols, temperature and water vapor", Proclim - 1st Global Swiss Change Day, Berne, (April, 2000).

[18] Collis, R.T.H., and Russell, P.B., "Lidar measurement of particles and gases by elastic backscattering and differential absorption", Laser monitoring of the atmosphere, E.D. Hinkley ed., Spinger-Verlag, 71-151, (1976).

[19] Baltensperger, U., Gäggeler, H.W., Jost, D.T., Lugauer, M., Schwikowski, M.., Weingartner, E., and Seibert, P. :"Aerosol climatology at the high-alpine site Jungfraujoch, Switzerland", J. Geo. Research, D102, pp. 19707-19715, (1997).

[20] Heimo et al. : "The Swiss atmospheric radiation monitoring network CHARM", proc. WMO Technical Conference on Meteorological and Environmental Instruments and Methods of Observation (TECO-98), Casablanca, Morocco, (1998).

[21] Delbouille L., Roland, G.:" High resolution solar and atmospheric spectroscopy from the Jungfraujoch high altitude station", Opt. Eng. , 34, pp. 2736-2739, (1995).

[22] De Mazière, M., van Roozendael, M., Hermans, C., Simon, P.C., Demoulin, P., Roland, G. :"Quantitative evaluation of the post-Pinatubo NO2 reduction and recovery, based on 10 years of FTIR and UV-visible spectroscopic measurements at the Jungfraujoch", Journal of Geophysical. Research, Vol. 103, pp. 10.849-10.858, (1998).

[23] Siegenthaler A., Lezeaux O., Feist D.G., and Kämpfer N. : "First water vapor measurements at 183 GHz from the high alpine station Jungfraujoch", in press, IEEE Transactions on Geoscience and Remote sensing.(2001).

[24] Arshinov Y., Bobrovnikov, S.M.., Zuev, V. E., and Mitev, V. M. : "Atmospheric temperature measurements using a pure rotational raman lidar", Applied Optics, 22, pp. 2984-2990, (1983).

[25] http://www.ndsc.ncep.noaa.gov/

[26] Schoulepnikoff, L., van den Bergh, H., Mitev, V., and Calpini, B. : "Tropospheric air pollution monitoring: Lidar", Encyclopedia of Environmental Analysis and Remediation, John Wiley & Sons, Ed. Robert A. Meyers, Vol. 8, pp 4873-4909, (1998).

[27] Newell, R.E., Thouret, V., Cho, J., Stoller, P., Marenco, A., and Smit, H.G. : "Ubiquity of quasi-horizonthal layers in the troposphere", Nature, Vol 398, pp. 316-319, (1999).

A qualitative assessment of the sources of uncertainty in climate change impacts assessment studies

DAVID VINER
Climatic Research Unit, University of East Anglia, Norwich, UK

Abstract: This paper, to be used as a guide by impacts assessors, identifies and then presents a qualitative assessment of the sources of uncertainty that are encapsulated in any climate change impacts assessment. The initial source, that of emissions scenarios derived from socio-economic projections is described and from here the cascade of the probable ranges of uncertainty associated with each step in an impacts assessment study is discussed. This assessment of the sources of uncertainty is coupled to the evolution of the science during the course of the Intergovernmental Panel on Climate Change (IPCC) process. This is used to explain (in specific cases) why certain *de facto* standards have been adopted in climate change research.

1. INTRODUCTION

For the basis of this assessment, it is proposed that there are two primary reasons why climate change impacts assessments are performed. First: science driven impacts assessments designed to study the sensitivity of a given system to changes in climate in order to understand linkages that may occur within a sector and or between sectors. Science driven studies have, to date, made up the bulk of reported impacts assessments. They have, however, often overlooked the range of uncertainties that exist at a particular step within the process and the end results are not able to be quantitatively incorporated as feedbacks. Second; policy driven impacts assessments have been performed to enable decisions to be made about what adaptive and /or

M. Beniston (ed.),
Climatic Change: Implications for the Hydrological Cycle and for Water Management, 139–149.

mitigative options areavailable for a particular stakeholder. Communicating the sources of uncertainty will enable the end-user/stakeholder properly utilise the results of any impacts assessment and treat these as scenarios rather than predictions or forecasts.

Science driven studies have not needed to incorporate uncertainty derived from the basic components of the climate to impacts to policy system. They were more concerned with the understanding of how a system functioned. Many of these impacts assessments took results from one single climate change integration and used these as the next step on from applying arbitrary changes in climate parameters, i.e., the estimates of change from a single grid box from one or more GCM integrations were used instad of arbritrary values. These climate change scenarios were then input into one impacts model and the results analysed. Often as there was little else to examine these results were used for informing stakeholders about future potential changes. There was as a result, therefore, no need or attempt to examine the actual sources and ranges of uncertainty that are present.

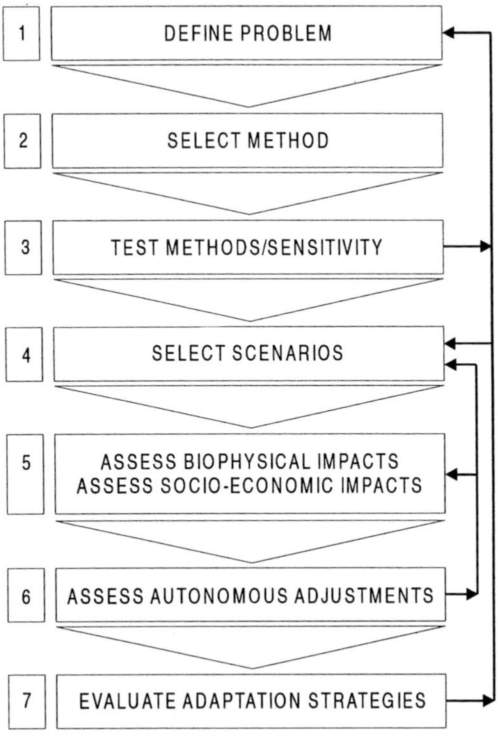

Figure 1: The seven stages of impacts assessment

The main climate change input used in impacts assessments are the results from climate change experiments and accompanying observed climate data. These have been viewed by the impacts community as being black box in nature. It was seen by impacts assessors that there was no need to understand how Global Climate Models (GCMs) operate, the range of forcing scenarios available and the sources of uncertainty associated with them. The process of climate change scenario data provision from the GCM community to the impacts community was often (especially during the First IPCC Assessment) seen as one way. The advent of interface projects, such as the Climate Impacts LINK Project (Viner and Hulme, 1998) which disseminates the results from the Hadley Centre have helped to alleviate this perception and encouraged an understanding of the characteristics of data from GCM experiment.

2. THE SOURCES OF UNCERTAINTY

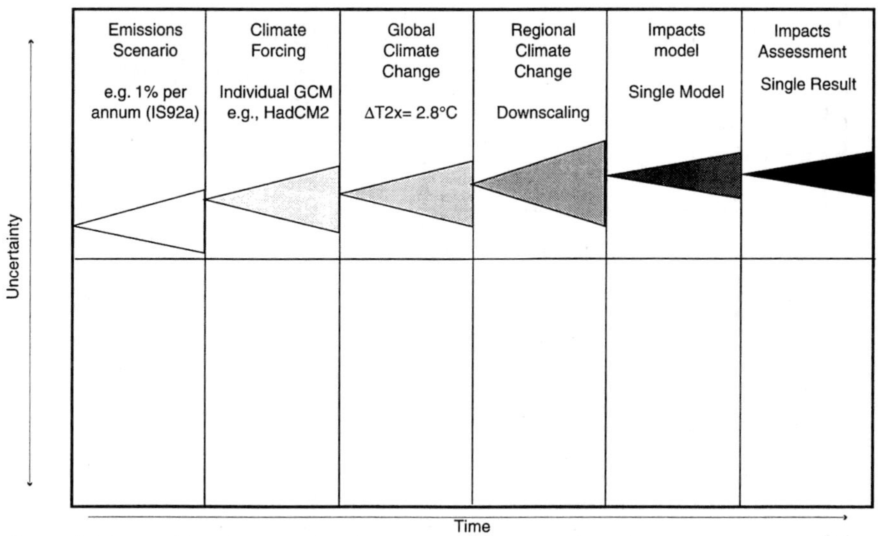

Figure 2 :The traditional approach to impacts assessments, where the uncertainties associated at each stage are often overlooked.

For the purposes of this qualitative assessment, the sources of uncertainty will be viewed from the traditional science driven approach to impacts assessment and then in more detail through the cascade of uncertainty that exits from emissions scenario through to the application of the results from a policy driven impacts assessment.

Figure 2 shows the process involved in a traditional science based impacts assessment. This process shows how to-date many impacts assessments have overlooked the uncertainties associated with the steps within a climate change impacts assessment. This is not problematic for the science, however, the results of such assessments can be mis-leading if used for policy decisions as they present more of a certain picture. In reality, however, we have a *cascade of uncertainty* that exists in each stage of the impacts assessment (Figure 3).

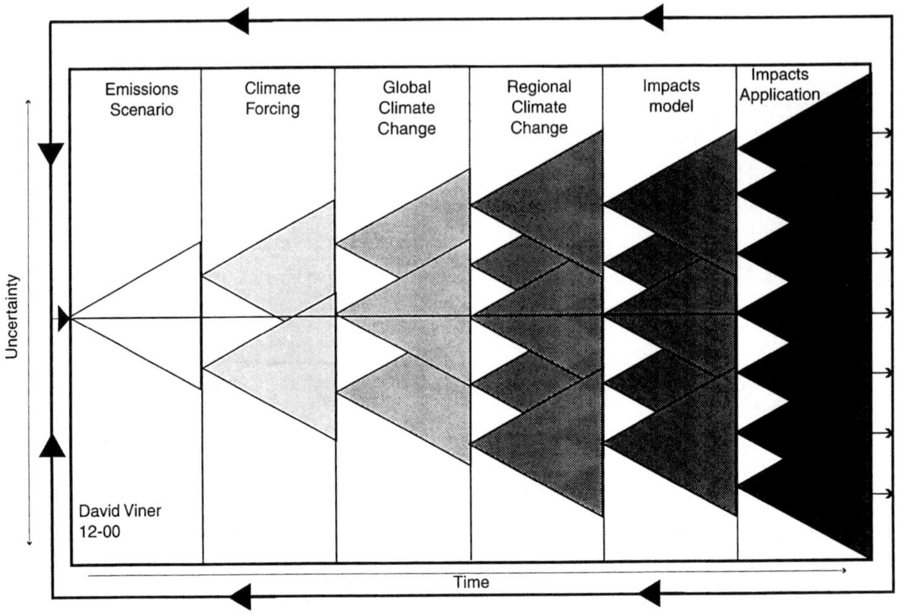

Figure 3 : A conceptual representation of the cascade of uncertainty in climate change impacts assessment

It is the purpose of this paper to present a qualatative description of the sources of uncertainty that exist in impacts assessments. The initial stage is the definition of an emissions scenario. For the First Assessment exercise of the IPCC this was a simple 2xCO2 integration. In 1992 the first consistent and plausible emissions scenarios were formulated (IPCC, 1992). These are widely known as the IS92 scenarios, in all six scenarios (IS92a-f) were described. At the same time the GCM community had developed the first coupled ocean-atmosphere GCMs. These could be used to examine the transient response of the climate system to a given forcing scenario. The initial transient experiments reported in Greco *et al*, 1995 were used as the first attempt to provide consistent scenarios to the "IPCC Community": UKTR (Murphy, 1995); GFDL89 (Manabe *et al.*, 1994), ECHAM1-A; (Cubasch *et al.*, 1992). These experiments were forced with a 1% per year compound increase in forcing. Conveniently this roughly equated to the IS92a scenario. The current literature often refers to the 1% per year forcing and IS92a as being "roughly equivalent". A characteristic that is used by the

IPCC Data Distribution Centre (http://ipcc-ddc.cru.uea.ac.uk) as a criteria for the results from a specific GCM to be included in its database.

2.1 Source 1: Emissions Scenarios

Uncertainties within this stage are derived from a number of sources. To construct an emissions scenario there is a need to define the way in which the socio-economic systems will develop in the future. This is possible, however the complex nature of these systems and in their contrast to the behaviour of natural physically based systems which obey scientific laws means that it is almost impossible to predict how the global socio-economic systems will develop in the future. It is from this basis alone that we must treat the results from further "down-the-line" in a climate change impacts assessment as being scenarios rather than predictions.

Through the Intergovernmental Panel on Climate Change's (IPCC) Special Report on Emissions Scenarios (IPCC, 2000) a range of world futures have been constructed. This process defined four "storylines" that describe possible future evolutions of the global socio-economic system:

SRES A1: a future world of very rapid economic growth, low population growth and rapid introduction of new and more efficient technology. Major underlying themes are economic and cultural convergence and capacity building, with a substantial reduction in regional differences in per capita income. In this world, people pursue personal wealth rather than environmental quality.

SRES A2: a very heterogeneous world. The underlying theme is that of strengthening regional cultural identities, with an emphasis on family values and local traditions, high population growth, and less concern for rapid economic development.

SRES B1: a convergent world with rapid change in economic structures, "dematerialization" and introduction of clean technologies. The emphasis is on global solutions to environmental and social sustainability, including concerted efforts for rapid technology development, dematerialization of the economy, and improving equity.

SRES B2: a world in which the emphasis is on local solutions to economic, social, and environmental sustainability. It is again a heterogeneous world with less rapid, and more diverse technological change but a strong emphasis on community initiative and social innovation to find local, rather than global solutions.

These futures are all very different, however they must be treated as being equally plausible. Unlike with the previous IS92 scenarios where the IS92a scenario was treated as the "best guess" or "central estimate" and therefore widely associated with the single (1% per year) forcing used by

GCM modelling centres, this should not be the case with the SRES scenarios.

2.2 Source 2: Climate forcing

From the range of future estimates for emissions of carbon dioxide and the other greenhouse gases are produced. For the SRES Scenarios these range from 500ppmv to 900 ppmv for 2100. If we take into account feedbacks from the carbon cycle this range becomes 550 to 1050 ppmv by 2100. Emissions are then converted by gas-cycle models to concentrations. At this stage the use of different gas cycle models produce different end result, thus adding to the cascade of uncertainty.

2.3 Source 3: Global Climate Change

The scenarios of future concentrations of greenhouse gases are then used to derive how the climate system will respond. At a global scale two types of climate model are used. Simple one-dimensional models (e.g. MAGICC, Wigley and Raper, 1992) and complex coupled ocean-atmosphere GCMs (e.g. HadCM3). The former due to their parsimonious computing requirements can be used to assess the full range of responses in the climate system as a result of different emissions scenarios. GCMs, on the other hand, require extensive computing requirements and as such only a limited range of forcing scenarios can be used in climate change experiments.

GCMs are the most powerful tool that we have available for assessing future global and regional climate change, they are, however, subject to uncertainties that need to be understood in order that the results of climate change experiments performed using them can be sensibly and appropriately applied. These errors are:

a) errors in the initial state of the GCM. The initial boundary conditions that are used at the start of a GCM experiment are partly responsible for how a climate change integration evolves through time and to what its final outcome will be. The method used to overcome this uncertainty is to perform an ensemble of integrations that have the same forcing but different initial conditions. Figure 4 shows the annual global precipitation rate for two ensembles of four different integrations forced with a 1% per year compound increase in greenhouse gas only and combined greenhouse gas and sulphate forcings performed with HadCM2. These two ensembles of four integrations produce a range of values at the global scale, this is greatly enhanced at the regional / grid box scale (see section 2.4). In order, however to capture a more complete picture of uncertainty there is a need to perform a larger number of integrations. This is, however is beyond the computing resources for any one given climate modelling centre.

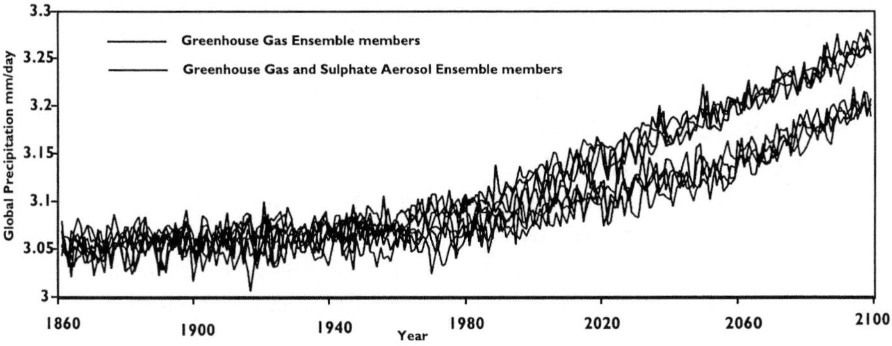

Figure 4 : Precipitation changes with and without sulphate forcings

b) errors related to poor representation of unresolved scales. Many of the physical processes of the climate system operate at scales below the resolution (e.g., 2°x3°) of GCMs. There is therefore, a need to parameterise many of these processes.

c) errors related to instability of the climate system to small perturbations. The climate system at all scales is chaotic and is subject to unforeseen states caused that can be as a result of small internal changes. For example,

d) Constraints in the design of climate change experiments do not take into to account volcanic effects, which on a moderate scale (e.g. Pinatubo, 1992) can have a sub-decadal impact upon the climate system and on a large scale can have profound multi century scale impacts upon socio-economic and natural systems at a continental scale.

e) There are a range of different GCMs that represent the state-of-the-art in climate modelling and the majority have been used to perform climate change experiments that use the adopted *de facto*standard of the 1% per annum increase in emissions. Most often within impacts assessments the results from one GCM integration that uses this forcing scenario are used. Each individual GCM, however, has its own climate sensitivity (the equilibrium temperature change associated with a doubling of CO2). The use of results from a single GCM will not capture this range.

2.4 Source 4: Regional Climate Change

From the global scale most impacts assessments require regional or local scale climate information. As we "zoom-in" on a region we start to see increasing differences in the changes described by one GCM integration compared to another. For example in the US National Assessmen. The results from two climate change integrations where used as the basis for wide ranging regional impacts assessments. These integrations, HadCM2GSa1 and CCCMa1GS, used the combined forcing of greenhouse gas and sulphate aerosols. As a result of the lower signal to noise ratio changes in these experiments there were contradictions in the climate change scenarios used. For example, over the central USA HadCM2GSa1 showed a wetting/drying and CCCMaGS showed the opposite. Ignoring the added uncertainties associated with incorporation of sulphate emissions scenarios, studies that use the results of just two GCM integrations have added uncertainty especially when there is a contradiction in the change fields , this is clearly shown within the US National Assessment.

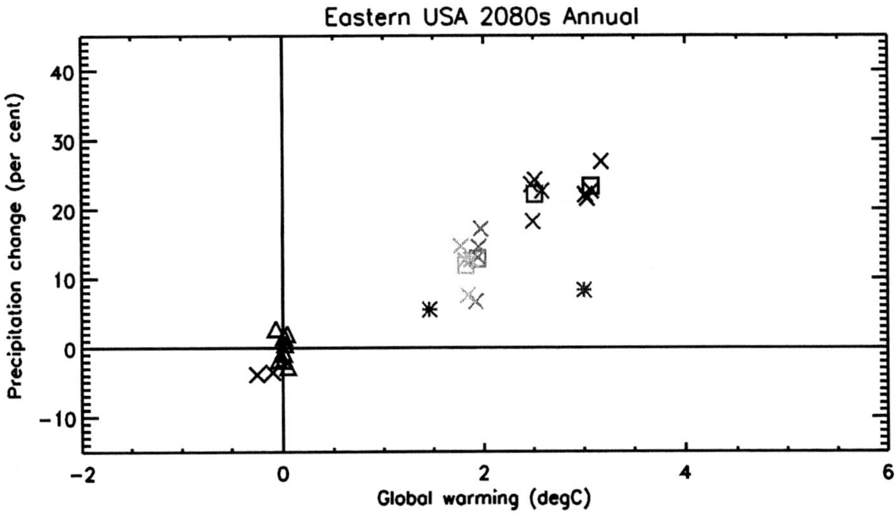

Figure 5: Changes in annual tempertaure and precipitation over the Eastern USA for the 2080s with respect to1961-90 for a range of climate change integrations perfomed with HadCM2.

Figure 5 shows a plot of annual-mean temperature changes against annual-mean precipitation changes for a range of climate change integrations performed with HadCM2 and HadCM3, the control integration of HadCM2 and two observational datasets for the Eastern USA. This figure shows the potential changes in temperature ranging from 1.5° to 3.2°C and precipitation changes ranging from 5 - 35%. When displayed in such a manner we are able to identify the possible range of future changes for a given region. This shows the advantage of using the results of a number of GCM integrations rather than for just one or two.

In order to obtain data at a higher spatial and or temporal resolution than which can be obtained directly from the results of a given GCM experiment there is a need to downscale. In order to employ a given downscaling technique there is often the requirement to have a high quality observational dataset. Observational datasets themselves have incorporated into them uncertainties due to data errors and imhogeneties of measurement within a given region. The higher the spatial resolution, the greater the uncertainty (New, 1999). There has, however, been no attempt to date to fully assess the performance, reliability and subsequent uncertainties associated with the various downscaling methods. There are three broad methods that are used within the impacts community:

a) Interpolation, applying an interpolation routine to the raw GCM data to produce a higher resolution version, or by applying the GCM resolution change fields to a high resolution baseline climatology.

b) Empirical, normally referred to as weather generators. There are two broad groups. First, circulation / weather typing, these use relationships between surface climate and synoptic circulation and then apply these to future climates derived from GCM experiments. Second, stochastic weather generators that are used to perturb time series of observed data with

c) Dynamical methods, mainly using Regional Climate Models (RCMs) driven by boundary conditions, these are used to produce higher spatial resolution data over a given region. The use of RCMs to date, has meant that one GCM has been used to drive one RCM as they are expensive on computing resources, this has introduced a new level of uncertainty, as differing RCMs forced by the same GCM will produce differing results and other uncertainties such as domain size create uncertainty.

2.5 Source 5: Impacts Model

Whilst not specifically covered by this paper, the use of a single impacts model (e.g. a hydrological model) for a given exposure sector does not cover the entire range of uncertainty. It is advised, therefore, where possible to force a number of comparable impacts models to investigate the range of sensitivity each has to climate data.

2.6 Source 6: Communication and Interpretation of impacts assessments into policy decisions

The majority of climate change impacts assessments that have been performed to date have followed the science driven model and many of the integrated assessments of climate change have undertaken bench studies that have drawn upon these results. Translation and application of the results, therefore, from impacts assessments through to a policy decision regarding a given adaptive of mitigative strategy will produce different interpretations. It is important that the results and associated uncertainties are communicated directly by the scientists to the policy makers, there is a need, therefore to:

a) Fully inform stakeholders about the uncertainties embodied in assessments;

b) Stress the robust aspect of the assessments;

c) Try to encompass probability.

3. DISCUSSION AND CONCLUSIONS

This paper attempts to describe in a qualitative manner the sources of uncertainty that exist in climate change impacts assessments. The results from any given climate change assessment are only as good as the weakest-link in the cascade of uncertainty. There is, therefore, a requirement and a need to identify and encapsulate the range of uncertainties that exist within the results of an impacts assessment.

It is recommended that a range of emissions scenarios should be used to force a suite of GCMs in order to take into account the range of possible socio-economic futures. To incorporate the range of potential changes that may occur for a given region it is advised that the results from a number of GCM experiments are used. The use of just two integrations within the US National Assessment has shown how contradicting changes for a given region enhance the range of uncertainty and thus dilute any impacts results. To date there has not yet been a quantitative assessment about the differing merits of adopting a certain climate change scenario construction / downscaling methodology. Such an assessment would be a valuable tool in guiding impacts assessors.

Communication of the results from impacts assessments to the policy community is difficult and as yet there appears no suitable framework available to incorporate the sources of uncertainty unless the above suggestions are taken into account when designing an impacts assessment.

4. REFERENCES

Carter T., Hulme M. and Viner D. 1999 *Representing Uncertainties in Climate Scenarios and Impacts Studies.* Proceedings of the EU Concerted Action Initiative ECLAT-2 Workshop 1, Helsinki, Finland April 16th - 18th, 1999, Climatic Research Unit, Norwich, UK. 128pp.

Cubasch U., Hasselmann K., Hock H., Maier Reimer E., Mikolajewicz U., Santer B.D. and Sausen R. 1992 Time-dependent greenhouse warming with a coupled ocean-atmosphere model. Climate Dynamics 8 55-69

Greco S., Viner D. Moss R. and Jenne R. 1994 *Climatic Scenarios and Socio-economic Projections for the IPCCÊWGIIÊAssessment.* Report Prepared for theÊIPCCÊWGII, Washington D.C., 12pp plus appendices.

IPCC 1992 Climate Change 1992: The Supplementary Report to the IPCC. (eds Houghton J.T., Callander B.A. and Varney S.K.) Cambridge University Press, 200pp.

IPCC, 1994: IPCC Technical Guidelines for Assessing Climate Change Impacts and Adaptations. Prepared by Working Group II [Carter, T.R., M.L. Parry, H. Harasawa, and S. Nishioka (eds.)] and WMO/UNEP. CGER-IO15-'94. University College -London, UK and Center for Global Environmental WMO/UNEP. CGER-IO15-'94. University College - London, UK and Centre for Global Environmental Protection

IPCC, 2000 Special Report of the Intergovernmental Panel on Climate Change Eds. Nakicenovic N. and Swart R. Cambridge University Press, UK. pp 570

Johns, T.C., R.E. Carnell, J.F. Crossley, J.M. Gregory, J.F.B. Mitchell, C.A. Senior, S.F.B. Tett, and R.A. Wood, 1997: The Second Hadley Centre coupled ocean-atmosphere GCM: Model description, spinup and validation. Climate Dynamics, 13, 103-134.

Jones, R.N., 2000a: Managing uncertainty in climate change projections - issues for impact assessment. Climatic Change Vol. 45 p.403-419.

Manabe S, Spelman M.J. and Stouffer R.J. Spelman M.J. and Bryan K. 1991 Transient responses of a coupled ocean-atmosphere model to gradual changes of atmospheric CO2: Part 1 Annual-mean response J. of Climate 5 105-126

Murphy J.M. 1995 Transient response of the Hadley Centre coupled ocean-atmosphere model to increasing carbon dioxide. Part I: Control climate and flux adjustment. J. of Clim. Vol. 8. 36-56

New M. 1999 Uncertainty in representing observed climate In *Representing Uncertainties in Climate Scenarios and Impacts Studies.* Eds. Carter T., Hulme M. and Viner D. Proceedings of the EU Concerted Action Initiative ECLAT-2 Workshop 1, Helsinki, Finland April 16th - 18th, 1999, Climatic Research Unit, Norwich, UK. 128pp.

Viner D. and Hulme M. The Climate Impacts LINK Project: Applying Results of the Hadley CentreÕs Climate Change Experiments in Climate Change Impacts Assessments. Report Prepared for the UK DETR, Climatic Research Unit, Norwich 16pp

Wigley T.M.L. and Raper S.C.B. 1992 The implications for climate and sea-level of revised IPCC emissions scenarios. Nature 357 293-300.

The GLOWA Volta project: Integrated assessment of feedback mechanisms between climate, landuse, and hydrology

NICK VAN DE GIESEN[1], HARALD KUNSTMANN[2], GERLINDE JUNG[2], JENS LIEBE[1], MARC ANDREINI[1], AND PAUL L.G. VLEK[1]

[1]*Center for Development Research (ZEF), Bonn University, Bonn, Germany*

[2]*Fraunhofer Institute for Atmospheric Environmental Research (IFU), Garmisch-Partenkirchen, Germany*

Abstract: Water management issues critical to the Volta Basin are discussed in this paper. An introduction to the Volta Basin and its hydrological characteristics provides the necessary background. Hydropower production for urban areas and irrigation development in rural areas compete for water resources. The paper introduces the GLOWA Volta project and presents details highlighting the project's integrative approach. In order to address water-policy issues, the GLOWA Volta Project seeks to develop a scientifically sound Decision Support System. The main premise of the GLOWA Volta investigation is that changes in landuse may have important, yet unknown, effects on the hydrological cycle of the basin. To predict future water availability, landuse change, driven by physical and social factors, needs to be predicted as well. In order to illustrate the different ways in which landuse affects the movement of water over the surface and through the atmosphere, three examples are given. The first example treats the indirect and direct effects of changes in soil moisture on the atmospheric circulation. In the second example, the high sensitivity of surface water availability with respect to rainfall is determined. The final example shows the increase in dam building activity in the basin which has a direct but rather limited impact on availability of water resources downstream.

M. Beniston (ed.),
Climatic Change: Implications for the Hydrological Cycle and for Water Management, 151–170.

1. INTRODUCTION: THE VOLTA BASIN

With a main channel of 1400 km, the Volta River drains 400,000 km² of the semi-arid and sub-humid savanna of West Africa (Figure 1). The Volta Basin lies mainly in Ghana (42%) and Burkina Faso (43%), with minor parts in Togo, Côte d'Ivoire, Mali, and Benin. The region is poor, also by African measures, with a per capita income around US$ 1100 per year. Population growth is almost 3% per year, implying a doubling every 25 years. The agricultural sector is by far the most important, followed at a distance by the service and mining sectors. Economic growth can only be achieved through increased agricultural productivity. Improved technologies such as high yielding varieties and phyto-chemicals are important, but their success depends on a reliable water supply. The average rainfall of 1000 mm per year would seem ample, but rainfall variability makes rainfed agriculture a risky enterprise. Integrated over one year and the whole basin, rainfall still looks rather stable. For any given point, however, the year-to-year variance is high. More important is the distribution of rain over the growing season. Especially the onset of the rainy season is very unreliable and the frequent short periods of drought (10-20 days) within the season cause significant crop damage (*Adiku et al.*, 1997). On top of this all, we find a decrease in rainfall in most of the region since the 1960's when compared to the 20th century average (*Nicholson*, 1998). Although technically and institutionally difficult, irrigation development is generally seen as an important condition for increased agricultural production.

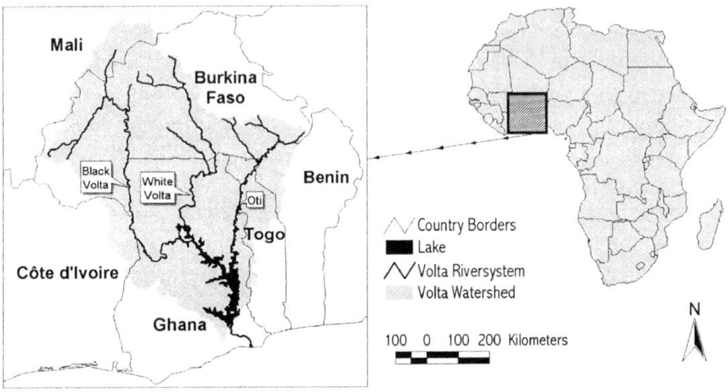

Figure 1. Overview of the Volta Basin

A dominating feature on the map of the Volta Basin is Lake Volta, which has the largest area of any man-made lake (*Lawson*, 1963). The purpose of the lake is hydropower generation at the dams at Akosombo and Kpong (total capacity 1060 MW). Between the lake and the ocean lies only 100 km of river, which limits the possibilities of re-use of hydropower water in downstream irrigation projects. The relatively cheap electric power produced by the dams is an important force behind the development of the urban-industrial centers in Southern Ghana, outside the basin. Drought induced power shortages, such as occurred in 1984 and 1998 disrupt urban life. At the macro-level, a water management issue emerges. Development of water resources for agriculture is needed to develop the poorer rural societies of Northern Ghana and Burkina Faso. Such development will have a yet unknown downstream effect on the availability of water for hydropower. In turn, reduced hydropower generation threatens industrial development in the South. Policies that discourage irrigation development to save water for hydropower, make rural life less attractive and urban life more attractive, thereby inducing migration to the cities. Upstream, Burkina Faso is more interested in irrigation development since they do not profit from hydropower generation and have no alternatives but agriculture. Ghana, however, appears to prefer electricity, as suggested by the on-going preparation for the construction of a third large dam in Bui gorge.

To address water policy issues as sketched above, the GLOWA Volta project seeks to develop a Decision Support System (DSS) for the sustainable development of water resources in the region. The project is part of the larger long-term GLOWA program, funded by the German Federal Ministry for Education and Research (BMBF). The GLOWA program examines the impact of global change on the hydrological cycle. The approach employs an integrated analysis of four larger watersheds in Europe and Africa. The GLOWA Volta research focuses on the modeling of the major processes that affect water supply and demand. Its strength lies in the integrated nature of the model, in which social and environmental processes are directly coupled. The multi-disciplinary research team includes meteorologists, hydrologists, agronomists, geographers, economists and legal scholars. The model will drive the DSS that, through policy gaming, will provide decision makers with quantified information on the impact of different water management strategies.

The objectives of this paper are twofold. First, we want to introduce the GLOWA Volta project and especially its conceptual framework. The conceptual framework will be presented as a set of variables and interactions to be researched in fourteen sub-projects. The second objective is to provide the reader with an impression of the intertwined nature of the different factors at play, especially the effects which landuse and land cover change may have on different parts of the hydrological cycle. Through three examples the sensitivity of water resources to human induced changes will be demonstrated. The examples are chosen to fit the theme of the BAHC workshop on "*Vulnerability of Water Resources to Changing Environmental Conditions*" which was part of the Wengen2000 conference.

2. FEEDBACK NETWORK AND MODELING FRAMEWORK

To understand supply and demand of water resources in the Volta Basin, one needs to address physical as well as social questions. The main scientific challenge of the GLOWA Volta project is to comprehensively analyze all relevant processes and their mutual interactions. The idea is not to model, for instance, water demand under different population and migration scenarios, but rather to couple population to water dynamics into one integrated model. It is not difficult to show that everything has an effect on everything and vice versa. What is difficult is to try to select a limited number of variables and processes that span the main dimensions of water resources management. Such a selection is necessary to maintain model sensitivity and, simply, to keep our data collection practically feasible. As a first step, project scientists identified eleven key-variables that need to be modeled over space and time: a) Precipitation, b) Evapotranspiration, c) Agricultural production, d) Landuse, e) Income, f) Population, g) Riverflow, h) Wateruse, i) Hydropower, *j) Health, k) Technological development, l) Institutional development*. Variables (a) through (i) will be treated as endogenous variables to be solved for simultaneously so that all feedback loops among these variables are accounted for. The variable *Health* will be calculated as an output variable because the feedback mechanisms are felt to be too complex to model meaningfully. *Technological* and *Institutional development* are treated as input variables as these are the variables over which to optimize for given objective functions within the DSS.

Predictions concerning the key-variables are based on models developed by different disciplines. The GLOWA Volta project is organized in fourteen sub-projects, grouped in three interdisciplinary clusters *Atmosphere*, *Landuse*, and *Water Use*. Each sub-project is associated with one Ph.D. project. Figure 2 gives a graphical overview and shows the exchange of information between the sub-projects. From top to bottom, the manageability of the processes increases together with the directness of the impact of human actions on the hydrological cycle. Without going into every detail of the diagram, we would like to point out three structural features which reflect the important operative principles of the project.

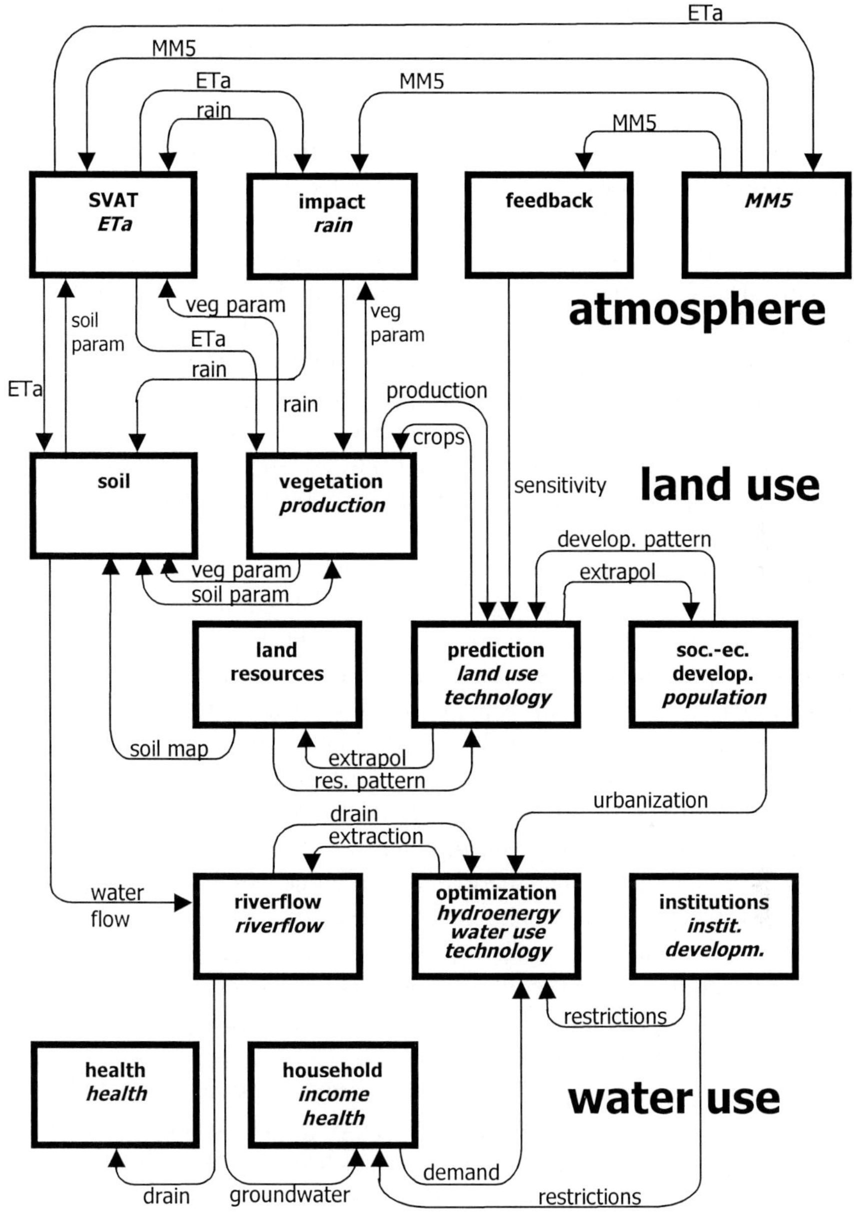

Figure 2. The three clusters and fourteen sub-projects of the GLOWA Volta project. The central theme is given in each box, together with the produced key-variables (in italics) where applicable. The labeled arrows denote the main information streams.

The first structural feature is the dense net of connecting arrows in the upper-left corner that models the land surface/atmosphere interface. We postulate that there is much, often non-linear, feedback played out over this interface. In addition, we encounter important scale problems here, where hydrological processes with characteristic lengths of 1 m to 10 km meet meteorological processes with characteristic lengths of 1 km to 1000 km (*Chehbouni et al.*, 1999). In order not to compound the difficulties associated with interdisciplinary information exchange, one scale of communications has been defined for the project. This scale was chosen to be optimal as a 9 km x 9 km grid. Given our CPU resources available at IFU, this is the finest resolution at which we can run the atmospheric model for regional climate simulations. Each discipline has to overcome its own scale problems through aggregation or disaggregation and deliver results to the other disciplines at this grid size. At this grid-size, social structures (villages and landuse patterns) as well as landscape elements (1st order watersheds) start to repeat.

A second noteworthy feature is the relatively central place of the *water optimization* sub-project. An optimal water distribution under simulated physical and given institutional constraints will be the main policy advice output of the project. The optimization project's central place also follows from the idea that integration over disciplines can more readily be realized between "kindred" scientists (such as hydrologists and meteorologists) and becomes more problematic between less related ones (say meteorologists and anthropologists). We see that in general, only scientists with similar backgrounds have to exchange information. For a truly integrated analysis, however, it is inevitable that there is at least one point where social and physical information functionally meet; somewhat analogue to the fixed-point theorem. In our case, we have two such points (*water optimization* and *landuse prediction*) each covered by scientists used to processing data from both domains: a resources economist for *water optimization* and a geographer for *landuse prediction*.

Finally, also the *landuse prediction* sub-project assumes a central place where information from physical as well as social science is used. At present, the main driving force behind most changes in the Volta Basin is population growth. Although it is recognized that population growth in itself is also an outcome of other processes, it is a convenient starting point for the feedback analysis. In general, reduced growth rates follow economic development and women's education and emancipation. This dependency implies that changes in population growth rate will not have much impact on the population within the next generation. The projected increase in rural population will probably result in both migration towards the cities and a more intense use of the land for agriculture. Since 80% of the population is active within agriculture, we see landuse change as the most important Global Change phenomenon within the Volta Basin. In the more populous areas (Upper East of Ghana, Central Burkina Faso), most land is already in continuous use and landuse will shift to more intense forms of agriculture. The project pays much attention to modeling landuse change. The project will build upon recent advances achieved within the IGBP/IHDP Land Use and Land Cover Change (LUCC) network, in which GLOWA Volta participates, to make spatially explicit predictions about the likelihood of transitions in landuse (*Lambin & Mertens*, 1997).

Within the Volta Basin, landuse and landuse change are a function of the availability of water resources, be it as rainfall or as irrigation water. In its turn, landuse change will have an effect on water resources. In the next section, we will present examples illustrating the three most important ways in which landuse change affects the movement of water through the watershed.

3. LANDUSE CHANGE AND THE HYDROLOGICAL CYCLE

As the project is still in the early stages of data gathering and analysis, it is not yet possible to give a quantified overview of all major interactions involved. Instead, we present some early results to give an idea about the different types of dynamics that may be encountered. The focus is on the effects of landuse and land cover change on the hydrological cycle describing *Atmospheric feedback*, *Rainfall/runoff relations*, and *Irrigation development*.

3.1 Atmospheric feedback

In this first example we apply an atmospheric model to assess the potential impact of changes in evapotranspiration on rainfall. The most highly developed model within the GLOWA Volta project so far is the community atmospheric model MM5 (*Grell et al.*, 1994). MM5 is a numerical mesoscale meteorological model coupled with the Oregon-State-University-Land-Surface Model (*Chen & Dudhia*, 2000). The atmospheric model serves three main purposes in the project. First, a daily hindcasting of the recent meteorological situation in the Volta Basin is performed. The hindcasting results are transferred to a central database and accessed by all project partners, especially the hydrologists, remote sensing specialists, agronomists and agricultural economists. Second, episode runs of up to 4 weeks are performed to investigate feedback mechanisms between soil moisture, landuse and precipitation. Finally, climate simulations are performed to assess the impact of global and regional change. In this final stage, global atmospheric change will enter through the lateral boundary conditions provided by the global climate model ECHAM 4. In this way, we assess the joint effect of global warming and regional landuse change on the long term water balance in the Volta Basin.

Here we present the results of an episode run in which the atmospheric water balance of the Volta Basin was simulated. MM5 was applied to three nested domains. The largest domain covers the area from 22°W to 22°E and from 14°S to 28°N with a grid size of 81x81 km^2. The embedded second domain runs from 8°W to 7°E and from 2°N to 17°N with a resolution of 27x27 km^2. The smallest domain covers the Volta Basin and immediate environment with a grid of 9x9 km^2. Each larger domain provides the lateral boundary conditions for the next smaller domain. The boundary and initial conditions for the largest domain is provided by the NCEP reanalysis data. The first domain is run with Four-Dimensional-Data-Assimilation (FDDA) allowing the incorporation of observations during the simulation period. The model has been optimally parameterized with respect to cumulus parameterization, microphysics, planetary boundary parameterization and radiation. Optimal parameterizations were obtained by comparing model outputs of episodic runs of 4 week duration in July/August 1998 with precipitation observations from the Ghanaian part of the Volta Basin and radiosonde information (wind speed, temperature and relative humidity at various pressure levels) from upper air measurements in West Africa. Meteorological data are sparse in West Africa. For example, there are daily data from roughly 10 upper air stations covering analysis domain 2 and 11 rain gauge stations within Ghana. The low density of observations implies that uncertainty with respect to the reliability and the performance of the obtained optimal parameterizations continues to exist.

We realize that regional models have certain problems such as potentially high sensitivity to initial and boundary conditions, but, as will become clear later, these risks are controlled by careful calibration and use of control and sensitivity runs. An alternative to using a regional model would be to use GCM output for the analysis but this would not have been sufficient to reproduce the observed small scale precipitation patterns that are created during the, mostly, convective precipitation events. GCM's have not been optimized to reproduce regional rainfall. Comparison of historical rainfall data with ECHAM4 data for the 1961-1990 period shows that ECHAM4 overestimates rainfall in the Volta Basin with 48%. This should come as no surprise because in the inter-tropical region almost all rain can be considered convective at the GCM gridsize. Correct precipitation predictions, therefore, depends completely on convective rain parameterization which has not been optimized for the region. Finally, the effect of land use changes in the Volta Basin on the regional water balance could not be investigated with a GCM leaving regional models as the only alternative.

To identify the feedback mechanisms between land surface (soil and vegetation) and atmosphere, an extensive numerical experiment was carried out. The methodology applied is based on (*Schär et al.*, 1999). The main research question was to what extent regional (intra-domain) evapotranspiration determines rainfall within the domains. If the amount of rainfall only depended on the amount of water vapor entering through the lateral boundaries, changes in the lower boundary conditions or the land surface would not affect regional rainfall.

We start out with the water balance for the atmosphere:

$$\Delta Q = I - O - P + ET \qquad \text{Eq. 1}$$

where ΔQ is change in storage, I inflowing and O outflowing water vapor, P precipitation, and ET evapotranspiration. We define a precipitation recycling ration β:

$$\beta = \frac{ET}{I + ET} \qquad \text{Eq. 2}$$

that denotes the fraction of moisture that originates from evapotranspiration within the analyzed domain. β is derived from splitting Eq. 1 into two fractions, one describing the water that comes from evapotranspiration within the analysis domain (Eq. 3) and one quantifying the atmospheric advection into the domain (*Schär et al.*, 1999):

$$\beta \cdot \Delta Q = -O \cdot \beta + ET - P \cdot \beta \qquad \text{Eq. 3}$$

$$(1 - \beta) \cdot \Delta Q = I - O(1 - \beta) - P(1 - \beta) \qquad \text{Eq. 4}$$

The higher β, the higher the contribution of locally evapotranspired water to precipitation in the interior of the analysis domain. Similarly, the precipitation efficiency χ:

$$\chi = \frac{P}{I + ET} \qquad \text{Eq. 5}$$

describes the fraction of water that enters the domain, either by evapotranspiration or atmospheric transport, and subsequently falls as precipitation within it.

The methodology applied assumes that the water molecules that derive from within and from outside the analysis domain are well mixed. This assumption is well justified in the vertical direction, since more than 90% of the precipitation in the analysis region is due to convective processes. In the horizontal direction, however, the assumption is weak especially when orographic contrasts or land-sea mask effects are involved. For this reason the calculated precipitation recycling rates and precipitation efficiency rates will later on be compared to results obtained from a more sophisticated approach based on evaporation tagging.

One objective of the atmospheric water balance study is to investigate the nature of the feedback between soil moisture, landuse, and precipitation. In particular the question arises whether the feedback is more of a direct or an indirect nature. In case of the soil-moisture feedback, the direct mechanism accounts for that part of the surplus precipitation that derives directly from extra evapotranspiration from the wet soils within the analysis domain. The indirect effect of soil moisture is that the state of the soil also co-determines the efficiency with which water vapor that is transported into the analysis domain precipitates (*Schär et al.*, 1999).

To examine the importance of both types of soil moisture-precipitation feedback, we simulated rainfall starting with three different initial soil moisture conditions, one control run (CTR), one with doubled (M×2) and one with half (M×0.5) the initial soil moisture of CTR. The differences in precipitation between control and adjusted runs (ΔP) are divided into three terms:

$$\Delta P = \chi(\Delta ET + \Delta I) + \Delta\chi(\Delta ET + \Delta I) + \Delta\chi(ET + I) \qquad \text{Eq. 6}$$

The first term describes the change in P due to changes in ET and I under a constant precipitation efficiency, and, thereby, the direct effect of the change in soil moisture. The third term gives the part of ΔP caused by the change in precipitation efficiency and thus tells us the indirect impact. The second-order middle term represents a mix of both effects.

The three simulations covered the four week period from 15 July through 15 August 1998. Table 1 summarizes the results for each domain.

Table 1. Values of precipitation efficiency χ and recycling ratio β for different initial moisture conditions (M×0.5, CTR, M×2), and percentages of change in precipitation ΔP explained by indirect (ind), direct (dir) and mixed (mix) mechanisms for the three domains D1, D2 and D3.

	χ			β			%ΔP explained M×0.5			%ΔP explained M×2		
	M×0.5	CTR	M×2	M×0.5	CTR	M×2	ind	dir	mix	ind	dir	mix
D1	0.31	0.34	0.37	0.37	0.41	0.44	59.7	44.4	-4.1	57.5	39.5	3.0
D2	0.194	0.190	0.191	0.108	0.126	0.132	35.5	66.0	-1.5	112.1	-12.2	0.1
D3	0.086	0.097	0.0914	0.057	0.068	0.07	30.0	70.0	0	60.0	40.0	0

Both β and χ are scale dependent, at global level they should be equal to 1. For all domains, we see increased recycling (β) with increased initial moisture. The changes in χ are only large for Domain 1, although even small changes in χ can have an important effect through the large $(ET+I)$ factor in the third term of Eq. 6. As can be seen from Table 1, the single terms of Eq. 6 can have negative values meaning they cause a change in precipitation contrary to the change in initial moisture. Ongoing research is investigating the physical reasons for the change of the 'bulk' parameters β and χ in terms of atmospheric stability factors such as equivalent potential temperature and moist static energy.

The changes in initial moisture induced rainfall changes of ±15%, ±2%, and ±6% for domains 1, 2, and 3, respectively. As will be shown later, such changes can have major effects on the availability of surface water and groundwater resources. In domain 3, which covers the Volta Basin, the indirect effect (third term in Eq. 6), originating from a change in precipitation efficiency, is responsible for 40% of the change in precipitation. The direct effect, caused by the local increase in evapotranspiration, is responsible for almost 60%. The second order mixed effects turn out to be negligibly small. The importance of the indirect effect complicates future analysis as it necessitates complete weather runs and makes the existence of simple relations between actual evapotranspiration and precipitation in the region highly unlikely.

What complicates matters further is the fact that impacts are spatially heterogeneous. In Figure 3, the normalized rainfall sensitivity with respect to initial soil moisture, $RS = \left(P_{M\times2} - P_{M\times0.5}\right)/P_{CTR}$, is mapped for the Volta Basin and surroundings. It turns out that RS is very variable over space and takes on negative as well as positive values. In most of the area, RS varies between –1 and 1 but extreme values up to 6 can be found, such as on the northern shore of Lake Volta. Continued sensitivity analysis will be undertaken to map not only those areas that are most affected by changes in land surface characteristics, but also those areas that, when changed, would have a large effect on weather and precipitation patterns. Given the high CPU demand of atmospheric models, no interactive climate simulations will be possible in the final DSS. Therefore, care has to be taken that in the climate runs, landuse change patterns are used that cover the entire range of possible impacts.

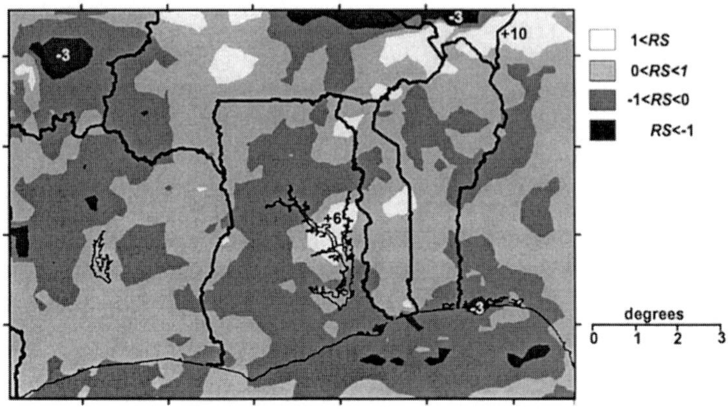

Figure 3. Differences in normalized rainfall sensitivity (*RS*) with respect to initial soil moisture. The numbers are the values of local maxima and minima.

In the control run the average boundary conditions did not change significantly over the simulation period. The period covered 31 days and therefore generated 31 realizations, originating from only slightly different boundary conditions. In this way, 31 realizations of the single dominant type of the usual daily tropical convective precipitation were created. We assume this ensemble of 31 realizations to be statistically sufficient in the sense that, within these 4 weeks, effects of noise in the boundary conditions on the episode precipitation pattern are cancelled out against the investigated effects due to initial soil moisture.

3.2 Rainfall/runoff relations

The second sample interaction between landuse and hydrological cycle concerns the partition of rainfall between evapotranspiration and runoff. The land surface (vegetation and upper-part of the pedosphere) is probably the single most important switch in the water cycle. Water can either be stored in the root zone to become later available for evapotranspiration or percolate beyond the reach of plants and recharge groundwater and, thereby, rivers. The land surface also divides the incoming radiation between latent and sensible heat fluxes, the importance of which was described by the atmospheric simulation above.

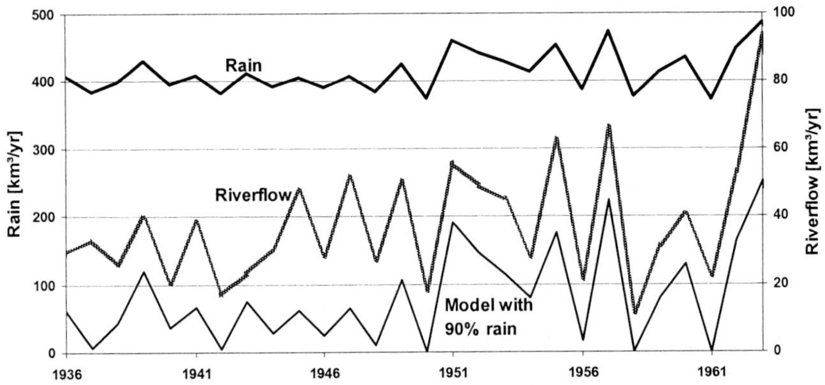

Figure 4. Annual water balance Volta. Sources *New et al.*, 2000 and *GRDC*, 1998.

As with the atmospheric model, we do not yet have a complete analysis of how the land surface affects water flow. We can, however, present a coarsely quantified sensitivity analysis for runoff or riverflow. The basis for this analysis is the annual water balance for the Volta Basin, as shown in Fig. 4 (*Andreini et al.*, 2000). For African standards, we have a relatively long time series available, mainly because the potential for a dam was recognized in the late 1920's. To facilitate direct comparison between years, we end the series in 1964 when the dam at Akosombo was closed. Both runoff and rainfall data are available from 1964 until 2000 as well but the runoff downstream from the dam over this period is completely determined by releases for power generation. For the present purpose of examination of the sensitivity of runoff with respect to rainfall, the series until 1964 suffices. The most salient feature in Figure 4 is the small variation of rainfall from year to year as compared with riverflow. Average annual rainfall over the 1936-1963 period was 414 km^3 with a range from 372 km^3 to 486 km^3 and a coefficient of variation of 0.073. Riverflow averaged 36.6 km^3 and ranged from 1.2 km^3 to 93.8 km^3 with a coefficient of variation of 0.524. One also sees that riverflow followed, in an amplified way, the variation in rainfall. In addition, monthly flows show that there was little or no memory from one year to the next. At the end of the dry season, riverflow had basically come to a halt and any storage was exhausted before the next rainy season started.

With relatively simple models, one can quite accurately simulate the yearly runoff as a function of rainfall on the basin (*Andreini et al*, 2000). For our present purpose, an elementary regression already provides important insights:

$$Q = 0.529 * (P - 342) \text{ [km}^3\text{]} \qquad \text{Eq. 7}$$

where Q is the yearly runoff [km^3] and P the yearly rainfall [km^3]. This regression has a surprisingly high r^2=0.80 and explains 70% of the year-to-year variance. Eq. 1 shows that there is a threshold of 342 km^3 which the rain has to exceed before runoff will occur. Once the threshold has been passed, more than fifty percent of the excess rain becomes runoff.

The large threshold at 85% of the average rainfall reflects the sensitivity of runoff with respect to rainfall. A decrease of 10% in the average annual rainfall would, *ceteris paribus*, result in a 64% decrease in runoff. The resulting yearly runoff for the 1936-1963 period is shown as the lower thin line in Figure 4. So even the small changes from the atmospheric simulations presented earlier would have large impacts on surface water availability.

One cannot directly associate the threshold value with a physical property of the basin. However, the value is clearly a measure of the dynamic storage within the basin from which water can be taken up by the atmosphere either through plants or through evaporation. If this storage is changed through changes in landuse, there will be a large effect on the riverflow. If the (further) disappearance of trees would decrease the average rooting depth and thus reduce the storage threshold by 10%, an average 54% increase in yearly runoff would be expected. If, on the other hand, dynamic storage is increased through the building of dams, which retain water in shallow lakes during the dry season, the threshold may rise and significantly reduce runoff downstream. Both increase in surface storage and decrease of the number of trees are definite trends observed in the Volta Basin. Detailed field work will be needed to see if these two counteracting trends will cancel each other out.

3.3 Irrigation development

The third and final early result regarding landuse and water flow concerns the most direct impact of human actions on surface water resources, namely the building of dams. An important premise underlying the project is that population increase will lead to increased urbanization and landuse intensification and that urban and rural populations have different interests when it comes to utilization of water resources. The potential water conflicts between city and countryside, between South and North, are still in an embryonic stage and may be defused through informed policies.

An important step is to quantify present developments with regard to the use and development of groundwater and surface water resources. Urban centers tend to have well defined locations from which water is extracted, either pumping stations on the rivers for drinking water or large dams for hydropower. Rural water use is much more diffuse. Extraction takes place through thousands of boreholes, wells, and small retinues that supply people and cattle with drinking water and feed small irrigation schemes. Large scale irrigation development is limited to the Vallée de Sourou and Bagré in Burkina Faso and Tono and Kpong in Ghana. Irrigation development is stagnating in Ghana but exponential in Burkina Faso (*FAO*, 2000). The growth in Burkina Faso is not only reflected in an increase in the total number of dams, but also in the conversion of drinking and cattle water storage reservoirs into irrigation supply basins.

To quantify this development, we evaluated the change of available surface water reservoirs from the mid-eighties until 1999 for the southern parts of Burkina Faso and the adjoining far north of Ghana. With the help of remote sensing we obtained spatially explicit objective information over the past two decades, as the following example shows. In 2000, the GLOWA Volta Landuse Workshop was held in Tamale. Here a panel of experts from the region discussed landuse changes over the past 25 years. Two of the landuse change "hot spots" that were identified were related to the development of surface water resources, namely irrigation development in the area to the Southeast of Ouagadougou in Burkina Faso and construction of small drinking water reservoirs in the Upper East Region of Ghana. Based on this information, a set of two Landsat 5 images (path195, row 52 from 18 November 1986 and path 194, row 52 from the 17 August 1984) was compared with corresponding Landsat 7 images from 13 October and 7 November 1999.

For the analysis of the scenes, a large training set containing surface reservoir pixels was compiled for each scene through on-screen digitization. Care was taken to include reservoirs of different depths and algae compositions. Through a maximum likelihood classification, the spectral information from Landsat bands 3, 4, and 5 was used to differentiate between water reservoirs and other landuse types (*Mather*, 1999). River water is the land cover class that has a response that is spectrally comparable to water reservoirs. Hardly any river pixels were classified as reservoirs, however, mainly due to the effect that sediment load had on the spectral response. After filtering the classified images with a 3x3 median filter to remove random noise and manual editing of some remaining river pixels, complete Boolean maps were produced by collating the scenes for '84/'86 and '99. Each contiguous set of reservoir pixels was vectorized and then the resulting polygons, together with their surface area, were entered in a database. The results were categorized by size and country, as summarized in Table 2. Small reservoirs (<5 ha) are mainly used for household purposes, cattle, and horticulture. Medium (5-100 ha) and large (>100 ha) reservoirs will, in general mainly have an irrigation function. The major dam in Burkina Faso, Bagré dam, is additionally used for hydropower generation.

Table 2. Number (Nr.) and total area (ha) per reservoir-size category for 1984/86 and 1999, and changes in that period, for Burkina Faso and Ghana

	Burkina Faso						Ghana					
	1984/86		1999		Increase (%)		1984/86		1999		Increase (%)	
	Nr.	ha	Nr.	ha	Nr.	ha	Nr.	ha	Nr.	ha	Nr.	ha
<5ha	109	145	377	421	268 (246%)	276 (191%)	96	115	113	170	17 (18%)	56 (48%)
5-100 ha	72	1704	154	3444	82 (114%)	1740 (102%)	18	197	48	534	30 (167%)	338 (172%)
>100ha	5	789	16	23421	11 (220%)	22632 (2868%)	2	1184	2	2218	-	1034 (87%)
Total	186	2638	547	27286	361 (194%)	24648 (934%)	116	1496	163	2922	47 (41%)	1427 (95%)

Comparative analysis of the development of surface water resources shows a significant increase in both number and area in southern Burkina Faso and northern Ghana. The increase in Burkina Faso was dramatically larger than in Ghana. In Burkina Faso, the categorized results show that the number of water bodies smaller than 5 ha more than triple as do the number of reservoirs larger than 100 ha. The medium sized dams (5-100 ha) mainly represent irrigation schemes at village level, which, given their size, can be concluded to have been mainly developed with outside construction support. The development in terms of numbers of dams and surface area covered are parallel for small and medium dams. The enormous increase in area for dams larger than 100 ha is due to the construction of Bagré Dam (20,957 ha in '99), which was finished in 1993.

In contrast to the development in Burkina Faso, the changes in northern Ghana were more moderate. The highest percentage increase in both number and area falls into the category of dams sized 5 to 100 ha. This development indicates an increase in the number of irrigation schemes between the mid-eighties and 1999 which runs counter to the national trend of stagnant irrigation development. The most likely explanation is that the population density, and thereby the need for intensive agriculture, is high in the Upper East Region. As in Burkina Faso, the largest dams contribute most to the total area and thereby to the expected downstream impact.

The increase in surface water bodies is not evenly distributed throughout the study area, but rather regionally clustered (Figure 5). Besides Bagré Dam in Boulgou Province, further major changes in Burkina Faso occurred south of Ouagadougou (Bazega Province). The changes that occurred in Sanguie Province in the NW section as well as in the Ganzourgou and Kouritenga Province in the NE parts of the study area were still significant although to a lesser extent. The edges of these major change clusters coincide with province boundaries and development of dams was hardly detected in Sissili,

Zoundweogo and Gourma Provinces, suggesting the importance of institutional arrangements. In Ghana, major changes are concentrated in the Upper East Region close to the border to Burkina Faso, where also Tono Dam (1,795 ha in '99) is located. In contrast to this, the Upper West Region features dams only sparsely and does not reveal a significant change.

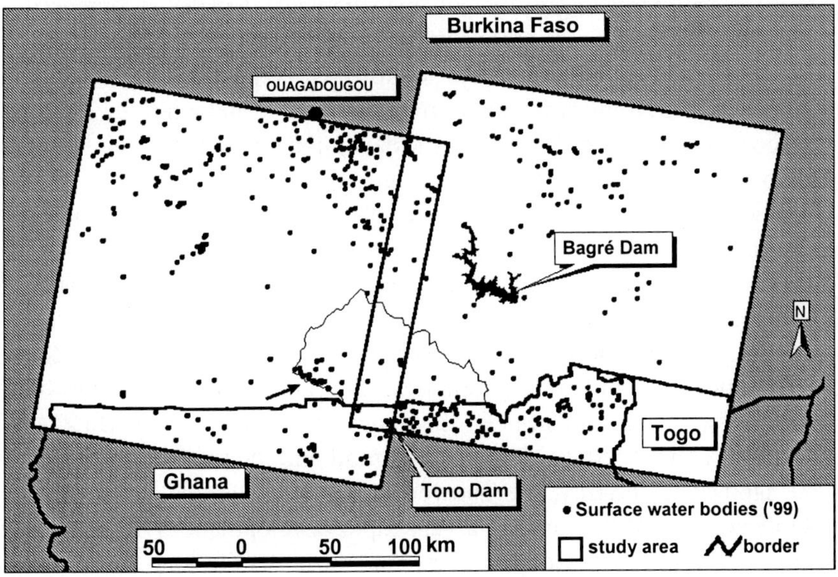

Figure 5. Distribution of dams in Burkina Faso, South of Ouagadougou, and Northern Ghana. The arrow points to the sharp transition in dam density between Naouri Province (outlined) and Sissili Province (no dams).

The total reservoir area in '99 was seven times that of '84/'86, which is a dramatic increase. Yet the total area is still only about 30,000 ha or less than 4% the surface area of Lake Volta. The topography of the basin is such that the reservoirs always have a rather shallow depth and a correspondingly small storage capacity. The total irrigated area can, therefore, not be very large. The total potentially irrigable area for Burkina Faso is estimated at 160.000 ha (*Sally*, 1997). If this area were to be completely developed and if we further make the generous assumption that irrigated areas evaporate at full potential rate throughout two growing season, we arrive at a total evapotranspiration from reservoirs and irrigation schemes of just 2.6 km^3 per year or less than 7% of total riverflow. We may conclude that although the rapid development of surface water resources for household and irrigation purposes may have the most direct effect on the downstream availability of water, the largest impact of landuse change on water resources may come indirectly through changes in atmospheric and recharge processes.

4. CONCLUSIONS AND OUTLOOK

The GLOWA Volta project addresses the complex of social and environmental processes that determine water fluxes through the Volta basin. The main goal is to provide a scientifically based Decision Support System for the increasingly difficult management of water resources in the region. Of special interest to the water supply side of the question is the impact of landuse change, the single most important global change phenomenon within the watershed. In terms of water demand, the main issue is the conflicting needs for irrigation development in the rural North and hydropower generation for the urban South.

As the project is still in its build-up phase, few quantitative conclusions can as yet be drawn. We have presented three sample interactions through which landuse change affects the hydrological cycle. The atmospheric feedback example showed that rainfall does indeed depend directly as well as indirectly on evapotranspiration within the region and may, therefore, be co-determined by landcover properties. The annual water balance of the second example demonstrated the high sensitivity of surface water availability with respect to changes in rainfall and the water storage capacity of the landscape. Finally, we showed that the increased landuse pressure does indeed precipitate translate a very rapid development of surface storage structures. Although these dams are the most visible impact of landuse change on water flows, the total potential impact is much less than that of the other two examples.

Given the complexity of the feedback network, it is clear that many interactions still remain to be quantified. It is no coincidence that our three examples concern physical models or observations. Social interactions are not only more complex -, they are also more difficult to measure and, unfortunately, - no spatially explicit historical databases exist. The social domain may yield the more interesting feedback since it is this domain that determines whether developments will be stable and sustainable or will enter into a downward spiral. We expect to be able to close some of the lower dimensional feedback loops relatively soon and proceed to the higher level loops as the project progresses. Specifically, the described effects of landuse change on water availability will be complemented in the near future by the analysis of the impact of water availability and climate on landuse. How to address the main water management question of irrigation development versus hydropower under changing rainfall reliability requires analysis of all key-variables involved and, thereby, the full life-span of the project.

5. ACKNOWLEDGEMENTS

We gratefully acknowledge financial support of the German Federal Ministry of Education and Research (BMBF) as main sponsor of the GLOWA project and the Ministry for Schools and Education, Science and Research of Northrhine-Westphalia. The GLOWA Volta project is *de facto* a research network and we express our thanks for the scientific inputs from our other partners: Remote Sensing Research Group (Bonn University), Institute for City Planning Rural Engineering (Bonn University), Institute for Tropical Medicine and Hygiene (Heidelberg University), Savanna Agricultural Research Institute (CSIR), Water Research Institute (CSIR), Remote Sensing Application Unit (University of Ghana), Population Impact Project (University of Ghana), Institute for Statistical, Social and Economic Research (University of Ghana), Institut de l'Environnement et des Recherches Agricoles (Burkina Faso).

6. REFERENCES

Adiku, S.G.K., Dayananda P.W.A., Rose C.W., Dowuona G.N.N., 1997: An analysis of the within-season rainfall characteristics and simulation of the daily rainfall in two savanna zones in Ghana, *Agricultural and Forest Meteorology*, 86, 51-62

Andreini, M., van de Giesen, N., van Edig, A., Fosu, M., Andah, W., 2000: *Volta Basin water balance,* ZEF Discussion papers, Nr 21, ZEF, Bonn, Germany

Chen F., and Dudhia J., 2000: Coupling an advanced land-surface/hydrology model with the Penn State/NCAR MM5 modeling system, Part 1: Model description and implementation, submitted to *Monthly Weather Review*

Chehbouni, A., Kerr, Y., Watts, C., Hartogensis, O., Goodrich, D., Scott, R., Schieldge, J., Lee, K., Shuttleworth, W., Dedieu, G., de Bruin, H., 1999: Estimation of area-average sensible heat flux using a land-aperture scintillometer during the Semi-Arid Land-Surface-Atmosphere (SALSA) experiment, *Water Resources Research*, Vol. 35, No. 8, 2505-2511

FAO, 2000: *FAOSTAT.* FAO, Rome, Italy, http://apps.fao.org

GRDC, 1998: *Global Runoff Data.* The Global Runoff Data Centre, D-56002 Koblenz, Germany

Grell G. A., Dudhia J., Stauffer D.R., 1994: *A description of the fifth-generation Penn State/NCAR mesoscale model (MM5), NCAR Technical Note*, NCAR/TN-397+STR

Lambin E. and Mertens B., 1997: Human ecology, remote sensing and spatial modelling of land-use changes in Africa, *LUCC Newsletter*, Nr. 2, 6-9

Lawson, G.W., 1963: Volta Basin Project. *Nature*, Vol. 199, No. 4896, August 31, 1963, 858-859

Mather P., 1999: *Computer processing of Remotely-Sensed Images. An Introduction.* 2nd Edition, Wiley&Son, Chichester, UK

New, M., Hulme, M., Jones P., 2000: Representing twentieth century space-time climate variability. II: Development of 1901-1996 monthly grids of terrestrial surface climate, *J. Climate*, **13** (13), 2217-2238

Nicholson S.E., 1998: Interannual and interdecadal variability of rainfall over the African continent during the last two centuries, in *Water Resources Variability in Africa during the*

XXth Century, (Servat E., Hughes D., Fritsch J.-M., Hulme M., eds) IAHS Publication 252, 107-116

Sally, H., 1997: *Améliorer les performances des périmètres irrigués: Les actes du séminaire régional du Projet Management de l'Irrigation au Burkina Faso,* IIMI, Colombo, Sri Lanka

Schär, C., Lüthi D., Beyerle U., Heise E., 2000: The soil-precipitation feedback: A process study with a Regional Climate Model, *Journal of Climate*, Vol. 12, 722-741

GLOWA Danube: A research concept to develop integrative techniques, scenarios and strategies regarding global changes of the water cycle

W. MAUSER AND R. LUDWIG
Institute for Geography, University of Munich, Munich, Germany

Abstract: Due to their complexity, sensitivity and high degree of integration, mountain areas and their forelands cannot be accurately described and analysed in integrated models with present capabilities. Yet, this is needed to determine the degree of sustainability of various management alternatives and to derive appropriate recommendations for public and commercial stakeholders and administration. Within the framework of the GLOWA (Global change of the water cycle)-Danube project, which is funded by the German Ministry for Research and Education (BMBF), a tool to determine water fluxes, their coupled fluxes of matter and matter turnovers is being created by a interdisciplinary group of researchers, combining expertise in the fields of hydrology, water resource engineering, meteorology, glaciology, geography, ecology, environmental economy, environmental psychology and informatics. It aims to develop and validate integration techniques, integrated models and integrated monitoring procedures and to implement them in the network-based Decision Support System DANUBIA. It will contain the essential physical and socio-economic processes required for realistic modelling of water and matter fluxes in mountain foreland situations. The large gradients in climate, vegetation and water supply, along with the good data coverage in both natural and social sciences, make the Upper Danube Catchment (≈ 78.000 km^2) an excellent prototype for an integrative research approach. On completion, DANUBIA will be able to simulate water related scenarios in environmental management and will examine the sustainability of the proposed solution scenarios in large, heterogeneous catchments. In order to develop DANUBIA as a mutual transdisciplinary communication platform of the participating disciplines, new information technologies will be utilised. Based on the industrial standard UML (Unified Modelling Language) as the common meta-modelling language, DANUBIA will be capable to integrate platform independent modules of all participating partners (e.g. climate model, SVAT model, ground water model, micro economic model, agent model describing behaviour. DANUBIA will be spatially distributed and raster-based at its core,

M. Beniston (ed.),
Climatic Change: Implications for the Hydrological Cycle and for Water Management, 171–188.

which simplifies the interdisciplinary description of the interactions between the considered processes. It will apply the concept of "proxels" (Process Pixel) as its basic building blocks, which has different dimensions depending on the viewing scale and connects to its environment through fluxes, such as runoff formation, lateral flow, evapo-transpiration, plant growth, economics or agents. A detailed description of the DANUBIA concept within GLOWA-Danube, its goals, scientific and technical structure, implementation and expected performance is presented.

1. INTRODUCTION

The scientific analysis of Global Change is characterized by a multitude of sectoral work in diverse research disciplines which have set global aspects as the centre of their research interests. Global Change is therefore characterized by a network of interactions, in which man himself acts as the cause, is acted upon, and is the mediator. Due to the regional character of the political and economical spheres of influence, these global changes have regional roots and causes as well as manifestations and impacts. A prerequisite for a sustainable development and management of natural resources, which usually takes place at the regional scale, is the possibility to plan and act with foresight. To accomplish this, the required straightforward, strategic and political concepts must be based upon farsighted analysis and evaluation of the interactions in the natural system, as well as in the relationship between man and nature. Farsighted analysis and evaluation requires the availability of suitable tools, which are based on process understanding and enable the prognosis of various possible future states in a deterministic manner and under consideration of all important natural and anthropogenic impacts. Thus, these impacts can be evaluated and alternative management plans can be developed.

Solving the conflict of interest between water demand and availability, with respect to water quantity as well as quality, can serve as a prototype for environmental management in this sense. It has lead to manifold technical and cultural solutions. However these solutions will be increasingly stressed in the near future, globally and regionally, by increasing population figures, changes in use and escalated conflicts over use, even to the extent that the usage will be pushed to or over the limits of their capabilities. The interaction between mountain areas as water suppliers and foreland areas as water consumers is an archetypical example of this. This type of interaction plays an outstanding cultural and economical role worldwide and can be found in different distinctions all over the world (e.g. Alps, Pyrenees, Himalayas, Andes, Caucasus Mountains, Atlas Mountains, Ethiopian Highland). A central element of the integrative global solution approach within the framework of GLOWA is to understand and explain this interaction.

For this purpose the Upper Danube catchment was chosen as a case study because it is extremely suitable as a catchment area consisting of high mountain regions and their forelands. One reason is the excellent natural science and socio-economic data base, covering the existing strong gradients in all natural (e.g. climate, hydrology, vegetation) and socio-economic (e.g. use, settlements, economy, tourism) factors. It is furthermore characterized by a strong usage of the natural resources connected with water through various economic activities, along with strong present and foreseeable conflicts in water use (with respect to quality and quantity), export of water to other catchments (mainly to the developing Eastern European economies) as well as a strong import/export of virtual water (used in the agricultural and industrial production).

2. THE AIM OF THE GLOWA-DANUBE PROJECT

The overall goal of GLOWA-DANUBE is to develop and validate integration techniques, integrated models, and integrated monitoring procedures and to implement them in the network-based Decision-Support-System DANUBIA. In order to be functional DANUBIA should contain the essential physical and socio-economic processes that are required for realistic modelling of water fluxes in mountain-foreland situations. Above all, the lateral flow, the relationships between the upper and lower river sections, the meteorological gradients as well as the special use of sensitive mountainous boundary areas will be taken into account. DANUBIA will be regionally transferable and thus applicable for the wide range of catchments within the GLOWA-(Global Change of the Water cycle)-transects. For more information on GLOWA see www.gsf.de/ptukf/glowa-e.pdf

The operative readiness of DANUBIA will be shown with various scenarios, that also contain the actual state. DANUBIA will be exemplarily applied with respect to the entire range of topics of the Global Change of the water cycle in the whole Upper Danube catchment regarding the actual state, and to develop various future scenarios and investigate these with regards to their sustainability. Scenarios, which will be developed within GLOWA-DANUBE and investigated with DANUBIA, will contain first of all climatic, political, economic, demographic, and technologic alternatives to the present state and comprise changes in the use and use intensity of the land and water resources.

Within the framework of GLOWA-DANUBE, local interest groups will be taken into account in defining the scenarios as well as in analysing and evaluating the results. Based on the outcome of the analysis DANUBIA will be refined. In the final stages, DANUBIA will be applicable to all parties concerned with water-resource-management (policy and administration, planning agencies, non-governmental organisations (NGOs), science and economy) as planning and management tool. In the first project phase, primarily climatic scenarios and alternatives to use and use intensity of the land and water resources will be developed and investigated with the DANUBIA prototype.

3. THE UPPER DANUBE CATCHMENT AREA

The Danube is the second largest river in Europe. 15 countries share its catchment area of 817,000 km². GLOWA-DANUBE is limited to the analysis of the Upper Danube to the discharge gauge Achleiten below Passau. Together with the catchment area of the Inn, it covers an area of 77,000 km². The altitudes in the catchment area range from 286 to 3,600 m a.s.l. This causes strong spatial, topographical and meteorological gradients (precipitation: 650 to >2000 mm/a, evaporation: 450-550 mm/a, discharge: 150-1600 mm/a, average annual temperature: -4.8 – +9 °C) as well as a strong differentiation in land cover and land-use. The Upper Danube is a region with a definite water surplus. The hydrology of the Upper Danube is characterized by strong impacts due to relief. All discharge regimens from straight nival through to pluvial occur. Floods also occur frequently. Generally they are triggered regionally by convective summer rains in the alpine foreland and in the Alps. However, characteristic large-scale weather patterns (e.g. Vb) also trigger floods, which have an impact upon the whole Upper Danube region and its tributaries (e.g. the Pentecost Flood of 1999).

The natural vegetation zones in the catchment area range from alpine grasslands, green alder and dwarf pine stands, mountain pine, mixed stands, to the dominating beech stands and oak-hornbeam woodlands. Alluvial flood plain forests are found in areas near rivers. The present land-use is mostly determined by human impacts, whereby agricultural use of different intensity (grassland, farmland) and forestry dominate. The high amount of precipitation in parts of the catchment area is disadvantageous for agricultural use.

The water resource management in the Upper Danube is complex and characterized by the different bordering countries: 73% of the Upper Danube is managed by the German states Bavaria and Baden-Württemberg, 24% by Austria and the rest by Switzerland, Italy and the Czech Republic. The Inn, as the most important "tributary", contributes up to 52% of the average discharge (MQ) of 1420 m³/s to the discharge measured at gauge Achleiten downstream of Passau.

The catchment area of the Upper Danube is densely populated with ca. 8 Mio. people. A large part of the water for the water supply of the larger cities and the industry originates in the pre-alpine region and the Alps. The most important industrial agglomeration areas are Munich (1.2 Mio inhabitants), Augsburg (260.000) and Ingolstadt (115.000) and the chemical triangle Burghausen.

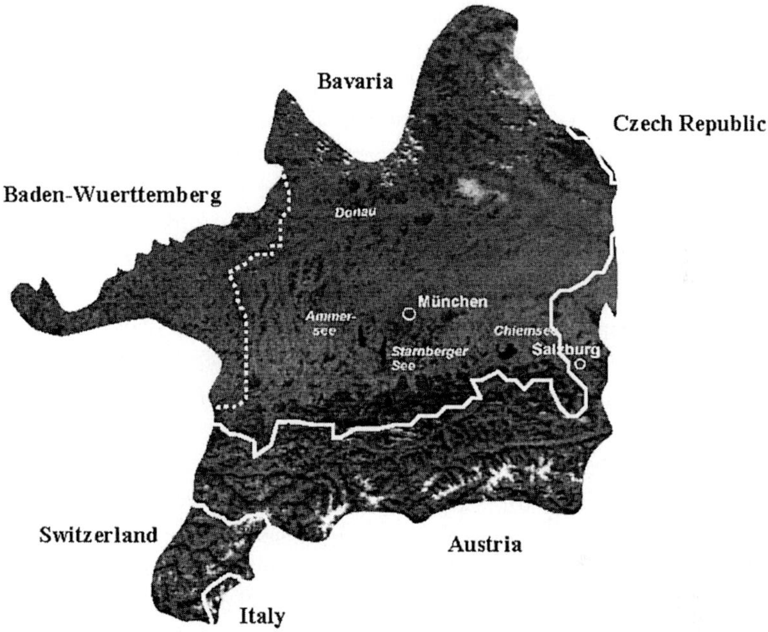

Fig.1: Catchment Area of the Upper Danube as seen from NOAA/AVHRR (3.7.1999, 1 km resolution)

For flood protection and for water-resource-management purposes, the discharge of all important tributaries of the Upper Danube has been regulated through reservoirs. Numerous power plants at rivers and reservoirs are used to produce energy. To a large extent, their management is determined by the dynamics of the snow and ice storage in the Alps. Reservoir management at present is largely uncoordinated between the different administrative entities involved in Alpine power production. Therefore a large potential for optimisation of the management practices exists. Parts of the Upper Danube are navigable and are part of an important waterway that connects the Black Sea with the North Sea. This waterway is already used to export water out of the catchment area of the Upper Danube into the Rhine-Main region and beyond. It is to be expected, that increasing demand for water during the course of a more intensive and more coordinated water use in Europe will put increasing pressure on the export of water out of the catchment area of the Upper Danube. Withal, the ecological and socio-economic effects as well as the environmentally sound capacity of the water resources are still largely unexplored.

3.1 Water Use Conflicts in the Catchment Area

The heterogeneity of the natural and anthropogenic conditions and the diversity (natural, economic and administrative) at different scales creates mutual dependency between the upstream and downstream areas. This results in important present and future water use conflicts in the Upper Danube. They can be considered typical for high mountain catchments and their forelands in developed, humid regions. This heterogeneity is caused by the altitude gradients and the resulting matter transport (water, fertilizers, pesticides, etc.) as well as the intensive and manifold anthropogenic uses. The fluxes of water and matter in the Upper Danube catchment are affected by human impacts to a degree, which is hardly exceeded by any other developed mountainous catchment. The diversity of uses as well as the abundance of the anthropogenic regulation measures lead to a dynamic and only apparent equilibrium. However, due to strategies to transfer and avoid conflicts, this apparent equilibrium is pushed far away from the natural equilibrium and therefore is not sustainable,. It is not regarded as a problem, that the pronounced, spatial dependencies which are typical for mountainous catchments and their forelands, result in a sensitive reaction of the water balance of the Upper Danube to changes in climate, land-use, and management at all temporal and spatial scales. The question arises though as to whether the presently implemented practice of a non-sustainable water and matter budget is economically feasible in the long run.

At present, the strongest water use conflicts in the catchment area of the Upper Danube exist between agricultural use and drinking water supply, due to fertilizer and pesticide drainage into the groundwater. In order to mediate this conflict, the states of Baden-Württemberg and Bavaria have adopted different strategies (water surcharge (Wasserpfennig), SchALVO (regulated land use in ground water protection areas), small-scale water protection concept vs. large-scale groundwater protection, compensation payments). Currently water supply in Bavaria is shared by app. 2200 companies. It is to be expected, that the water supply structure and drinking water exploitation in the catchment areas will be centralized during the upcoming years in the course of the catchment related water management of the European Water Resources. The basis for such a strategy of common usage and balancing of interest beyond the catchment boundaries has not yet been reached among the participants involved. Balancing the interests is aggravated by the diversity of the participants involved and by existing structural dependencies which are due to the fact that changes in the upstream catchment areas have an impact upon the downstream catchment areas and must be taken into account in the management of the catchment.

Strong conflicts of use also exist in the Upper Danube area between agricultural use and forestry as well as between tourism and environmental protection. Southern Bavaria is the centre of tourism in Germany. Tourism is a decisive and employing factor in Bavarian economy. Agriculture, forestry, tourism and environmental protection compete against each other in shaping the landscape and the industrial structure of the alpine forelands and in the Alps. In this context, especially the artificial snow production and the ski runs, the avalanche protection forests and the water quality of the lakes and rivers must be mentioned.

Especially the alpine area of the Upper Danube is being increasingly limited in its development possibilities due to spontaneous mud flows and avalanches caused by the relief. A combination of the factors water, land-use changes as well as the settlement dynamics in the valleys of the Alps plays an essential role in both cases. A steady decrease in the groundwater levels in the alpine foreland has also been observed for almost 30 years. The causes for this and the significance with respect to a sustainable water resources management are still largely unclear.

It is to be expected, that climate changes will lead to strong land-use changes. However, these changes are also affected by other factors, that are not related to climate change. Among these are the creation of cultivated plants with a higher resistance to cold, precipitation, and parasites and their changed yield structure, changes in the vegetation growth and the water use efficiency due to increased CO_2 concentrations, especially at higher altitudes changes in agricultural production goals (quality vs. quantity) and the structure of agricultural industry in Germany, and the conspicuous deterioration of the prerequisites for agricultural production in the Mediterranean area due to decreased precipitation. As a result, the Bavarian alpine foreland may be more strongly used to maintain the European food production. The ecological, economical, and societal effects of this bundle of foreseeable changes have not yet been assessed. There are also no strategies available for sustainable environmental management under such circumstances.

With respect to its Global Change relevance, the Upper Danube catchment is characterized less by a lack of water as by a lack of substantiated definitions of the various future functions, which catchments of this type must meet as important and sustainable global water supply and water export areas, particularly with regards to sustainable management of the natural resources, even beyond the borders of the catchment area.

4. INTEGRATIVE METHODS FOR THE DECISION SUPPORT SYSTEM DANUBIA

Up to now, there is no commonly recognized method available to both integratively describe the interactions between the partially very different natural and anthropogenically determined processes over the wide range of considered climates and catchments, and to simulate scenarios about the future development. The main obstacle regarding the integration is the large difference in the way that the various disciplines formalize and describe their understanding of the respective processes. This results in differences in terms and concepts, differences in comprehension of the involved scales and dominant processes, as well as different methodical approaches. Due to these reasons, numerous investigations exist, in which the respective disciplines developed their separate contributions for solving parts of this task, which however can only be solved as a whole through an integrative approach. Thus the combination of these disciplinary approaches often led to failure. Using new networked modelling technologies though opens new possibilities of integrating various disciplines.

The integrating factor within GLOWA-Danube is the *common* development and use of the Decision-Support-System DANUBIA. The main themes covered by DANUBIA are water use, land use, drinking water quality, quantity, and availability, water export, possibilities and limits of the economical and political controls, prices, and political acceptance. At the end of the first project phase, it shall quantitatively describe the interactions of the different disciplines concerned with water fluxes for the complex mountainous catchment and it shall also cover simple scenarios about the future developments and their influence upon water quantity and quality in the Upper Danube catchment. Simple scenarios are characterized by a manageable number of future changes that they cover. In the simplest case, this is one factor and is e.g. either climate or land-use change. In contrast, complex scenarios have several changing factors that interact, and in the most complex case, these factors change during their development as a result of mutual interactions.

In the second phase, procedures shall be developed and applied for the common evaluation of the sustainability of the various complex scenarios of the future global and regional changes. These procedures will include local stakeholders, integrating and formulating their respective needs and considerations of future possibilities. This will lead to a refinement and improvement of DANUBIA. In the final phase of the project, DANUBIA shall be prepared for practical applications. DANUBIA describes the involved processes with a spatial differentiation. Thus it allows the explicit determination of lateral redistributions based on fluxes of water, energy, and matter as well as migration and capital within the involved catchments and beyond. In the fully developed stage, DANUBIA will cover water, carbon, nitrogen, phosphorus, capital and human migration as the "lead-substances", whose fluxes will be integratively modelled:.

4.1 Methodology

The methodology used to develop DANUBIA is characterized by applying integrative numerical, network based model development methods and integrative monitoring (e.g(. remote sensing). In the area of integrative numerical modelling, the necessity of improving the optimisation of the industrial production cycle across all areas of the production processes has led to two methodologies in computer sciences during the last years. These shall serve as the basis for developing a new approach for integration of the disciplines in GLOWA-DANUBE:

1. Formal languages, which will enable the involved disciplines to model sub-models of very complex systems as well as their interactions in a common way. These Meta-Modelling languages describe the essence of the modelling in a way that is independent of the respective discipline. It is thus very well suited to formulate the interaction and communication between the various processes and to check for completeness and functionality. The Unified Modelling Language (UML, (Booch et al. (1999)) is such a language and has developed within a few years to the industry standard. UML shall be used in the project as the common Meta-Modelling language for all groups.
2. The networked communication of distributed objects (i.e. independent, self-contained parts of a model compound). Hereby, the possibilities of the increased degree of networking are used, in order to enable the various elements of the whole model or model compound (called objects in the rest of the text) to be able to run in a network on different computers in different locations (namely the places with the most expertise). The communication consists of an exchange of data and methods and in the synchronization of the various model elements. Furthermore, the possibility to subdivide complex and voluminous applications onto a large number of different networked computers theoretically enables an almost inexhaustible computer capacity. The standardized communication infrastructure, which is needed for the distribution of the objects in the network, shall be used in order to integrate the different model elements of various groups, which were developed in different places and with different programming languages. A prerequisite for this is that only selected, standardized communication procedures will be used by all groups. The industry standard in the area

of communication between distributed networked objects is the Common Object Request Broker Architecture (CORBA) (Henning & Vinoski (1999)). CORBA is used in this project in order to allow for communication of the networked objects from the various groups. On this basis, with DANUBIA, a system will be developed that consists of distributed networked objects and which can communicate through CORBA in the net.

The mutual networking of the objects is described with the Meta-Modelling language based on UML independent of the discipline. Each discipline contributes its part of the complex model compound as an object. An object is an encapsulated unit, which completes a distinct function in the Decision-Support-System and carries out the data exchange and the synchronization through defined interfaces. Hereby, the object can be implemented in any desired language. This selected concept of a network-based, distributed model compound for integrative modelling is new. However it is foreseeable, that along with the expansion of network technology and the internet, this concept will prevail and provide the basis for interdisciplinary work in the future. In this sense, the project plans to provide basic research in the development of interdisciplinary integration techniques.

4.2 Modelling Concept

Beyond the use of industry standards, the following characteristics and concepts make certain that the set-up of DANUBIA is feasible: As shown schematically in Fig.2, DANUBIA is *raster based* at its core. Although this may not be the optimal spatial representation for all use cases, it simplifies the interdisciplinary description of the interactions between the considered processes. The spatially distributed, raster based modelling of the core processes (run-off generation, evaporation, plant growth, economics, human behaviour through agents) are expanded at first by one-dimensional modelling of the processes in flumes and water bodies with regards to quantity, matter turnover, and management.

DANUBIA uses the *concept of "Proxels (process pixels)"*. Proxels are the basic building blocks of DANUBIA and consist of a pixel (picture element) in the form of a cube, in which processes occur (Tenhunen et.al., (1999), Mauser (2000)). This cube, which has different dimensions depending on the viewing scale, connects to its environment (other proxels) through fluxes. An object of DANUBIA can thus be e.g. a surface proxel, that describes the water flow on the surface. Encapsulation allows the

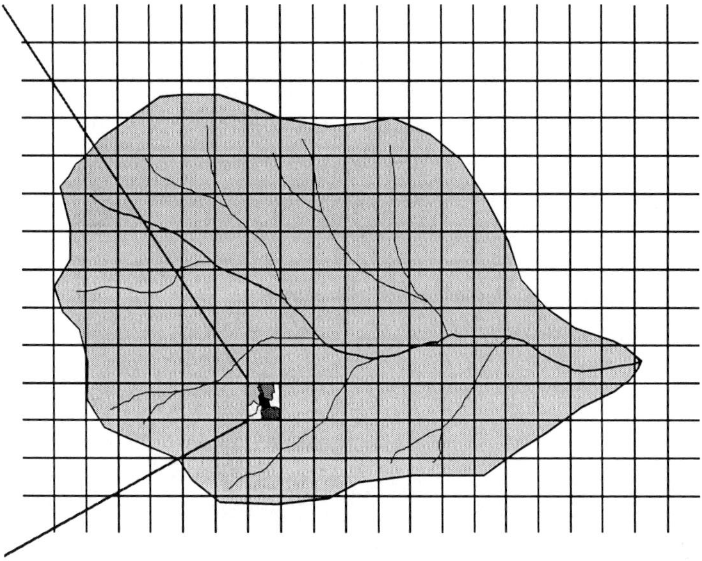

*Fig.2:*Schematic raster based modelling in DANUBIA on the pixel basis. The focus connects to Fig.3

management of the involved processes, in this case evaporation, surface run-off, lateral run-off in the unsaturated zone, and percolation into groundwater. It also enables the management of the necessary parameters within the object and to carry out data imports and exports via defined and standardized interfaces. For further development and improvement, all characteristics of a proxel can be handed down to the later versions of the programs, i.e. are available for reuse, scaling, and refinement. In the same way, this is realized for natural science processes as objects. Socio-economic processes (migration, behaviour, market situation, perception, etc.) are also described on the basis of proxels. This concept is to be used for the integrative tasks in GLOWA-DANUBE and is to be expanded at the appropriate places.

In socio-economics, DANUBIA uses the concept of *agent based modelling (Suleiman 2000)*. The core idea of an agent model is that *every single* agent involved is modelled in a domain of individuals. Thus, this procedure differs fundamentally from all summative modelling approaches, in which the input-output-functions of a whole system are described. Agent based modelling can be subdivided into shallow and deep models, according to the theoretical content. *Shallow models* specify the input and output connections of the behaviour of single agents with little or no theoretical assumptions. This can be implemented through straight forward mathematics or also through neuronal networks. *Deep models* generate the behaviour out of a theory about the agents and enable semantically meaningful statements about the motives of the agent. Methods used here are e.g. symbolic modelling, in which decision making is described as a sequence of rules. Agent models can be used successfully, if a meaningful standardization of the agents can be carried out. Modelling of the agents enables not only a detailed extrapolation of behaviour (which is important in the case of DANUBIA, with respect to the agent types as well as the spatial distribution), but is also easily expanded and refined. A deep model enables mapping and interpretation of the dynamic phenomena of adaptation, learning, and interaction.

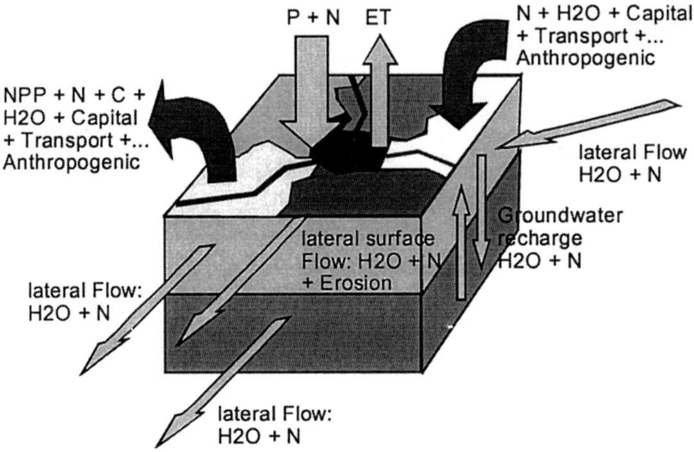

Fig.3: Simplified schematics of a proxel as a basic element in DANUBIA using a land surface proxel as an example.

At first, agent models will be used in the DANUBIA model compound psychology. In the second project phase, they will also be developed in water resources management/groundwater and in political science. At first, shallow models will be used. Especially in modelling political agents, water supply companies and farmers, it is obvious that a deep, i.e. theoretically based modelling of the agent behaviour, is a very demanding task. Besides this, in a dynamic and highly interactive domain, prognoses based on such a model will initially have a rather exemplary character.

4.3 Use of remote sensing for integrative monitoring of the environment

Using remote sensing in GLOWA-Danube-like projects is not yet common, but from the viewpoint of the project partners very important especially with regard to the transferability of the developed methods to other regions. Research in the field of information retrieval from remote sensing data has made great advances in recent years. Thus, stable algorithms to determine a multitude of land surface and atmospheric parameters now exist. These shall be applied in the project. At the same time, new and improved possibilities will become available during the course of the project with the new sensors of the next generation (TERRA, AQUA, ENVISAT, METEOSAT 2nd Generation). Field campaigns, accompanying the aquisition of remote sensing data with in-situ measurements, will be performed for further algorithm development and validation purposes. The risk of a false start of these mentioned systems exists. In such a case, the data from the existing sensors (AVHRR, SeaWifs, MOS, METEOSAT) will be used. Even with these data, new procedures to assimilate remote sensing data into integrative models can be developed, even if the expanded possibilities cannot be used at first. Thus, the expectation of success is rated good in this area.

5. THE GLOWA-DANUBE RESEARCH GROUP

In the course of setting-up, validating, and applying DANUBIA, models and methods from the following natural and socio-economic disciplines are involved:

Core Groups:

- *Hydrology*, Institute for Geography, University of Munich
- *Remote Sensing and GIS*, Institute for Geography, University of Munich
- *Meteorology*, Institute for Meteorology, University of Munich
- *Water Resources Management / Groundwater*, Institute for Water Resources, University of Stuttgart
- *Water Resources Management / Surface Waters*, Institute for Applied Water Resources, Ottobrunn
- *Plant Ecology*, Institute for Plant Ecology, University of Bayreuth
- *Environmental Psychology, Institute for Psychology, University of Freiburg*
- *Environmental Economics, IFO-Institute, Munich*
- *Information Sciences, Institute for Informatics, University of Munich*

Bridge Groups:

- *Glaciology, Institute for Meteorology and Geophysics, University of Innsbruck*
- *Remote Sensing and Meteorology, Institute for Geography, University of Marburg*
- *Tourism Research, Institute for Geography, University of Regensburg*

Each participating core group will develop an object based on their available disciplinary model approach and knowledge and partially with the help of the bridge groups. The objects communicate through a standardized mechanism, which will ensure that for every time step in the model calculations, the spatial data (e.g. temperature, radiation, biomass, income, etc.) and parameters (e.g. water-import, suggested subsidy of corn, etc.) required by a partner-object will be provided. This mechanism is thus comparable with a market place, in which data are offered and exchanged. An object can simultaneously be the supplier and receiver of data. This "Market place concept" was developed and refined through intensive discussions within the project groups. It is shown in Fig.4.

The "Market place" in this figure is the sum of all boxes outlined in black. They show the essential parameters, that will be exchanged between the objects, generally as grid data. DANUBIA's structure provides for the synchronization of the data exchange. This kind of interaction structuring between the groups enables the respective points to be clearly identified, in which scenarios regarding the future development are entered into the system. These scenarios are external factors to the catchment, such as climate change (meteorology), EU-political boundary conditions (politics), life styles or decision structures in the area of psychology (e.g. agent based decision, life-style changes), or economics (e.g. changes in prices, supply, technologies, industrial structure, etc.). In this sense, economics is of prime concern in the selected structure. For instance regarding the introduction of a newly cultivated plant and its impact upon the water balance, the optimisation of the cost structure of the agricultural production is of prime concern and requires an answer to the question of the production costs and impacts upon water quality and environmental influences. At first, the property-structure of the plant is determined by the economical viewpoints, whereby the ecological consequences of the introduction can only be foreseen with great difficulties. The groups plant ecology, hydrology/remote sensing, groundwater and water bodies react to this change by searching for an optimal region to cultivate these plants, by setting-up fertilizer plans, by letting these plants grow according to meteorological data (today's and future data), by quantifying their influence upon the water cycle and sending feedback to the disciplines of psychology and economy about the resulting data regarding the fluxes of water and the water quality. In a further step, psychology has the possibility to react unsatisfied, which induces politics to check the regulations.

Since the various data have to be available in different time steps, the mutual synchronization of the objects is a central task. This can be carried out with UML relatively easily and independent of the discipline. These methodical approaches are breaking new ground. The possibility to reuse the developed code (through object-oriented programming), the easy serviceability of the interfaces (through standardization) and the inherent explicit documentation through the use of the Meta-Modelling Language in this approach, create new integrative structures between the participating scientists. They also easily allow the introduction of refined and improved algorithms into the structure, as implied in the ellipses in the core groups. The core groups start with a simple implementation of its functionality (phase 1) and continue by gradual improvements of the methods (phase 2 and 3) (for details see the individual project descriptions). Based on this, a prototype of DANUBIA shall be developed, that will already contain all basic characteristics of the final system. Building upon this, validation, refinement, and application of the created objects and interfaces will follow with a wide selection of Global Change tasks and scenarios in the Upper Danube catchment. Basic approaches for their formulation are given in the description of the catchment areas. After the initial phase, regional stakeholders and interest groups will be included in the validation and improvement as well as in the development and handling of scenarios, in order to assure the relevance and applicability of DANUBIA.

6. PERSPECTIVES

GLOWA-DANUBE is trying to break new ground in many areas. The integration of the expert disciplines plays the key role for the success of the project, for which the following aspects are of central importance:

1. the agreement to a common raster-oriented model structure between all involved project partners
2. the clear identification of common interfaces and a common meta-modelling language, which is independent of disciplines,
3. the establishment of a hierarchical communication structure between the groups with bilateral cooperation, working groups, workshops, and colloquia.

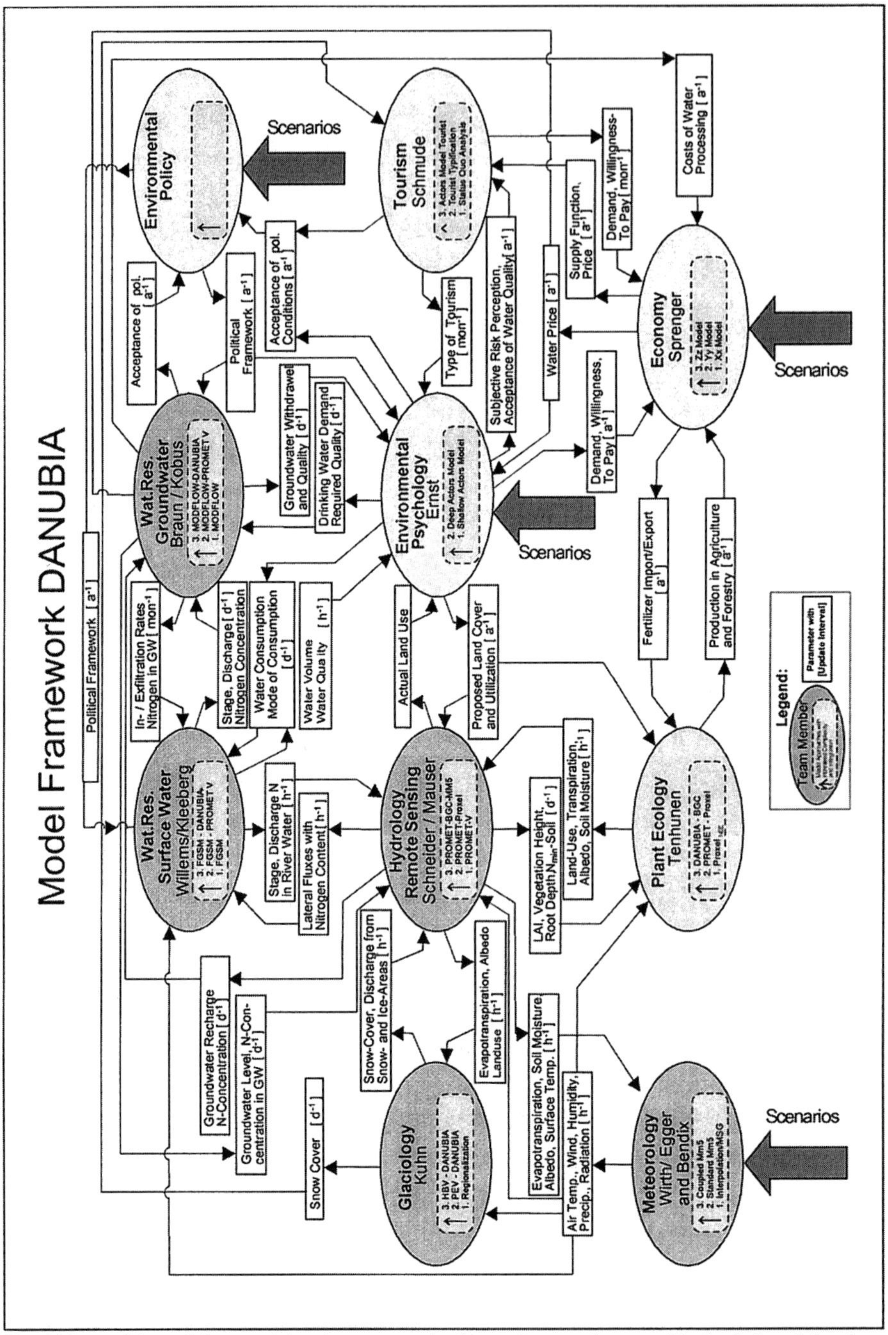

Fig.4:Networking of the participating groups in the project as shown with the help of the data exchange between the groups and gates for the development of scenarios

The development of integration methods and techniques in modelling water flow based on distributed objects in the network with CORBA and formalization of the multi-disciplinary modelling with the Meta-Modelling Language UML is essentially the only possible approach, if the project is to be conceived as a distributed expert network and developed on the basis of re-useable, refineable, and documented sub-models. The applied industry standards are open, exactly designed and coded, excellently documented and expandable. Extensive experience about their applications in industry is available (Jones, C. (1996)). These show that successful inter-disciplinary software projects in the magnitude such as GLOWA-Danube, are characterized by items such as agreements as to interfaces, determined mutual check-ups, exactly defined quality-checks and clearly assigned areas of responsibility. The beginning of the networked development will take place in the areas of hydrology, plant ecology (where it is easiest) and groundwater. The network modelling prototype created, will serve as a fundamental platform, from which further developments can descend.

In order to develop mesoscale models and to interpret the results at the microscale, transition in the process description from microscale to mesoscale and vice versa becomes necessary. This is common to almost all sub-projects as a cross-cutting theme. The complex nature of the investigated objects is supported to a large degree by models of sub-processes already developed in the group, by successes achieved in the group in upscaling from microscales to mesoscale and by the already mentioned commonly agreed upon raster-based model structure, that make successful scaling-techniques relatively easily transferable. Due to this, it is perspected that a large step towards a common description of the processes in the natural and social sciences will finally be possible.

7. REFERENCES

Booch, G., Rumbaugh, J, Jacobson, I.: The Unified Modelling Language User Guide. Addison-Wesley, Reading, 1999

Chen, F., Dudhia, J.: Coupling an advanced land-surface/hydrology model with the Penn State/NCAR MM5 modelling system. Part I: Model description and implementation. [submitted to the Monthly Weather Review], 2000

Chiang, W.-H. & W. Kinzelbach, W.: Processing Modflow - A simulation system for modelling groundwater flow and pollution . Scientific Software Group: Washington, DC, 1996

Henning, M., Vinoski, S.: Advanced CORBA Programming in C++. Addison-Wesley, Reading, 1999

Jones, C: Patterns of Software System Failures and Success, Int. Thomson Computer Press, 1996

Mauser, W., J.D. Tenhunen, K. Schneider, R. Ludwig, T. Stolz, R. Geyer, and E. Falge: Assessing Spatially Distributed Water, Carbon, and Nutrient Balances at Different Scales in Southern Bavaria; in: Tenhunen, J.D. et al. (eds.) Ecosystem Approaches to Landscape Management in Central Europe. Ecological Studies, Vol. 147. Springer Verlag. Berlin, 2000

Michalakes, J., 2000: The Same-Source Parallel Implementation of MM5, Journal of Scientific Computing. Vol. 8, No. 1. pp. 5-12.

Suleiman, R., K.G. Troitzsch, N. Gilbert (Ed.): Tools and Techniques for Social Science Simulation, Physica Verlag, Heidelberg, 2000

Tenhunen, J.D., P. Kabat (Ed.): Integrating Hydrology, Ecosystem Dynamics, and Biogechemistry in Complex Landscapes, John Wiley & Sons, Chichester, 1999

Climate change and water resources in Sweden – analysis of uncertainties

M. GARDELIN, S. BERGSTRÖM, B. CARLSSON, L.P. GRAHAM AND G. LINDSTRÖM
Swedish Meteorological and Hydrological Institute, Norrköping, Sweden

Abstract: Simulations of the impacts of climate change on water resources in Sweden are produced within the Swedish Regional Climate Modelling Programme, SWECLIM. The impact studies are based on a combination of global climate models (GCMs), a regional climate model and a hydrological runoff model. The two different GCMs used so far are the UKMO HadCM2 from the Hadley Centre and the ECHAM4/OPYC3 of the Max Planck Institute for Meteorology. The regional climate model, RCA, was developed at the Rossby Centre of the Swedish Meteorological and Hydrological Institute (SMHI) and is a modified version of the international HIRLAM meteorological forecast model. The RCA model performs downscaling from GCM scenarios on a time horizon of 50 to 100 years. Based on the RCA scenarios, water resources scenarios were produced with the HBV hydrological runoff model developed at the SMHI. Two different methods for estimation of evapotranspiration in the hydrological model were used. Neither of the methods takes into account the possible feedback from changing land-use, vegetation dynamics or changing plant use of water at increasing CO_2 concentrations in the atmosphere. The impacts on water resources were simulated from differences between control runs and scenario runs of the RCA model for a number of selected test basins covering the major climate regions in Sweden. Changes in runoff totals, runoff regimes and extreme values were analysed with focus on the uncertainties introduced by the choice of global climate model, routines for estimation of evapotranspiration in the hydrological model and methods applied in the interface between the models. It was further analysed how these choices affect the statistical return periods of future extremes in a design situation.

M. Beniston (ed.),
Climatic Change: Implications for the Hydrological Cycle and for Water Management, 189–207.

1. INTRODUCTION

The Swedish Regional Climate Modelling Programme (SWECLIM) produces Nordic regional climate scenarios on a time horizon of 50 to 100 years in the future. In the programme are included studies of the impact of climate change on water resources in the region. Earlier water resources scenarios within SWECLIM for basins in Sweden are described by Bergström et al. (2000) and Bergström et al. (2001). Studies for Swedish basins have so far mainly addressed issues related to hydropower production, dam safety and flooding. Graham (2000) and Graham et al. (2001b) describe the hydrological impact of SWECLIM scenarios on the entire Baltic Sea drainage basin. Other regional impact studies for the Nordic countries have previously been reported by Vehviläinen and Lohvansuu (1991), Roos (1996), Vehviläinen and Huttunen (1997), Saelthun et al. (1998), Lemmelä and Helenius (1998) and Saelthun et al. (1999).

This paper gives a detailed description of the methods used for the hydrological impact studies within SWECLIM. Results are presented from the full ensemble of water resources scenarios for Swedish basins that has so far been performed within SWECLIM. The paper includes results from the latest high-resolution water resources scenarios and the analysis focuses on different sources of uncertainty that are associated with the impact study.

2. METHODS

Studies of the impacts of climate change on water resources are produced within SWECLIM by the use of a combination of global climate models (GCMs), a regional climate model and a hydrological runoff model. The regional climate model performs dynamical downscaling of GCM output to produce climate scenarios on a regional scale. The regional modelling makes it possible to obtain higher spatial resolution and to provide more accuracy for certain processes. Of particular importance for the Nordic region is that the treatment of the Scandinavian mountains and regional water bodies, such as the Baltic Sea and the numerous Nordic lakes, can be included in the regional climate model.

For studying the climate change impact on water resources the scenarios furthermore have to be interpreted by a hydrological model as the regional climate models at present lack, among others, detailed representation of snow in mountainous areas and river routing functions. A commonly used strategy in the hydrological modelling of climate change effects is to transfer changes of meteorological variables in the climate model to off-line hydrological simulations (Arnell, 1999; Kaczmarek et al., 1996; Lemmelä and Helenius, 1998; Lettenmaier et al., 1999; Saelthun et al., 1998 and 1999; Vehviläinen and Huttunen, 1997). This method, also used within SWECLIM (Bergström et al., 2000 and Bergström et al., 2001), implies that an interface must be used between the climate model and the hydrological model. Differences in meteorological variables between present climate simulations and future climate scenario simulations are extracted and processed in the interface before use in the hydrological model. Direct use of meteorological output from regional climate models is currently not applicable, as there is at present a bias in both temperature and precipitation in most models. As hydrological models are particularly sensitive to the seasonal distribution of precipitation for example, such biases will strongly affect a hydrological simulation. Previous attempts to use climate model output directly have lead to large discrepancies in modelled river discharge for the present climate (Graham and Jacob, 2000). The indirect use of the meteorological changes does not however exclude uncertainties related to the bias in the present day simulations in the climate models. A future aim within hydrological modelling is therefore to do away with the interface as the climate models develop and improve their ability to accurately represent the variables of the hydrological cycle.

Within SWECLIM, combinations of different GCMs, different model versions of a regional climate model, and variations of a hydrological model are being used to produce a range of various water resources scenarios. The use of this ensemble of simulations, which to date comprises 8 samples, made it possible to analyse the importance of different sources of uncertainty involved in the impact study. The results have been analysed with respect to average conditions (as annual runoff totals and runoff regimes) and with respect to extreme values. Frequency analyses of peak floods in present day climate and under scenario conditions were carried out separately for spring floods (January-July) and for summer or autumn floods (August-December). The analyses were based on the Gumbel probability function estimated by the method of moments (see e.g. Chow et al., 1988).

2.1 Climate models

Two GCMs have thus far been used within SWECLIM to provide driving boundary conditions for the regional climate model. The simulations used are 10-year time slices from the UKMO HadCM2 of the Hadley Centre in Reading (Johns et al., 1997) and the ECHAM4/OPYC3 GCM of the Max Planck Institute for Meteorology in Hamburg (Roeckner et al., 1996; Oberhuber, 1993a;b). The scenario time slice represents a future climate affected by changes in greenhouse gases. The control time slice represents a reference to which the scenario is compared. For HadCM2, the control is the

GCM simulation representing atmospheric CO_2 from the 1950's. The scenario is a transient greenhouse gas simulation. The control and scenario time slices differ by an increase in the atmospheric equivalent CO_2 concentration of 150 percent. For ECHAM4/OPYC3, both of the time slices are from a transient greenhouse gas simulation. The control time slice corresponds to present day conditions (around 1990) in the sense that it is preceded by the forcing changes due to increases in greenhouse gases since 1860. The scenario time slice is taken from the time of doubling of atmospheric equivalent CO_2 concentration compared to the control. Consequently, the forcing from these two sets of GCM simulations are not directly comparable. However, the mean global warming between control and scenario time slices is virtually the same in both cases, i.e. 2.6°C.

The regional climate simulations are made with the Rossby Centre Regional Atmospheric Climate Model (RCA), (Rummukainen et al., 2001), developed at the Rossby Centre of the Swedish Meteorological and Hydrological Institute. The model is based on the international HIRLAM (High Resolution Limited Area Modelling) forecast model (Källén, 1996; Eerola et al., 1997). The RCA model hosts new parameterisation schemes for surface, soil and snow (Graham et al., 2001a) and an improved treatment of regional water bodies, i.e. the Baltic Sea and the Nordic lakes. The model domain covers the Nordic region and extends over most of Europe and the North Atlantic. The model simulations used up to now for water resources simulations have a horizontal resolution of 0.8° (~ 88 km) or 0.4° (~ 44 km). The simulations based on HadCM2 are referred to as RCA88-H and RCA44-H, whereas the ECHAM4 based simulations are referred to as RCA88-E and RCA44-E.

2.2 Hydrological model

The hydrological model used within SWECLIM is the conceptual semi-distributed HBV runoff model (Bergström, 1995; Lindström et al., 1997). The model was originally developed for operational runoff forecasting at the Swedish Meteorological and Hydrological Institute. It has since then been used in various applications in more than 30 countries. Its use in climate change studies has mainly occured in the Nordic countries as documented by Vehviläinen and Lohvansuu (1991), Vehviläinen and Huttunen (1997), Saelthun et al. (1998 and 1999) and Graham (2000).

The HBV model includes routines for snow accumulation and melt, soil moisture accounting, groundwater response and river routing. The snow routine is basically a degree-day approach with a liquid water holding capacity of the snow. Sub-basin variability is provided for by elevation zones, vegetation zones and a statistical distribution of snow to account for patchiness and snowdrifts above the tree line. The soil moisture routine is based on a variability parameter, representing the distribution of storage capacity within a basin, which determines the contribution to runoff at varying soil moisture deficit. The response routine includes two storage boxes with a gradually increasing recession coefficient in the upper box. Major lakes and reservoirs are modelled explicitly by a storage-discharge relationship. Further spatial distribution and smoothing of runoff is achieved by the use of several sub-basins.

Evapotranspiration in the model is calculated as a function of soil moisture deficit from estimates of potential evapotranspiration. In general, either monthly mean values calculated with the Penman formula (Penman, 1948) or a simple temperature index method is used to estimate potential evapotranspiration. In the SWECLIM climate change simulations, the evapotranspiration in the future climate has been estimated by two different methods. In the first method (HBV-a) a simplified version of the Thornthwaite formulation (Thornthwaite, 1948) was applied. Potential evapotranspiration was in this method (Lindström et al., 1994) proportional to air temperature by a seasonally dependent coefficient (Saelthun et al., 1998). The method was applied both for today's climate and for scenario conditions. Since temperature index methods do not take into account possible changes in humidity, radiation or other significant variables and can be questioned in climate change applications, an alternative method (HBV-b) was introduced. For today's climate in the HBV-b simulations, potential evapotranspiration was based on long term monthly estimates with the Penman formula. For scenario conditions, the future changes in actual evapotranspiration from the regional climate model were transferred to the HBV model simulations. This was done by adjusting the monthly potential evapotranspiration input to the HBV-b model simulations so as to produce the same relative changes in actual evapotranspiration as given by the RCA model. Neither of the two methods applied in the study takes into account the possible feedback from changing land-use, vegetation dynamics or changing plant use of water at increasing CO_2 concentrations in the atmosphere.

2.3 Model interface

Changes to meteorological variables, between present climate simulations (also referred to as control simulations) and scenario simulations, were in the model interface extracted from the regional climate model to be transferred to off-line hydrological model simulations.

In the RCA88 simulations the meteorological variables were extracted as an average over grid-boxes covering the specific test basin. Monthly average changes in precipitation and temperature between the scenario and the control simulation of the RCA model were extracted and smoothed to 3-monthly average running mean values. In this way, standard sets of 12 monthly values were created that were then applied to all years of the hydrological simulation representing scenario conditions. The monthly changes in temperature were added to the daily temperature values of an observed climate database used in the hydrological simulations. The average monthly changes in precipitation were applied as a percentage change from the present climate. This method thus assumes the same relative changes for all years and for extreme values as well as for average conditions. It further assumes that the number of days with precipitation stays the same in the future climate. In the HBV-b simulations the relative changes in evapotranspiration were transferred to the hydrological model as a percentage change on a yearly basis. Further descriptions of the methods used in the RCA88 simulations are given by Bergström et al. (2001).

To reduce the influence of local divergence, the analyses of meteorological variables in the RCA44 simulations were made over larger regions. All meteorological variables in these simulations were integrated over three geographical regions that represent 1) the northern Scandinavian Mountains, 2) northern Sweden excluding the mountains and 3) southern Sweden. In the RCA44 simulations another important modification was introduced in the interface to account for the variation of temperature change with the magnitude of temperature. Analysis of the RCA model scenarios shows that changes in temperature are strongest at low temperatures and less pronounced at higher temperatures. Therefore, a temperature dependent linear function for applying the temperature change to the climate database was introduced. The resulting mean annual temperature change in the HBV model was set to equal the average temperature change of the RCA model over the specific region. This modification has an important impact on the hydrological response as it affects snow accumulation and snowmelt. A similar analysis of changes to precipitation frequency in the climate model was considered too uncertain to be quantified in the hydrological simulation. Therefore, the precipitation changes in the RCA44 scenarios were transferred to the HBV model in the same way as in RCA88, except regarding the region size for data extraction.

3. DATABASE

The hydrological simulations were carried out for six test basins of the size 1700-6000 km^2, which represent different climate regions in Sweden (Figure 1). Hydropower plants are located in all the selected rivers, but impact studies in the four northernmost basins are of particular interest for the hydropower industry. The Suorva reservoir in test basin 1, with a storage capacity of 6 km^3, is the most important reservoir in the Swedish hydropower system. The HBV model was calibrated automatically, with the routine developed by Lindström (1997), against an observed database of river discharge for the period 1967-97. This period was also used for the HBV simulations representing today's climate and as the initial meteorological database for the simulations representing scenario conditions.

The use of two different GCMs for external boundary forcing (H and E), two different horizontal resolutions of the RCA model (RCA88 and RCA44) and two variations of the HBV model (HBV-a and HBV-b) resulted in an ensemble of eight different water resources scenarios (RCA88-H & HBV-a, RCA88-E & HBV-a, RCA44-H & HBV-a, RCA44-E & HBV-a, RCA88-H & HBV-b, RCA88-E & HBV-b, RCA44-H & HBV-b and RCA44-E & HBV-b). The RCA model average changes in temperature and precipitation that were used as input to the hydrological simulations are summarised in Table 1.

Table 1. Changes in annual mean temperature and precipitation over Sweden in the different RCA scenarios used for hydrological simulations.

	RCA88-H	RCA88-E	RCA44-H	RCA44-E
Temperature (°C)	+3.2	+3.5	+3.7	+3.8
Precipitation (%)	+13	+7	+18	+8

4. WATER RESOURCES SCENARIOS

4.1 Average conditions

Figure 1 summarises results from the eight different water resources scenarios for the Swedish test basins as described above. The curves illustrate the effects of the climate change scenarios on the seasonal variation of runoff. A general tendency, which can be seen in all the impact simulations in northern Sweden, is the shift in the runoff regime towards decreasing spring flows and increasing flows during autumn and winter. Only in the northernmost basin will the magnitude of runoff in spring be unchanged since snow accumulation will not be significantly affected in this region. The timing of snowmelt will however be affected in all the northern basins. A decrease in spring flows can also be noticed in southern Sweden, while the impact simulations in this region give diverging results for winter and autumn.

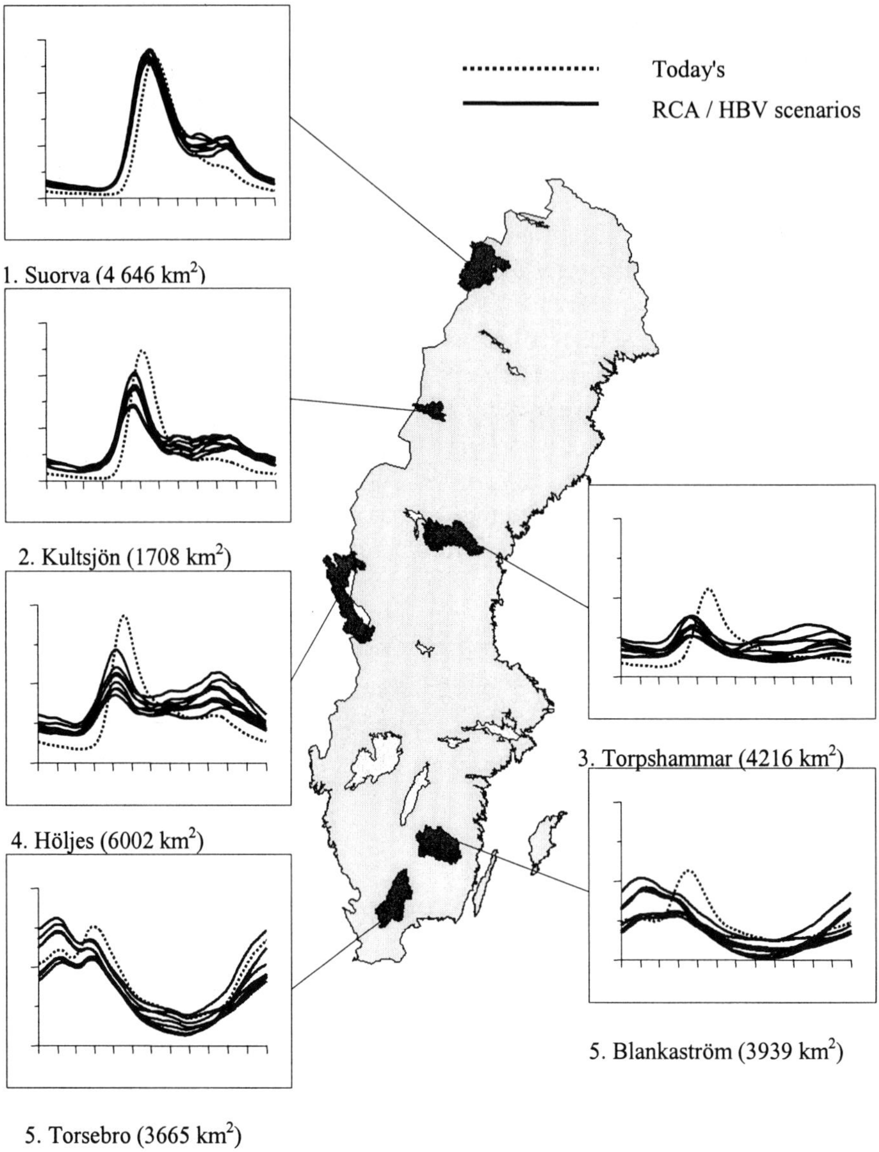

Figure 1. Average calculated daily runoff in six test basins in Sweden for today's climate and according to eight different climate change impact simulations. Daily runoff has been smoothed over 30 days.

The total range of the different runoff simulations illustrates some of the uncertainties in the impact study. Generally, the spread of the runoff curves is larger towards the south of the country. The simulations of summer and autumn runoff are particularly unstable in all basins. The changes to annual runoff totals, summarised in Table 2, indicate the same tendency concerning the geographical distribution of impact uncertainties. The simulations show on average increasing annual runoff in the 4 northern basins (~10-25%), whereas annual runoff on average decreases in the two southernmost basins (~10-20%). The spread of the runoff totals in the different simulations is however significantly smaller in the northernmost basins. The largest annual runoff change is generally found in the HBV-b simulations based on the RCA44-H scenarios.

Table 2. Changes in annual runoff (%) for six test basins in Sweden. Climate change impact simulations based on boundary forcing from two different GCMs (H=HadCM2; E=ECHAM4), two different resolutions in the RCA model and two versions of the HBV model.

		RCA88		RCA44	
		HBV-a	HBV-b	HBV-a	HBV-b
1. Suorva	H	+25	+30	+29	+33
	E	+23	+28	+18	+28
2. Kultsjön	H	+13	+23	+22	+32
	E	+11	+22	+14	+24
3. Torpshammar	H	+10	+38	+5	+41
	E	-19	+2	-12	+14
4. Höljes	H	+1	+14	+16	+40
	E	-1	+12	+5	+29
5. Blankaström	H	-41	-10	-21	+16
	E	-38	-21	-34	-22
6. Torsebro	H	-21	+3	-15	+14
	E	-26	-11	-26	-13

The different sources of uncertainty are set in focus by reviewing a selection of scenarios within the full ensemble. The influence of the choice of global climate model used for boundary conditions is illustrated in Figure 2. The graphs show two different water resources scenarios based on identical hydrological and regional climate models, but with boundary forcing from two different GCMs.

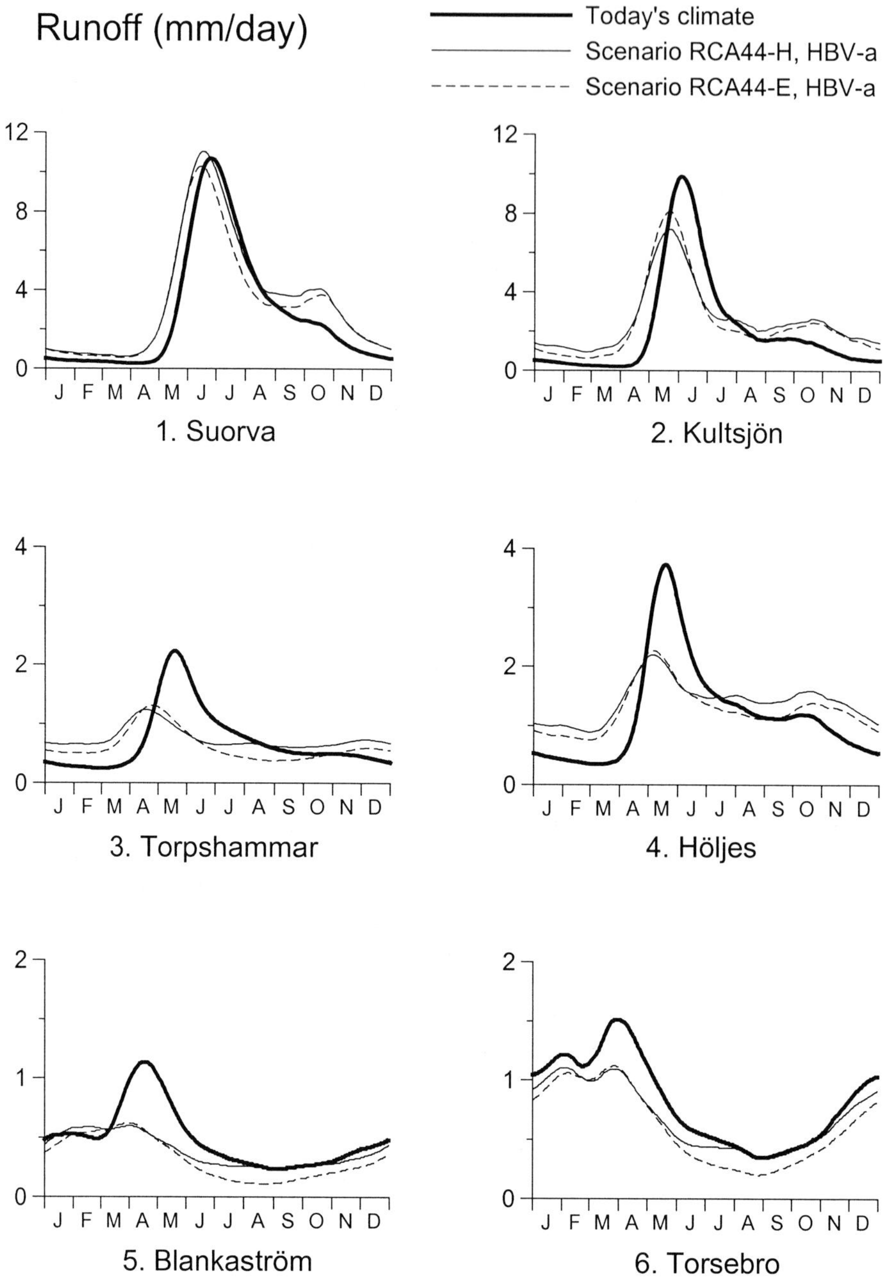

Figure 2. Average calculated daily runoff in six test basins in Sweden for today's climate and according to climate change impact simulations based on boundary forcing from two different GCMs (H=HadCM2; E=ECHAM4). Daily runoff has been smoothed over 30 days.

The impact on runoff differs particularly during summer and autumn, although snowmelt during spring is also affected by the choice of GCM. In southern Sweden the ECHAM4 scenario results in a significant decrease of low flows during summer, which can not be observed in the HadCM2 scenario. The differences in the precipitation increase and distribution in the global scenarios thus have an obvious impact on runoff totals.

Figure 3 shows two water resources scenarios based on identical hydrological and global climate models, but using different regional climate models and different methods in the model interface. The RCA44 has a higher horizontal resolution and a further developed parameterisation scheme than RCA88. In this example, the most significant difference in the impact on runoff can be observed as less pronounced decreases of spring flows in the RCA44 simulation as compared to RCA88. The results illustrate the apparent effects of changes to the model interface, which for this case had an important impact on snow accumulation and snowmelt in the northern basins. The differences in summer runoff in the southern basins were however more related to the diverging changes to precipitation in the RCA88 and RCA44 scenarios.

The impact on runoff from different formulations of evapotranspiration in the hydrological model is indicated in Figure 4. The graphs show two different simulations based on identical global and regional climate models, but with different versions of the hydrological model. The model version based on the temperature index method (HBV-a) yields a considerable decrease of runoff as evapotranspiration is strongly affected during scenario conditions. In the example based on HadCM2 (Figure 4), the temperature index method (HBV-a) resulted in a decrease of runoff of approximately 15-20% in southern Sweden, while HBV-b resulted in an increase of approximately 15%. The influence of the choice of hydrological model version on the annual cycle of runoff is thus strong in most basins and affects summer and autumn flows even in the northernmost basins.

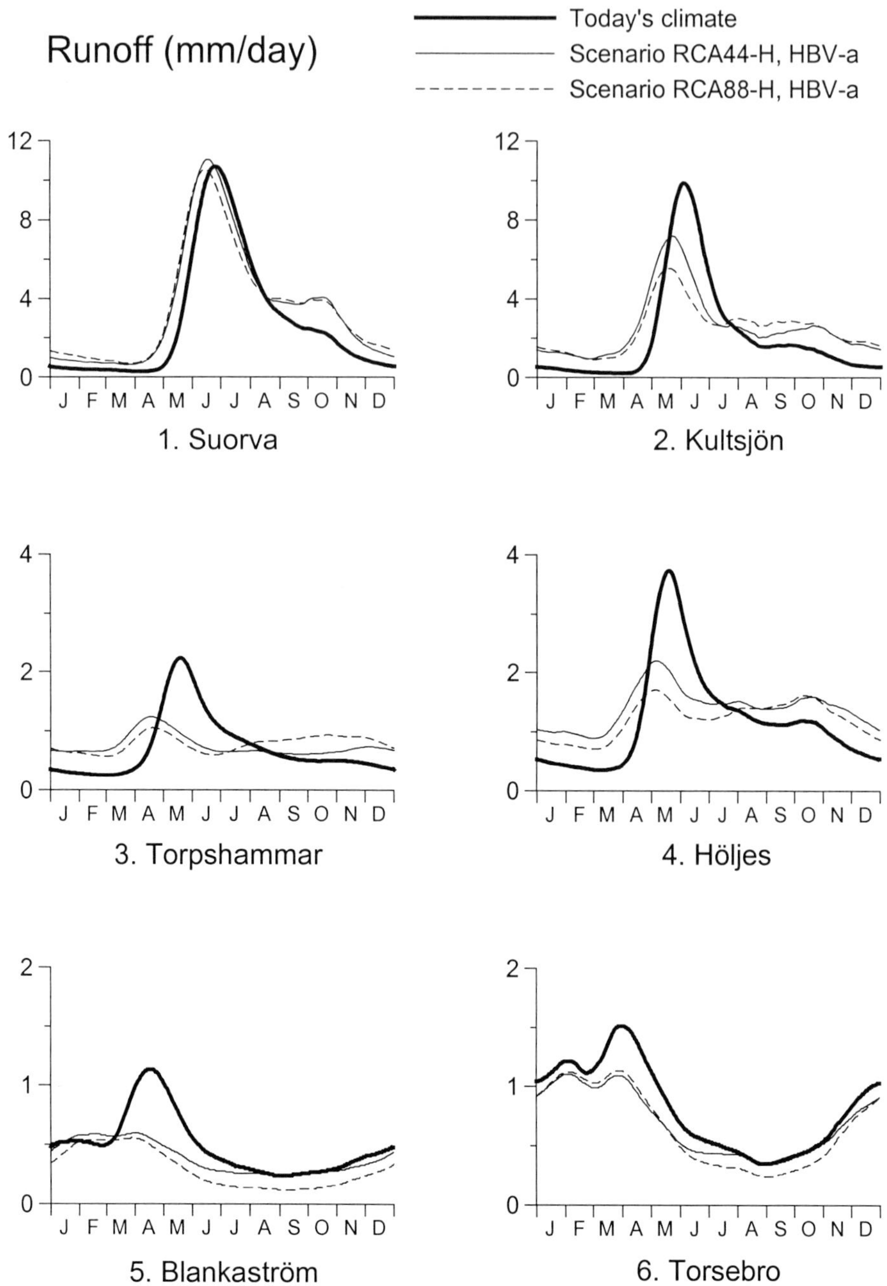

Figure 3. Average calculated daily runoff in six test basins in Sweden for today's climate and according to climate change impact simulations based on two different resolutions of the RCA model and using different methods in the model interface. Daily runoff has been smoothed over 30 days.

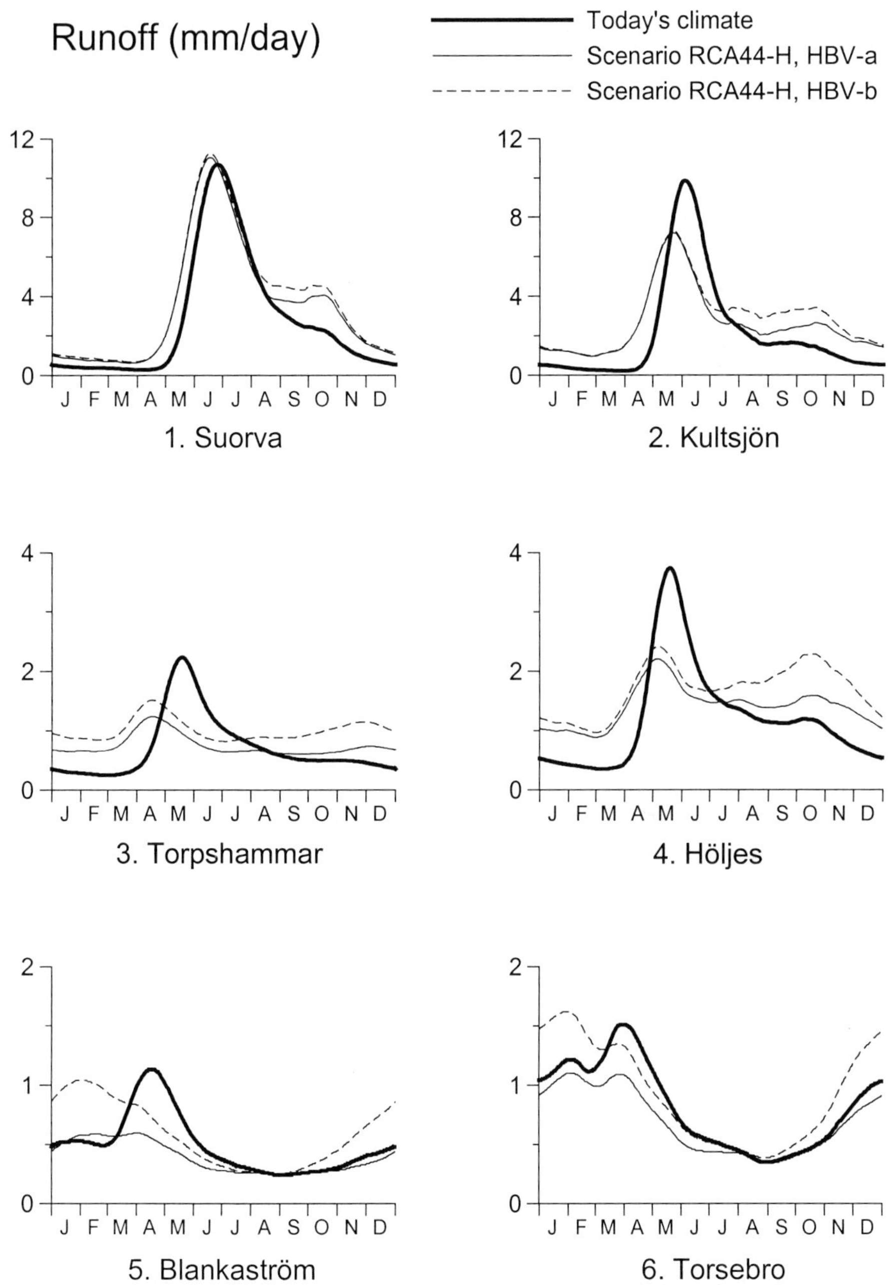

Figure 4. Average calculated daily runoff in six test basins in Sweden for today's climate and according to climate change impact simulations based on two different versions of the HBV model. Daily runoff has been smoothed over 30 days.

4.2 Extreme values

Table 3. Changes in 100-year floods and return periods for six test basins in Sweden. Scenarios based on two different GCMs (H=HadCM2; E=ECHAM4), two different resolutions in the RCA model and two versions of the HBV model.

		Change in 100-year flood (%)				Return period of today's 100-year flood under scenario conditions			
		RCA88		RCA44		RCA88		RCA44	
		HBV -a	HBV -b	HBV -a	HBV -b	HBV -a	HBV -b	HBV -a	HBV -b
Spring (January-July):									
1. Suorva	H	0	0	+2	+2	100	100	80	80
	E	-1	-1	+1	+5	110	110	90	60
2. Kultsjön	H	-28	-25	-20	-17	>1000	>1000	710	500
	E	-24	-22	-15	-12	>1000	870	400	300
3. Torp-shammar	H	-44	-37	-47	-37	>1000	>1000	>1000	>1000
	E	-51	-46	-43	-36	>1000	>1000	>1000	>1000
4. Höljes	H	-48	-44	-42	-36	>1000	>1000	>1000	>1000
	E	-46	-43	-43	-31	>1000	>1000	>1000	>1000
5. Blankaström	H	-51	-35	-49	-28	>1000	>1000	>1000	>1000
	E	-46	-32	-48	-36	>1000	>1000	>1000	>1000
6. Torsebro	H	-21	-7	-29	-10	700	170	>1000	230
	E	-20	-8	-27	-13	640	190	>1000	320
Summer and autumn (August-December):									
1. Suorva	H	+31	+36	+45	+49	20	10	10	10
	E	+27	+32	+23	+42	20	20	30	10
2. Kultsjön	H	+36	+48	+44	+56	20	10	10	5
	E	+13	+28	+17	+31	50	20	40	20
3. Torp-shammar	H	+58	+82	+41	+88	10	5	20	5
	E	-12	+7	0	+41	240	70	100	10
4. Höljes	H	+27	+40	+50	+92	30	20	10	5
	E	+21	+35	+29	+74	30	20	20	5
5. Blankaström	H	-33	-5	-11	+21	>1000	140	240	30
	E	-22	-2	-27	0	550	110	>1000	100
6. Torsebro	H	-23	-8	-21	+1	600	170	570	90
	E	-24	-12	-27	-10	670	230	>1000	210

Table 3 shows the relative changes of the 100-year spring floods and summer/autumn floods in the future climate according to the different impact simulations. The analyses indicate decreasing spring floods (approximately between 10-50%) in all basins except in the northernmost one. Floods during the rest of the year generally increase in the four northern basins. The largest increases (up to approximately 90%) are found in the HBV-b simulations based on RCA44-H. The analysis shows very diverging results for southern Sweden, however summer/autumn floods decrease in most scenarios for this region. The shift in northern Sweden from spring floods towards increased floods during the rest of the year can be critical for reservoirs and dams in regulated rivers. Reservoirs are often low in spring but autumn and winter floods may occur when reservoirs are full and without capacity for flow dampening. Table 3 also shows expected return periods of the present-climate 100-year flood in the future climate. These estimates are particularly uncertain as they involve a long extrapolation of the frequency curve. In general, the return period for today's 100-year spring flood far exceeds 100 years under scenario conditions (the Suorva basin excluded). The return period for floods during the rest of the year generally decreases in the four northern basins and increases in the southern basins.

5. DISCUSSION AND CONCLUSIONS

The use of ensembles of differently structured impact simulations illustrates some of the many uncertainties involved in this type of climate change study. The application including different global and regional climate models, different methods of estimating evapotranspiration and different techniques in the model interface shows a possible range of the impacts, even though the importance of different sources of uncertainty is difficult to quantify.

A fundamental source of uncertainty in the process lies in the global climate scenarios, which are affected by the emission assumptions, how these result in atmospheric concentrations, the resulting radiative forcing changes and the climate sensitivity to these. An important issue is the procedure for extraction of time slices to represent climate change in different models. Use of 10-year time slices also results in poor representation of natural climate variability. Different global scenarios further differ more in their description for particular regions than for global mean results. There are, for instance, considerable differences in the distribution of changes to precipitation over the Nordic region between different global scenarios. The different parameterisations used in both global and regional models, and how well they represent the present climate, add further uncertainties. To thoroughly address the question of uncertainty in climate models, additional simulations with forcing from several different models should be used in the hydrological simulations to produce a wider range of potential outcomes.

Besides the uncertainties in the climate models, further uncertainties are generated in the hydrological modelling. The calibration of the hydrological model parameters to present conditions is for example affected by the choice of reference period. Furthermore, the same parameters are assumed to represent future conditions, although changes in both climate and landscape will occur. Extrapolation outside the calibration range is thus inevitable in climate change studies. The uncertainty due to parameter identification even for the present climate is another factor, as discussed by Uhlenbrook et al. (1999). The structure of the hydrological model itself and its representation of different processes, as for example evapotranspiration, proved to have a large impact in the present study. Another fundamental source of uncertainty is the possible feedback from changing land-use or from changing plant use of water at increasing CO_2 concentrations in the atmosphere. The processing of the hydrological modelling results for user impacts studies, such as frequency analysis of floods, introduces further uncertainties.

The technique used in the interface between climate models and hydrological models forms yet another source of uncertainty. Direct use of output from climate model scenarios in hydrological simulations is not currently advisable, as the control runs of the climate models still do not accurately match seasonal climatic observations, primarily precipitation. A future aim within hydrological modelling is to be able to use meteorological variables directly from the climate models and thus reduce smoothing that may occur through the interface processing. In the meantime, improvements of the technique in the model interface are essential. Such modifications should attempt to account for eventual changes in extreme values or changes in precipitation frequency.

The aim within SWECLIM is to communicate the existing uncertainties directly to the users of the impact studies and to thereby include uncertainties as a base for decision making. The applicability as well as the limitations of the climate change impact studies could be better understood through a close dialogue with the users. The SWECLIM impact studies are therefore regularly presented to representatives of companies, authorities and other organisations in affected sectors, such as hydropower and flood protection.

6. ACKNOWLEDGEMENTS

The global climate simulations were provided by the Hadley Centre in Reading and the Max-Planck Institute for Meteorology in Hamburg. Staff at the Rossby Centre carried out the regional climate simulations. SWECLIM is funded jointly by the Swedish Foundation for Strategic Environmental Research (MISTRA) and the Swedish Meteorological and Hydrological Institute. Additional financial support was supplied by the Swedish Electrical Utilities R&D Company (Elforsk), the Swedish dam safety authority (Svenska Kraftnät) and the Swedish Rescue Services Agency (Räddningsverket).

7. REFERENCES

Arnell, N.W. (1999) The effect of climate change on hydrological regimes in Europe: a continental perspective. Glob Environ Change 9, 5-23

Bergström, S. (1995) The HBV model. In: Singh VP (ed) Computer Models of Watershed Hydrology. Water Resources Publications, Highlands Ranch, Colorado, p 443-476

Bergström, S., Carlsson, B., Gardelin, M., Lindström, G. and Pettersson, A. (2000) Water resources simulations based on climate change scenarios within SWECLIM. XXI Nordic Hydrological Conference. NHP Report No. 46, 315-322

Bergström, S., Carlsson, B., Gardelin, M., Lindström, G., Pettersson, A. and Rummukainen, M. (2001) Climate change impacts on runoff in Sweden - assessments by global climate models, dynamical downscaling and hydrological modelling. Climate Research 16, 101-112

Chow, V.T., Maidment, D.R. and Mays, L.W. (1988) Applied Hydrology. McGraw-Hill, New York, USA

Eerola, K., Salmond, D., Gustafsson, N., Garcia-Moya, J.-A., Lönnberg, P. and Järvenoja, S. (1997) A parallel version of the HIRLAM forecast model: Strategy and results. In: Hoffmann G-R, Kreitz N (eds) Making its mark. Proceedings of the seventh ECMWF Workshop of the use of Parallel Processors in Meteorology, Reading, UK, November 1996, p134-143

Graham, L.P. (2000) Large-scale hydrological modeling in the Baltic basin. Division of Hydraulic Engineering, Dept of Civil and Environmental Engineering, Royal Institute of Technology, PhD thesis TRITA-AMI PHD 1033, Stockholm

Graham, L.P. and Jacob, D. (2000) Using large-scale hydrologic modeling to review runoff generation processes in GCM climate models. Meteorologische Zeitschrift 9, 49-57

Graham, L.P., Lindström, G., Bringfelt, B., Gardelin, M., Gollvik, S., Bergström, S. and Samuelsson, P. (2001 a) Using conceptual hydrological modelling to develop better subgrid variability in the Rossby Centre regional atmospheric model. Contribution to the 6th Scientific Assembly of the IAHS, Maastricht 18-27 July

Graham, L.P., Rummukainen, M., Gardelin, M. and Bergström, S. (2001 b) Modelling climate change impacts on water resources in the Swedish regional climate modelling programme. In: Detecting and Modelling Regional Climate Change and Associated Impacts, Springer-Verlag (in press).

Johns, T.C., Carnell, R.E., Crossley, J.F., Gregory, J.M., Mitchell, J.F.B., Senior, C.A., Tett, S.F.B. and Wood, R.A. (1997) The second Hadley Centre coupled ocean-atmosphere GCM: model description, spinup and validation. Clim Dyn 13, 103-134

Kaczmarek, Z., Strzepek, K.M., Somlyódy, L. and Priazhinskaya, V. (Eds) (1996) Water Resources Management in the Face of Climatic/Hydrologic Uncertainties. Water Science and Technology Library. Kluwer Academic Publishers, Dordrecht, 395 pp.

Källén, E. (ed) (1996) HIRLAM documentation manual. System 2.5, Norrköping

Lemmelä, R. and Helenius, N. (eds) (1998) Proceedings of the Second International Conference on Climate and Water, Espoo, Finland, 17-20 August

Lettenmaier, D.P., Wood, A.W., Palmer, R.N., Wood, E.F. and Stakhiv, E.Z. (1999) Water resources implications of global warming: A U.S. regional perspective. Climatic Change 43, 537-579

Lindström, G., Johansson, B., Persson, M., Gardelin, M. and Bergström, S. (1997) Development and test of the distributed HBV-96 hydrological model. J Hydrol 201, 272-288

Lindström, G., Gardelin, M. and Persson, M. (1994) Conceptual modelling of evapotranspiration for simulations of climate change effects. SMHI RH 10, Norrköping

Lindström, G. (1997) A simple automatic calibration routine for the HBV model. Nord Hydrol 28, 153-168

Oberhuber, J.M. (1993a) Simulation of the Atlantic circulation with a coupled sea ice-mixed layer-isopycnal general circulation model. Part I: Model description. J Phys Oceanogr 23, 808-829

Oberhuber, J.M. (1993b) The OPYC ocean general circulation model. Tech. Rep., 7, Deutsches Klimarechenzentrum GmbH, Hamburg

Penman, H.L. (1948) Natural evapotranspiration from open water, bare soil and grass. Proc. R. Soc. London, Ser. A, 193, 120-145

Roeckner, E., Arpe, K., Bengtsson, L., Christoph, M., Claussen, M., Dümenil, L., Esch, M., Giorgetta, M., Schlese, U. and Schulzweida, U. (1996) The atmospheric general circulation model ECHAM-4: Model description and simulation of present-day climate. Max-Planck-Institut für Meteorologie, Report No 218, Hamburg

Roos, J. (ed) (1996) The Finnish Research Programme on Climate Change, Final report. Publications of the Academy of Finland 4/96, Helsinki

Rummukainen, M., Räisänen, J., Bringfelt, B., Ullerstig, A., Omstedt, A., Willén, U., Hansson, U. and Jones, C. (2001) A regional climate model for northern Europe - model description and results from the downscaling of two GCM control simulations. Clim Dyn (in press)

Saelthun, N.R., Aittoniemi, P., Bergström, S., Einarsson, K., Jóhannesson, T., Lindström, G., Ohlsson, P.-E., Thomsen, T., Vehviläinen, B. and Aamodt, K.O. (1998) Climate change impacts on runoff and hydropower in the Nordic countries. Final report from the project "Climate Change and Energy Production" Tema Nord 1998, 552, Oslo

Saelthun, N.R., Bergström, S., Einarsson, K., Jóhannesson, T., Lindström, G., Thomsen, T. and Vehviläinen, B. (1999) Potential impacts of climate change on floods in Nordic hydrological regimes. In: P. Balabanis, A. Bronstert, R. Casale and P. Samuels (Editors), Proceedings from the Ribamod - River Basin Modelling, Management and Flood Mitigation Concerted Action - Final Workshop. Wallingford, UK, 26-27 February 1998, 103-115

Thornthwaite, C.W. (1948) An approach toward a rational classification of climate. Geogr Rev 38, 55-94

Uhlenbrook, S., Seibert, J., Leibundgut, C. and Rodhe, A. (1999) Prediction uncertainty of conceptual rainfall-runoff models caused by problems to identify model parameters and structure. Hydrol Sci J 44, 779-797

Vehviläinen, B. and Lohvansuu, J. (1991) The effects of climate change on discharges and snow cover in Finland. Hydrol Sci J 36, 109-121

Vehviläinen, B. and Huttunen, M. (1997) Climate change and water resources in Finland. Boreal Environment Research 2, 3-18

Hydrological cycle changes in China's large river basin: The Yellow River drained dry

CHANGMING LIU AND HONGXING ZHENG
Institute of Geographical Sciences and Natural Resources Research, Chinese Academy of Sciences, Beijing Normal University, Beijing, China

Abstract: The Yellow River is one of the most important rivers in China. In the recent years, the hydrological cycle of the basin have changed greatly. Particularly, the main course of the lower reaches has been drying up which has caused a series of ecological, environmental and socioeconomical problems. In this paper, we have tried to explain the reasons for the hydrological cycle changes regarding the physical aspects and human activities. On the basis of the research on the characteristics of hydrological cycle and water resources in the basin, we find that the climate change impact may be one of the reasons. However, the prevailing impacts on the water resources were from human activities associated with land use and land cover change in the river basin. According to the research, some major countermeasures were suggested to enhance water management as the adaptive approach. Finally the authors addressed the highlights of hydrological cycle regarding water resources development and major conclusion remarks.

1. GEOGRAPHICAL BACKGROUND

It is well known that the Yellow River, or Huang He, is one of the most important rivers in China. Throughout history the river can be seen as the cradle of Chinese civilization. Yellow River is located in the north part of China (fig.1). It flows from west to east, passing through nine provinces or automatic regions as Qinghai, Sichuan, Gansu, Ningxia, Inner Mongolia, Shaanxi, Shanxi, Henan and Shandong. As the outlet, the Yellow River reaches the Bohai Sea in Kenli County of Shandong province.

M. Beniston (ed.),
Climatic Change: Implications for the Hydrological Cycle and for Water Management, 209–224.

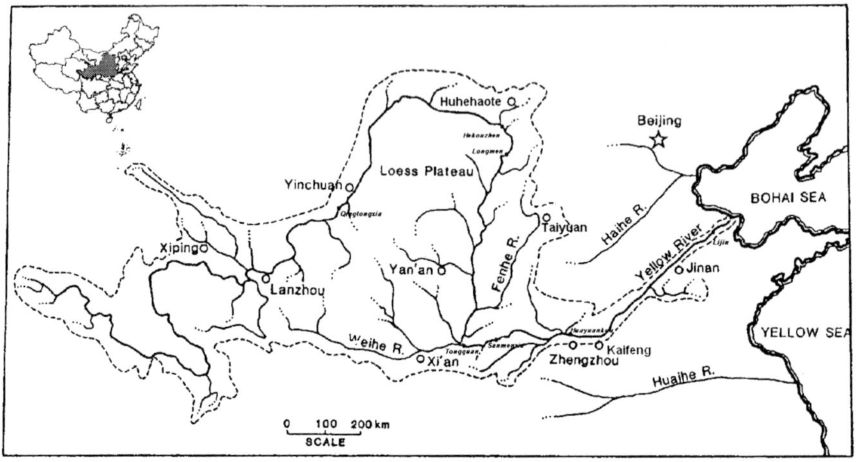

Figure 1. A Sketch Map of the Yellow River

The basin is located in the geographic region of 96^0E-119^0E, 32^0N-42^0N, the east to west length is about 1,900 km, and the north to south width is 1,100 km (Chen Xiande, 1996). The length of the Yellow River and the area of the basin keep changing since the sedimentation process, which also makes the river mouth keep extending year by year in the past. According to the survey result of 1973, the Yellow River is 5,464 km with area of 752,443 km^2, the watershed area may also be considered as 794,712 km^2 if the Erdos inner flow area is included (fig.1).

The population in the basin is 107 million (1997) with an average population density of about 143 per km^2. The area of farmland totals 12.6 million ha, of which about 40% are under irrigation using the Yellow River's water resources. Per capita grain production reached 352 kg in 1997. About 70% of the grain yield produced by irrigated lands. The urban population of about 25.1 million comprises 23.4% of the total. Per capita GDP was 4,530 Yuan equivalently to about $550 USD (Yellow River Commission, 1996). Water resources in the basin total 71.9 billion m^3 including annual average runoff of 58 billion m^3 and groundwater of 13.9 m^3 (Xi Jiazhi, 1996), while the Yangtze River has an annual average runoff of around 1,000 billion m^3. The Yellow River is characterized by its hyper-concentrated sediment-load (Liu Changming and Liang Jiyang, 1988). The measured annual sediment runoff during 1919-1996 averaged 1.69 billion tons, which mainly comes from the Loess Plateau of the Yellow River's middle reaches (fig.1).

Recently, heavy industries have been developed in many cities in the basin and highly productive oilfields at Zhongyuan and Shengli have been exploited in the river's lower reaches and estuary, respectively. Thus, the management of the Yellow River is significant in the nation's economy.

As for hydropower exploitation, the Yellow River is fairly well developed. There are several high dams with hydropower facilities on the river's main course distributed in its upper and middle reaches. Among seven hydropower projects the Longyangxia is the largest one, with a dam height of 175 m, storage capacity of 24.7 billion m^3 and installed capacity of 1,280 MW (Liu Changming, 1989).

2. A GREAT CHANGE IN THE LOWER YELLOW RIVER RUNOFF

2.1 The Lower Yellow River Drained Dry.

Since 1970s, the flow of the Yellow River has broken down frequently in its lower reaches. During the 27 years from 1972 to 1998, the drying up course had appeared 21 times (Liu Changming, Chen Li, 2000). It means four times every five years (see Table 3). The condition has become more austerity since 1990s. During the 1990s, the channels of the lower Yellow River dried up every year. From year to year, it appeared earlier and lasted longer (Yellow River Commission, 1997). Moreover the river channel drying up segment has continuously extended towards the upper part (Table 3). The average channel length of drying up segments in the lower reaches during 1970s was 242km, it increased to 256km in 1980s while it has been extended to 392km in 1990s. In 1997, parts of the river dried up almost throughout the whole year and flow in the lower river channel had dried up 13 times, which had added up to 226 days. In the serious drought year in 1997, by the hydrological observation at Lijin Station in Shandong Province, it had lasted for 330 days without any freshwater flowing into the sea. This year the dried up segments of river course extended near to Liuyuankou in Kaifeng City, Henan Province. The length of the drying up segments was 704 kilometers and accounted for 90 percent of the river course in the lower reaches. Furthermore, in the middle reaches of the river, the flows of the primary tributaries such as Qinhe River, Yihe River, Fenhe River, Dawenhe River, Yanhe River and Weihe River were also drained dry subsequently. The Yellow River is shrinking.

According to the bearing capacity of the Yellow River and the social and economic development of the provinces those fetch water from it, we may concluded that it is urgent time to control the water diversions along the upstream river courses. Otherwise, with further developing economy, the demand of water will keep increasing, which may cause drying up river channels in downstream reaches continuously. Moreover, as a result of the sharpening conflicts among all kinds of water utilization, such as economic development, sediment transition, water-and-soil conservation and ecological protection in river's estuary and delta areas, the extension of drying up channel may be enlarged and the situation may be worsened in space and time. In short, the drying up lower Yellow River is in a very austere condition.

Table 1. Records of the drying up channels in the Lower reaches of the Yellow River at Lijin Hydrological Station

Year	Times	Lasting Time (d)	Length (km)	Year	Times	Lasting Time (d)	Length (km)
1972	3	19	310	**1986**	0	0	0
1973	0	0	0	**1987**	2	17	216
1974	2	20	316	**1988**	2	5	150
1975	2	13	278	**1989**	3	24	277
1976	1	8	166	**1990**	0	0	0
1977	0	0	0	**1991**	2	16	131
1978	4	5	104	**1992**	5	83	303
1979	2	21	278	**1993**	5	60	278
1980	3	8	104	**1994**	4	74	308
1981	5	36	662	**1995**	3	122	683
1982	1	10	278	**1996**	7	136	579
1983	1	5	104	**1997**	13	226	704
1984	0	0	0	**1998**	16	142	(490)
1985	0	0	0	**1999**	6	42	(300)

2.2 Changes of the Annual Runoff Coefficient

The annual runoff can be seasonally divided into two parts, namely rainy period in a year (from June to September) and dry period (from October to May across years). Thus, we obtained three time series including series of rainy periods, series of dry period and the annual series. By using the data of 48 years from 1950 to 1997, the three series of runoff formed three annual curves illustrating changing processes of the river runoff (fig. 2).

From Figure 2 we can see that the runoff (R) measured at Huayuankou Hydrological station in the lower reaches of the Yellow River for last 48 years since 1950 has been apparently decreasing.

The annual runoff coefficient(C), which links precipitation (P) and runoff(R) and calculated according to the expression C=R/P, may be accepted as an index to describe the character of hydrological cycle in a certain region. Changes of the annual runoff coefficient may means a change of hydrological cycle.

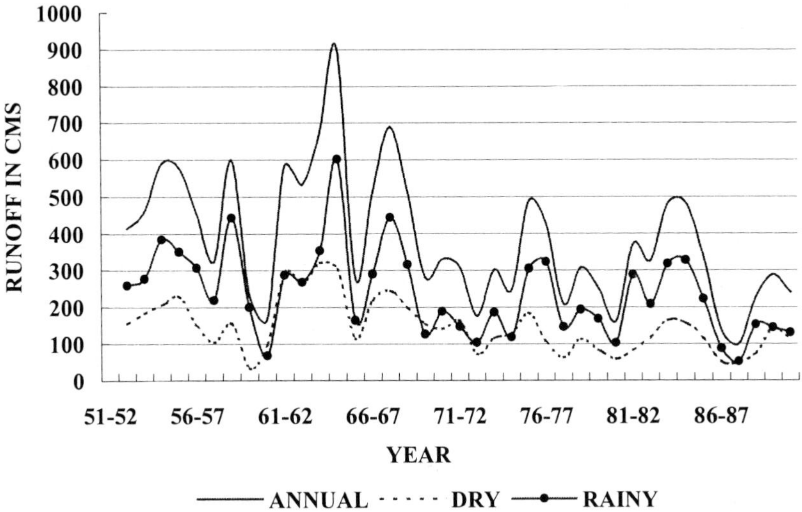

Figure 2. Changes in Yellow River Runoff Measured at Huayuankou Station of Lower Yellow River in Last Decades (1950-1997)

According to the meteohydrological records of the basin, we have chosen two hydrometric stations (drainage area control section) to calculate the annual runoff coefficient of two special regions. One of the station is the Lanzhou Station (Fig.1), which represents the upper basin, the other is the Huayuankou station (near Kaifeng, Fig.1), which used to calculate the coefficient of the region between Sanmenxia and Huayuankou. The results have shown that the coefficients of both two regions all tend to decrease, while that of the Sanmenxia-Huayuankou region is in a more obvious trend (Fig.3). For the whole region, it is evident that the hydrological cycle feature has been changed by even prevailing evapotranspiration (E). The ten years (decade) mean value of runoff coefficient has decreased with an increase of evapotranspiration coefficient (E/P) (Table 2). The decrease of the annual runoff coefficient has implied an increasing in evapotransplation which may be caused by climatic change or land use and land cover change.

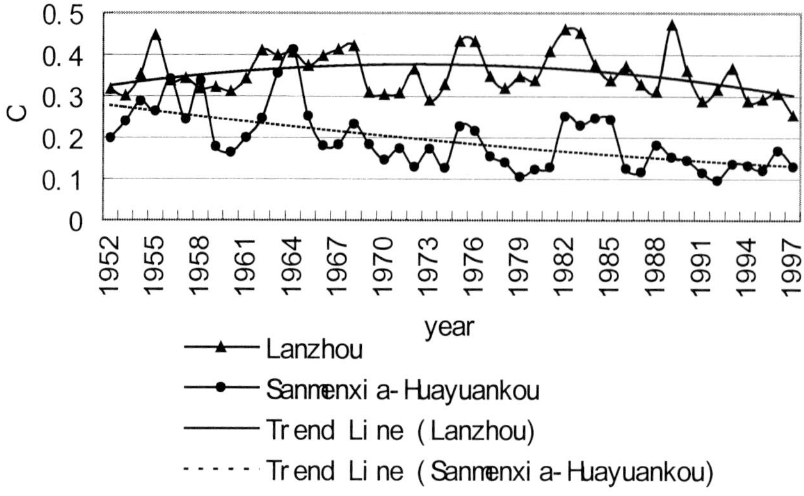

Figure 3. Change of Annual Runoff Coefficient in the Yellow River

Table 2. Measured Changes in Hydrological Cycle of the Yellow River Basin from 1950 to 1997

Drainage Area	Elements	1950s	1960s	1970s	1980s	1990s	1950-97
730,036 km^2 at Huayuankou Hydrological Station	Precipitation (mm)	460.8	472.3	449.0	443.8	417.1	449.9
	R/P (%)	14.4	14.7	11.6	12.7	8.8	13.7
	E/P (%)	85.6	85.3	88.4	87.3	91.2	86.3

3. MAIN REASONS OF HYDROLOGICAL CYCLE CHANGES

3.1 Physical Aspects

3.1.1 Main Features of Basin Hydrological Cycle

Climatically, most of the river basin lies in semiarid and arid regions. Its hydrological cycle features high evaporation and low runoff. The annual precipitation of the river basin averages only 475.9 mm while the mean annual evapotranspiration is 388.3 mm. The annual runoff is only 87.6 mm, accounting for just 18.5% of the precipitation (Fig. 4).

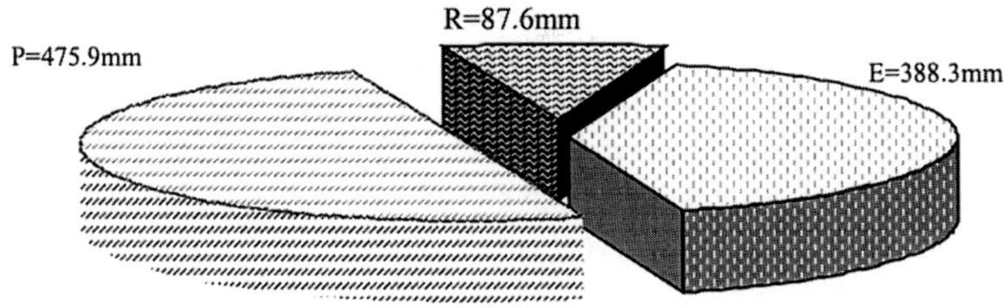

Figure 4. Annual Hydrological Cycle Elements of the Yellow River Basin

The Yellow River water is featured by its uneven distribution spatially. In the area above Lanzhou (fig.1), 30% of the total watershed yields 56% of the entire water. This region has a large water-retaining capacity and stable runoff and is regarded as the main source area of clean water in the Yellow River basin. The annual runoff at Hekouzhen (fig.1), outlet of the next reach, is less than that at Lanzhou because of losses in the intermediate arid region in addition to less precipitation and high evapotranspiration. The middle reach can be divided into three sections from Hekouzhen to Longmen (fig.1), a region covered predominantly by hilly and gully loess, contributes 12% of the runoff of the overall drainage basin. The second section from Longmen to Sanmenxia (fig.1) comprises 19% of the water runoff. The third section from Sanmenxia to Huayuankou (fig.1) yields 10% of the total runoff.

The temporal distribution of runoff is also uneven, 60% of the annual runoff and 85% of the sediment discharge takes place in the rainy season. According to the records, the total runoff in wet years is 3~4 times of that in dry years in general for stations on the main course of the river. The interannual variability is large too. The lowest recorded annual runoff between 1919 and 1997, was only 27.5 billion m^3 in 1928, while the highest annual runoff reached 93.3 billion m^3 in 1964. As an important feature, the sustained periods of successive dry years was found in the Yellow River. There were successive low-water years in the periods of 1922-1932, 1969-1974, and 1990-1997. In the record from 1922 to 1932, successive droughts occurred as all-year-long dry period. Average annual runoff in this 11 years' period is about 70% of the perennial average value. The variability of the Yellow River runoff, which is one important component of the hydrological cycle, may contribute to water resources vulnerability.

3.1.2 Water Resources Vulnerability

According to the analysis of long-term historical records of runoff, the average annual runoff of the Yellow River amounts to 58 billion m^3, equaling to only 2.1% of that of the whole country. The volume of water per capita in the basin is 542 m^3 per annum corresponding to only 25% of the nation's average. Therefore, in comparison with other large rivers in China, it is obvious that the Yellow River is deficient of water resources.

Since 1980s especially since 1990s, the economy of the regions along the riverside has been growing at a high speed. Concurrently, the demand of water has been increasing fast. However, water quality management against pollution and environment prevention is lagged (the average disposal rate of industrial drainage is about 21%). The amount of sewage directly to the river is rising up, for instance, in 1993, the total sewage amount of the basin reached to 4.17 billion m^3. As a result, the river is more seriously polluted. According to the Bulletin on Water Environment of the Country, during 1991 to 1997, water quality of the main rivers was still deteriorating. Among the main river basins in China, the pollution of the Yellow River especially of its middle reaches is aggravated most obviously and quickly.

Table 3. Water Quality Assessment Segment from Tongguan to Xiaolangdi located at the end of Middle Yellow River Surveyed on Jan. 26,1999

Section name	COD_{Cr}			COD_{Mn}		
	Concentration	Times above □	Times above□	Concentration	Times above □	Times above□
Tongguan	122.0	7.1	3.9	8.8	0.5	
Sanmenxia	50.5	2.4	1.0	12.4	1.1	0.2
Xiaolangdi buffer	52.3	2.5	1.1	10.2	0.7	0.0
Xiaolangdi bridge	47.5	2.2	0.9	9.1	0.5	
Mengjin bridge	47.3	2.2	0.9	9.6	0.6	
Huayuan kou	33.0	1.2	0.3	6.8	0.1	
Gaocun	37.0	1.5	0.5	4.6		
Likou	33.9	1.3	0.4	4.2		

Section name	Oil		Non-iron		Integrated assessment
	Concentration	Times above □	Concentration	Times above □	Above □
Tongguan	0.98	18.6	0.027	0.4	Above □
Sanmenxia	0.38	6.6	0.029	0.5	Above □
Xiaolangdi buffer	0.28	4.6	0.041	1.1	Above □
Xiaolangdi bridge	0.23	3.6	0.060	2.0	Above □
Mengjin bridge	0.15	2.0	0.051	1.6	Above □
Huayuan kou	0.62	11.4	0.046	1.3	Above □
Gaocun	0.00		0.018		Above □
Likou	0.00		0.015		Above □

Its raising rate of the comprehensive index COD_{Mn}, which represents organic pollution, had been the higher range in China. Moreover, it is quite obvious that the value of NH_4-N is increasing too, which represents the aggravating pollution caused by fertilization. In China, water quality assessed by the classified grades respectively representing water quality from high to low. The annals on surface water resources quality of Yellow River in 1997 showed that the surveyed lengths of the segments with grade of water quality were IV, V or less than V had occupied 86.2% of the whole river (the whole lengths of the surveyed river network was 6631km). It still occupies 82.3% of the surveyed river length even in the flood season. A latest water quality assessment for the segments from the middle reaches on the 26th of January in 1999 is shown in Table 3. In such case, water deficit would be aggregated by very poor water quality, which means a reduction of the available water.

3.1.3 Climate Change

According to China's long term climatic records from 1950 to 1997 and GIS mapping, we can see that there is geographical differentiation of precipitation corresponded with temperature. In the Yellow River basin, the general of the annual temperature anomaly is positive since 1970s, influenced by global warming. On the contrast, the annual precipitation in the basin has shown a decrease tendency with continuously negative anomalies in the same period. In 1980s, the negative anomaly was obviously shown in the Ningxia Hui Autonomous Region of the middle Yellow River areas. Moreover, during 1990-1997, the annual precipitation showed a more obvious decreasing tendency with a maximum negative value. –0.1 (-10%).

Annual precipitation has been decreased with annual temperature rising in the Yellow River basin, consequently, a reduction of the river runoff has been formed (Fig.5).

3.2 Human Activities

3.2.1 Water Withdrawing

The amount of withdrawal water has exceeded the availability of the natural runoff. The Yellow River is very important in water supply for Northwest China and North China. With the fast developing economy in these regions, the demand of water resources is strongly increasing.

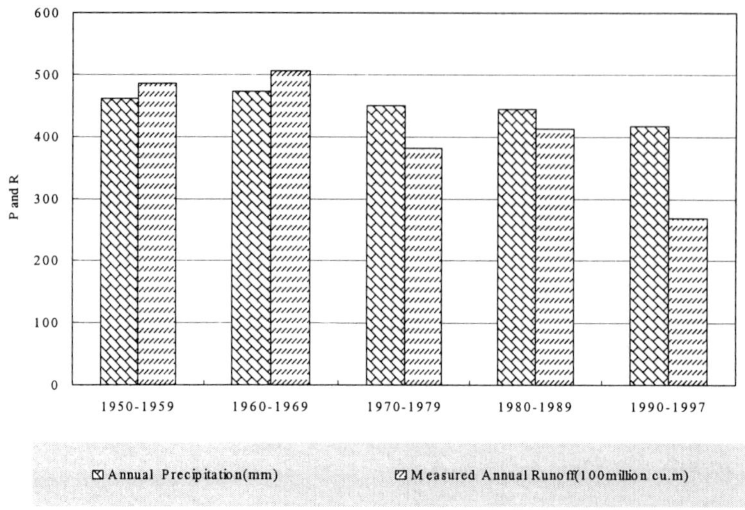

Figure 5. Changes in the measured annual precipitation (P) and runoff (R)

As far as in the 1990s, water fetched from the river for domestic use and production had amounted to about 30 billions m^3, which was 1.6 times more than that of 1950s (Table 4). For these regions, agricultural irrigation is the main sector of water utilization. It occupies 92% of the total water consumption in the basin. The irrigated area of the basin had been changed from 1,402,700 hectare in 1950s to 4,870,700 hectare in 1990s. The irrigated area had enlarged by 93%, 215% and 641% in upper reaches, middle reaches and lower reaches respectively. As a result the amount of utilized water increased by 80%, 100% and 470% respectively. In the whole basin especially in the lower reaches, water diverted from the Yellow River for irrigation and industry production increased fast which had sharpened the conflicts between water supply and demand. In addition that, parts of the runoff had been preserved by the reservoirs on tributaries and the main course thus the downstream runoff in the rivers had been reduced. As more water was needed for water-and-soil conservation and ecological rehabilitation, it can be said that the amount of fetched water might exceed that was available water in a ordinary dry year. Therefore the increasing water consumption is the most direct reason of Yellow River drying up.

Table 4. Annual Average Amount of Utilized Water and Irrigation Areas during Different Periods of Different Segments

Items	Utilized Water $10^8 m^3$				
Periods	1950s	1960s	1970s	1980s	1990s
Upper Reaches	73.4	95.2	102.9	121.1	131.7
Middle Reaches	30.0	49.4	63.4	62.1	60.2
Lower Reaches	18.9	33.1	83.5	112.9	107.8
Whole River	122.3	177.7	249.8	296.1	299.7

Items	Irrigation Areas 10^4ha.				
Periods	1950s	1960s	1970s	1980s	1990s
Upper Reaches	68	88.46	112.7	130.3	131.5
Middle Reaches	42.29	80.22	114.4	130.1	133.3
Lower Reaches	30.0	33.3	100.0	146.81	222.3
Whole River	140.29	201.98	327.1	407.21	487.1

3.2.2 Land Use and Land Cover Changes

It is well known that the Yellow River is characterized by its hyperconcentration of sediment load. The sediment comes mainly from vast area of the Middle Yellow River reaches, where is the Loess Plateau (fig.1) with loose soil and a lack of vegetation cover. In order to control soil erosion from these areas, the water and soil conservation work has been conducted for last several decades. Since the 1950s, soil conservation work, such as field terracing, afforestation, grassland development, building check dams (sediment control dams), water sediment warping, and small scale reservoirs, have been carried out in the Loess Plateau of middle reaches. It is estimated that the proportion of the controlled area for which erosion is ameliorated via water and soil conservation work has reached about 35 per cent (1995). The effect of conservation measures upon sediment reduction was fairly good. From 1970 to 1984, the annual sediment retention of terraces check dams and warping with irrigation accounted for above 50 per cent. So the sediment retention resulting from soil conservation works would amount to about 300 million tons, or 25 per cent of total sediment discharge of the river. Soil conservation has also benefited agricultural production. For instance, grain production of terrace fields can be three times higher than that of hilly slopes. According to the watershed experiment in terms of the rehabilitation of forest, soil erosion reduction shows even better effect.

However, the land use and land cover change has obviously resulted in reduction of stream runoff, though it increases the underground water in a certain degree. It is estimated that the reduction of water yield amounts to more than 1 billion m^3 in the target area which land use and land cover have changed greatly. Therefore, change in land use and land cover is favorable to controlling the river's sediment, but has some influence on water quantity of downstream reaches in terms of decreasing water yield.

4. MAIN COUNTERMEASURES CONCERNING WATER MANAGEMENT

4.1 Forming unified management systems and mechanism

Unified water resources management means a context of unified planning and unified allocation for the whole Yellow river drainage basin, which takes total control of water quantity and water quality as a target. Meanwhile it strengthens monitoring financed through the collection of water resource fee based on the National Water Law.

Enforcing supervision of water resource is the base of effective management. The Yellow River Commission (YRC) should conduct the total control on water input, output and quality according to resource allocation scheme approved by the national government. The united management should be conducted on hydrological and water quality stations of inter province mainstreams and main tributaries. Moreover, the management and development of large irrigation districts should be supervised by YRC.

4.2 Setting up the Yellow River water resource administrative and protection authority at national level.

The leadership of the State Council is regarded as the authority coordinating relevant government departments, the YRC and local governments along the river should participate. While YRC serves as representative agency of the central government in the Yellow River basin. Defining the legal status of Yellow River basin administrative agencies is key to conduct the unified management of Yellow River basin water resources. Therefore it is necessary to recommend to the government to authorize the YRC to manage and to protect Yellow River water resource. Under the power assigned by the national government, the local governments along the river should support and subordinate to the unified management.

4.3 Recommending the legislation of the "Yellow River Law"

The "Yellow River Law" is an urgent need for managing such large basin. This law should be put into forward to realize efficient administration and economic benefits in terms of rational use of the Yellow River, and to promote the social and economic sustainable development and the improvement of the eco-environment. It is suggested that in the "Yellow River Law", the legal status of the administrative agency should be defined and its responsibility in controlling water resources should be made clear; The explicit contents of the unified management and protection of the Yellow River water resources should be formulated; The relations between local water manager and basin water management agency should be solidified consistently. And the relations between basin water administrative agency and water resource development department should be also defined clearly.

4.4 Developing water saving on irrigation regions

Water consumption on the agricultural sector accounts for about 92%, so the potential amount of water saving is in close relation to water-saving agriculture. According to field study on the natural and economic conditions, the suitable method, which can be applied quickly, is to combine water convergence saving such as lining ditch. And to adopt underground low-pressure pipe with field saving method such as leveling ground, smaller ridge field, field mulching, and to adopt the combination of well and ditch irrigation assisted by reasonable water price policy. All those would conduct to ensure the success of water saving.

Irrigation areas in the Yellow River basin are lack of the utilization of groundwater and have the problem of "emphasis on irrigation regardless discharge". The combination of well and ditch irrigation is the most effective method for rationally using water resources against subsurface water loss in evaporation in the surface irrigated area of Yellow River basin. Therefore, conjunctive use of both surface water and subsurface water should be fully developed in this area (Liu Changming, 1992). Comprehensive control of water utilization should be conducted to increase effectiveness of water resources. Local water resource and limited irrigation quota on basis of agronomic techniques should be fully developed to increase water use efficiency (WUE).

4.5 Recollecting water resources for Yellow River as water transfer

Although the serious condition of the water supply can be relieved to some extent by the enforcement of the unified management, and the comprehensive water saving some other measures must be taken to conserve the available water quantity. This will relieve the conflict between the congenital shortage of Yellow River water resources and the increasingly water demand from the eco-environmental improvement and economic development. The water transfer project diverting water from the Yangtze River to the upper Yellow River is likely to be advisable. This project is involving in so-called "South-to-North Water Transfer" planned by Chinese government, which includes three routes. The western route would supplement water to the Yellow River (Liu Changming, 1998). For western route water transfer there are major points to be addressed as follows:

(1) The feasibility study about the western route of water transfer project must be conducted accordingly. In the long term□it would relieve the water shortage of the Yellow River basin. The earlier stage of the western route of water transfer project must be enforced to carry out, so that the planned water transfer project can be realized earlier.

(2) Enlarging the water regulation storage must be considered. While the western route project would speed up, the harness of the Yellow River itself in terms of water regulation should be included. In addition to this, the water storage reservoirs along the river courses should be constructed to regulate water and to increase water resources availability.

5. DISCUSSION AND REMARKS: HIGHLIGHT OF HYDROLOGICAL CYCLE STUDY

The Yellow River basin has vast area, equivalent to a regional scale in global change study. It is good for us to do detailed research. Because the river basin lies in between semiarid and humid region, it is sensitive to both climate change and human activities. In respect of water resources, it is characterized by the vulnerability. To carefully study water resource changes in the Yellow River basin will be of great highlight in better understanding of after effects of climate change and human activities on the hydrological processes.

The hydrological cycle is a core issue of water resources management. In general, there are three basic aspects of hydrological cycle study in close regard with water resources development.

Firstly, hydrological cycle study will provide a scientific base for water resources assessment. Water resources can be divided into different types of reserves. The hydrological states include the waters in air, land surface, soil layer or unsaturated zones, underground aquifers or saturated zones as well as water content in the plant bodies (the author called them "five waters"). All these water can be utilized by human beings as the water resources. They are always interacting with and exchanging between each other. The driving force is the hydrological cycle. Considering such relationship, the author presents a phrase like "five water interactions", namely "transformation of hydrological states". Hydrological cycle links all water bodies in continents and oceans, and studies in different scales in global extent and in a local area. It is essential for scientifically assessing available fresh water resources.

Secondly, a better understanding of hydrological cycle is of great significance in utilizing and managing water resource rationally. It is well known that in water resource exploitation, an optimal method is towards conjunctive use of surface water and ground water. In particular, for agricultural production, such conjunctive use can be expanded to more kinds of water including precipitation (rainwater catchment) and soil water (effective rainfall in farmlands). Obviously, the study on hydrological cycle is highly necessary for designing and planning of water conjunctive use project, and effectively administrating water allocation. Resourcefully, hydrological cycle study will lead the way to keeping the water renewable.

Thirdly, eco-environmentally, hydrological cycle plays as a very important role in joining climate change processes as heat transportation and regulation, landform evolution as erosion and sedimentation constitution, bring along with dissolved matters (chemical elements and pollutants) movement as well as joining plant (crop) physiological processes through transpiration etc. Thus the hydrological cycle is one of the actors taking part in all the physical geographical and ecological processes. Its mechanism will create deepened knowledge of the eco-environmental protection and rehabilitation. In other words, study on hydrological cycle is of great importance for human beings to constitute the coordination between the eco-environment and socio-economy in terms of sustainable development.

As far as the water resource problems in the Yellow River basin are concerned, the following major conclusion remarks can be addressed:

1) The study on the river basin's hydrological cycle is an essential topic for renewably the water resource development and management;
2) The Yellow River water is characterized by its vulnerability in terms of less water with hyper concentration of sediment load and the relationship between water and sediment must be carried on carefully;

3) The river water resources variation in time and space is great and sensitive to climate change and human activities. The influence of human activities is the key issue for further study;
4) Drying up along the course of the lower Yellow River has been found in last three decades since 1972 and its duration has become longer with earlier appearance. To mitigate such situation needs to enhance water management;
5) The depletion of water in the lower Yellow River was a result from both the reduction of precipitation and the enlargement of upstream water withdrawal. The latter is a prevailing factor which may be supposed by about 80% of the influence;
6) The enhancing water management associating with adaptive measures for climate variation impact is likely to be highly advisable.

6. ACKNOWLEDGEMENTS

This study is supported by China's key project: 973-G19990436-01. Also the author wishes to thank Dr. Zhang Shifeng for the assistance in preparing the paper.

7. REFERENCES

(1) Liu Changming, Chengli, 2000, Analysis on Runoff Series with Special Reference to Drying Up Courses of Lower Huang He River, Acta Geographical Sinica, Vol.55, No.3
(2) Liu Changming, 1998, Environmental Issues and the South-North Water transfer Scheme, The China Quarterly.
(3) Yellow River Commission, 1998, Yellow River Harness and Water Resources Development, A series of 21 books, Published by Yellow River Water Conservancy Press.
(4) Yellow River Commission, 1997, River Drying Up and Countermeasures Research□Published by Yellow River Water Conservancy Press. Vol.19(10),
(5) Chen Xiande (ed.), 1996, Yellow River Hydrology, Published by Yellow River Water Conservancy Press.
(6) Xi Jiazhi (ed.), 1996, Yellow River Water Resources, Published by Yellow River Water Conservancy Water Press.
(7) Liu Changming, 1992, Conjunctive Use of Agricultural Water in the Lower Researches of the Yellow River", Proceedings of Workshop on Agricultural Water Use Efficiency, Co-edited with J. Shalhevet, Israeli Academy's Office in Beijing (English version published by Israeli Academy Press) and Chinese version published by Science Press.
(8) Liu Changming, 1989, Problems In Management of the Yellow River, Regulated Rivers: Research &Management, Vol.3, 361-369 (by John Wiley & Sons, Ltd.)
(9) Liu Changming and Liang Jiyang, 1988, Capricious Yellow River, the Geographical Magazine, London, February, pp. 41-46.

Floods in the context of climate change and variability

Zbigniew W. Kundzewicz
Research Centre of Agricultural and Forest Environment
Polish Academy of Sciences, Poznan, Poland
and Potsdam Institute for Climate Impact Research, Potsdam, Germany

1. INTRODUCTION

Flood protection systems have been designed and operated based on the assumption of stationarity of hydrological processes of river stage or discharge. If this assumption is incorrect then the existing design procedures for embankments, dams, reservoirs, relief channels, polders, etc. will have to be revised. Without revision, the flood protection systems can be over- or under-designed and either not serving their purpose adequately or being overly costly. Studies of trend detection are also of importance because of our need to understand the changes of the "natural" world. The process of river flow has been directly influenced by changes caused by man (e. g. land-use changes: urbanisation, deforestation, changes in agricultural practices, and engineering works: drainage systems, dam construction, river regulation, etc.). Other changes may have been caused by man in an indirect way, e. g. through enhanced emissions of greenhouse gases resulting in the global warming and the related effects. Also natural changes (e.g. in channel morphology, solar activity, ENSO cycle) can play a role. All in all, detection of trends in long time series of flood data is of paramount scientific and practical importance.

According to the Red Cross, floods in 1971-1995 affected more than 1.5 billion people worldwide. In this number are included: 318 thousand killed and over 81 million homeless (IFRCRCS, 1997).

Berz (2001) examined inter-decadal variability of great flood disasters in the period 1950-1998. Great disasters are understood as those where the

M. Beniston (ed.),
Climatic Change: Implications for the Hydrological Cycle and for Water Management, 225–247.

ability of the region to help itself is distinctly overtaxed, making international or interregional assistance necessary. Based on the data presented by Berz (2001) one could state that the number of great flood disasters has grown considerably worldwide in the last decades (six cases in 1950s, seven in 1960s, eight in 1970s, 18 in 1980s, and 26 in 1990s). The number of great flood disasters in the last decade was higher than in the three decades 1950-1979.

In eight of disastrous floods in the last decade of the twentieth century, the number of fatalities exceeded a thousand and in two dozens flood events the total losses were in excess of one billion US dollars each, with the most fatal storm surge in Bangladesh in April 1991 (140 000 fatalities) and the most costly flood in China in summer 1998 when material losses exceeded 30 billion dollars (cf. Munich Re, 1997 and Kundzewicz & Takeuchi, 1999).

2. CHASING A GREENHOUSE COMPONENT IN RIVER FLOW DATA IS NOT AN EASY TASK

Finding a significant change in time series of river flow data by statistical testing is not difficult if a change results from a major human intervention in the river regime, such as, for instance, dam construction. It is far more difficult to find a gradual change (e. g., related to climatic impacts) in the behaviour of the extremes of flow, amidst strong natural variability.

When studying climatic impacts, it is very important to understand the difference between climate variability and climate change. The former can be understood as a natural variation in the climate from one period to the next (e. g. inter-annual or inter-decadal variability). The latter refers to a long term alteration in the climate. Climate variability appears to have two important effects for trend detection. On the one hand, it can cause an apparent trend. On the other hand, it can obscure other changes (Robson *et al.*, 2000). Climate variability is likely to give rise to apparent trends when records are short. Such trends would be expected to disappear once more data have been collected. Because climate variability is typically large, it can effectively obscure any underlying changes either due to climate change or due to anthropogenic causes, such as urbanisation.

The very issue of detecting a climate change signature in river flow data is complex. It is well established that increasing concentrations of greenhouse gases in the atmosphere cause global temperature rise. This, in turn, enhances evapotranspiration and precipitation in most areas, accelerating the hydrological cycle. Runoff is a difference between precipitation and evapotranspiration (whose annual means are increasing in most cases), so the net effect on their difference is not intuitively clear. In order to detect a weak, if any, climate change component, it is necessary to eliminate other influences. Using data from pristine / baseline river basins is recommended. In case of a strongly modified (e. g. dammed) river, conceptual reconstruction of the natural flow (e. g. by calculating the flow, which would have occurred in the absence of an existing reservoir).

Apart from the inherent complexity of the issue of detecting a greenhouse component in flow records, there are serious problems with the data with which to work, and also with the methodology to detect changes.

Data should consist of long time series of good quality records. Because of strong climate variability, records of 30 years or less are almost certainly too short for detection of climate change. It is suggested that *at least 50 years of record is necessary* for climate change detection (Kundzewicz & Robson, 2000), but even this may not be sufficient (cf. Chiew & McMahon, 1993).

Data should be quality controlled *before* commencing an analysis of change. There are further problems related to data. Perhaps the major one is nonhomogeneity: baseline conditions are rare, typically human influence is strong (river regulation, deforestation, urbanisation, dams and reservoirs, instability of rating curves). Possible further sources of heterogeneity (e.g. due to changes in instruments and observation techniques) should be also identified and corrected. A great deal of uncertainty results from the need of extrapolation of rating curve (stage-discharge relationship) to high values, where no direct flow measurements exist. Missing values and gaps are further complicating factors. It is difficult to give a general advice as to how to deal with them: whether or not to fill missing values and gaps, and if so, in what way?

But, even if the data are perfect, it is worthwhile to re-state a tautology: extreme (hence rare) events are rare. They do not happen frequently, so even having a very long time series of instrumental records one deals with a small sample of truly extreme floods, of most destructive power.

One practical approach to detection of change is a comparison of sub-series of the variable in question by an appropriate division of the period of observation in two (or more) sub-periods. Sometimes, this is done in an arbitrary way ("by eye"), sometimes in a more objective way, such as segmentation (e. g., Hubert, 2000).

In order to detect rigorously a weak greenhouse signal in a noisy, and highly variable, hydrological record, one needs an appropriate advanced methodology. Are trustworthy methodological tools available? The existing methods are based on three types of assumption commonly made when carrying out statistical tests: the form of the distribution, the constancy of it and the independence.

Many statistical tests are founded on an assumption that the data being tested are normally distributed. In such a case, and for data following a strongly non-normal distribution, test results cannot be trusted. A normal-score approach can then be used (Kundzewicz & Robson, 2000). Most basic statistical tests assume that the distribution of the data does not change. This assumption is violated if there are seasonal variations, other cycles in the data, or if there is an alteration over time in any other feature of the data that is not part of the test (e. g. variance). If there are seasonal cycles in the data, then the options are either to (1) deseasonalize the data, i.e. estimate the seasonal structure and remove it from the data series (cf. Radziejewski *et al.*, 2000), or (2) to use a testing approach that allows for seasonality. Data values can be said to be independent if they are completely unrelated to one another. Many hydrological data sets show correlation from one time value to the next and therefore data values are not independent. Knowing the flow in the river now may say quite a bit about what tomorrow's flow is likely to be – so these data values are not independent. However, knowing today's flow does not usually say much about what the flow will be in a year's time – thus these values can be treated as independent.

Most common statistical tests do not allow for serial correlation in the data. If it is present then possible options are (cf. Robson *et al.*, 2000):

- Use block permutation or block bootstrap methods
- Decrease the frequency of the data series (e.g., by calculating monthly or annual averages)
- Use other methods, such as, for instance, phase randomisation (cf. Radziejewski et al., 2000) that build in serial correlation.

It is very important to check that test assumptions are approximately met by the data. The violation of a test assumption can result in surprisingly inaccurate significance levels. For example, data that is assumed to be independent when it is not, could result in a significance level of 5% when in reality it should only be 25%. Most test statistics can be tested using resampling approaches in order to minimize assumptions (Robson *et al.*, 2000). Assumptions can be checked by visual techniques, such as:

- Histograms and normal probability plots – to examine distribution
- Time series plots - to spot time dependence patterns or possibly changes in variance
- Autocorrelation plots.

If visual methods are not sufficient, the formal tests are also available for assumptions checking, such as tests for normality and independence of data.

Radziejewski *et al.* (1998) compared performance of different tests for generated data contaminated by artificially introduced, and fully controlled, trends. All methods considered could detect stronger changes, in form of a gradual trend or abrupt jump, yet they could not detect weaker changes. The results of detection for short-lasting change (analogous to climate variability effects) were different for different tests. Beyond the "strength" of the trend or step-change, a duration of occurrence of a trend is important (cf. Pittock, 1980, Chiew & McMahon, 1993). It is unlikely to detect a trend that has not continued for a long time – the run-up phase must be appropriately long.

A number of tests for change detection exist, some of which have been developed to find an abrupt jump, while others can detect a gradual trend. Often, the observed change does not clearly fall into either of these categories. Therefore, it is advisable to use a number of available tests (Kundzewicz & Robson, 2000).

3. WHAT HAVE OBSERVED DATA BEEN TELLING US?

It is a well established fact that the effects of human pressure have aggravated flood risk. Development of low lying areas at risk has continued. The storage capacity of catchments has decreased and runoff and river flow rates have accelerated. As noted by Bronstert (1996), direct urbanization effects are particularly visible in small or middle size floods, which often constitute a substantial contribution to flood losses in a longer term. The urbanized area in West Germany more than doubled from 6% in 1950 to approximately 13% in 1995. An additional factor affecting flood risk has possibly been a climate change.

Search for climate change signature in hydrological data has been an area of much recent interest. It is driven by the existing prognostication of the possibility of increasing frequency and severity of floods in the future warmer world with accelerated hydrological cycle. A question can be posed: how do these prognostications fare in the context of the evidence provided by the already observed data?

3.1 Precipitation

According to IPCC (2001, Chapter 2), instrumental records of land surface precipitation continue to show an increase of 0.5 to 1 % per decade over much of mid- and high latitudes of the Northern Hemisphere. Furthermore (IPCC, 2001, Chapter 2), it is very likely (judgmental estimate of confidence: 90-99% chance) that in regions where the total precipitation has increased, there have been even more pronounced increases in heavy and extreme precipitation events. The converse is also true. Moreover, increases in heavy and extreme precipitation have also been documented even in the regions where the total precipitation has decreased or remained constant.

3.1.1 Europe

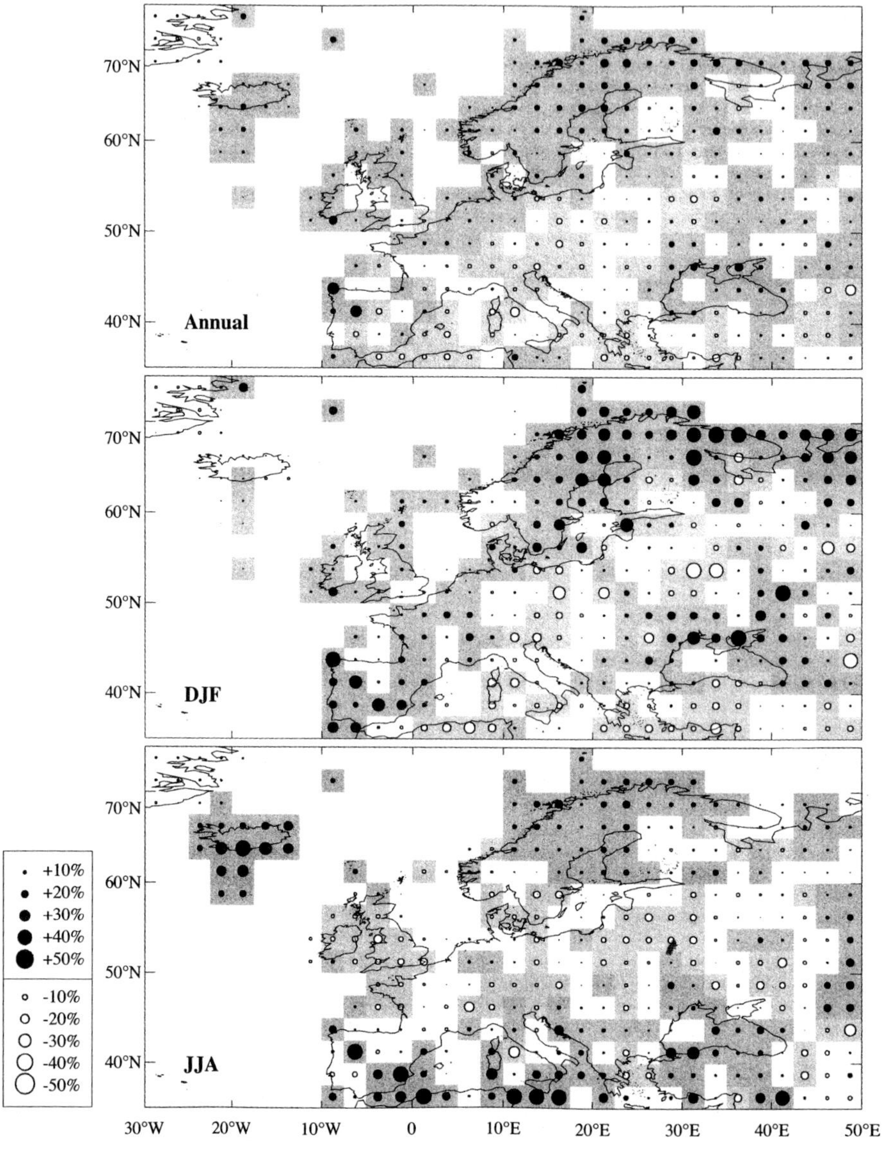

Figure 1. Trends in annual, winter, and summer precipitation in Europe (grid 2.5o), expressed as % per century. Notation: wetting – black circles, drying – white circles. Trend magnitude is related to circle size, shaded trends – significant at 90%. (From IPCC, 2001a, Chapter 13, based on data from New et al., 1999).

Although the trends are by no means uniform, about 20% of the stations analyzed worldwide show statistically significant increase of both the proportion of total annual daily precipitation within the upper five percentile and the maximum consecutive 5-day precipitation totals.

Increases in heavy precipitation events can arise from a number of causes, e. g. changes in atmospheric moisture, thunderstorm activity and large-scale storm activity. Trends in precipitation in the 20th century show indeed marked increase of precipitation, especially in winter, in much of Europe, in particular North and Northeast (Fig. 1). There are numerous studies restricted to a single drainage basin or a country, corroborating these findings. There is evidence that the frequency of extreme rainfall has increased in the UK (IPCC, 2001a, Chapter 4.3.2) and a greater proportion of precipitation is currently falling in large events than in earlier decades (Osborn *et al.*, 2000).

In their studies of *Grosswetterlagen* (synoptic-scale weather patterns), Bárdossy & Caspary (1990) noted a rise of frequency and persistence (measured by the time intervals of occurrence) of some "wet" patterns (in particular Wz, i.e. West cyclonic) in catchments in Southwest Germany during the fall. A similar tendency of precipitation was detected by Engel (1997), who compared climatological standard normals of precipitation over the intervals 1931-1960 and 1961-1990 in the Rhine basin up to Cologne, Germany. He found increased precipitation during the fall (November to January) and spring (March to June). The precipitation growth was also detected over the time period 1891-1990.

3.1.2 North America

Karl *et al.* (1995) noted that within the United States, the proportion of total precipitation contributed by extreme one-day events has increased significantly during the 20th century. The incidence of intensive precipitation events has steadily increased at the expense of moderate events.

Observations confirm that atmospheric moisture is increasing in many places. For example, growth at a rate of about 5% per decade was observed in the USA (Trenberth, 1998). Increased atmospheric moisture contents favours more intensive precipitation events thus increasing the risk of flooding.

3.1.3 Australia and Oceania

As stated in IPCC (2001a, Chapter 12), Australian annual mean rainfall has increased by a marginally significant amount over the last century.

However, increases in heavy rainfalls have been observed over many parts of Australia in the 20th century (IPCC, 2001, Chapter 2).

After 1877, increases (some statistically significant) have been noted in mean rainfall for New Zealand's west coast. This is partially explained by the increase in El Niño conditions over recent decades. There is some evidence of long-term variations in the Australasian region in storm frequency and tropical cyclones (IPCC, 2001a, Chapter 12).

3.2 River flow

There have been a plethora of studies of time series at a single stream gauge, reported in the literature. Several reports of significant changes detected in flow records at a single gauge encouraged researchers to extend the analysis into a truly spatial domain, to check whether or not a pattern observed at a single gauge has been reproduced in the neighbouring locations.

Yet, it would be a gross oversimplification to say that, in general, floods have exhibited growing trends worldwide. Only some series show a significant trend and out of those only some (yet, typically more than half) feature a positive trend, while others exhibit negative trends. It is not uncommon that neighbouring gauges behave in a different way. The time series of flood data show a complex response (due to other, non-climatic factors), whose behaviour is not necessarily in tune with gross climate-related prognostications.

The finding in IPCC (2001a, Chapter 8) is that the costs of extreme weather events have exhibited a rapid upward trend in recent decades and yearly economic losses from large events have increased ten-fold between 1950s and 1990s (in inflation-adjusted dollars). The insured portion of these losses has grown even stronger. Demographic and socio-economic trends are increasing society's exposure to floods and part of the observed upward trend in weather disaster losses is linked to socio-economic factors, such as increase in population, wealth, and developing settlements in vulnerable areas, but these factors alone cannot explain the observed growth. As stated in IPCC (2001a, Chapter 8), a part of losses is linked to climatic factors, such as the observed changes in precipitation and flooding events. Even if precise attribution is complex, the growth in manmade losses has been far lower than of weather-related events.

3.2.1 Europe

Engel (1997) analyzed annual maxima and 9-day-means of flow of the river Rhine at Cologne from 1890 to 2000 and detected a positive trend.

Major floods observed during the last decade in Southwest Germany occurred during the Wz (West cyclonic) pattern of atmospheric circulation in winter, whose increased frequency of occurrence was detected (Bárdossy & Caspary, 1990). Caspary (2000) analyzed time series of discharge in four rivers in Germany. After having smoothed the year-to-year oscillation of annual peak discharge, he found a marked recent increase in the amplitude of floods. He also compared floods of different recurrence intervals for two consecutive sub-periods. The 100-year-flood determined from the older data in the first sub-period corresponds to much lower return periods (between 5 and 30-year-flood) for the more recent data. Large flows are therefore becoming more frequent. However, no space-covering study placing these results in a truly regional perspective has been available yet.

Nobilis & Lorenz (1997) analyzed the flood trends in Austria. They considered different periods of observation (40 year-interval: 1952-1991 and parts thereof). Only in a portion of cases, a significant trend was detected. The quantitative results depended on the sub-period and the characteristics studied (whether annual maxima, or number of floods per year, or partial duration series). The portion of cases for which a significant trend was detected ranged from 4.3% to 31.5%. Among those cases where a significant trend was detected, there were more examples of positive trend (64.3%) than of negative trend (35.7%). Analysis of the full 40-year period results in detecting a positive trend in 66.3% of the cases with significant trend.

Gilvear & Black (1999) found that the number of flood-induced embankment failures on the river Tay in Scotland in the time period 1988/9 to 1996/7 was nearly five times higher than in 1978/9-1987/8; i. e. 26.1 and 5.4, respectively.

Mansell (1997) presented evidence of a general increase in river flow (including the maximum and the percentile, Q5%) in four catchments in Scotland in the last 30 years, being significantly stronger than the increase in rainfall over the same period.

A comprehensive study of flood records has been conducted in the UK by Robson & Reed (1996). Using a data base consisting of ca. 600 stream gauges with long data series (from 15 to over 100 years), they presented a map of gauging stations in the UK exhibiting significant non-stationarity in annual maxima and peak-over-threshold (POT) variables. Figure 2, stemming from Robson & Reed (1996), shows a summary measure (trend gradient) plotted at the geographical location at each site, with type of trend and its intensity noted. Some regional features are visible in the results. There are more incidences of increased flooding than decreasing flooding, particularly in Scotland and in South East of England.

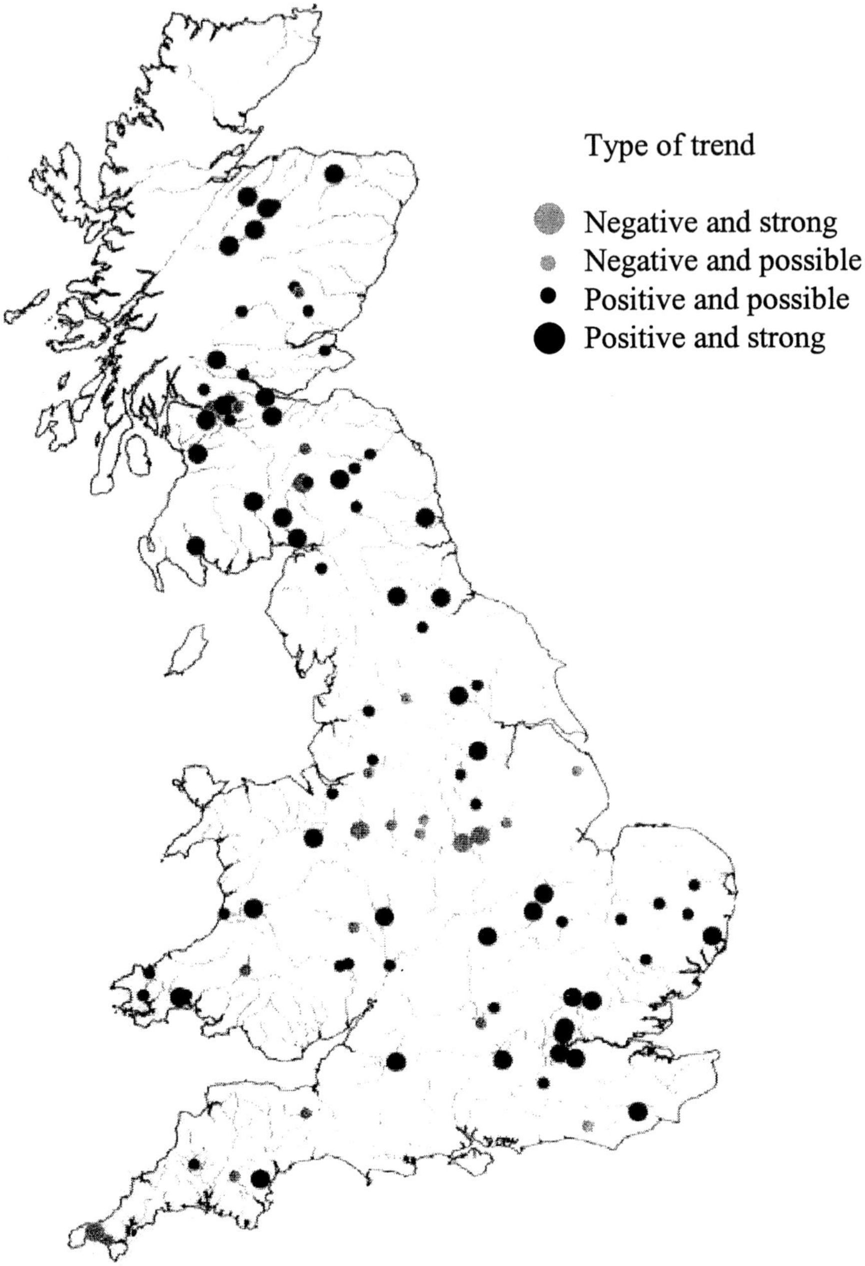

Figure 2. - Non-stationarity in British flow records. Increases in annual maxima or any peak-over-threshold (POT) variables (except POT intervals) are marked as positive; decreases as negative. (From Robson & Reed, 1996).

3.2.2 North America

Olsen *et al.* (1999) looked into the distribution of long series (up to nearly 120 years) of flow records in the Upper Mississippi, Lower Missouri and Illinois rivers and their relationship to climate indices. In many gauges, large and statistically significant upward trends were detected. Past-to-present and present-to-past analysis of subsets of data (between 10 and 100 years of length) showed several significant correlations (with significance level of 1% or better in many cases), typically corresponding to growing trends.

Table 1 (based on results of Lins & Slack, 1999) – Aggregate statistics illustrating changes of selected quantiles of streamflow.

	Years of record					
	30	40	50	60	70	80
Beginning year	1964	1954	1944	1934	1924	1914
No. of stations tested	395	395	395	193	70	34
Annual minimum (daily mean) discharge						
No. of significant trends (% of total)	112 (28%)	177 (45%)	163 (41%)	85 (44%)	34 (49%)	13 (28%)
No. with increasing trend	74	145	127	76	32	10
Annual 30th percentile of daily discharge						
No. of significant trends (% of total)	109 (28%)	160 (41%)	135 (34%)	81 (42%)	28 (40%)	9 (26%)
No. with increasing trend	76	148	125	79	27	8
Annual 70th percentile of daily discharge						
No. of significant trends (% of total)	59 (15%)	130 (33%)	64 (16%)	58 (30%)	19 (27%)	9 (18%)
No. with increasing trend	55	124	61	58	19	6
Annual maximum (daily mean) discharge						
No. of significant trends (% of total)	37 (9%)	53 (13%)	35 (9%)	20 (10%)	9 (13%)	4 (12%)
No. with increasing trend	12	31	14	11	5	2

Lins & Slack (1999) studied secular streamflow trends, using long series of daily data from 395 climate-sensitive stream gauging stations in the conterminous United States. When studying quantiles of discharge, they found two significant patterns of change. Trends were most prevalent in the annual minimum (Q_0) to medium (Q_{50}) flow categories, while they were least prevalent in the annual maximum (Q_{100}) category. For all, but the highest quantiles, streamflow has increased across broad areas of the US. These results were summarized as "getting wetter, but less extreme" (Lins &

Slack, 1999), since streamflow has increased except for the highest quantiles.

In order to evaluate interdecadal streamflow variability, quantile trends were calculated for 30-, 40-, 50-, 60-, 70-, and 80-year periods, all ending in 1993.

The principal results of Lins & Slack (1999) are summarized in Table 1 and Fig. 3. Table 1 shows the aggregate statistics illustrating changes of selected quantiles of streamflow. Figure 3 presents results of spatial studies of change in flow data, showing trends in annual maximum daily value of flow.

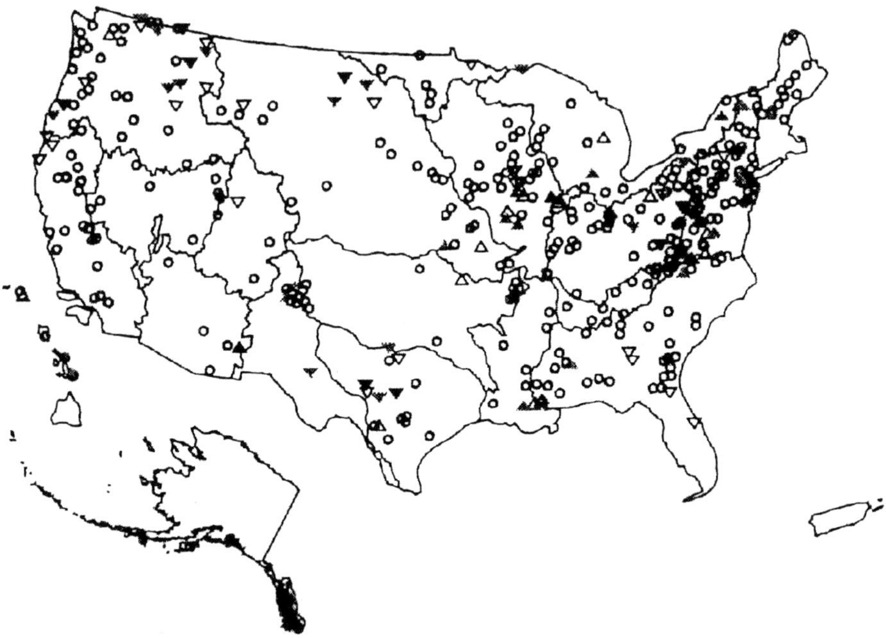

Figure 3. Results of spatial studies of change in observed flow data in the USA. Trends ($p \leq 0.05$) in annual maximum flow (daily observations). Upward pointing triangles indicate increasing discharge (downward-pointing – decreasing). Delta-like solid triangles denote stations exhibiting a trend in three or more time periods analysed (see Table 1), an inverse delta-like solid triangles – a trend in two time periods, open triangles – in one time period. Circles denote no trend in any period. (From Lins & Slack, 1999).

Pielke & Downton (2000) studied the rates of change in flood characteristics and socio-economic indicators in the USA in the time period from 1932 to 1997. They found that the total annual flood damage, adjusted for inflation, has grown in the average with the rate of 2.92% per year, that is more strongly than population (+1.26%) and tangible wealth per capita, in inflation-adjusted dollars (+1.85%) but less strongly than the net stock of fixed reproducible tangible wealth (+3.13%). They also found significant correlations of flood damage measures with several precipitation indices.

3.2.3 Australia

Chiew & McMahon (1993) stated that with the current data set, there is no clear evidence to suggest that the greenhouse signal is impacting on Australian streamflow. They showed that the detectability of change in the mean depends more on interannual variability and less on the length of data available. As the interannual variability of Australian streams is high, being twice as high as that of the Northern Hemisphere, the detection threshold is also high. If scenarios predicted by GCMs could be reached, then significant trends would be detected. Chiew & McMahon (1993) analyzed percentage changes in the means required in the future data set of 25 and 50 years to be considered as statistically different from the historic mean. They studied relationships between the historic data length, length of future data (since the trend commences), percentage change (strength of the trend), and coefficient of variation. For high values of the variation coefficient, long data records are needed to detect an existing trend; e. g., for $C_v = 1.48$; 76 to 88 years.

3.3 Changes in seasonality

An important change observed in flow data refers to seasonal characteristics. River flow regimes, i. e. temporal distributions of flow, have considerably changed. It was reported from much of Europe that high flows come earlier in the year due to earlier snowmelt (sometimes in winter rather than spring) and less snow cover may reduce the severity of spring snowmelt floods. During warmer and wetter winters with less water storage in snow, increased flows are observed. It seems that, where the rivers freeze, milder winters lead generally to thinner ice cover and shorten persistence and reduce severity of ice jams. Ice-jam floods are not a major problem anymore in much of Europe, where the rivers freeze less often in the warming climate (with industrial waste heat playing also a role in many locations).

Beltaos & Prowse (2001) found that in Canada the trends in timing of freeze-up and breakup are consistent with concomitant changes in average temperature. Most stations show later freeze-up and earlier breakup. It is not only spring breakup but also winter thaws, which can lead to severe flood destruction, especially if a re-freeze follows soon. Increased incidence of mid-winter breakup events and higher freshet flows in certain parts of Canada could enhance the frequency and severity of ice jams. Destructive premature breakup, associated with rapid runoff (rapid melt and heavy rain) is a phenomenon of growing concern.

3.4 Sea level rise

The global average sea level rose about 15 cm during the 20th century, at least partly due to the thermal expansion of ocean water (steric rise) and the widespread loss of land ice associated with the 20th century warming. The average rate of the sea level rise in the 20th century was about ten times larger than the average rate over the last 3000 years (IPCC, 2001, Technical Summary). Rise of the mean state of the sea level clearly impacts on extremes. Increased frequency of flooding in Venice accompanied 30 cm relative sea level rise in the 20th century (IPCC, Chapter 13). After Cecconi *et al.* (1999), this relative sea level rise (approximately half due to subsidence, half due to an absolute sea level rise) has led to an increase in the number of floods (exceedences of the water level of 80 cm) per year. It grew from a few to ten in the early 1920s to over 30 in 1990s with the maximum exceeding 100 in 1996.

Storm surge data obtained from hourly tide gauge records in the US Atlantic coast provide a unique quantitative measure. A considerable increase has been detected over the 20th century in the number of hours of anomalous high water level (surges greater than two standard deviations) per year at Atlantic City, N. J. (Zhang Keqi *et al.*, 1997). Yet, no discernible secular trend was found in storm activity or severity during the 20th century.

3.5 Links with climatic variability

Search for links of hydrological extremes with climatic variability is a challenging study area (e. g. oscillations in the oceanic-atmosphere system, such as El Niño–Southern Oscillation (ENSO) and North Atlantic Oscillation (NAO)). The frequency and intensity of ENSO have been unusual since the mid 1970s, as compared with the previous 100 years, in the context of the instrumental record. Warm phase of ENSO episodes has become more persistent and intense (IPCC, 2001, Chapter 2). In some regions, intensive precipitation (and floods) occur more frequently in the El Niño phase than in the La Niña phase (e. g. in the Atlantic side of Central America, Northwest Peru, Equador, in western, central-western and Pampas region of Argentina, and Chile). In other locations (e. g. Colombia, Northern Amazonia and Northeast Brazil, and also Australia and Southern Africa), it is the other way round. There seems to exist a link between the frequency of extreme flood events and the SOI anomalies, while no clear connections for the magnitude of extreme floods have been detected.

Kaczmarek (2002) showed that, on average, the peak snowmelt-induced floods in Polish rivers decrease with the increase of the value of the NAO index.

Kuhnel & Coates (2000) studied fatalities from natural perils in eastern Australia in 1876-1991. They found that the probability of reaching the mean annual number of flood fatality event days is over four times higher during La Niña seasons (80%) than the corresponding probability associated with El Niño periods (18%).

The Sun plays a role of external forcing to the Earth system and search for links between flood indices and solar parameters has been continuing. Part of the studies refer to sunspot numbers. Correlations were found (Tomasino & Dalla Valle, 2000) between the discharges of the river Po and the characteristics of the solar activity cycle. Landscheidt (2000) interpreted the frequency of 8.7 years, determined by Tomasino & Dalla Valle (2000) as corresponding to the cycle of the Sun's motion about the centre of mass of the solar system.

4. EXTREME EVENTS – PROJECTIONS FOR THE CHANGING CLIMATE

A significant portion of the increase of flood hazard has been caused by human pressure, including land-use change and river engineering, and the increase in population and wealth accumulated in endangered areas. These drivers will continue to be active in the future.

However, there are also several climate change related mechanisms, which may significantly influence flood characteristics. Characteristics of extreme climatic phenomena related to floods, can change in the future climate, in the light of scenario analyses (cf. IPCC, 2001 and 2001a, Technical Summaries). Among climate-related impacts relevant to floods are: increased magnitude of precipitation events of high intensity in many locations, more frequent wet spells in mid / high latitude winters, more intense mid-latitude storms, and more El Niño-like mean state of ENSO. Some models show increase of mean and peak intensity of precipitation and peak wind intensity of tropical cyclones (IPCC, 2001, Technical Summary). No conclusive evidence as to the paths of depressions and storms has been established.

As stated by Cubasch & Kasang (2001), in regions with increasing precipitation, heavy precipitation will increase as well. The reason being that atmosphere's capacity to absorb moisture, and thus its absolute water content, increase with temperature. Thus, the potential for intensive precipitation, and likewise floods, will increase.

Based on global model simulations and for a wide range of scenarios, global average water vapour concentration and precipitation are projected to increase during the 21st century.

Precipitation extremes are projected to increase more than the mean. The frequency of extreme precipitation events is projected to increase over many areas (IPCC, 2001, Technical Summary).

However, as noted by Walsh & Pittock (1998), confidence in the ability of models to predict changes in climate for specific regions of the globe is not strong. Moreover, since the evaluation of climate models has focused on their ability to simulate the mean climate, less is known about their capacity to simulate events of shorter durations, such as floods.

Walsh & Pittock (1998) state that there is considerable evidence that the frequency of extreme rainfall events may increase in the tropics, which would cause decreases in flood return periods. Yet, assessment of regional impacts remains uncertain and awaits improvements in model resolution and parameterization.

Arnell (1999) compared the percentage changes in precipitation across the world land areas, by the 2020s, 2050s and 2080s, under six scenarios considered). In all those scenarios, precipitation is expected to increase, globally, depending on the scenario; from 0. 6 to 3.4% in the 2020s, and from 2.1 to 7.5% in the 2080s, as compared to the climatological standard normals, 1961-1990. He also noted an amplification effect in change in annual runoff, versus annual rainfall. Arnell (1999) showed that many regions will feature considerable changes in 10-year return period maximum monthly runoff. Based on HadCM3 GGa1 analysis for the time horizon of 2050 he noted increases, in particular, in the northwest north America and in the east Asia.

4.1 Europe

Regional precipitation scenarios for winter developed under the ACACIA project (Parry, 2000), for the ACACIA A2-high scenario, show growth across most of Europe, while the median of annual precipitation grows in particular grids by up to 14% in 2020s, through up to 35% in 2050s and up to 53% in 2080s. Wetter winters are predicted throughout the continent, with two regions of highest increase: the Northeast and northwest Mediterranean coast, including northern Italy. Figure 4 shows the future climatology of Poland and Spain, indicating possible growth of winter flood hazard.

Climate change is likely to increase the risk of river flooding, and loss and damage generally in Europe. As stated in Chapter 13 of IPCC (2001): "[i]n more continental and upland areas, where snowfall makes up a large proportion of winter precipitation, a rise in temperature would mean that more precipitation falls as rain and therefore that winter runoff increases and spring snowmelt decreases. The timing of streamflow therefore alters significantly." "In some areas the time of greatest risk would move from spring to winter." In particular, winter flood hazard is likely to rise for many catchments under many scenarios. However, global warming may not necessarily reduce snowmelt flooding. Winter precipitation is going to increase, but in the areas where the temperature will still be below 0^0C, the snow cover may increase. Increased urban flooding from overflow of storm drains is also predicted. Frequency of floods in the mountains is likely to grow.

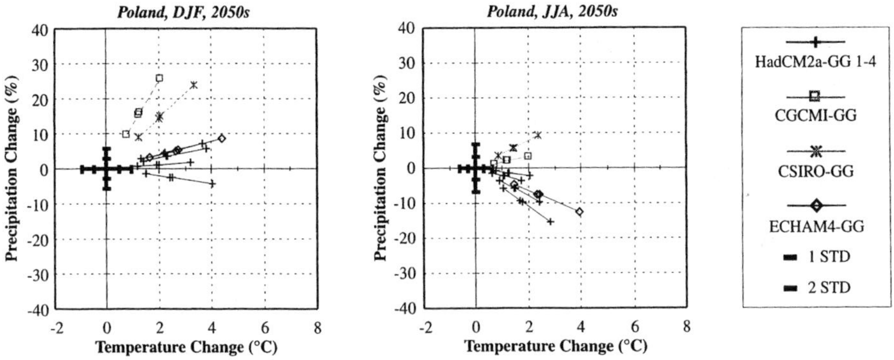

Figure 4 Scaled mean winter (December to February) temperature and precipitation change over land grid boxes representing Poland and Sweden. Seven lines represent seven GCM simulations, where each point refers to one of four ACACIA scenario. One- and two-standard-deviations limits (from 1400-year unforced Hadley CM2 simulation) indicate natural multi-decadal variability (From IPCC, 2001, Chapter 13, based on material produced by Dr Tim Carter for ACACIA, Parry, 2000)

Bárdossy & Duckstein (2001) studied extreme daily areal winter precipitation for the Lemme catchment in Germany under $1xCO_2$ and $2xCO_2$ conditions. They found that the intensive precipitation with the return period of 20 years under $1xCO_2$ is lower than the 10-year-intensity under $2xCO_2$ (44.5 and 46.5 mm, respectively).

Reynard *et al.* (1998) detected changes in floods of different return periods for 2050s for the river Severn in the U.K., ranging, for different scenarios, from 6.0 to 21.5%, if greenhouse gases rise only is considered and from –1.8 to 11.9%, if both greenhouse gases and the aerosols are taken into account. The results for the Thames are slightly less pronounced. For the 2050s, the climate change scenarios produce an increase in both frequency and magnitude of flooding events, and the number of days with high flow.

Arnell (1998) studied changes in floods over Britain for the time horizon of 2050. Even if the annual rainfall may change only slightly, still there will be a considerable change in seasonality of precipitation – moderate to large increases in winter and considerable decreases in summer. As a result of seasonal changes in precipitation and in potential evapotranspiration, there could be considerable changes in annual runoff over Britain – increase in the north and decrease in the south and south west.

The study of Arnell (1998) shows that it is difficult to estimate the effects of climate change on extreme high flows. Increased winter rainfall implies an increase in winter flooding, whilst less frequent but more intense summer rainfall may also cause increase in the occurrence of summer floods. The effects of a change in rainfall on flood depends on catchment characteristics. A small catchment underlain by clay is very sensitive to changes in short-term intense rainfall, while in a larger catchment, changes in accumulations over several days would be more significant.

Naden *et al.* (1996) found that floods in three large British catchments (of the rivers Severn, Trent and Thames), with return periods between 5 and 50 years, could increase in magnitude by 8-11% by 2050s.

4.2 North America

After IPCC (2001a, Chapter 15), "[p]ossible changes in runoff patterns, coupled with apparent recent trends in societal vulnerability to floods in parts of North America suggest that flood risks may increase as a result of anthropogenic climate change. Seasonal changes in flow are also expected, caused by changes in snowmelt accumulation and timing of snowmelt." "Canadian rivers in northern areas may begin to experience winter ice break-ups and regional flooding."

Loukas & Quick (1999) studied, in a comprehensive way, changes in floods in two largely different watersheds in British Columbia, Canada, under a single and a double carbon dioxide scenarios. The results for the two catchments were strikingly different. The floods in the maritime watershed were found to increase, on average, in magnitude by 14%, in volume by 94%, in frequency by 119% and in duration by 44%. In the interior mountain catchment, all these flood characteristics would decrease, except for a marginal increase in duration.

4.3 Australia and Oceania

Even if "drier conditions are anticipated for most of Australia over the 21st century", "an increase in heavy rainfall is also projected, even in regions with small decreases in mean rainfall. This is due to a shift in the frequency distribution of daily rainfall toward fewer light and moderate events and more heavy events. This could lead to … more floods." (IPCC, 2001a, Chapter 12)

4.4 Asia

"Enhanced hydrological cycle and increase in mean annual rainfall over most of Asia", but highest in Boreal Asia is expected (IPCC; 2001a, Chapter 11). The frequency and severity of floods is projected to increase in many countries. "Increasing precipitation intensity, particularly during the summer monsoon, could increase vulnerability of many climate-dependent sectors affecting the economy in the Boreal Asia region". "The intensity of extreme rainfall events are projected to be higher in a warmer atmosphere suggesting thereby a decrease in return period for extreme precipitation events and the possibility of more frequent flash floods in parts of India, Nepal and Bangladesh."

5. LINKS WITH OCEAN-ATMOSPHERE SYSTEM

Perhaps the sea level rise is the most important single climatic factor, which contributes to increasing (coastal) flood risk. In this case, a change in the mean level may exert significant impacts on the extremes. Also higher peak wind intensities could contribute to increasing severity of coastal flooding. Needless to say, that sea level rise itself is a dangerous occurrence, which, after reaching some stage, may cause permanent inundations and the need of massive relocation.

Man-induced climate change has the potential to trigger large-scale singular events that could have important consequences for flood risk. An example is a disintegration of the West Atlantic and Greenland Ice Sheets, in a long time scale of multiple centuries, which would cause a significant sea level rise (of the order of metres) leading to a permanent inundation of large, now densely populated areas. The probability of triggering such events in the near future is low, and is poorly understood, but should not be ignored given the severity of its consequences (IPCC, 200, Technical Summary). As summarized in IPCC (2001, Technical Summary), no significant trends of Antarctic sea ice have been observed so far since the 1970s, except for the Antarctic Peninsula. Greater melting is projected in 21st century.

6. CONCLUSIONS

Destructive floods observed in the 1990s all over the world have led to record high material damage. The immediate question emerges, as to the extent in which this sensible rise of flood hazard and vulnerability can be linked to climate variability and change.

Some recent studies show that the flood hazard is likely to rise in the future and that plausible climate change scenarios result in future increases of both amplitude and frequency of flooding events. Yet, there has been no conclusive and general proof as to how climate change affects the flood behaviour, in the light of the data observed so far.

Discussion is offered of results reported in the literature for both analysis of the past data and analysis of scenarios for the future, emphasizing the existing uncertainties, controversies and inconsistencies. The general statement that high floods are becoming more frequent is supported by several studies (e. g. local flood frequency changes). Yet, this statement has also been challenged based on contradictory evidence, when a non-stationarity could not be distinguished, or when the finding was: "wetter but less extreme".

A regional change in timing of floods has been observed in many areas, with increasing late autumn and winter floods and less ice-jam-related floods. This has been a robust result. Yet, intensive and long-lasting precipitation episodes happening in summer have also led to disastrous recent flooding (e.g. the Odra / Oder deluge in 1997), cf. Kundzewicz *et al.* (1999).

It is difficult to disentagle the climatic component in the flood data subject to strong natural variability and influenced by man-made environmental changes: urbanization, deforestation, human occupying hazardous areas, reduction in storage capacity and increase in runoff coefficient.

All in all, the response of flood risk to climate forcing in the future will be complex. In many places flood risk is likely to grow, due to a combination of anthropogenic and climatic factors. Yet, quantification of flood statistics is difficult and subject to high uncertainty. As stated in IPCC (2001, Technical Summary, "[t]he analysis of extreme events in both observations and coupled models is underdeveloped" and "the changes in frequency of extreme events cannot be generally attributed to the human influence on global climate."

7. ACKNOWLEDGEMENTS

The work reported in the paper was carried out within the grants no. 6P 04 G 03 814 and 6 P 04 E 05 520 of the Scientific Research Committee, Republic of Poland. Constructive comments of Dr A. Barrie Pittock are gratefully acknowledged.

8. REFERENCES

Arnell, N. W. (1998) Climate change and water resources in Britain, *Climatic Change,* 39: 83-110.

Arnell, N. W. (1999) Climate change and global water resources, *Global Environ. Change* 9: S31-S49.

Bárdossy, A. & Caspary, H.-J. (1990) Detection of climate change in Europe by analysing European Atmospheric Circulation Patterns from 1881 to 1989. *Theor. and Applied Climatol.* 42, 155-167.

Bárdossy, A. & Duckstein, L. (2001) Hydrological risk under nonstationary conditions changing hydroclimatological input. In: Bogardi, J. J. & Kundzewicz, Z. W. (Eds) *Risk Reliability, Uncertainty and Robustness of Water-Resources Systems.* Cambridge Univesrity Press, Cambridge, New York.

Beltaos, S. & Prowse, T. D. (2001) Climate impact on extreme ice-jam events in Canadian rivers, *Hydrol. Sci. J.* 46(1) *in press.*

Berz, G. (2001) Climatic change: Effects on and possible responses by the insurance industry. In: Lozán, J. L., Graßl, H. & Hupfer, P. (Eds) *Climate of the 21st Century: Changes and Risks.* Office: Wissenschaftliche Auswertungen, Hamburg, 392-399.

Bronstert, A. (ed.) (1996) Hochwasser in Deutschland unter Aspekten globaler Veränderungen. Bericht über das DFG-Rundgespräch am 9. Oktober 1995 in Potsdam, PIK Report No. 17, 59 pp.

Caspary, H.-J. (2000) Increased risk of river flooding in southwest Germany caused by changes of the atmospheric circulation across Europe. In: Bronstert, A., Bismuth, Ch. & Menzel, L. (Eds) Proc. Eur. Conf. on Advances in Flood Research, PIK Report No. 65, Vol. 1, 212-223.

Cecconi, G. *et al.* (1999) Climate record of storm surges in Venice. In: RIBAMOD – River Basin Modelling, Management and Flood Mitigation, Concerted Action, Proc. of the final Wallingford Workshop, European Commission, Luxembourg, EUR 18287 EN, p. 149-156.

Chiew, F. H. S. & McMahon, T. A. (1993) Detection of trend and change in annual flos of Australian rivers, *Int. J. of Climatology*, 13: 643-653.

Cubasch, U. & Kasang, D. (2001) Extremes and climate change. In: Lozán, J. L., Graßl, H. & Hupfer, P. (Eds) *Climate of the 21st Century: Changes and Risks.* Office: Wissenschaftliche Auswertungen, Hamburg, 256-261.

Engel, H. (1997) The flood events of 1993/1994 and 1995 in the Rhine River basin. In: Leavesley, G. H., Lins, H. F., Nobilis, F., Parker, R. S.,. Schneider, V. R. &. van de Ven, F. H. M. (Eds) *Destructive Water: Water-caused Natural Disasters, their Abatement and Control*) IAHS Publ. No. 239, 21-32.

Gilvear, D. J. & Black, A. R. (1999) Flood-induced embankment failures on the River Tay: implications of climatically induced hydrological change in Scotland, *Hydrol. Sci. J.,* 44(3) 345-362.

Hubert, P. (2000) Segmentation. Chapter 10 in: Kundzewicz, Z. W. & Robson, A. (Eds) *Detecting Trend and Other Changes in Hydrological Data*; World Climate Programme – Water, World Climate Programme Data and Monitoring, WCDMP-45, WMO/TD – No. 1013, Geneva, May 2000, 113-119.

IFRCRCS (International Federation of Red Cross and Red Crescent Societies) (1997) *World Disasters Report 1997.* Oxford University Press.

IPCC (Intergovernmental Panel on Climate Change) (2001) *Climate Change 2001: The Scientific Basis.* Contribution of the Working Group I to the Third Assessment Report of the Intergovernmental Panel on Climate Change, Cambridge University Press, Cambridge.

IPCC (Intergovernmental Panel on Climate Change) (2001a) *Climate Change 2001: Impacts, Adaptation and Vulnerability*. Contribution of the Working Group II to the Third Assessment Report of the Intergovernmental Panel on Climate Change, Cambridge University Press, Cambridge.

Kaczmarek, Z. (2002), contribution to this volume.

Karl, T. R., Knight, R. W. & Plummer, N. (1995) Trends in high-frequency climate variability in the twentieth century, *Nature*, 377: 217-220.

Kuhnel, I. & Coates, L. (2000) El Niño-Southern Oscillation: Related probabilities of fatalities from natural perils in Australia, *Natural Hazards* 22: 117-138

Kundzewicz, Z. W. & Robson, A. (ed.) (2000) *Detecting Trend and Other Changes in Hydrological Data*; World Climate Programme – Water, World Climate Programme Data and Monitoring, WCDMP-45, WMO/TD – No. 1013, Geneva, May 2000.

Kundzewicz, Z. W., Szamałek, K. & Kowalczak, P. (1999) The Great Flood of 1997 in Poland, *Hydrol. Sci. J.*, 44(6): 855-870.

Kundzewicz, Z. W. & Takeuchi, K. (1999) Flood protection and management: Quo vadimus? *Hydrol. Sci. J.*, 44(3): 417-432.

Landscheidt, T. (2000) River Po discharges and cycles of solar activity. Discussion of "Natural climatic changes and solar cycles: an analysis of hydrological time series" by Tomasino, M. & Dalla Valle, F. *Hydrol. Sci. J.* 45(3) 491-493.

Lins, H. F. & Slack, J. R. (1999) Streamflow trends in the United States, *Geoph. Res. Letters* 26(2): 227-230.

Loukas, A. & Quick, M. C. (1999) The effect of climate change on floods in British Columbia, *Nordic Hydrology*, 30: 231-256.

Mansell, M. G. (1997) The effect of climate change in rainfall trends and flooding risk in the West of Scotland, *Nordic Hydrology*, 28: 37-50

Munich Re (1997) Flooding and Insurance, Munich Re, Munich, Germany.

Naden, P. S., Crooks, S. & Broadhurst, P. (1996) Impact of climate and land use change on the flood response of large catchments, Ministry of Agriculture, Fisheries and Food, Conference of Rivers and Coastal Engineets, Keele, Englamd, July 1996, p. 16.

New, M. Hulme, M. & Jones, P. D. (1999) Representing twentieth century space-time climate variability, part 1: development of a 1961-90 mean monthly terrestrial climatology, *J. of Climate* **12**, 829-856.

Nobilis, F. & Lorenz, P. (1997) Flood trends in Austria. In: *Destructive Water: Water-caused Natural Disasters, their Abatement and Control* (ed. by G. H. Leavesley, H. F. Lins, F. Nobilis, R. S. Parker, V. R. Schneider & F. H. M. van de Ven) IAHS Publ. No. 239.

Olsen, J. R., Stedinger, J. R., Matalas, N. C. & Stakhiv, E. Z. (1999) Climate variability and flood frequency estimation for the Upper Mississippi and Lower Missouri rivers, *J. AWRA*, 1509-1522.

Osborn, T. J., Hulme, M., Jones, P. D. & Basnet, T. A. (2000) Observed trends in the daily intensity of United Kingdom precipitation, *Int. J. of Climatology*, 20, 347-364.

Parry M. L. (ed.)(2000) *Assessment of Potential Effects and Adaptations for Climate Change in Europe: The Europe ACACIA Project*. Jackson Environment Institute, University of East Anglia, Norwich, UK, 320 pp.

Pielke, R. A. & Downton, M. (2000) Precipitation and damaging floods. Trends in the United States, 1932-1997, *J. of Climate*, 13: 3625-3637.

Pittock, A. B. (1980) Monitoring, causality and uncertainty in a stratospheric context, *Pageoph*, 118: 643-660.

Radziejewski, M., Bardossy, A. & Kundzewicz, Z. W. (1998) Intecomparison of tests for detection of changes in long time series of river flow. Proceedings of the Second

International Conference on Climate and Water, Espoo, Finland, 17-20 August 1998, Vol. 3, 1120-1129.

Radziejewski, M., Bardossy, A. & Kundzewicz, Z. W. (2000) Detection of change in river flow using phase randomization, *Hydrol. Sci. J.,* 45(4) 547-558.

Reynard, N. S., Prudhomme, C. & Crooks, S. (1999) Impact of climate change on the flood characteristics of the Thames and Severn rivers. In: Balabanis, P., Bronstert, A., Casale, P. & Samuels, P. (Eds) *The Impact of Climate Change on Flooding and Sustainable River Management.* RIBAMOD – River Basin Modelling, Management and Flood Mitigation, Concerted Action, Proc. of the final Wallingford Workshop, European Commission, Luxembourg, EUR 18287 EN, p. 49-63.

Robson, A., Bardossy, A., Jones, D. & Kundzewicz, Z. W. (2000) Statistical methods for testing for change. Chapter 5 in: Kundzewicz, Z. W. & Robson, A. (ed.) *Detecting Trend and Other Changes in Hydrological Data*; World Climate Programme – Water, World Climate Programme Data and Monitoring, WCDMP-45, WMO/TD – No. 1013, Geneva, May 2000, pp. 49-85.

Robson, A. J. & Reed, D. W. (1996) Non-stationarity in UK flood records. Flood Estimation Handbook Note 25, Institute of Hydrology, Centre for Ecology and Hydrology, Wallingford, UK, October 1996.

Tomasino, M. & Dalla Valle, F. (2000) Natural climatic changes and solar cycles: an analysis of hydrological time series, *Hydrol. Sci. J.* 45(3): 477-490.

Trenberth, K. E. (1998) Atmospheric moisture residence times and cycling: implications for rainfall rates and climate change, *Climatic Change,* 39: 667-694.

Walsh, K. & Pittock, A. B. (1998) Potential changes in tropical storms, hurricanes, and extreme rainfall events as a result of climate change, *Climatic Change*, 39: 119-213.

Zhang Keqi, Douglas, B. C. & Leatherman, S. P. (1997) East Coast storm surges provide unique climate record, *Eos*, Sept. 16, 1997, 389 and further pages.

Climate change impacts on river flooding: A modelling study of three meso-scale catchments

LUCAS MENZEL[1], DANIEL NIEHOFF[1], GERD BÜRGER[2], AND AXEL BRONSTERT[1,3]

[1] *Potsdam Institute for Climate Impact Research (PIK), Global Change and Natural Systems Department, Potsdam, Germany*

[2] *Potsdam Institute for Climate Impact Research (PIK), Climate System DepartmentPotsdam, Germany*

[3] *University of Potsdam, Institute for Geo-Ecology, Potsdam, Germany*

Abstract: Whether the high number of river flood disasters in recent years in different European countries and many other areas of the globe are triggered or worsened by human activities has been the subject of a great deal of debate. Possible anthropogenic activities leading to increased flood risk include river regulation measures, intensified land use and forestry, and emissions of greenhouse gases causing a change in the global climate. This article discusses the latter by presenting modelling studies of three meso-scale catchments in Germany. These catchments represent different conditions of land-use, landscape morphology and climate type, therefore showing different dominating flood generation processes. The results of the case studies show how changes of *temperature* and *rainfall regime* can lead to significant changes in flood risk. A development towards both an increase or a decrease of the frequency and/or magnitude of flood events is possible, depending mainly on the altered timing of snow accumulation and snowmelt and of a possible shift in rainfall seasonality and intensity.

M. Beniston (ed.),
Climatic Change: Implications for the Hydrological Cycle and for Water Management, 249–269.

1. INTRODUCTION

Whether the high number of river flood disasters in recent years in different European countries and many other areas of the globe are triggered or exacerbated by human activities has been the subject of much debate (e.g., Bronstert, 1996, Bronstert et al., 1999, Kundzewicz & Takeuchi, 1999; Kundzewicz & Kaczmarek, 2000; Longfield & Macklin, 1999). Anthropogenic activities which might lead to increased flood risk include river regulation measures, intensified land-use and forestry, and anthropogenic climate change.

This paper discusses the latter by presenting the results of three modelling studies investigating the link between climate change scenarios and the subsequent possible change of flooding conditions in river systems. The decisive questions for evaluating the effects of anthropogenic climate change on possibly altered flooding conditions of river systems are the following:

- how will the regional - or even local - climate conditions *relevant for flood events* look like if the global climate changes ? This question addresses the value of information about the future climatic conditions available at the global or continental scale (usually derived from General Circulation Models) for the derivation of climate scenarios at the scale which is relevant for the generation of river floods (*climate downscaling*).
- what are the effects of an altered regional - or even local - climate on the flood generation processes ? This question addresses the hydrological processes which are relevant for the generation of flood in a specific catchment *(hydrological origin of a flood)*. This shows the need to account jointly for possibly altered climate conditions (as mentioned before), catchment morphology, soil surface conditions and vegetation.
- What are the effects of climate change on flooding in comparison to other anthropogenic impacts and in comparison to the "normal" natural variability ? This is important if one aims at quantifying the change of flood risk. Other anthropogenic activities relevant here are such as land-use change, river training measures, and an intensified settlement in flood plains.

In addressing these questions it is useful to differentiate different categories of floods according to the relevant weather type, specific flood generation processes involved, and to the spatial and temporal scales of the flooding events investigated.

Extensive, long-lasting floods ("plain floods") occur in larger catchments (approx. 1,000km² - 300,000km²) and are generally caused by rainfall lasting several days or weeks, often associated with the melting of snow and with high antecedent soil saturation. The inundations caused by this flood type occur mostly in plain areas when the river dikes can no longer contain flood discharges. This can lead to flooding over wide areas, as occurred, e.g., during the flood of the rivers Rhine and Maas in December 1993 and in early 1995, the flood of the Oder/Odra in July 1997, and in Yorkshire in autumn 2000.

Local, sudden floods ("flash floods") occur in small catchments (e.g., those of less than 100 km^2 - 1,000 km^2) and are mainly caused by intense localised precipitation (e.g. thunderstorms or hurricanes). Flash floods occur primarily in hilly or mountainous areas because of prevailing convective rainfall mechanisms, sometimes intensified by thin soil layers and high runoff velocities. In general this type of flood event is short in duration, but is nonetheless frequently connected with severe damage.

2. A SUMMARY OF THE REQUIREMENTS FOR MODELLING CLIMATE CHANGE IMPACTS ON RIVER FLOODS

A *quantitative* analysis of the impacts of climate change on flooding conditions requires simulations of the climatological-hydrological system. The models on which the simulations are based must give an adequate representation of the system dynamics *relevant for flood generation.* This means that both the *relevant* internal processes of the climatological-hydrological system and the *relevant* external forces (boundary conditions) must be part of – and remain recognisable within – the system to be modelled. This calls for an *integrated (or coupled) approach* of climatological and hydrological model applications.

For the analysis of flooding conditions, it is sufficient to realise the integration by means of a one-way-coupling, i.e. the climate model is coupled to the hydrological model by prescribing the climatological forcing of the hydrological model. A two-way-coupling (meaning that the climatological model also depends on feedback information from the hydrological model) is not necessary at the temporal and spatial scales which are relevant for flooding analysis.

In this section we summarise the features of climate and hydrological models and the requirements for impact analysis of climate change on river flooding.

2.1 Requirements for climate modelling

The most powerful tool available today for climate research are coupled global atmosphere-ocean circulation models (GCMs) which simulate the three dimensional climate system. An assessment, however, of possible changes of flood characteristics resulting from climate change requires models which give reliable information on much smaller scales than the global or continental scale. The following information is required:

- A realistic description of changes in precipitation regime. This includes changes of the average value, of the statistical features in space and time, and of frequency or seasonality. Scenarios which only give changes of the average value are hardly sufficient.
- A realistic description of changes in temperature. This is particularly important for catchments where floods can be composed of *both* rainfall and snowmelt events, which is the case, e.g., in many central and northern European catchments.

A variety of techniques has been developed to derive the climate forcing required for assessing the hydrological, basin-wide impacts of climate change. This includes dynamical downscaling techniques (i.e. the use of regional climate models); empirical downscaling (circulation type techniques); stochastic weather generators; and simple interpolation approaches. The most important of these are dynamical and empirical downscaling.

Regional climate models (RCMs) have been applied for some time. In contrast to general circulation models, they cover only a section of the globe, which can be modelled at a finer spatial resolution (e.g., using grid widths of 0.5°). The climatic conditions at the boundaries of the regional sections are predetermined by the results from GCM experiments. In comparison to a GCM resolution, the spatial resolution of an RCM is more adequate for the estimation of flood-relevant precipitation, particularly with regard to weather conditions connected with large-scale precipitation fields. However, to obtain accurate information *on the location, quantity and intensity of precipitation* and on *changes in precipitation characteristics* necessary for the analysis of flood generation processes, the models are not yet sufficiently spatially detailed and accurate (see, e.g., Xu, 1999).

Empirical downscaling bridges the two different scales by establishing (statistical) relationship between large-scale features simulated reliably by the GCMs (such as geopotential height fields) and regional or local climate variables (such as temperature and precipitation at a certain location). While the dynamic approach utilises physical principles only, the empirical approach is anchored in the observed fact that weather phenomena are often caused by the conditions of the prevailing large-scale atmospheric circulation. The method of GCM-downscaling, in its various forms, has by now been well established as an appropriate and necessary tool for impact assessment studies, so it shall be enough to refer to the review article of Wilby & Wigley (1997). Empirical downscaling techniques have been applied in a series of studies (for an overview, see, e.g., Wilby et al., 1999), but only a minority of these studies have been carried out in the context of climate change and flooding.

2.2 Requirements for hydrological models

Transferring climatic conditions into river runoff requires the application of an appropriate hydrological model. An overview of some popular hydrological models was put together by, e.g., Singh (1995). It can be of high importance that certain features of the changed climatic condition (e.g. changed seasonality or rainfall intensity) may lead to a relevant change of the hydrological system, i.e. to a different relevance of some hydrological processes for storm runoff composition. If this mechanism is to be analysed, it is important to apply a hydrological model which includes the relevant runoff generation processes. It goes beyond the scope of this paper to review the existing hydrological models as regards to their ability to describe the hydrological processes relevant for flood generation. However, it is possible to derive some criteria on which the selection of hydrological models for this kind of impact analysis should be based.

2.2.1 Representation of the soil zone

The behaviour of the soil surface and the unsaturated zone is regarded as crucial for the quick rainfall-runoff process. Models which lump different runoff generation processes are not advisable, in particular if a change of the hydrological processes is included in the analysis. If the climate conditions have a *second order impact on the runoff generation conditions*, e.g. by causing crusting of the soil, altering the characteristics of the vegetation cover etc., this needs to be taken into account, especially for long-term projections.

2.2.2 Spatial distribution

The model applied should operate in a spatially *distributed manner*, with approximately the same resolution as the one used for representing the climatic and catchment conditions. A distributed approach is essential if the flood generation processes are highly variable in space, especially if this variability can be attributed to soil and vegetation characteristics.

2.2.3 Temporal resolution

If rainfall intensity is relevant for flood generation, this should be reflected in the temporal resolution of both the climate and the hydrological modelling time step. This is typically the case in small catchments and/or if infiltration excess induced overland flow (Hortonian overland flow) contributes significantly to flood generation.

3. CASE STUDIES

3.1 Modelling the impact of climate change on the flood situation in a forest dominated catchment: The Selke study

This study analyses the possible impact of anthropogenic climate change on flood generation by means of a modelling study in the rural meso-scale catchment of the Upper Selke. This catchment is located in the Harz Mountains in Northern Germany and covers an area of 105 km². The topography is characterised by moderately steep mountains, with altitudes ranging between 330 m asl. and 600 m asl. The dominating land-cover types are forest (58.2%) and pasture and arable land (38.2%). The study focuses on climatological conditions with heavy precipitation, their possible changes due to anthropogenic climate change and on the corresponding flooding characteristics.

3.1.1 The floods in the hydrological year 1994

The Selke catchment has been affected by several severe floods during recent decades. In particular during the hydrological year 1994 (November 1993 until October 1994) a series of severe floods struck this catchment. The most extreme flood occurred on 13 and 14 April 1994, with a peak discharge measured at the catchment outlet of 74 m³/s. The return period of this event is estimated to be above 100 years. This flood caused huge damage in this and neighbouring catchments, causing total damage estimated at about € 125 Million.

The extreme flood in April 1994 was caused by a combination of extreme rainfall (over 80 mm in one day) and wet antecedent soil moisture conditions. It was a purely rainfall-induced flood, i.e., there was no snow cover left in the catchment at the beginning of the storm rainfall period. The other flood events during the hydrological year 1994 (19 March 1994 and 31

December 1993 / 1 January 1994) were caused by a combination of heavy rainfall and melting of the snow cover in the catchment.

The occurrence of three severe floods, of which the one in April was a catastrophic event, was the reason that the hydrological year 1994 was chosen as a reference simulation period for this study. It was the purpose to simulate this extreme hydrological year and to investigate the possible influence of a changed climatic forcing on such extreme events.

3.1.2 Derivation of climate scenarios

The climate scenario method derived for this study (see below) was applied as a basis for the following thought experiment: Looking only at the climate *statistics* excludes the possibility of basing the analysis on a specific, historic hydrological event or period. However, it is of particular interest to use an extreme, historic storm period as a basis and to speculate on how this record would look if 'transferred' to a different climate. Taking the specific observed weather record for the Upper Selke during the hydrological year 1994, how would this record change if 'transferred' to a future climate such as, say, that of the 2050s? We assume here that the specific sequence of events that led to the weather process (in 1994) remains the same, but takes place with modified climatic boundary conditions.

The weather transfer is based on two fundamentals: First, a long historic record defining the base climate of the station and, second, a downscaled climate simulation performed with the ECHAM4/OPYC3 general circulation model (Roeckner et al. 1996), which was driven by the IS95a CO_2 emission scenario 'business as usual' of the IPCC (Houghton et al., 1995). The climate scenario method applied for this study is based on the Expanded Downscaling (EDS) method developed by Bürger (1996), which can be regarded as a compromise between deterministic, linear regression and stochastic methods based on weather type. EDS renders, for each variable from the observational record, a corresponding daily time series representative for the IS95a scenario, that is, for the period 1860-2100. Please note that the transfer is not specific to EDS; we could have used any other daily climate scenario to drive the weather transfer. The transfer itself is carried out along the following lines:

- From the historic record one defines a set of so-called probit parameters, ψ_0, that define the monthly base climate of the station. These parameters can be viewed as a generalisation of 'mean' and 'variance', applicable equally to normally and non-normally distributed variables, such as, e.g., precipitation. In the same way as using mean and variance for normal variables one can transform, in a one-to-one way, any given time series with the help of the probit parameters into one that is N(0,1)-distributed.

- The ψ_0-transformation is applied to the hydrological year 1994, rendering a 'normalised' series ξ. ξ contains all temporal information and relative amplitudes of the hydrological year 1994, including covariability of the station variables.
- Another set of probit parameters, ψ_S, is determined for the climate of the 2050s from the IS95a scenario by using the period 2035-2065. The series ξ is back-transformed using the parameters ψ_S, resulting in a 1-year precipitation series that represents the events of the hydrological year 1994 transferred into the climate of the 2050s.

In other words, it is assumed here that the specific sequence of events that led to the weather process remains the same, but takes place with modified climatic boundary conditions. This thought experiment makes it possible to compare directly the different hydrological responses on storm weather periods under present and under future climatic boundary conditions. More details of this particular downscaling procedure are explained in Bronstert et al., 1999. The resulting time series for the years 1994 and 2050 and their differences are shown in Fig. 1a (temperature values) and Fig. 1b (precipitation values).

The difference series looks rather erratic. This is meaningful even for the much simpler case of temperature T (which is almost Gaussian). In this case the mean, μ, the variance, σ^2, and the anomaly ξ itself determine this difference according

$$T_s - T_o = \mu_s - \mu_o + (\sigma_s - \sigma_o)\xi$$

As μ and σ vary monthly and ξ varies daily a rather complicated interplay takes place. For example, a positive increase in the mean and a comparable decrease in variance lead to a difference whose sign depends on the actual ξ. Therefore, the interpretation must be carried out very carefully and should take into account all of the above constituents. A dominant feature is the persistent positive temperature anomaly from November to March, and another one in June and July. It seems that extreme temperature anomalies are damped, as indicated by the positive and negative signals in February and May, respectively. This might be caused, for example, by a systematic negative bias in GCM variability. For precipitation, the main signal is the cluster of heavy events in March that become heavier under the 2050s climate. The maximum event on April 13/14 is weaker by about 10mm, similar to the May-June clusters. Again, the reduction of heavy rainfall for this time of the year might well be attributable to a GCM deficiency. Another source of error might arise from the downscaling method for which extreme values are very critical. On the other hand, we might be seeing climatic change. To reach to a more informed judgement it would certainly be advisable to apply the weather transfer to more than one GCM, and preferably to the control runs, too.

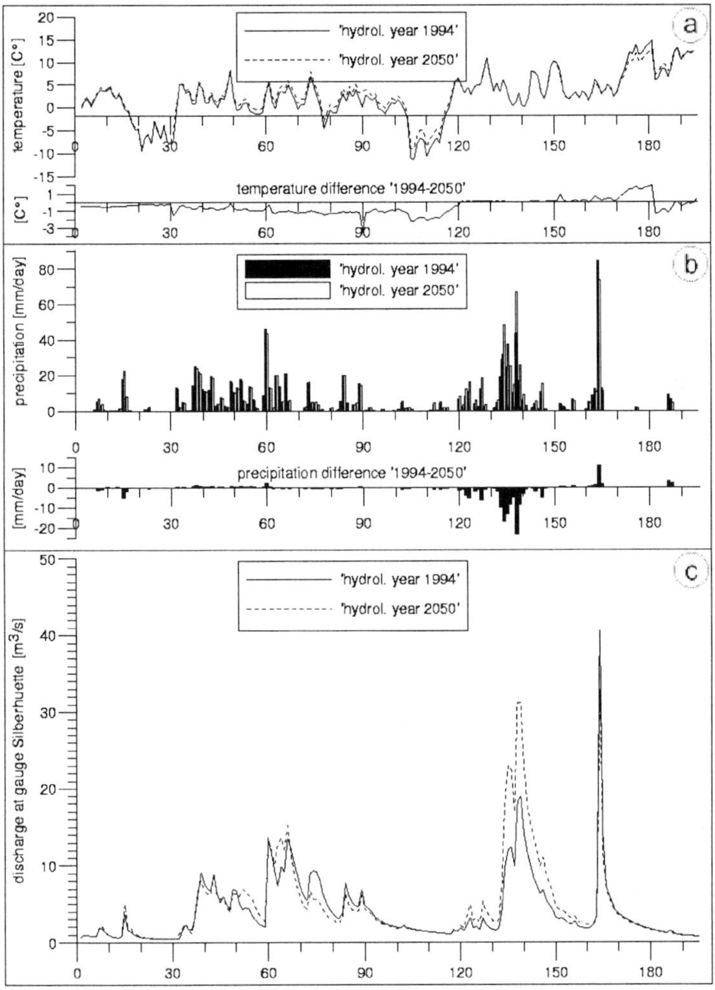

Figure 1. Temperature, precipitation and discharge hydrographs for the Upper Selke Catchment for the hydrological year 1994 and scenario conditions 2050: a) Temperature values (measured values and climate change scenario) at the climate station Harzgerode and the differences; b) Precipitation values (measured values and climate change scenario) at the climate station Harzgerode and the differences; c) Simulated hydrographs for the Selke at the gauge Silberhütte for the hydrological years 1994 and 2050

3.1.3 Results of the hydrological modelling

The hydrological model chosen for this study is BROOK (Federer & Lash, 1983), which has been designed especially for small forested catchments in mountainous landscapes. BROOK is a semi-distributed hydrological model, which contains most of the vertical and lateral hydrological fluxes and storage processes. Storm runoff generation is represented by saturation excess and snowmelt processes. Infiltration excess (Hortonian overland flow) is not included, subsurface stormflow is approached by a simple linear storage concept. We considered these assumptions about the storm runoff generation processes to be appropriate for this catchment, because infiltration excess overland flow is not relevant in this environment. More details of the model and the calibration applied are given in Bronstert et al. (1999).

Fig. 1c shows the resulting hydrographs gained from the different meteorological input ('hydr. year 1994' = present conditions; 'hydr. year 2050' = scenario conditions). The differences in the calculated discharges are significant. There is a considerable increase in the peak discharge during the winter floods, particularly in March (day 139) from 19.0 m^3/s to 31.2 m^3/s (64% increase), which is due to both an increase in temperature and higher precipitation for the scenario conditions. The increase of this early spring flood –generated by high rainfall *and* snow melt – highlights the importance of snow accumulation and melting conditions and snow seasonality for flood generation. On the other hand, there is a reduction in the peak discharge in April (day 164) from 40.7 m^3/s to 33.0 m^3/s (23% decrease), which is due to the reduced rainfall for the scenario conditions.

This example demonstrates the importance of the climatic boundary conditions ("climate scenarios") for the flood simulation. The difference in meteorological input was small (in the range of -1°K and +2°K or +15 mm/day and –10 mm/day), but the differences in the hydrological response were rather significant. The method of deriving the scenarios is very important, in particular for studies of climate change impacts on floods. The use of "simple scenarios" (e.g., „general increase of 2°K", or „general increase of precipitation by 10%") is not an appropriate method. Tests for the Selke catchment have shown that the discharge results obtained with simple climate scenarios as climate forcing are rather arbitrary, i.e. one can produce any hydrograph change by choosing different simple climate change assumptions, see Bronstert & Köhler (2000).

3.2 Modelling the impact of climate change on flood generation in an agriculture dominated catchment: The Lein study

The impact of climate change on storm-runoff generation in an agriculturally used landscape has been investigated by means of a modelling study for the Lein catchment, which is a meso-scale tributary to the Rhine, situated in south-western Germany. The mainly loess-covered Lein catchment is dominated by agricultural land-use (68 % arable land, 25% forest, 7% residential area) and covers an area of approximately 115 km^2. For this typical agricultural catchment, climate scenarios have been generated by the help of the same "transfer approach" which was explained above for the Selke catchment: Taking the specific observed weather record for the Lein during the hydrological year 1994, we assume that for the climate scenario for the years from 2070 to 2097, the specific sequence of events remains the same as during the actual weather process in 1994, but takes place with modified climatic boundary conditions.

3.2.1 Hydrological modelling technique

The hydrological response to the scenarios is being simulated using a modified version of the process-oriented and spatial distributed hydrological model WaSiM-ETH (Schulla, 1997). WaSiM-ETH has been extended in several ways (Katzenmaier et al., 2000), now including descriptions for macropore infiltration, siltation effects and the retention capacity of urban sewer systems. Hence, this model is well suited to investigate the effects of a possible change of runoff generation mechanisms, which might be a second order consequence of land-use changes or climatic change.

Due to their high spatial and temporal variability, global circulation models are not capable of reproducing *convective* storm events. Correspondingly, the same is true for the *expanded downscaling* approach. Therefore, hydrological modelling has been restricted to the hydrological winter period lasting from November to April. During this period, rainfall in this region is mainly bound to *advective/cyclonic* precipitation events.

3.2.2 Results of the hydrological simulations

The upper diagram in Fig. 2 shows the *differences in daily precipitation* between the conditions in 1994 and the climate scenario for the end of this century. The scenario suggests a decrease in precipitation for November and December as well as a pronounced increase mainly in March and April. This pattern is partly a product of the actual weather sequence in 1994, but also an indication for a general *shift in the precipitation regime* for the future climate. As the lower diagram in Fig. 2 illustrates, this shift is also reflected in catchment response, i.e. by an increased flood event in April.

Another result of the climate scenario is a *higher variability* in the range of *extreme precipitation*. One manifestation of this behaviour can be detected round the 20th of December 1994, where the maximum of future daily precipitation exceeds the amount observed under present conditions – despite a general decrease of rainfall in December. In this case, the consequence is a scenario flood peak which is as high as the one simulated for the actual year 1994.

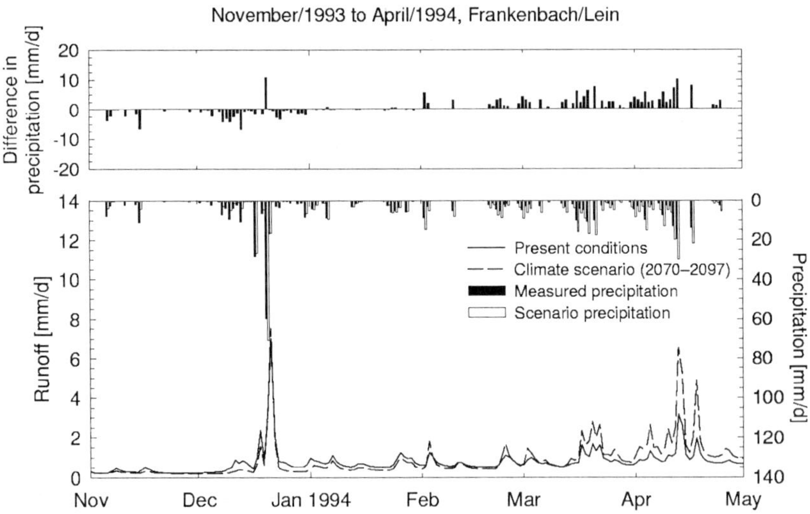

Figure 2. Simulated hydrographs for the Lein Catchment at the gauge Frankenbach for the hydrological year 1994 and scenario conditions 2100

3.3 Modelling the impact of climate change on the flood situation in a large catchment: The Mulde study

Within the framework of large-scale hydrological modelling of the Elbe river and its tributaries (approximately 150 000 km² both in Germany and the Czech Republic) the Mulde river has been selected for a detailed analysis of climate change impact on flooding (Menzel et al., 2000). The Mulde catchment covers an area of 6,105 km² at the transition zone between the highlands at the Czech-German border (ranging up to 1200 m asl.) and the lowlands of north-eastern Germany. Forest predominates in the mountainous areas while agriculture and urban areas represent the main land cover types in the plains. Fig.3 gives an overview of the runoff regime in the years 1981 – 1988 at gauge Bad Düben, the outlet of the Mulde catchment. The discharge reflects the influence of winterly snowpack and snowmelt in combination with rainfall in the early springtime. Therefore, the highest flood events usually occur between January and April. Strong convective rainfall causing significant floods is relatively rare (e.g., see the flood event in summer 1983 in Fig. 3).

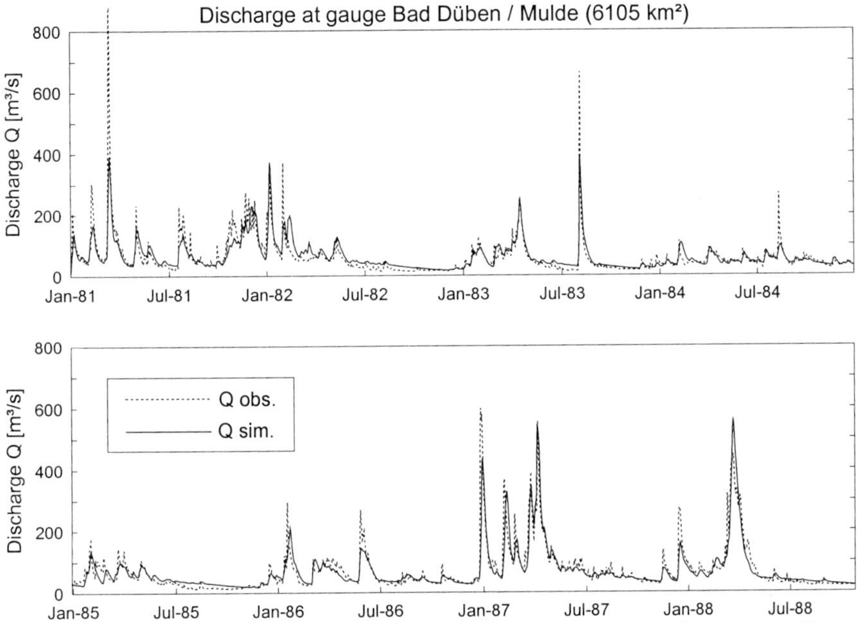

Figure 3. Measured and simulated daily discharge, using HBV-D with measured climate variables. Mulde at gauge Bad Düben, years 1981–1988.

The hydrological model applied is the distributed version of the conceptual rainfall-runoff model HBV (Bergström, 1995), called HBV-D (Krysanova et al., 1998, 1999). Based on a sound physical description, HBV-D combines relative simplicity and low data demand to cover the most important runoff generating processes by using robust model structures. HBV-D consists of three main components: 1) snow accumulation and snow melt, 2) the simulation of soil moisture and runoff and 3) a response and river routing routine. Continuous simulations of river discharge by use of HBV-D were carried out for the Mulde on a daily time step over the eight successive years 1981 – 1988 (Fig. 3).

3.3.1 Results of the hydrological and climate modelling

HBV-D was repeatedly applied to simulate discharges with different climate input. First, observed daily data from 75 precipitation and 4 climate stations have been used. In contrast to the case studies described in sections 3.1 and 3.2 the analysis of the impact of climate change on flooding in the Mulde catchment is not based on a specific, historic hydrological period transferred to modified climatic boundary conditions. For the Mulde study EDS has been applied to observed Atmospheric Circulation Fields (ACFs) for the reference period 1961 – 1990. This enables a direct comparison of simulated discharge using climate input from measurements and from downscaled ACFs, thus allowing to test the performance of the selected method for regional hydrological modelling. Finally, EDS was applied to deliver regional climate data from ACFs, derived from climate change simulation runs over the period 1860 – 2100. This allows a continuous hydrological simulation over longer periods, indicating possible trends in modelled discharge due to climate change.

Generally, the application of HBV-D for the Mulde basin with measured climate forcing gave good results (Fig. 3) in terms of reflecting the observed discharge. Except for 1981 the spring floods were well simulated. The exceptional and strong convective event in summer 1983 was poorly reproduced by the hydrological model. This is probably due to the limited areal reach of this single event, which may have caused its poor representation in the measured rainfall patterns used for the simulations.

In the next step HBV-D has been run with climate input as determined by the application of EDS and observed ACFs. Fig. 4 shows a comparison between observed discharge and the simulated discharge using HBV-D and the local climate from observed and downscaled ACFs. Fig. 4 illustrates that the temporal occurence between measured and simulated flood discharges are in good agreement. Nevertheless, the simulated hydrograph using EDS seems to reflect a lower tendency for high runoff (both in frequency and in peak volume), especially during winter/spring times. This is probably due to a lower capability of EDS to produce and downscale heavy rainfall events, implying that small scale heavy events can not be predicted from large scale circulation patterns. In some cases reduced runoff appears to be coupled to a simulation of relatively low values of snow accumulation and related, shorter melting periods.

For the future runoff conditions it is worth to analyse the development of temperature and precipitation in the Elbe catchment and its tributaries over the last 30 years. Menzel et al. (2000) showed that for the Mulde catchment both the analysis based on measured climate data and the downscaled GCM simulations show a clear tendency towards increased areal temperatures.

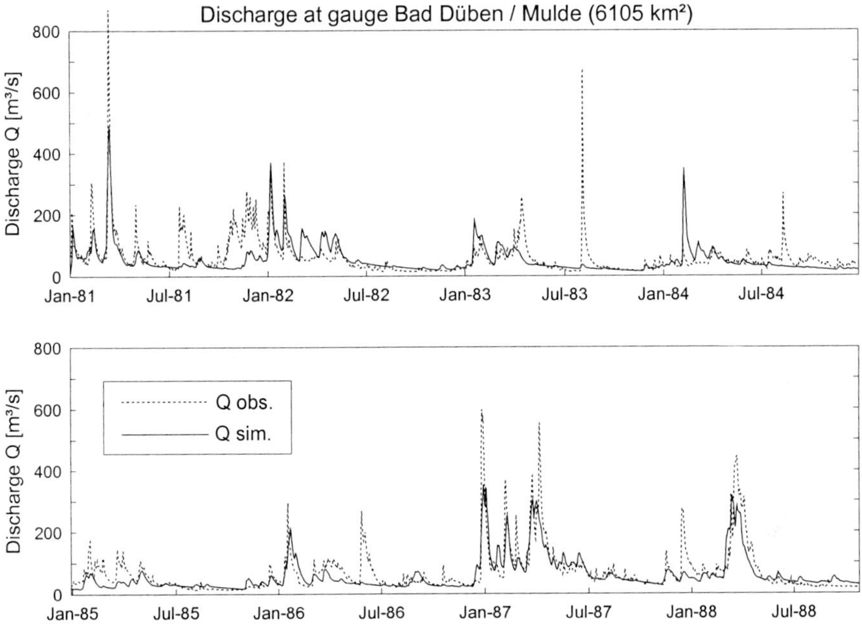

Figure 4. Measured and simulated daily discharge, using HBV-D with local climate from observed and downscaled atmospheric circulation fields. Mulde at gauge Bad Düben, years 1981–1988.

This tendency is predicted to be maintained and even to be amplified within the next 100 years. Due to its high temporal variation the future development of areal precipitation is less clear, but one can assume a general tendency towards reduced values within the next 100 years in the investigated area.

A decrease in annual areal precipitation and a simultaneous increase in temperature doesn't automatically imply a decrease of peak discharges. To test this hypothesis, annual peak discharges HQ have been derived from the simulated, daily discharge data.

In Fig. 5 HQ-values based on measurements are opposed to the respective simulations for the reference period 1961–1990. It is important to state that the two data series are completely independent since the simulated discharge is based on the climate scenario run. Nevertheless, in spite of the high variations in peak discharge from year to year it appears that the ranges of the two curves are in the same order. The development of the simulated HQ-values within the next 100 years reveals a slight tendency towards reduced HQ's. This is especially true for the second half of the 21st century when the simulated HQ-values clearly fall below the line of mean peak discharge (Fig. 5). Nevertheless, it is difficult to draw general conclusions since the year to year variations of HQ are quite high. The application of extreme value statistics reveals high uncertainty. However, there is a clear tendency towards reduced peak discharges in the period 2001–2100 for the Mulde. Similar results were obtained for other investigated sub-catchments of the Elbe basin.

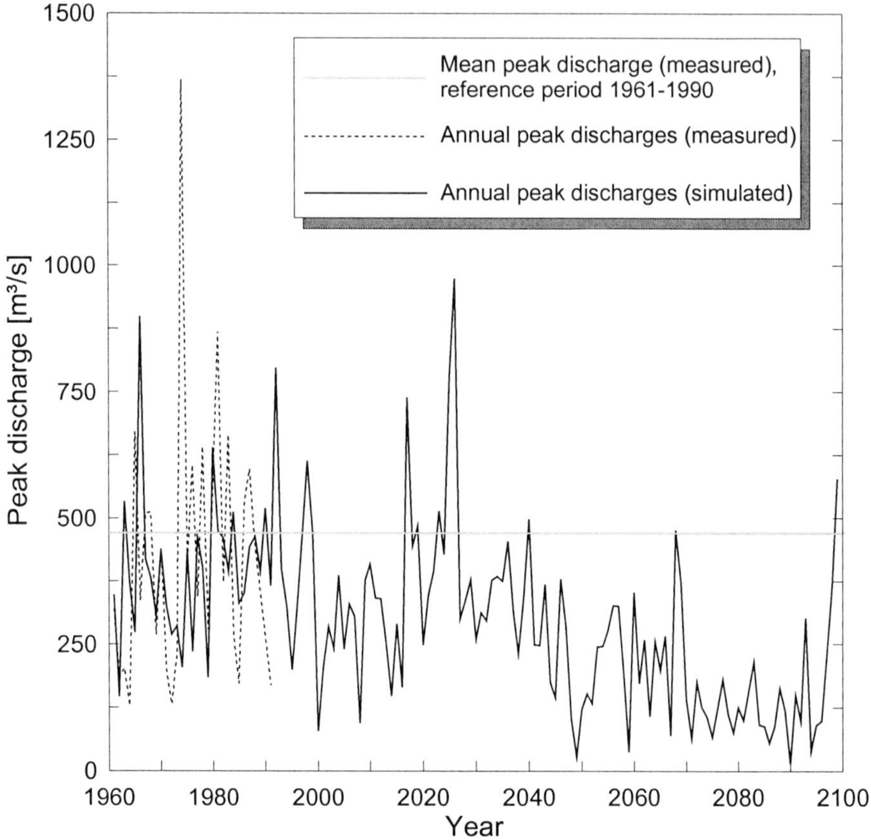

Figure 5. Measured annual peak discharges at gauge Bad Düben / Mulde (dashed line; reference period 1961-1990) and simulated annual peak discharges using downscaled climate input from the GCM scenario run (solid line).

4. SUMMARY OF RESULTS

- The case study of the Selke catchment took as its starting point a small, fairly natural catchment in a mountainous area in combination with the hydrological year 1994, during which three major flood events (all winter or early spring floods) occurred. Its purpose was to simulate this extreme hydrological year and to investigate the possible influence of changed climatic boundary conditions on such extreme events. The results indicate that anthropogenic climate change can have an important impact on flood generation. Even relatively small changes of the meteorological forcing can cause much stronger changes in catchment runoff. It also became clear that a change in climate can have both aggravating and mitigating effects, depending on the change of the snow melt season and on the change in precipitation patterns.

- The selection of the appropriate climate scenario approach is very important in studies of climate change impacts on flood generation, and it has to be recommended that sophisticated methods should be applied, taking into account the general climatic conditions of the region studied and possible future changes of these conditions (average values *and* variations).
- The simulations which have been carried out for the Lein catchment underline the findings obtained for the Selke, especially concerning possible consequences of an altered precipitation regime. In addition to that, the climate scenario implies an increase in the variability of extreme precipitation for the Lein region. A comparison of catchment response to the *actual* weather sequence of the year 1994 and the *altered* weather sequence under scenario conditions demonstrates, how this growing variability can lead to an aggravation of flood risk despite a general decrease in rainfall.
- The continuous simulation of discharge over the next 100 years using a climate scenario for the Mulde catchment reveals a high uncertainty about the future development of flood risk for this case study. The simulations based on the regionalised GCM scenario predict a weak tendency towards reduced peak discharges. This seems to be valid for a major part of the large-scale Elbe catchment. Based on the applied climate change scenario, the simulations indicate that a future problem might arise from reduced water availability in the Elbe catchment.

5. CONCLUSIONS AND OUTLOOK

The possible impacts of climatic change on river flooding are of high relevance both to the public and to the scientific community. It was the purpose of this paper to demonstrate how a combination of refined climate downscaling and hydrological modelling can contribute to a quantitative analysis of the impacts of climate changes on river flooding.

5.1 Outcome of the case studies

The climate impact study conducted for the meso-scale Selke catchment shows how changes of *temperature* and *rainfall regime* can lead to significant changes in flood risk. In the case of the Selke, the altered timing of snow accumulation and snowmelt combined with a time shift in the rainfall regime may result in an unfavourable superposition of snowmelt and high precipitation in early spring. In the simulation example, a considerable increase of 64% of the peak discharge during a spring flood in March was calculated, which was due both to an increase in temperature and higher precipitation for the scenario conditions.

The importance of the *meteorological boundary conditions* and *process-representation* within hydrological models was demonstrated by simulations in the Lein catchment. These results indicate a rising flood risk towards the end of this century. In accordance with the results obtained for the Selke, this is due to a *shift in the rainfall regime* from early winter to spring. Furthermore, the hydrological simulations illustrate how an increase in *rainfall variability* can lead to higher flood risk despite a general decrease in monthly precipitation. For large catchments such as the Rhine basin, where extreme floods can be caused by a *superposition* of a snowmelt-originated flood in a high altitude sub-catchment *and* a rainfall-originated flood in a lower lying sub-catchment, the shift of rainfall seasonality and temperature increase leads to a higher probability of a *coincidence* of extensive rainfall and snowmelt runoff. The relevance of this hydro-climatological feature for floods in the Rhine catchment has been mentioned already in the report of Grabs (1997).

The example for the third case study, the southern part (Mulde) of the large Elbe catchment, shows that the same procedure to produce regional climate scenarios, i.e. the application of ECHAM4/OPYC3 and EDS leads to a *decrease of flood peak discharges* within the next 100 years. This might be interpreted as a result of uncertainty of the climate model and downscaling. On the other hand this may also be due to a different trend in future rainfall conditions. This is an example for the value of more profound research on the regional variations of climate change impacts.

5.2 What are the implications of these studies for climate change impact assessment?

If the impact of changes to a hydrological system is being simulated, the *relevant processes* need to be recognisable within the models used for this purpose. Regarding the impact of *climatic changes*, this demand refers particularly to the representation of *meteorological boundary conditions* within both the climate model and the hydrological model.

The derivation of *climate scenarios* is limited by the quality of the information available about the future development of our earth's climate. In many regions of the world, including large parts of Europe, the results delivered *directly* by GCMs are inconsistent. Therefore, it is indicated to apply *sophisticated regionalization methods* to the climate scenarios, taking into account the specific climatic conditions of the study area. Bronstert & Köhler (2000) have demonstrated for the Selke catchment that the use of simple scenarios, such as a simple linear trend in temperature or precipitation change, is not an adequate way to cope with this problem.

The hydrological model should take into account those meteorological features which are most important for flood generation, such as high intensity rainfall and snowmelt conditions. This calls for an *adequate temporal resolution of the model* and for a *proper treatment of snowmelt processes*. If the climate conditions have a *second-order impact on runoff generation conditions*, e.g. by causing siltation or crusting of the soil, this needs to be accounted for, especially for long term projections.

5.3 Coping with uncertainty

Though the authors believe that combined climate and hydrological models are an adequate tool with which to investigate climatic change impacts on flooding conditions, it has to be emphasised that there are limitations to such modelling studies, since climatic and hydrological modelling are accompanied by a high degree of uncertainty. This uncertainty is the consequence of insufficient data availability and quality (data uncertainty), high spatial heterogeneity and/or temporal variability (uncertainty due to sub-scale heterogeneity and variability), insufficient knowledge on the physical appearance and the stochastic features of the processes involved, in particular during extreme precipitation periods (process uncertainty), and simplifications inherent in the model structure (model uncertainty).

For non-complex hydrological models (small number of parameters), a rigorous uncertainty estimation can be carried out, such as described by Freer et al. (1996). This procedure, however, is not easily transferable to detailed process-oriented hydrological models like, for instance, the one applied for the Lein catchment (see section 3.3). Due to the large number of parameters and long computing times involved, the uncertainty estimation techniques developed so far are hardly applicable in that context.

One must be aware of the fact that a considerable amount of uncertainty remains unresolved regarding the climate situation of extreme flood events (long return periods). This is mainly caused by the immense natural variability of the climate system, which requires long estimation periods for the statistical derivation of long return periods (= very rare events). With the current setting of climate downscaling approaches, using the information of - relatively short - observed and simulated circulation fields, one is bound to return periods in the order of decades; everything beyond belongs to the realm of speculation. But even for the shorter return periods one must expect that, for a given global emission scenario, different climate models coupled to different subsequent downscaling methods generate a whole spread of local responses. Currently, there exists no universally accepted procedure which allows for the quantification of this residual uncertainty. For a thorough impact assessment it is suggested, at least, to apply as many scenarios and downscaling methods as possible, which should provide a rough picture of the significance of simulated results.

6. ACKNOWLEDGEMENTS

The work reported here has been funded by the European Union, DG XII (EUROTAS-project), the German Federal Environmental Agency (UBA), and the EU DG XVI (IRMA-Programme).

7. REFERENCES

Bergström, S., 1995: The HBV model. In: Singh, V.P. (Ed.): Computer Models of Watershed Hydrology. Water Resources Publications, Highlands Ranch, Colorado, pp 443-476.

Bronstert, A., 1996: River Flooding in Germany: Influenced by Climatic Change? Physics and Chemistry of the Earth, 20(5-6), pp 445-450.

Bronstert, A., Bürger, G., Heidenreich, M., Katzenmaier, D. and Köhler, B., 1999: Effects of climate change on storm runoff generation: Basic considerations and a pilot study in Germany. In: Balabanis et al. (eds.): The Impact of Climate Change on Flooding and Sustainable River Management. Official Publications of the EU, Luxembourg, pp 325-340.

Bronstert, A. and Köhler, B., 2000: Simulation der Einflüsse anthropogener Klimaänderungen auf die Hochwasserentstehung: Eine Fallstudie in einem kleinen ländlichen Einzugsgebiet im Ostharz, Z. f. Kulturtechnik und Landentwicklung 41 (5), pp 1-7.

Bürger, G., 1996: Expanded downscaling for generating local weather scenarios. Clim Res 7, pp 111-128.

Federer, C.A. and Lash, D., 1983: BROOK: A hydrologic simulation model for eastern forests. University of New Hampshire, Research Report No 19, Durham/USA.

Freer, J., Beven, K.J., Ambroise, B., 1996: Bayasian estimation of uncertainty in runoff prediction and the value of data: an application of the GLUE approach. Water Resources Research, 32, 2161-2173.

Grabs, W. (Edt., 1997): Impact of climate change on hydrological regimes and water resources in the Rhine basin. CHR-Report No. I-16, Lelystad, 172 pp.

Houghton, J.T., Meira Filho, L.G., Callender, B.A., Harris, N., Kattenberg A. and Maskell, K. (Eds.), 1996: The Science of Climate Change. Contribution of Working Group I to the Second Assessment of the IPCC. Cambridge University Press, 572pp.

Katzenmaier, D., Fritsch, U. and Bronstert, A., 2000: Influences of land-use and land-cover changes on storm-runoff generation. In: Bronstert, A., Bismuth, C., Menzel, L. (Eds.):

European Conference on advances in flood research, PIK-Report No. 65, Potsdam Institute for Climate Impact Research, pp. 276-284.

Krysanova, V., Mueller-Wohlfeil, D.I., Bronstert, A. and Fleckenstein, J., 1998: Modelling runoff dynamics of the Elbe drainage basin: An application of the HBV model. In: Hydrology in a Changing Environment, Vol. 1: Global hydrological processes, ecological and hydrological interactions. John Wiley & Son.

Krysanova, V., Bronstert, A. and Menzel, L., 1999: Comparison of data requirements and performance of two semi-distributed hydrological models of different complexity. In: Quality, management and availability of data for hydrology and water resources management; Proc. of the International Conference. Koblenz, 22–26 March 1999, pp. 161-164.

Kundzewicz, Z.W.and Takeuchi, K., 1999: Flood protection and management: quo vadimus? Hydrological Sciences Journal. 44(3) 417-432.

Kundzewicz Z.W. and Kaczmarek Z., 2000: Coping with hydrological extremes. Water International, 25(1) 66-75.

Longfield S.A. and Macklin M.G., 1999: The influence of recent environmental change on flooding and sediment fluxes in the Yorkshire Ouse basin. Hydrological Processes, 13(7) 1051-1066.

Menzel, L., Bronstert, A, Bürger, G., and Krysanova, V., 2000: Environmental change scenarios and flood responses in the Elbe catchment (Germany). In: Bronstert, A., Bismuth, C., Menzel, L. (Eds.): European Conference on advances in flood research, PIK-report No. 65, Potsdam Institute for Climate Impact Research, pp. 139-148.

Roeckner, E., Oberhuber, J.M., Bacher, A., Christoph, M. and Kirchner, I., 1996. ENSO variability and atmospheric response in a global coupled atmosphere-ocean GCM. Climate Dynamics, 12, 737-754.

Schulla, J. (1997): Hydrologische Modellierung von Flußgebieten zur Abschätzung der Folgen von Klimaänderungen. Zürcher Geographische Schriften, Heft 69. Zürich.

Singh. V.P. (Ed.), 1995: Computer Models of Watershed Hydrology. Water Resources Publications, Highlands Ranch, Colorado, 1144pp.

Wilby, R. L. and Wigley, T.M.L., 1997. Downscaling general circulation model output: a review of methods and limitations. Prog. in Phys. Geogr. 21, 530-48.

Wilby, R.L., Hay, L.E. and Leavesley, G.H., 1999: A comparison of downscaled and raw GCM output: implications for climate change scenarios in the San Juan River basin, Colorado. Journal of Hydrology, 225, 67-91.

Xu C.Y., 1999: Climate change and hydrologic models: A review of existing gaps and recent research developments. Water Resources Management. 13(5) 369-382.

Impact of climate change and higher CO_2 on hydrological processes and crop productivity in the state of Brandenburg, Germany

VALENTINA KRYSANOVA AND FRANK WECHSUNG
Potsdam Institute for Climate Impact Research (PIK), Potsdam, Germany

Abstract: The paper: (1) describes an analysis of climate change impacts on hydrological processes and crop yield for the state of Brandenburg, Germany, and (2) suggests a methodology for climate change impact studies at the regional scale, which implies (a) the application of an integrated ecohydrological model for simulating the impacts of climate change on hydrological regime and crop yield simultaneously, and (b) studying the physiological effects of higher CO_2 in parallel with the 'climate change only' scenarios.

The analysis of climate change impacts was performed using the regional ecohydrological model SWIM. Two transient and two equilibrium scenarios of climate change, assuming 1.5 and 3.0 °C increase in temperature were applied. The scenarios were developed from ECHAMT21 GCM results using a statistical downscaling method. Three periods were compared: 1980 - 1990 (control period A), 2020 - 2030 (period B), and 2040 - 2050 (period C). The crop spectrum was restricted to winter wheat, winter barley and silage maize. The physiological effects of increased CO_2 concentrations on plants were studied considering: adjustment of the potential plant growth rate per unit of intercepted photosynthetically active radiation by a temperature dependent correction factor alpha; and assuming a CO_2 influence on transpiration at the regional scale (factor beta), which is coupled to the factor alpha.

According to the simulation results, actual evapotranspiration is expected to increase in four scenario periods of eight. Despite of the higher average temperature, it decreased in the other four scenario periods due to the low precipitation. Runoff and groundwater recharge always decreased, whereas groundwater recharge responded more sensitively to the anticipated climate change. The crop yield of cereals was decreased in all four scenarios considering the "climate change only" case. For maize, changes were not significant. The impact of higher atmospheric CO_2 (alpha factor) compensated partly for climate-related crop yield losses. The assumption that in addition a stomatal reduction in transpiration is taking place at the regional scale (alpha and beta factors) lead to further increases in crop yield, which were larger for

M. Beniston (ed.),
Climatic Change: Implications for the Hydrological Cycle and for Water Management, 271–300.

maize than for barley and wheat. Reduction in regional transpiration due to higher CO_2 may partly compensate for the decrease in runoff and groundwater recharge (feedback of vegetation on hydrological flows). Across the scenario runs, a high sensitivity of the main model outputs (three hydrological flows and crop yield) to precipitation was found.

1. INTRODUCTION

Over the last decade, there have been many studies of the potential impact of climate change on hydrological processes and agriculture at the regional scale. They are reported and reviewed by the Intergovernmental Panel on Climate Change (IPCC, Watson *et al.*, 1996 & 1998) and elsewhere (Reilly and Schimmelpfennig, 1999; Arnell, 1999; Parry *et al*, 1999; Krysanova *et al.*, 1999a). However, in most studies climate change impact was investigated for hydrological regime and agriculture separately, using different tools. One of the major challenges for current research in the frame of the IPCC is the adequate description and modelling of the complex interactions between climate parameters, hydrological and ecological processes at different scales. The ecohydrological model SWIM (Krysanova *et al.*, 1998a) has been developed as a tool to serve this purpose in the mesoscale river basins and at the regional scale. This paper aims to demonstrate the ability of SWIM to explore the potential changes due to climate change in major components of the hydrological cycle and crop yield simultaneously at the regional scale.

Agricultural crops are directly and indirectly affected by higher CO_2. The indirect effects result from the CO_2-induced climate shift, which influences crop growth and development. In addition, higher atmospheric CO_2 concentrations affect plants directly. The higher CO_2 level stimulates photosynthesis and increases stomatal resistance to transpiration. An increase in production for annual C3 crops averaging about 30% at doubled CO_2 level has been confirmed in many studies (Lawlor *et al.*, 1993; Kimball *et al.*, 1993; and Kimball *et al.*, 1995). This is less pronounced for C4 crops, and variations among crop varieties of the same crop and over years are high (Rötter and Geijn, 1999). Therefore, uncertainties concerning the magnitude of these direct effects and their consequences at the regional scale remain unresolved. Several feedback processes can actually diminish the initial effect at the leaf and plant level. Photosynthesis can be suppressed by down regulation processes resulting from temporal and spatial imbalances between source and sink processes. The increased stomatal resistance can be counteracted by the increased vapour pressure gradient between the leaf and the atmosphere. However, the amount of water transpired by crops under higher CO_2 determines not only their sensitivity to stress, but also the components of the regional water balance, as groundwater recharge and runoff.

The uncertainties related to the climatic and plant physiological effects of higher CO_2 limit our ability to predict consequences of future climate change for crop yield and regional hydrology. However, scenario techniques can be used to reveal the most sensitive and rather stable processes. Therefore, in our case study the effects of climate change were studied in parallel with the potential physiological effects of the elevated atmospheric CO_2 concentrations.

The case study region is the state of Brandenburg, Germany (see Fig. 1). The region essentially overlaps with the lowland part of the Elbe River basin. The primary reason for selecting Brandenburg for the impact assessment is its vulnerability to water stress in dry periods. Due to the position of the Elbe River basin around the boundary between the relatively wet maritime climate of Western Europe and the more continental climate of eastern Europe with longer dry periods, the annual long-term average precipitation is relatively small, less than 600 mm yr^{-1} in the lowland part of the basin. During the last years, integrated studies on impacts of Global Change on the environment have been initiated for the Elbe River basin and for the state of Brandenburg (Krysanova *et al.*, 1999a; Wechsung *et al.*, 2000).

The main objective of the study was to investigate vulnerability of water resources and agriculture to expected climate change for the state of Brandenburg. The sensitivity of the main hydrological components (evapotranspiration, runoff and groundwater recharge) and three major crops (silage maize, winter barley and winter wheat) to four regional warming scenarios (two transient and two equilibrium scenarios, assuming +1.5 °C, +3.0 °C increase in temperature until 2050, respectively) was investigated. In addition, differences in precipitation and temperature between the scenarios were used to explore the sensitivity of crop yields and hydrology to changes in climate parameters. The paper also includes a discussion on uncertainties associated with the climate change impact assessment. The suggested approach can be transferred to other regional studies of climate change impact.

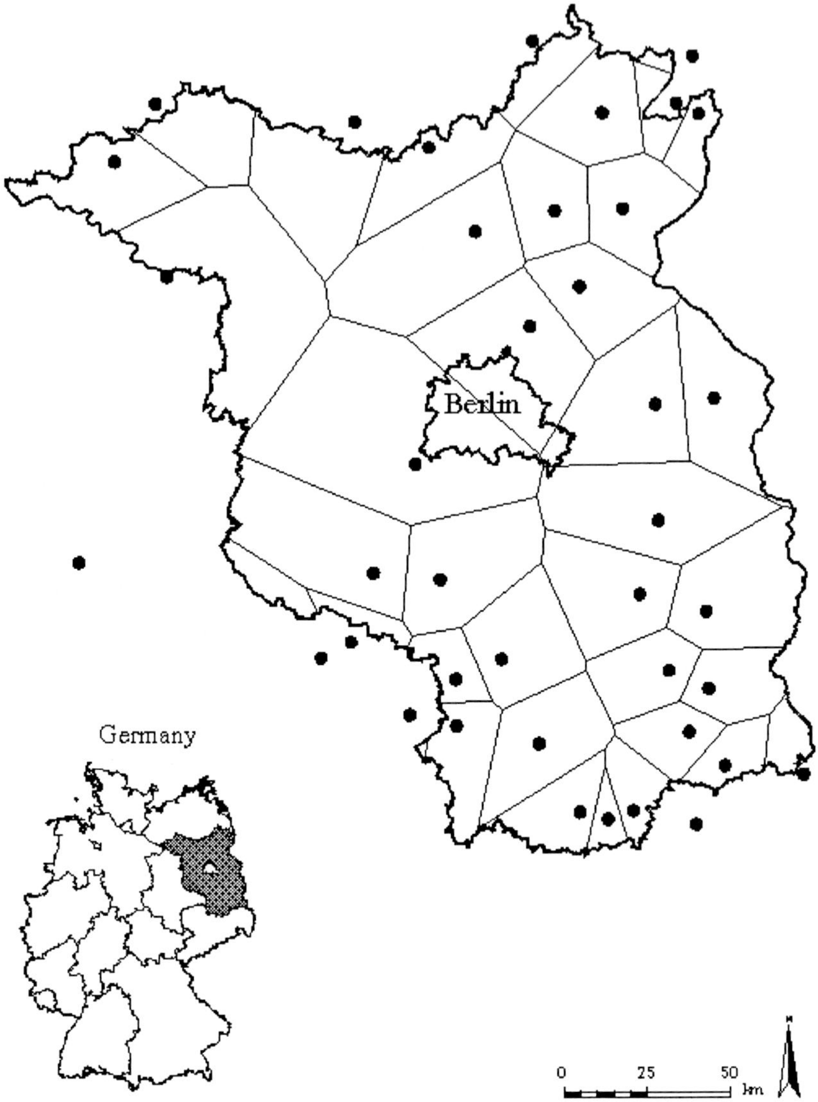

Fig. 1. The state of Brandenburg and Berlin, climate stations and Thiessen polygons for the climate station used in the study

2. METHODOLOGY

2.1 The modelling approach

Prior to the climate impact study, hydrological validation was performed in three mesoscale river basins in the area. As a next step, the crop module was validated regionally for Brandenburg, using crop yield data for districts (see Krysanova *et al.*, 1999a), considering three major crops in Brandenburg: winter wheat, winter barley, and silage maize.

After that, the climate change impact study was performed, considering hydrological processes and crop yield. The climate change impact assessment in general followed the standard methodology outlined by Parry and Carter (1998). The model SWIM was run under current climate conditions (the reference scenario SB) and under altered climates, and the model outputs compared. Four climate scenarios derived from GCM outputs as described below, were applied. The 11-yr period 1980 - 1990 was extracted from the reference period 1951 - 1990 and considered as the control. Three periods were compared in the simulation study: the control period A, 2020 - 2030 (the scenario period B), and 2040 - 2050 (the scenario period C). The atmospheric CO_2 concentrations for the control period A and two scenario periods B and C were set to 346, 406 and 436 ppm, respectively (extrapolation of a linear trend independent from the temperature change across the scenarios).

The crop spectrum was restricted to the three above-mentioned crops. All cereals currently cover about 73% (winter wheat and winter barley being the major cereals), and maize covers only about 15% of arable land in Brandenburg. Three crops were assumed to cover the arable land in equal proportions for the calculation of impacts on hydrological processes, which were averaged over three simulation runs with the single crop cover. The 'equal proportions' assumption approximately corresponds to the real C3:C4 relationship in the cropland of Brandenburg. The impacts on crop yield were based on simulation results assuming that the whole arable land is covered by one crop. Crop yield was averaged over 11 years (maize) or 10 years (winter crops) in order to exclude inter-annual variability. The dates of sowing and harvesting were the same for the control period and for the climate scenarios in order to avoid this kind of adaptation to the changed climate. Other possible adaptation options for agriculture crops were also excluded from the study.

2.2 A short description of the model

The modelling system SWIM (Soil and Water Integrated Model) (Krysanova *et al.*, 1998a) includes as its core a continuous-time spatially distributed model, integrating hydrology, vegetation, nutrients (nitrogen, N and phosphorus, P) and sediment transport at the river basin scale. In addition, the system includes the interface to the Geographic Information System GRASS (Geographic Resources Analysis Support System), which allows to extract spatially distributed parameters of elevation, land use, soil types, and the routing structure for the basin under study. The model can be applied for mesoscale and large river basins with an area up to 25,000 km^2 or, after validation in representative sub-basins, for regions of similar size.

A three-level scheme of spatial disaggregation „basin – sub-basins – hydrotopes", plus vertical subdivision of the root zone into a maximum of 10 soil layers are used in the model. The hydrotopes are delineated by overlaying the sub-basins (or climate zones), land use and soil maps. At first, water and nutrient balances are calculated for every hydrotope, and the outputs from hydrotopes are subsequently integrated to estimate the sub-basin outputs. Finally, the routing procedure is applied to the sub-basin lateral flows of water, nutrients and sediments, taking into account transmission losses.

The simulated hydrological system consists of four control volumes: the soil surface, the root zone, the shallow aquifer, and the deep aquifer. The soil column is subdivided into several layers in accordance with the soil database. The water balance for the soil column includes precipitation, surface runoff, evapotranspiration, percolation and subsurface runoff. The water balance for the shallow aquifer includes groundwater recharge, capillary rise to the soil profile, lateral flow, and percolation to the deep aquifer.

Surface runoff is estimated as a non-linear function of precipitation and a retention coefficient, which depends on soil water content, land use and soil type (modification of the SCS curve number method). Lateral subsurface flow (or interflow) is calculated simultaneously with percolation. Interflow appears when the storage in any soil layer exceeds field capacity after percolation and is especially important for soils having impermeable or less permeable layer(s) below permeable ones. Potential evapotranspiration is estimated using the method of Priestley-Taylor. Actual evaporation from soil and actual transpiration by plants are calculated separately.

The module representing crops and natural vegetation is an important interface between hydrology and nutrients. A simplified EPIC approach (Williams *et al.*, 1984) is included in SWIM for simulating arable crops (like wheat, barley, rye, maize, potatoes) and aggregated vegetation types (like 'pasture', 'evergreen forest', 'mixed forest'), using specific parameter values for each crop/vegetation type. It is simplified in comparison to EPIC mainly in the description of phenological processes in order to decrease the requirements of input information. This enables crop growth to be simulated in a distributed modelling framework at the regional scale. A number of parameters are specified for 74 crop/vegetation types in the crop database attached to the model, i.e. biomass-energy ratio, harvest index, base and optimal temperature for plant growth, maximum leaf area index (LAI), fraction of growing season when LAI declines, maximum root depth, potential heat units required for maturity of crop, etc.

The nitrogen and phosphorus modules include the following pools: nitrate nitrogen, active and stable organic nitrogen, organic nitrogen in the plant residue, labile phosphorus, active and stable mineral phosphorus, organic phosphorus, and phosphorus in the plant residue, and flows: fertilisation, input with precipitation, mineralisation, denitrification, plant uptake, leaching to groundwater, losses with surface runoff and erosion. Regarding the lateral transport, the runoff and leaching are more important for nitrogen than for phosphorus. The latter is mainly transported with erosion.

Vegetation in the model affects the hydrological cycle by the cover-specific retention coefficient, which influences runoff, and indirectly - the amount of evapotranspiration, which is simulated as a function of potential evapotranspiration and LAI. The interaction between vegetation and nutrient supply is modelled by the plant consumption of nutrients and using nitrogen and phosphorus stress functions, which affect the plant growth.

Prior to this study, SWIM was tested and validated for hydrology in several mesoscale and large river basins (with 100 to 24000 km^2 drainage area), for nitrogen dynamics, crop growth, and erosion (Krysanova *et al.*, 1998a & b, Krysanova *et al.*, 1999a & b, Krysanova & Becker, 1999). All these model tests and applications have shown that the model is able to describe with a reasonable accuracy the basic hydrologic processes, including the spatial and temporal variability of the main water balance components; the cycling of nutrients in soil and their transport with water; vegetation growth (especially agriculture crops); the dynamical features of soil erosion and sediment transport under different environmental conditions in the Elbe basin. This provides a justification for studying the effects of changes in climate and land use on all these interrelated processes and characteristics.

3. CLIMATE AND PHYSIOLOGICAL EFFECTS OF HIGHER CO_2

3.1 Climate change scenarios

It is well known that current General Circulation Models (GCM) work at a spatial resolution that is too coarse for representing hydrological processes. Therefore GCM outputs represent rather average weather conditions than real weather dynamics for the regional scale. In other words, climate change scenarios derived from GCMs have inadequate spatial resolution for regional impact studies and have to be regionalised.

During the last decade a number of statistical methods for downscaling GCM climate scenarios to the regional scale were developed. In this study we applied four regional climate change scenarios created in the Potsdam Institute for Climate Impact Research by Gerstengarbe and Werner (see description in Lasch *et al*., 1999) using the results of the ECHAMT21 GCM of the Max Planck Institute for Meteorology (Roeckner & Cubasch, 1995). The statistical algorithm for regionalisation was based on the stability principle of the main statistical characteristics (variability, frequency distribution, annual cycle, persistence) and used a special cluster analysis method (Werner and Gerstengarbe, 1997). The algorithm ensures the realization of an imposed temperature trend and maintains auto-correlative characteristics of single climate parameters and correlative relationships between them. The scenarios are site-specific and differ within the region because of differences in the historical weather records and their statistical characteristics.

The reference scenario SB represents the observed climate over the period 1951 - 1990. To reflect the uncertainties in the prediction of global warming by current GCMs, two equilibrium scenarios, SE15 and SE30, and two transient scenarios, ST15 and ST30, assuming temperature increase by 1.5 and 3.0 °C, respectively, were used in our study. Every scenario was created for 40 meteorological stations (Fig. 1), using the observed data to estimate the necessary statistical characteristics. The following meteorological parameters were used in our study: average, maximum and minimum daily air temperature, daily net solar radiation, and daily precipitation. Climate data were considered as homogeneous inside 40 polygons. Spatial disaggregation in SWIM was done by overlaying Thiessen polygons for climate stations (Fig. 1), land use and soil maps.

Comparison of average climate characteristics for the scenario SE15 and the reference scenario, as well as distribution functions for daily maximum temperature and daily precipitation are given in Lasch *et al.* (1999) for a number of stations in Brandenburg. Here average annual precipitation and average annual temperatures (average daily, maximum and minimum) are listed for the reference scenario SB and four climate change scenarios (Table 1). One can see that the differences in temperature between the control period and the scenario periods are very close to the assigned values for the equilibrium scenarios, and for the transient scenarios in period C. Precipitation is lower than in the control period for all scenario periods. It is significantly lower for the scenario SE15 in periods B and C (-15.2%, -15.9%), and in period C for both transient scenarios (-13.4% and –12.1%).

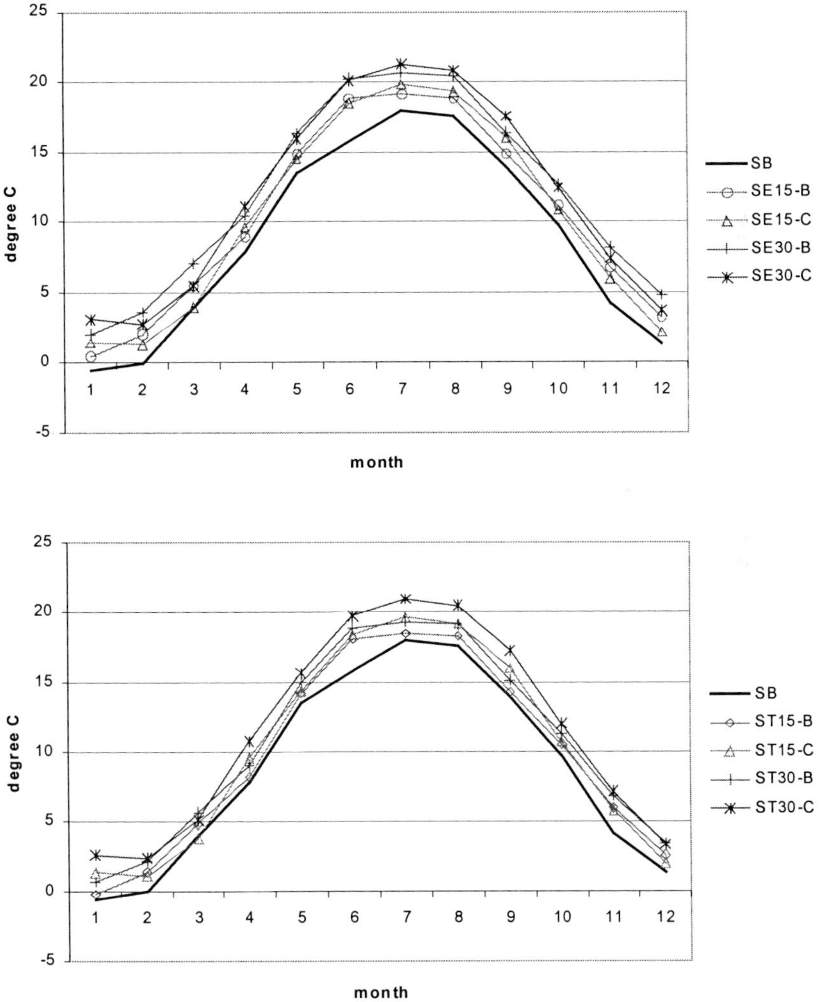

Fig. 2.: Average annual variations in temperature for the reference scenario SB and for four scenarios of climate change.

Table 1. Average characteristics of climate scenarios in Brandenburg for the reference scenario and four climate change scenarios

Scenario	Time period					
		A	B		C	
	1951-1990	1980-1990	2020-2030		2040-2050	
	Precipitation					
	mm	mm	mm	change in % to SB	mm	change in % to SB
SB	560	560				
SE15*			475	-15.2	471	-15.9
SE30			544	-2.8	524	-6.5
ST15			531	-5.2	485	-13.4
ST30			546	-2.4	492	-12.1
			Average temperature			
	°C	°C	°C	Difference to SB in °C	°C	difference to SB in °C
SB	8.6	8.7				
SE15			10.4	+1.7	10.3	+1.6
SE30			11.9	+3.2	11.8	+3.1
ST15			9.7	+1.0	10.1	+1.4
ST30			10.5	+1.8	11.5	+2.8
			Maximum temperature			
	°C	°C	°C	Difference to SB in °C	°C	difference to SB in °C
SB	12.9	13.1				
SE15			14.7	+1.6	14.7	+1.6
SE30			16.2	+3.1	16.2	+3.1
ST15			14.0	+0.9	14.5	+1.4
ST30			14.8	+1.7	15.8	+2.7
			Minimum temperature			
	°C	°C	°C	Difference to SB in °C	°C	difference to SB in °C
SB	4.6	4.9				
SE15			6.5	+1.6	6.2	+1.3
SE30			8.0	+3.1	7.7	+2.8
ST15			5.8	+0.9	6.0	+1.1
ST30			6.6	+1.7	7.3	+2.4

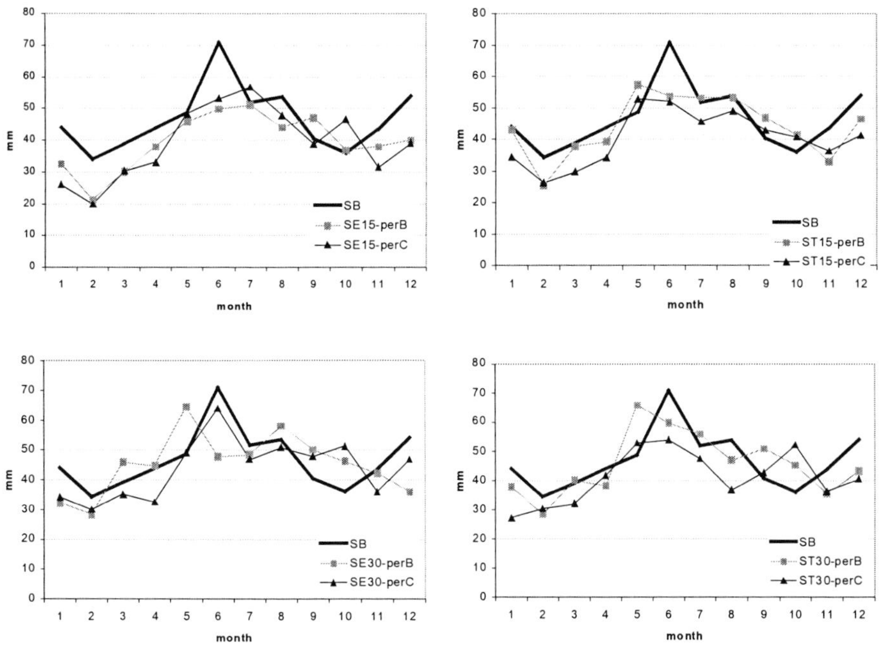

Fig. 3. Average annual variations in precipitation for the reference scenario SB and for four scenarios of climate change.Average annual variations of temperature (Fig. 2) demonstrate that the increase in average monthly temperature is more or less constant over the whole season and corresponds to the declared temperature increase over the whole season. For the transient scenarios the increases in average annual and average monthly temperatures are lower in period B than in period C (Table 1 and Fig. 2).

Average annual variations of precipitation in the scenario periods can be compared to that in the control period in Fig.3. Precipitation in winter months is almost always decreasing, the summer peak observed in June in the 80-ies is shifted to May for scenarios SE30 and ST30 in period B, and it is significantly decreased for scenarios SE15, ST15 and in period C for scenario ST30.

3.2 Adjustment of photosynthesis and transpiration to higher CO_2

Different approaches for the adjustment of net photosynthesis and transpiration to higher atmospheric CO_2 have been used in modelling studies (Goudrian & Ketner, 1984; Rotmans & den Elzen, 1993). Detailed results about the interaction of higher CO_2 and water use efficiency are described in Easmus (1991) and Kimball *et al.* (2001). In our study the adjustment of net photosynthesis and transpiration to altered atmospheric CO_2 concentrations was studied considering two additional factors:

- *the factor alpha* to account for the increase in potential plant growth rate per unit of intercepted photosynthetically active radiation (PAR) under higher CO_2; and
- *the factor beta* to account for the reduction of potential transpiration under higher CO_2.

The factor beta is coupled to the direct CO_2 effect on photosynthesis (the factor alpha). In our study these effects were restricted only to cropland.

A semi-mechanistic approach for the adjustment of net photosynthesis (see full description in Krysanova *et al.*, 1999a) derived from a mechanistic model for leaf net assimilation (Harley *et al.*, 1992) was used. The method takes into account the interaction between CO_2 and temperature:

$$\alpha_{C3} = \left(\exp(a + bT)\right)^{0.6} \tag{1}$$

$$\alpha_{C4} = \left(\exp(a + bT)\right)^{0.36}, \tag{2}$$

where α_{C3} and α_{C4} - the alpha factors for C3 and C4 crops, a and b – the empirical coefficients, dependent on the current and future CO_2 levels, and T – the temperature.

The reduction of potential leaf transpiration due to higher CO_2 was derived directly from the enhancement of photosynthesis in the form:

$$\beta = \alpha \cdot \frac{(CO_2)^0}{(CO_2)^*}, \tag{3}$$

where indices ^{0} and * correspond to the current and scenario periods.

Fig. 4 depicts the factors alpha (at 17° C) and beta for wheat and maize as functions of atmospheric CO_2 ranging from 330 to 660 ppm, calculated in accordance with equations 1, 2 and 3.

Altogether, three variants were studied for every scenario:

(1) *Climate change only' (CCO),*
(2) *climate change + factor alpha' (CC + α), and*
(3) *climate change + factors alpha and beta' (CC + α + β).*

In this way we accounted for the current uncertainty regarding the significance of higher CO_2 effects for regional evapotranspiration and photosynthesis.

Several feedback processes can actually diminish the initially observed effect at the leaf and plant level. Photosynthesis can be suppressed by down regulation processes resulting from temporal and spatial imbalances between sources and sinks (Adam *et al.*, 2000; Wall *et al.*, 2000). The increased stomatal resistance can be counteracted by the increased vapour pressure gradient between the leaf and the atmosphere. Several simulation studies (Jarvis and McNaughton, 1986; Field *et al.*, 1995) have shown that the stomatal control of regional transpiration decreases with a higher stomatal conductance and is less effective for aerodynamically smooth (crops) than for aerodynamically rough (forest) canopies. However, under drier climate conditions the stomatal control of regional transpiration becomes more significant. The beta factor, which is used in our simulations to reduce the potential transpiration in the case 'CC + α + β', assumes total stomatal control of regional transpiration. The real effect of higher CO_2, if climate scenario conditions were to be realized, will most probably lie between those simulated for the variants 'CC + α' and 'CC + α + β'. For the C4 crop maize (with its relatively low stomatal apparture) and the drier climate conditions the real effect will be closer to the 'CC + α + β', whereas for the C3 crops (relatively wide stomatal apparture) in wetter climate scenarios the real effect will be closer to the variant 'CC + α'.

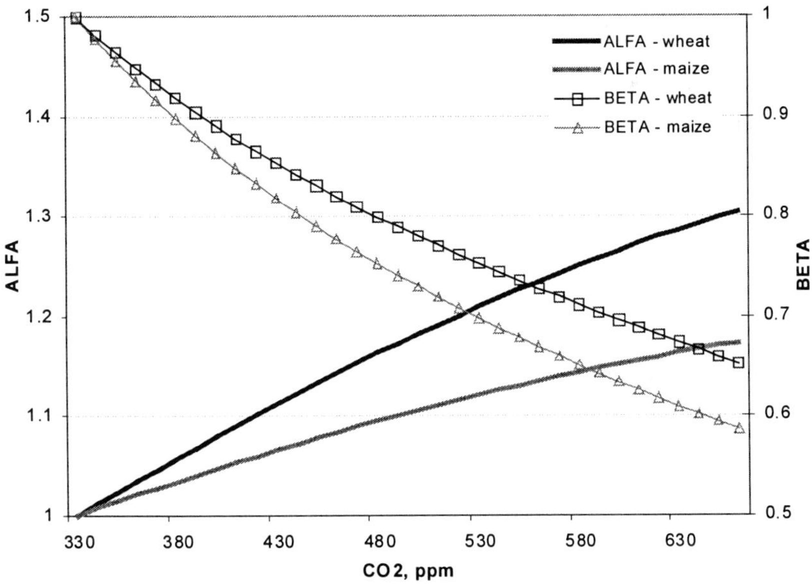

Fig. 4. The physiological factors alpha (at 17° C) and beta for wheat and maize as functions of atmospheric CO_2 concentrations

4. RESULTS

4.1 Impact on hydrological processes

Table 2: Average hydrological flows in Brandenburg for the basic scenario and changes in hydrological flows under climate change scenarios (case 'climate change only')

Hydrological flow	Units	*Basic scenario SB* 1980-1990	
Evapotranspiration	mm	414.2	
Runoff	mm	25.8	
Groundwater recharge	mm	119.5	
		2020-2030 B	2040-2050 C
		Scenario SE15	
Evapotranspiration	%	-3.7	-4.7
Runoff	%	-21.1	-18.5
Groundwater recharge	%	-51.7	-51.5
		Scenario SE30	
Evapotranspiration	%	+6.3	+1.3
Runoff	%	-4.8	-7.1
Groundwater recharge	%	-29.4	-30.7
		Scenario ST15	
Evapotranspiration	%	+2.1	-2.8
Runoff	%	-4.0	-20.4
Groundwater recharge	%	-29.5	-46.0
		Scenario ST30	
Evapotranspiration	%	+4.4	-1.3
Runoff	%	-2.3	-24.3
Groundwater recharge	%	-24.0	-45.0

The hydrological validation performed in three mesoscale river basins in the area and the validation of the crop module were described earlier (see Krysanova *et al.*, 1999a).

The results of our climate change impact study are summarised in Tables 2, 3 and 4 as areal average values of change (in %) compared to those for the control period A. Table 2 reports average hydrological flows for the reference scenario SB in period A and changes in hydrological flows for the scenario runs (periods B and C) considering the total area of the region. Here 'runoff' means the sum of direct surface runoff and interflow. Hydrological flows were calculated under the 'equal proportions' assumption for crop distribution (see Section 2.1).

Evapotranspiration is notably decreased for the driest scenario SE15 in periods B and C, and for both transient scenarios ST15 and ST30 in period C. In other cases it is increased. Runoff and groundwater recharge are always decreased, and especially significant for the scenario SE15 and in period C for the both transient scenarios. Groundwater recharge responds most sensitively to the anticipated climate change. This is caused by a notable precipitation decrease in the climate scenarios during winter (Fig. 3), when potential evapotranspiration is minimal and potential groundwater recharge is therefore relatively high.

It is worth mentioning that the relatively small percentage of change for evapotranspiration (in comparison to those for runoff and groundwater recharge) is not so small in absolute units, because evapotranspiration itself is high in comparison to other hydrological flows in the region. For example, for scenario SE15 in period C the following changes were simulated: evapotranspiration –19.6 mm, runoff –4.8 mm, groundwater recharge –61.5 mm, and for scenario SE30 in period B: evapotranspiration +26.0 mm, runoff –1.2 mm, groundwater recharge –35.1 mm.

Table 3 presents results for hydrological processes restricted to the cropland area, where in addition to the climate effect of higher CO_2 also the physiological effect on leaf photosynthesis and transpiration, represented by the response factors alpha and beta, was considered. The same tendencies of change for the hydrological flows as in Table 2 are observed in Table 3 in the 'climate change only' case. When only the effect of higher CO_2 on photosynthesis is added (the factor alpha), a very minor effect on hydrological flows is observed; therefore it is not reported in Table 3. However, adding factor beta – reduction in potential transpiration for arable crops – leads to a notable decrease in evapotranspiration ranging from 1.2 to 2.8% for different scenarios, a 7 –15% increase in runoff, and a 5-12% increase in groundwater recharge. However, as mentioned above, this feedback on hydrological processes under higher CO_2 has to be treated with caution, because it assumes a total stomatal control of regional transpiration. Therefore, the real effects would be probably lower than the simulated ones.

Table 3: Change of hydrological processes for cropland in Brandenburg in the cases 'climate change only' (CCO) and 'climate change + factors alpha and beta' (CC + α + β)

Hydrological flow	CCO		CC + α + β	
	2020-2030 B	2040-2050 C	2020-2030 B	2040-2050 C
		Scenario SE15		
Evapotranspiration	-3.8	-4.6	-5.0	-6.4
Runoff	-35.0	-30.0	-28.4	-19.6
Groundwater recharge	-68.4	-68.7	-63.4	-60.9
		Scenario SE30		
Evapotranspiration	+6.9	+2.4	+4.7	-0.4
Runoff	-8.3	-13.2	+3.9	+2.2
Groundwater recharge	-43.4	-46.1	-33.7	-33.6
		Scenario ST15		
Evapotranspiration	+1.9	-2.0	0.0	-3.8
Runoff	-8.0	-37.7	+3.3	-29.2
Groundwater recharge	-41.4	-63.7	-33.4	-55.3
		Scenario ST30		
Evapotranspiration	+4.5	-0.9	+2.4	-3.0
Runoff	-6.1	-47.7	+6.1	-37.7
Groundwater recharge	-36.2	-61.7	-27.1	-52.1

Fig. 5 shows average annual changes in hydrological flows. The changes in runoff peaks correspond to the changes in precipitation peaks. In all scenarios with 1.5 °C increase in temperature the runoff peaks are shifted to August, whereas for the scenarios with 3 °C temperature increase they occur in May – June. The level of runoff is always lower in winter months. The dynamics of groundwater recharge shows significant decrease in winter months for all scenarios, on response to the decreased winter precipitation (compare with the results reported in Table 2). Evapotranspiration is reduced in summer and slightly increased in winter.

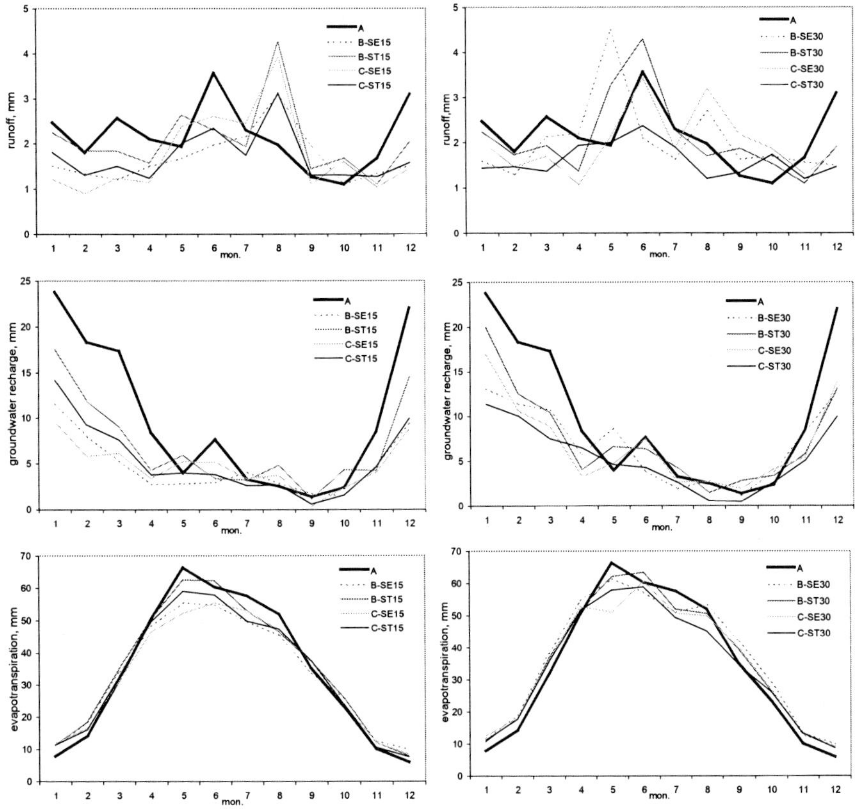

Fig. 5. Average annual changes in hydrological flows for the reference scenario SB and for four scenarios of climate change

4.2 Impact on crop yield

Table 4 summarises climate change impact on three investigated crops, including the case 'climate change only', as well as the cases with factor alpha and with both factors alpha and beta for all four scenarios and two periods. In general, winter wheat is the most negatively affected crop, with yields more than 15% lower in period C for all four scenarios in the case 'climate change only'. Winter barley is less affected, though the yield is notably (by 6 to 14%) decreased in period C for all scenarios. Maize is least affected among the considered three crops. The higher yield decreases of winter wheat compared to those of winter barley are mostly due to the later sowing of wheat. Winter barley finishes its growth before winter break at a more advanced stage, and it is therefore less sensitive to drought in spring.

Table 4: Change of crop yield in Brandenburg in the cases 'climate change only' (CCO), 'climate change + factor alpha' (CC + α), and 'climate change + factors alpha and beta' (CC + α + β)

Crop	CCO		CC + α		CC + α + β	
	2020-2030 B	2040-2050 C	2020-2030 B	2040-2050 C	2020-2030 B	2040-2050 C
			Scenario SE15			
w. barley	-7.8	-13.8	-1.8	-5.5	+6.5	+6.8
w. wheat	-20.6	-22.3	-15.2	-14.5	-8.1	-4.3
maize	+0.1	-1.1	+4.4	+5.4	+15.9	+23.1
			Scenario SE30			
w. barley	+1.6	-6.2	+8.1	+2.9	+13.2	+12.2
w. wheat	-11.8	-18.7	-6.2	-10.9	-0.5	-1.9
maize	+7.3	+1.5	+12.1	+8.3	+25.0	+26.7
			Scenario ST15			
w. barley	+0.5	-7.5	+7.0	+1.3	+14.0	+14.0
w. wheat	-8.3	-15.9	-2.4	-7.6	+4.5	+3.6
maize	+1.2	-0.4	+5.5	+6.0	+16.5	+23.8
			Scenario ST30			
w. barley	+3.1	-8.3	+9.9	+0.3	+16.1	+10.4
w. wheat	-10.4	-21.6	-4.5	-13.7	+2.1	-4.0
maize	+8.2	-5.3	+12.9	+1.0	+24.7	+18.8

When the physiological effect of higher CO_2 concentrations on plant growth was taken into account ('CC + α 'case), the negative effects of climate change were partly compensated. As could be expected, this improvement was more noticeable for the C3 cereals than for the C4 crop maize. This corresponds to the model assumptions as described by equations 1-2. However, still winter wheat yield is more than 10% lower for three out of four scenarios in period C in comparison to the reference scenario.

When in addition to the alpha factor the reduction in potential transpiration was included (case 'CC + α + β'), a further increase in crop yield was observed. Thus, winter wheat yield losses with 'climate change only' were almost compensated (now practically in all cases the changes in wheat yield do not exceed ±5%). Winter barley yield was more than 10% higher, except in scenario SE15. The most significant increase was found in the case 'CC + α + β' for maize: from 16 to 27% increase in comparison with the reference scenario. Because of the direct coupling between the alpha and beta factors, the lower stimulation of photosynthesis for C4 than C3 plants at higher CO_2 concentrations goes along with a higher reduction in potential leaf transpiration due to higher CO_2. Under the chosen scenario conditions, the latter effect is more significant for yield formation than the direct effect of CO_2 on photosynthesis. Therefore, the increase in yield for maize is the highest, if both alpha and beta factors are considered. However, as mentioned earlier, caution should be maintained because of the not fully understood regional realisation of this effect.

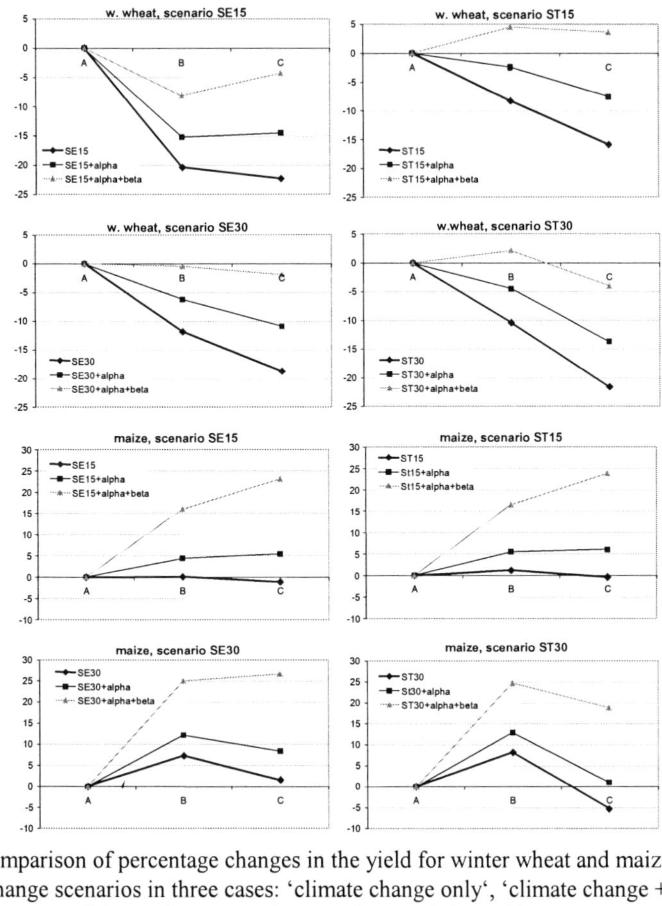

Fig. 6. Comparison of percentage changes in the yield for winter wheat and maize under climate change scenarios in three cases: 'climate change only', 'climate change + factor alpha' and 'climate change + factors alpha and beta'

Fig. 6 depicts a comparison of the crop yield changes for winter wheat and maize in three cases: 'CCO', 'CC + α', 'CC + α + β' for all four scenarios (see also Table 4). It gives an impression of the crop yield gains when the factors alpha and beta are taken into account, and the differences between periods B and C.

4.3 Regression analysis

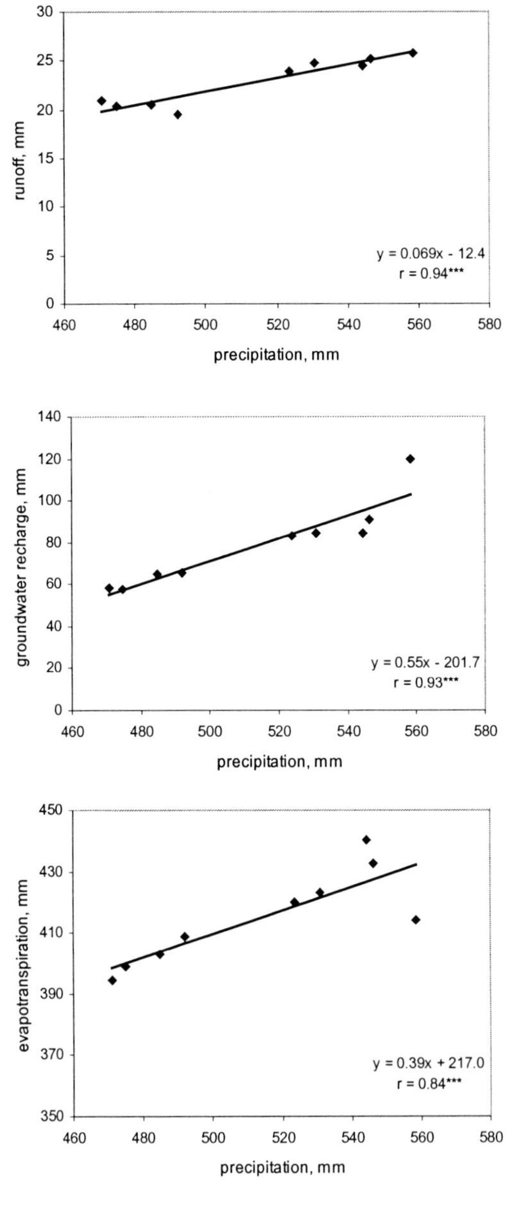

Fig. 7. Regression lines, equations, and the corresponding coefficients of correlation for the dependence between hydrological flows and precipitation taken as regional average values for the total area for the reference period and 8 scenario periods (*** means the significance at the 0.01 level)

After the comparison between scenarios was done, the obtained results were analysed statistically. Regression analysis was performed, considering the main model outputs (three hydrological flows and crop yield for winter wheat, winter barley and maize) and the main climate parameters (temperature and precipitation) as regional average values for the control period and eight scenario periods. This was done in order to study the general sensitivity of the model outputs to climate change. The regression analysis results could be further used for comparison with other climate change impact studies.

The correlation of the average model outputs with temperature was always low and not significant, whereas all of them were significantly correlated with precipitation in the 'climate change only' case. The significance of correlation coefficients was estimated applying F-test.

Fig. 7 shows regression lines and coefficients of correlation for the dependence between hydrological components and precipitation, when all values were averaged over the total area. The correlation is higher for runoff and groundwater recharge (r = 0.94, 0.93), and it is lower for evapotranspiration (r = 0.84), all correlation coefficients are significant at the 0.01 level.

Table 5: Coefficients of correlation between the main model outputs (hydrological flows & crop yield) and precipitation taken as regional average values for the cropland area for the control period and eight scenario periods in the cases 'climate change only' (CCO), 'climate change + factor alpha' (CC + α), and 'climate change + factors alpha and beta' (CC + α + β)

Model outputs	CCO	CC + α	CC + α + β
runoff	0.90 ***	0.89 ***	0.86 ***
groundwater recharge	0.89 ***	0.90 ***	0.94 ***
evapotranspiration	0.82 ***	0.82 ***	0.90 ***
winter wheat	0.85 ***	0.85 ***	0.55 NS
winter barley	0.91 ***	0.72 **	0.07 NS
maize	0.62 **	0.30 NS	0.29 NS

The significance of the correlation coefficients estimated using F-test is indicated beside as follows:

*** significant at the 0.01 level, ** significant at the 0.05 level, NS not significant

Fig. 8 allows to compare the dependence between hydrological flows and precipitation for cropland, when in addition the factors alpha and beta were taken into account. The corresponding coefficients of correlation are collected in Table 5. All correlation coefficients for hydrological components are statistically significant at the 0.01 level. The high sensitivity of hydrological flows to precipitation is preserved also with the alpha and beta factors. When only the alpha factor is added, there are practically no changes in hydrological components, therefore the corresponding regression lines are not shown in Fig. 8. When in addition the factor beta is considered, the hydrological components are changed, but the regression lines have practically the same slope, and the coefficients of correlation are still high and significant.

The most diverse effects take place for the regression lines expressing the dependence of crop yield on precipitation (Fig. 9 and Table 5). The correlation is higher for winter cereals than for maize in the 'CCO' case ($r = 0.85, 0.91, 0.62$ for winter wheat, winter barley and maize, respectively), and they are all significant (see Table 5). In the case '$CC + \alpha$' the high correlation is preserved for winter wheat, it is decreased for winter barley, and becomes not significant for maize. And finally, in the case '$CC + \alpha + \beta$' the sensitivity of yield to precipitation is very low for all the crops. Coefficients of correlation in this case are low ($r = 0.55, 0.07, 0.29$ for winter wheat, winter barley and maize, respectively) and not significant. This reflects the fact that precipitation is no longer the most limiting stress factor in this case.

The spatial patterns of change in the model outputs under climate scenarios differ between scenarios and reflect the changed spatial patterns of precipitation. Fig. 10 shows the histogram of the areal difference map for groundwater recharge between scenario SE15, period C (the driest scenario period) and the reference scenario SB, period A. A notable decrease ranging from –45 mm to –90 mm is expected to occur over a significant part of the region (68%). A difference in excess of –45 mm was simulated for 77% of the area in this case.

Fig. 11 depicts changes in crop yield for the same scenario SE15 in period C. The decrease is most important for winter wheat, and is the smallest for maize. For all the crops the response to climate change is more negative in the southern part of the region. This reflects the changing precipitation patterns.

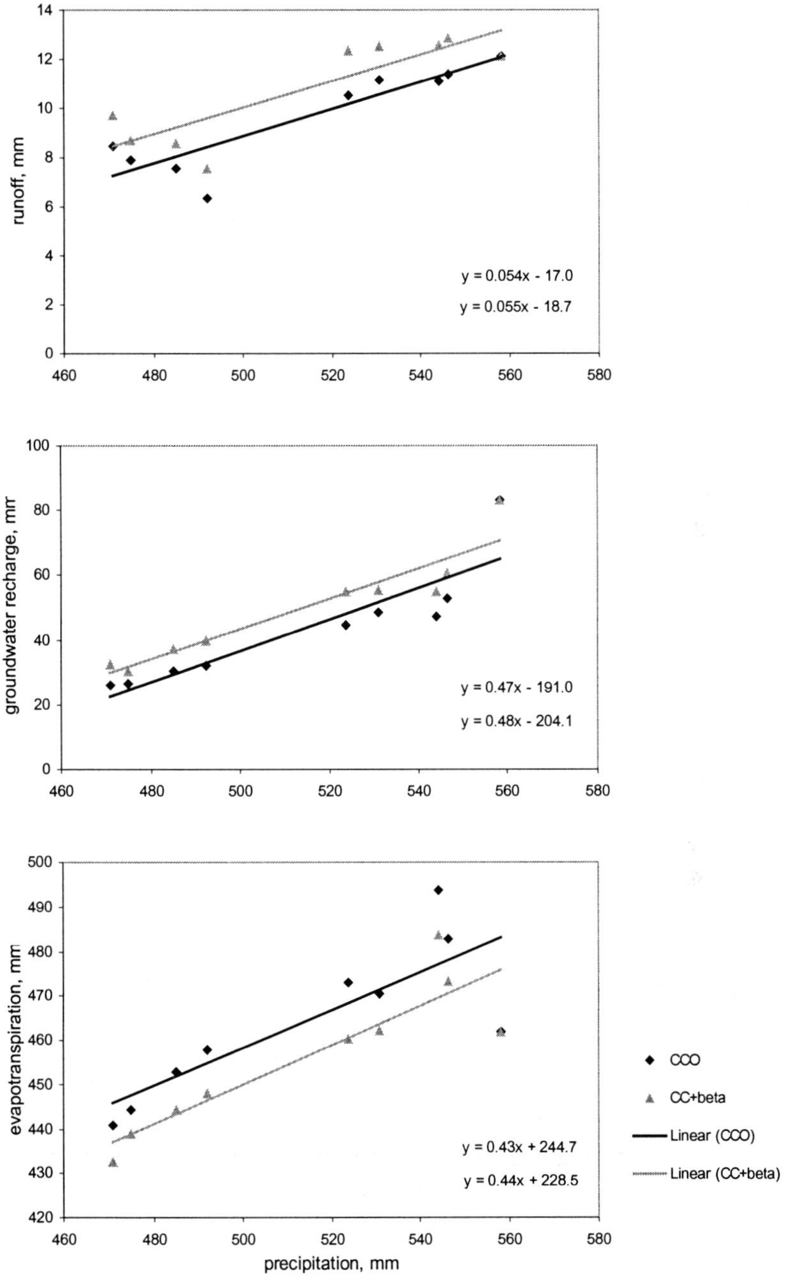

Fig. 8. Regression lines and equations (the upper equation corresponds to the upper line) for the dependence between hydrological flows and precipitation taken as regional average values for the cropland area for the reference period and 8 scenario periods in two cases: 'climate change only' and 'climate change + factors alpha and beta'. The corresponding coefficients of correlation and their significance are indicated in Table 5.

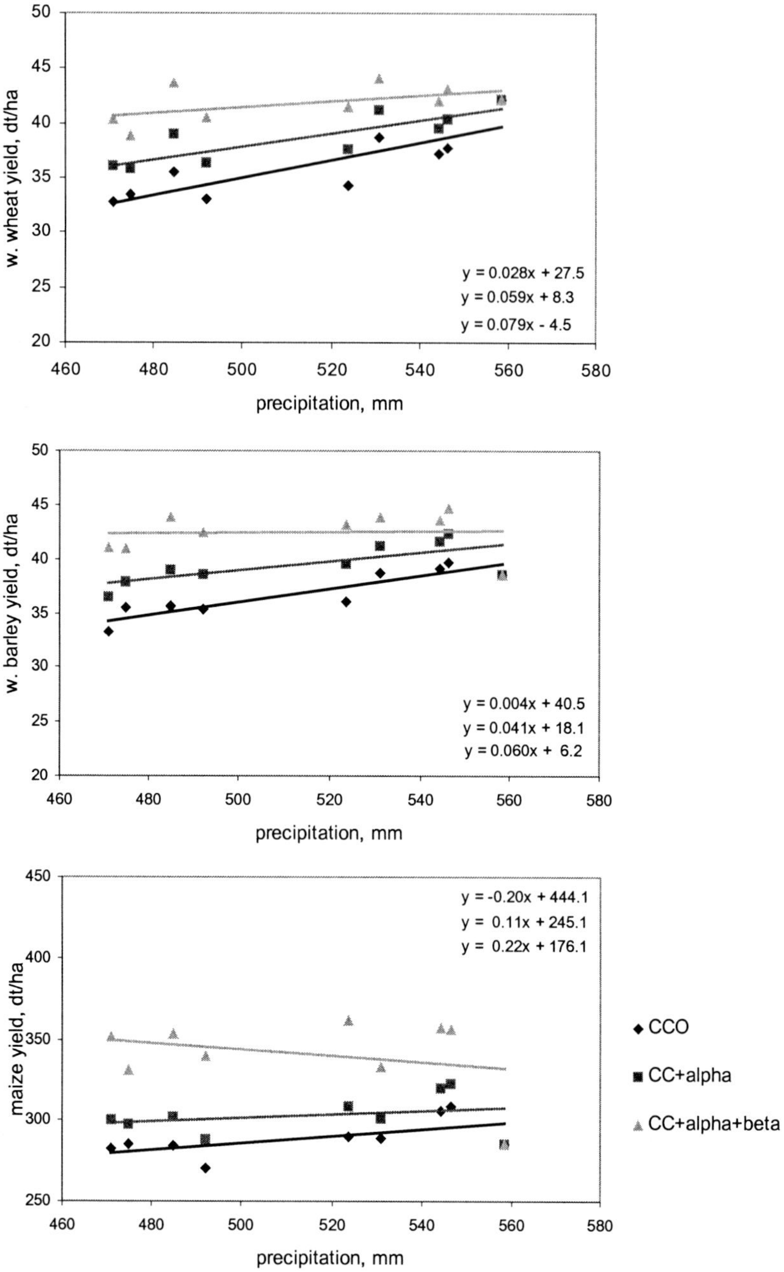

Fig. 9. Regression lines and equations for the dependence between crop yield and precipitation taken as regional average values for the reference period and 8 scenario periods in three cases: 'climate change only', 'climate change + factor alpha' and 'climate change + factors alpha and beta' (the equation are placed in the same order as the lines). The corresponding coefficients of correlation and their significance are indicated in table 5.

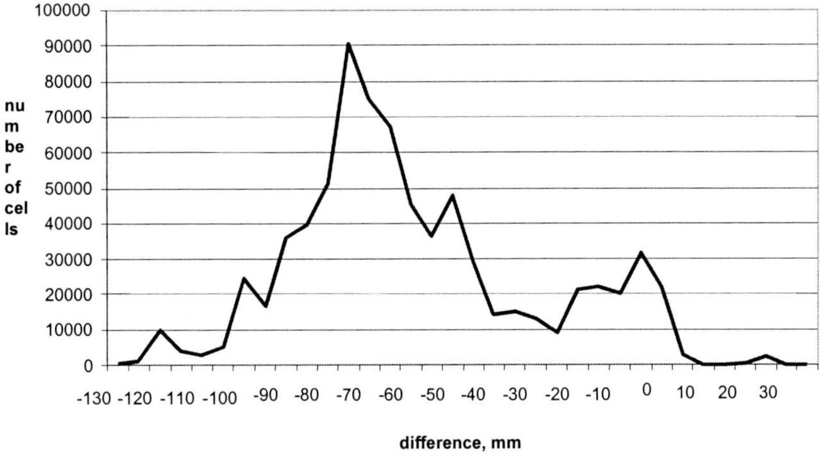

Fig. 10. The histogram of the difference map for groundwater recharge between scenario SE15, period C and the reference scenario SB

4.4 Uncertainties involved in the impact study

The IPCC tried to summarise the uncertainties in hydrological impact assessment. According to their conclusions (Watson *et al.*, 1996), the largest uncertainty is connected with weaknesses of current GCMs with their spatial resolution in coupling the land surface and atmospheric hydrological cycles. This results in a high uncertainty regarding the simulation of precipitation.

There are five main sources of uncertainties in estimating the effects of climate change at the regional scale in our case study:

- *climate scenarios based on GCM simulations (especially precipitation),*
- *conversion of 'the signal' from GCM scenarios to the regional scale using statistical downscaling,*
- *possible feedbacks from climate impact on vegetation to the regional climate,*
- *taking into account the stomatal control of regional evapotranspiration, and*
- *regional ecohydrological model itself, which simulates hydrological processes and crop growth taking the regionalised climate as an input.*

We tried to address these uncertainties by (a) considering four different scenarios, (b) including/excluding the physiological effects of higher CO_2, and (c) validation of the regional model in advance for hydrological processes and crop yield. Still, the conclusions and results summarised in the Tables and Figures have to be treated with caution.

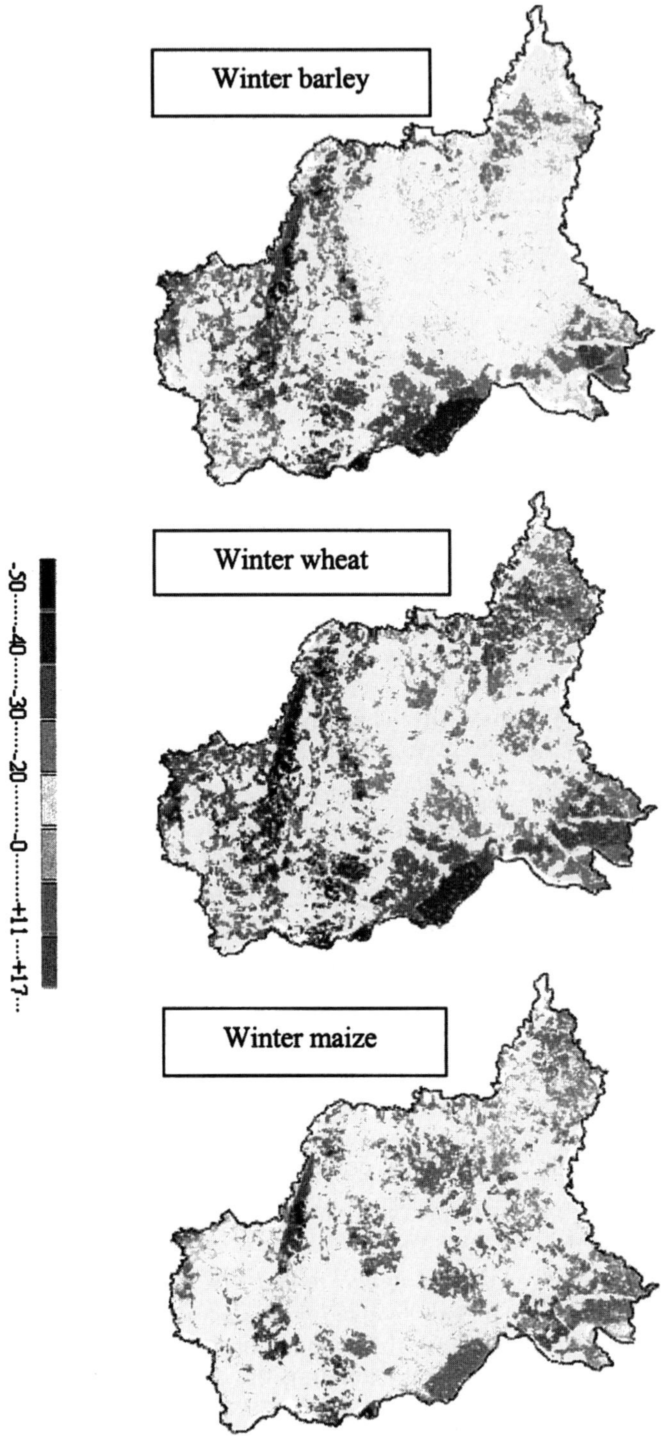

Fig. 11. Changes in crop yield for scenario SE15, period C in comparison with the reference scenario SB.

5. CONCLUSIONS

Main conclusions from our regional climate change impact study are the following:

- *a warmer (1.5 – 3 °C higher temperature) and drier (12-16% lower precipitation) climate in Brandenburg may lead to a significant decrease in the regional runoff and groundwater recharge,*
- *a warmer (1.5 – 3 °C higher temperature) and drier (12-16% lower precipitation) climate in the region would strongly affect agricultural crops, especially wheat and barley, if the physiological effects of higher CO2 on photosynthesis and transpiration are not realised at the regional scale,*
- *maize yield would be 16 - 27% higher, if the physiological effects of higher CO2 on photosynthesis and transpiration would be realised at the regional scale; in this case maize expansion on arable land may be considered as a mitigation measure.*

In general, hydrological processes and crop production in Brandenburg are more sensitive to precipitation than temperature change for the considered scenarios with their ranges of change in temperature and precipitation. A drier climate would significantly decrease the regional runoff and groundwater recharge. Taking into account a high water demand in Brandenburg, and a high share of groundwater used for municipal water supply (88%), the effect of warming scenarios may be very critical. Current attempts to increase the deciduous forest share (which has lower average annual transpiration and higher groundwater recharge in comparison to pine forest) can be considered as an effective mitigation measure to decrease region's vulnerability to climate change.

Hydrological components on cropland were insensitive to the direct effect of higher CO_2 on photosynthesis. However, they were notably altered in the case when the CO_2 impact on regional transpiration (the factor beta) was considered. For the considered crop cover, the simulated decrease in groundwater recharge in this case was by 10-15% lower than in the 'climate change only' case. This could be treated as an important feedback of vegetation on hydrological processes.

Warmer and drier climate will increase the suitability of maize cropping in the region. Yield increase in the range between 16-27% for the regional climate change scenarios related to 1.5 and 3.0 °K temperature increase are possible, even if a lower CO_2 effect on transpiration at the regional scale compared with our simulations would be realized.

Regression analysis between the main model outputs (evapotranspiration, runoff, groundwater recharge and crop yield) and climate parameters (temperature, precipitation) was done in order to study the general sensitivity of the model outputs to climate change. It demonstrated that the correlation with temperature is always low, while the dependence of almost all model outputs on precipitation is strong. All three hydrological flows are highly correlated with precipitation in all variants – with and without factors alpha and beta. Two grain crop yields were highly correlated with precipitation in the case 'climate change only', and the correlation was very low in the case 'climate change + alpha + beta' for all three crops. The regression analysis results can be used for reviewing the recent climate impact studies for different regions.

The suggested methodology: (a) application of an integrated ecohydrological model for simulating the impacts of climate change on hydrological regime and crop yield simultaneously, and (b) studying the physiological effects of higher CO_2 in parallel with the case 'climate change only' can be recommended for transfer to other regional studies.

6. REFERENCES

Adam, N.R., Wall, G.W., Kimball, B.A., Pinter, P.J., Jr., LaMorte, R.L., Hunsaker, D.J., Adamsen, F.J., Thompson, T., Matthias, A., Leavitt, S.W. and Webber, A.N., 2000. Acclimation response of spring wheat in a Free-air CO_2 enrichment (FACE) atmosphere with variable soil nitrogen regimes. 1. Leaf position and phenology determine acclimation response. *Photosynthesis Res.* 66:65-77.

Arnell N., 1999. The effect of climate change on hydrological regimes in Europe: a continental perspective. *Global Environmental Change,* 9, p. 5-23

Easmus D., 1991. The interaction of rising CO_2 and temperatures with water use efficiency: commissioned review. *Plant, Cell and Environment*, Special Issue: Elevated CO_2 levels, 14 (8), p. 843-852.

Field C.B., R.B. Jackson & H.A. Mooney, 1995. Stomatal responses to increased CO2: implications from the plant to the global scale. *Plant, Cell and Environment*, 18, p. 1214-1225.

Goudrian J. and P. Ketner, 1984. A simulation study of the global carbon cycle including man's impact on the biosphere. *Climatic Change*, 6, p. 167 - 192.

Harley P.C., R.B. Thomas, J.F. Reynolds & B.R. Strain, 1992. Modelling photosynthesis of cotton grown in elevated CO_2. *Plant, Cell and Environment*, 15, p. 271-281.

Jarvis P.G. and McNoughton K.G., 1986. Stomatal control of transpiration: scaling up from leaf to region. *Adv. Eco. Res.*, 15, p. 1-49

Kimball B.A, Mauney, J.R., Nakayama, F.S. & Idso, S.B., 1993. Effects of elevated CO_2 and climate variabilities on plants. *Journal of Soil and Water Conservation*, 48, p. 9-14.

Kimball B.A., P.J. Pinter Jr., R.L. Garcia, R.L. Lamorte, G.W. Wall, D.J. Hunsaker, G. Wechsung, F. Wechsung & T. Kartschall, 1995. Productivity and water use of wheat under free-air CO_2 enrichment. *Global Change Biology,* 1, p. 429-442.

Kimball B.A., LaMorte R.L., Pinter Jr. P.J., Wall G.W., Hunsaker D.J., Adamsen F.J., Leavitt S.W., Thompson T.L., Matthias A.D. & Brooks T.J., 2001. Free-Air CO_2 enrichment (FACE) and soil nitrogen effects on energy balance and evapotranspiration of wheat. *Water Resources Research* (in press)

Krysanova V., D.I. Müller-Wohlfeil and A. Becker, 1998a. Development and test of a spatially distributed hydrological / water quality model for mesoscale watersheds. *Ecological Modelling*, 106, p. 261-289.

Krysanova V., A. Becker and B. Klöcking, 1998b. The linkage between hydrological processes and sediment transport at the river basin scale. In: W.Summer, E.Klaghofer, W.Zhang (eds.) *Modelling Soil Erosion, Sediment Transport and Closely Related Hydrological Processes*. IAHS Publication no. 249, p. 13-20.

Krysanova V., Wechsung, F., Becker, A., Poschenrieder, W. & Gräfe, J., 1999a. Mesoscale ecohydrological modelling to analyse regional effects of climate change. *Environmental Modelling and Assessment,* 4, 4, p. 259-271.

Krysanova V., Gerten, D., Klöcking, B. & Becker, A. 1999b. Factors affecting nitrogen export from diffuse sources: a modelling study in the Elbe basin. In: L. Heathwaite (ed.) *Impact of Land-Use Change on Nutrient Loads from Diffuse Sources*, IAHS Publications no. *257*, p. 201-212.

Krysanova V. & Becker, A., 1999. Integrated Modelling of Hydrological Processes and Nutrient Dynamics at the River Basins Scale. *Hydrobiologia*, 410, p. 131-138.

Lasch P., M. Lindner, B. Ebert, M. Flechsig, F.W. Gerstengarbe, F. Suckow and P.C. Werner, 1999. Regional impact analysis of climate change on natural and managed forests in the Federal State of Brandenburg, Germany. *Environmental Modelling and Assessment,* 4, 4, p. 273-286.

Lawlor D.W. and A.C. Mitchell, J. Franklin, V.J. Mitchell, S.P. Driscoll & E. Delgado, 1993. Facility for studying the effects of elevated carbon dioxide concentration and increased temperature on crops. *Plant, Cell and Environment*, 16, p. 603-608.

Parry M.L. & T. Carter, 1998. Climate Change Impact and Adaptation Assessment. Earthscan, London.

Parry M., I. Carson, T. Rehman, R. Tranter, P. Jones, D. Mortimer, M. Livermore, and J. Little, 1999. Economic Implications of Climate Change on Agriculture in England and Wales, Jackson Environment Institute, University College London, 114 pp.

Reilly J.M., and D. Schimmelpfennig, 1999. Agricultural impact assessment, vulnerability, and the scope for adaptation. *Climatic Change*, 43, p. 745-788.

Roeckner E. & Cubasch, U., 1995. Das Klimamodell ECHAM4. DKRZ, Max-Planck-Institut fuer Meteorologie, Hamburg.

Rotmans J. & M.G.J. den Elzen, 1993. Modelling feedback mechanisms in the carbon cycle: balancing the carbon budget. *Tellus* 45 B, p. 1-20.

Rötter R. & S.C. Van de Geijn, 1999. Climate change effects on plant growth, crop yield and livestock. *Climatic Change*, 43, p. 651-681.

Wall, G.W., Adam, N. R., Brooks, T.J., Kimball, B.A., Pinter, P.J., Jr., LaMorte, R.L., Adamsen, F.J., Hunsaker, D.J., Wechsung, G., Wechsung, F., Grossman-Clarke, S., Leavitt, S.W., Matthias, A.D. and Webber, A.N. 2000. Acclimation response of spring wheat in a free-air CO_2 enrichment (FACE) atmosphere with variable soil nitrogen regimes. 2. Net assimilation and stomatal conductance of leaves. *Photosynthesis Res.* 66, p. 79-95.

Watson R.T., M.C. Zinyowera & R.H. Moss, 1996. Climate Change 1995. Impacts, adaptations and mitigation of climate change: scientific-technical analyses, Contribution of Working Group II to the Second Assessment Report of the Intergovernmental Panel on Climate Change, Cambridge University Press, Cambridge.

Watson R.T., M.C. Zinyowera & R.H. Moss, 1998. The regional impacts of climate change. An assessment of vulnerability. Intergovernmental Panel on Climate Change, Cambridge University Press, Cambridge.

Wechsung, F., Krysanova, V., Flechsig, M., and Schaphoff, S., 2000. May land use change reduce the water deficiency problem caused by reduced brown coal mining in the state of Brandenburg? *Landscape and Urban Planning,* 51, p. 177-189.

Werner P.C. & F.-W. Gerstengarbe, 1997. A proposal for the development of climate scenarios. Climate Change, 8, 3, p. 171-182.

Williams J.R., K.G. Renard & P.T. Dyke, 1984. EPIC - a new model for assessing erosion's effect on soil productivity. *Journal of Soil and Water Conservation,* 38 (5), p. 381-383.

Local climate and water availability changes due to landscape modifications

A numerical experiment in Southeastern Spain

N. HASLER[1], I. IORGULESCU[1], A. MARTILLI[2], G.E. LISTON[3], AND R. SCHLAEPFER[1]

1Laboratory of Ecosystem Management, Swiss Federal Institute of Technology, Lausanne, Switzerland

2Laboratory of Air and Soil Pollution, Swiss Federal Institute of Technology, Lausanne, Switzerland.

3Department of Atmospheric Science, Colorado State University, Fort Collins, Colorado, USA

Abstract: Water availability is a crucial problem for many countries and will become more important in the context of global change, particularly for arid and semiarid regions. The water cycle is linked to natural processes but is also strongly influenced by human activities, in particular land-use changes. Since evapotranspiration is highly dependent upon vegetation characteristics and landscape structure, different land-use distributions are expected to have a significant influence on evapotranspiration rates. The fundamental paradigm on which this analysis is based is that landscape pattern, functional interactions among landscape elements, and natural and anthropogenic changes in structural relationships are intimately related. In order to assess the impact of landscape structure on local climate and water cycle variables, we set up a series of numerical experiments for July 1997 over a domain that broadly corresponds to Castilla-La Mancha, a region of southeastern Spain (364 x 280 km^2). We use a regional climate model (ClimRAMS) with nested grids of 14 km and 3.5 km cell size, implemented with a USGS global database of elevation and land cover, and the FAO soil classification database. Landscape changes are applied on a 112 x 84 km^2 sub-domain (the 3.5 km grid). We developed different landscape structure scenarios by considering changes in landscape composition (type and extent of ecosystems) and configuration (spatial arrangement of ecosystems). Changes as large as 1K in daily average air temperature and 9.5K in daily temperature range take place when replacing mixed forest with semi-desert over the whole sub-domain. For analyzing impacts within each landscape scenario, we concentrate on grid cells that have the same land cover between two scenarios. For example, we looked at a forested cell in two scenarios with same land composition (50% mixed

M. Beniston (ed.),
Climatic Change: Implications for the Hydrological Cycle and for Water Management, 301–328.

forest – 50% semi-desert) but with a different spatial configuration of the land-cover types. Differences in evapotranspiration between the two scenarios in that particular gird cell reached 12.5 mm of evapotranspiration (15% of total amount). The number of water stress days also differ in these kinds of grid cells. This study demonstrates that changes in landscape structure can have a significant impact on moisture budgets in arid regions.

1. INTRODUCTION

Water availability is a crucial problem for many countries and may become more important in the context of global change, particularly for arid and semiarid regions. Crudely modelled, the water cycle is a balance between incoming quantities associated with precipitation processes, and outgoing fluxes associated with evapotranspiration, infiltration, and runoff. The partitioning between these quantities is driven by atmospheric and surface conditions and processes, and is also strongly influenced by human activities such as land-use changes. A host of studies and reviews support the hypothesis that land use affects hydrology (Calder, 1993) and climate (Cotton and Pielke, 1995; Pielke, 2000). Most hydrological studies have been catchment based experiments, and focus on the effects of land-use change on water yield (see Bosch and Hewlett, 1982). They usually deal with small to medium scales (0,1-1000 km^2) and do not account for interactions between vegetation and the atmosphere. On the other hand, atmospheric scientists have focused on understanding landscape processes in order to identify their influence on the planetary boundary layer and the so-called non classical mesoscale circulations (Segal et al., 1988; Avissar and Pielke, 1989; Jia and Ye, 1996; Giorgi and Avissar, 1997). They have also developed parameterisations of subgrid-scale processes for use in general circulation models (GCMs) (Klaassen and Claussen, 1995; Noilhan and Lacarrère, 1995; Raupach and Finnigan, 1995; Laval, 1997, among others). Lynn et al. (1995) varied soil moisture conditions in different chessboard-configured patches as part of mesoscale atmospheric simulations; they considered soil moisture to be the most important factor influencing energy fluxes. Differences between their mesoscale experiment and the corresponding GCM-resolution result was as large as changes predicted by GCMs for doubling the atmospheric CO^2 concentration. Mahfouf et al. (1987) found that land-surface induced circulations may be as strong as sea breezes for the case of a transpiring vegetated area adjacent to dry, bare land. Marengo et al. (1996) noted that the importance of feedbacks between the atmosphere and the surface hydrology in the drought of the Sahel region is widely accepted.

Apparently, landscape characteristics not only affect regional atmosphere, but they also have a strong impact on local climate. Pielke and Avissar (1990) reveal, for example, that severe thunderstorms are more likely over an irrigated area. McNaughton (1989) argue that if a uniform canopy covers an area of some hundreds of square kilometres then "the effect of the vegetation will be felt throughout the whole of the turbulent planetary boundary layer, up to a kilometre or so". Rainfall also appears to be a key determinant in boundary layer behavior, and land-use affected atmospheric fluxes may result in the formation of convective clouds and precipitation (Liu and Avissar, 1996). It is also hypothesised that less evapotranspiration may ultimately reduce the moisture content of the atmosphere, thereby decreasing the rainfall. But, on the other hand, decreases in sensible heat flux through an albedo increase can cause a decrease in convective potential energy, which in turn limits the number of times convection is triggered (Dolman et al., 1997).

Some studies in arid regions like Israel (e.g., Anthes, 1984; Ben-Gai et al., 1993), emphasise land-use induced climatic change. De Ridder and Gallée (1998) compared simulation runs over southern Israel, before and after the start of an irrigation program and found that irrigation may reduce the diurnal amplitude of wind speed and temperature, and increase the potential for moist convection. Nováky et al. (1985) suggested that in arid and semiarid areas the impact of land factors on evaporation can even exceed those of climatic variation. Other studies have focused on historical landscape changes and their simulated corresponding climatic effects (e.g., Copeland et al., 1996; Pan et al., 1999). In general, these studies suggest that land-use change effects on weather have to be included in climate-trend analyses to avoid erroneous conclusions on the reasons for climate change. In his recent review, Pielke (2000) concluded that vegetation and soil processes and change directly affect the surface energy and moisture exchanges with the atmosphere. In the context of climate, soil and vegetation dynamics are as much a part of the climate system as are the atmospheric variables.

Landscape impacts on atmospheric circulations (at mesoscale or regional scales) have been studied either as theoretical sensitivity studies (e.g., Yang et al., 1999), usually focusing on one specific variable, or in a historical perspective, by analysing changes in climate that occurred between the potential climax vegetation (considered to be the landscape before human impact) and actual human-modified landscape effects (e.g. Pielke et al., 1999; Chase et al., 2000; Heck et al., 2000). The theoretical studies stress important factors that can modify atmospheric circulations, and the historical studies more realistically present the probable reasons for actual climate changes, at least in some regions or at some extent. However, for sustainable land-use planning, it's important to know the magnitude of future management impacts on ecosystems, the water cycle, and climate. Therefore, there is a need for more precise and realistic sensitivity studies and their implications for atmospheric, water resources, and ecosystems, at more local scales. The present study begins to address these aspects by performing a series of sensitivity studies for one month of a normal-climate-condition year, over a semiarid area. A similar study has been performed for tropical conditions, where small-scale deforestation impacts on rainfall and clouds were assessed (Wang et al., 2000).

Water and atmosphere are key elements of ecosystem function, and a better understanding of the interactions between soil, vegetation, and atmosphere could lead to important conclusions, especially for sustainable ecosystem management and land-use planning. Moreover, the growing problem of water availability, particularly in arid and semiarid regions, will lead planners to consider these questions of interaction between land use and water cycles. This perspective implies that effects have to be analysed at the landscape scale, a scale at which stakeholders act (corresponding to usual atmospheric gamma-mesoscale) and on ecologically relevant variables, such as near-surface air temperature, atmospheric humidity, soil moisture, energy fluxes, and precipitation rates. These variables have to be analysed at the landscape level and within the landscape itself.

The fundamental paradigm on which the analysis described herein is based, is that 1) landscape pattern, 2) functional interactions among landscape elements, and 3) natural and anthropogenic changes in structural relationships, are intimately related (Forman, 1995). In order to analyse the magnitude of landscape changes on the landscape itself, we set up a series of numerical experiments, over a domain that broadly corresponds to Castilla-La Mancha, a region of southeastern Spain (364 x 280 km^2). We uses a regional climate model (ClimRAMS) with nested grids of 14 km and 3.5 km grid increments. Very simple landscape change scenarios were applied over a 112 x 84 km^2 sub-domain. These scenarios separate the effects of changes in landscape composition (type and extent of ecosystems) from effects of changes in landscape configuration (spatial arrangement of ecosystems).

2. MATERIAL AND METHOD

2.1 Study area

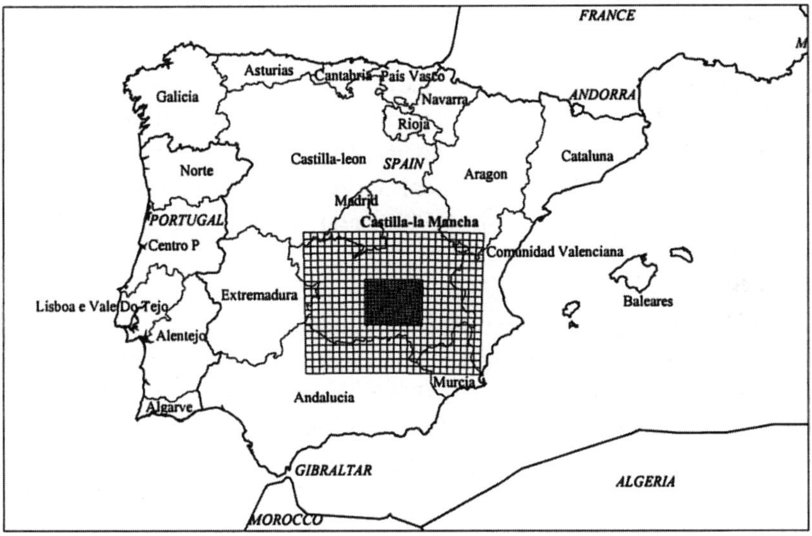

Figure 1. . Study area, model domain and grid configurations. The coarse and fine grid intervals are 14 and 3.5 km, respectively.

To assess the impact of landscape structure on local climate and moisture variables, model simulations were performed for south-eastern Spain. This region is chosen for its semiarid climate; it therefore has a high sensitivity to water availability, and the summer precipitation is largely convective and thereby strongly influenced by surface properties and processes. A previous vegetation-climate feedbacks study for all of Europe outlined that, for human-induced landscape modifications during the last 2000 years, the Mediterranean region had the major climate change (Heck, 1999). In that study, this change was partly due to the larger land-cover changes that had occurred in this area during the period, but also due to the intrinsic atmospheric sensitivity of this area to land-surface processes. Our region of interest is centred in Castilla-La Mancha. This Spanish area has an annual-average precipitation of 360-400 mm, and summer-average air temperatures of 20-25 °C (Garcia de Pedraza and Rieja Garrido, 1994). The model domain and grid configurations are given in Figure 1.

2.2 Mesoscale model

ClimRAMS (Liston and Pielke, 2000) is a climate version of the widely used CSU-RAMS model (Regional Atmospheric Modelling System developed at Colorado State University in Fort Collins, USA). The model is based on equations of motion, heat, moisture, and continuity in a σ_z terrain-following coordinates system. It includes a land-surface sub-model briefly described hereafter, and cloud, precipitation, and radiation sub-models representing physical processes associated with atmospheric energy and moisture interactions. More theoretical aspects of RAMS and ClimRAMS can be found in Pielke et al. (1992) and Liston and Pielke (2000), respectively.

The land-surface sub-model that represents the storage and exchange of heat and moisture associated with the atmosphere-terrestrial interface is based on the multi-level prognostic soil model of (McCumber and Pielke, 1981) and (Tremback and Kessler, 1985) and on (Avissar and Mahrer, 1982) for the vegetation part. Each ClimRAMS surface grid cell has a land-cover type value and a land-water percentage value. Each vegetation class is assumed to have a bare ground part under the vegetation canopy, given by the "vegetation fraction" parameter. Thus, each grid cell has three parallel energy exchange processes corresponding to water, bare soil and vegetation surfaces. For the bare soil part, soil surface temperature is computed from the surface energy balance equation and below the soil surface, only vertical diffusion is permitted. Similarly, soil moisture is linked to a one-dimensional vertical flux, with the deepest soil level moisture held constant to initial value. Soil surface albedo, thermal conductivity, hydraulic conductivity and diffusivity, and soil moisture potential are temporally-evolving as a function of soil moisture. Vegetation temperature is, as for bare soil temperature, computed from a basic surface energy budget equation. The transpiration process is a soil moisture uptake by roots, controlled by atmospheric conditions through the leaf stomatal resistance. Vegetation characteristics like albedo, emissivity, roughness length, displacement height, root profile, are defined based on climatologic temperatures and vegetation type, and kept constant during the run. Leaf area index, canopy transmissivity to solar radiation, and fractional coverage are defined using a cosine functionof the climatologic temperatures, to account for a seasonal variation and are also vegetation type-dependent. In this way, in terms of interactions, energy and water budgets depend on soil and vegetation types, soil parameter vary with moisture and vegetation react to atmospheric conditions through stomata closure.

2.3 Initial and boundary conditions and model set-up

A 364 x 280 km^2 domain is considered for the numerical experiments. This area is covered by a regular mesh in the ClimRAMS polar stereographic projection using 26 x 20 grid elements, each 14 x 14 km^2. A second grid, of 34 x 26 grid elements of 3.5 x 3.5 km^2 grid increment, is two-way nested in the centre of the first grid. The model is run in non-hydrostatic mode. The precipitation scheme is a dump-bucket parameterisation for large-scale precipitation, and the ClimRAMS-implemented convective parameterisation (see Liston and Pielke (2000) for details).

In these simulations, GTOPO30 was used as the topography data set[1]. It is a global digital elevation model with a horizontal grid spacing of 30 arc seconds (approximately 1 kilometre). The land-cover data used for this study (land cover class and land percentage) is taken from the 1-km resolution Global Land Cover Characteristics database[2]. The land-cover type is defined with a block-majority method, assigning the dominant land-cover type to the entire cell. The principal characteristics of the encountered land-cover types are given in table 2. Sea-surface temperatures are derived from the global one-degree weekly database provided by the United States National Oceanic and Atmospheric Administration (NOAA). The values are interpolated to fit to the ClimRAMS grids and transformed to daily data. Atmospheric initial and lateral boundary conditions, in the form of 6-hourly air temperature, horizontal wind components, relative humidity and geo-potential height, are provided by the United States National Centres for Environmental Prediction (NCEP) reanalysis data[3]. They are gridded on 17 pressure levels and on a global 2.5° * 2.5° latitude-longitude grid. The variables are interpolated in time and space to the coarsest ClimRAMS grid. Nudging is performed on the two outer-boundary grid cells of the coarser grid. Soil types[4] are derived from Food and Agriculture Organisation (FAO) soil classification. Soil moisture is initialised with a constant 50% of saturation value for every soil

[1] GTOPO30, Global topography 30arc second, a digital elevation model available through the United States Geological Survey's (USGS) Earth Resources Observation System (EROS) Data Center, http://edcdaac.usgs.gov/gtopo30/gtopo30.hml

[2] Global Land Cover Characteristics database, done on IGBP-DIS initiative, version 1.2, 1-km resolution, Biosphere-Atmosphere Transfer Scheme (BATS) classification (19 classes), available through the United States Geological Survey's (USGS) Earth Resources Observation System (EROS) Data Center, http://edcdaac.usgs.gov/glcc/glcc.hml

[3] NCEP Reanalysis data provided by the NOAA-CIRES Climate Diagnostics Center, Boulder, Colorado, USA, from their web site at http://www.cdc.noaa.gov/

[4] FAO soil database, available from their web site at http://www.fao.org/

type and every scenario. Table 1 summarises the different initial and boundary conditions.

Table 1. Initial and boundary conditions and model set-up.

Condition	*Value*
Period and length of simulation	Entire month of July 1997 (considered as having normal climate conditions)
Atmospheric boundary conditions	NCEP 4-daily reanalysis data
Current land cover	USGS – 1km resolution IGBP v.1.2
Elevation	GTOPO30 – 30 arc second resolution
Soil texture	FAO classification – 1:5'000'000
Initial soil moisture	Constant profile at 0.5 of saturation value
Centre of the domain	39.11°N; 2.85°W
Horizontal resolution	14 x 14 km^2 / 3.5 x 3.5 km^2
Vertical resolution	σ_z terrain-following coordinate of first 250 m and a stretch of 1.2
Top of the model	21'895 m
Nudging condition	On the 2 outer grid cells of the coarser grid, above 19'000m at the top of the domain
Precipitation scheme	Dump-bucket for large-scale precipitation and usual ClimRAMS convective parameterisation

Table 2. Land-cover types characteristics.

Land-cover type	Leaf area index*	Vegetation albedo	Vegetation fraction*	Roughness length	Displacement height [m]	Root in top 5 cm / 12.5 cm soil layers [%]
Mixed forest	5.99	0.18	0.80	0.80	20.	83 / 16
Semi-desert	5.99	0.25	0.09	0.10	0.	80 / 18
Crop	5.99	0.20	0.85	0.06	0.75	63 / 24
Tall grass	5.99	0.16	0.79	0.05	0.	80 / 18

*values on 1 July 1997. Maximum value of leaf area index is fixed to 6 on July 18 for all kinds of vegetation.

2.4 Numerical experiments

To assess the magnitude of changes in climate and water cycle variables due to landscape structure, a set of 6 numerical experiments were performed with different scenarios defined by varying landscape composition and configuration. Simulations were done during the entire month of July 1997; 1997 is an average year in terms of synoptic-scale meteorological conditions. Changes are applied on the finer grid, corresponding to the 112 x 84 km^2 sub-domain.

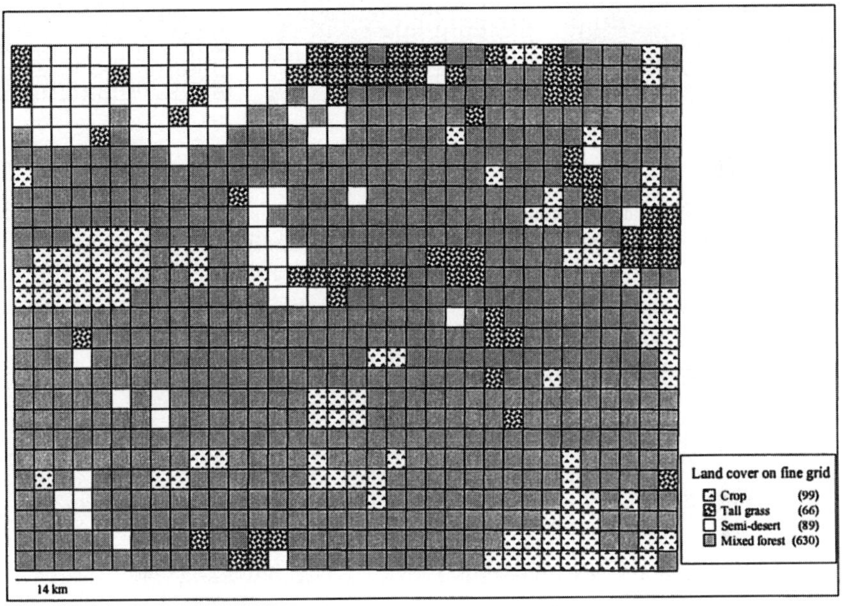

Figure 2. Land cover on fine grid. Number of cells in each land-cover type is given in brackets.

First three scenarios represent changes in landscape composition. As a base simulation, Scenario 1 is the current landscape, given by the USGS land cover (Figure 2). The second two scenarios consider homogeneous landscapes, on the finer grid only, with a single land-cover type of mixed forest for Scenario 2, and semi-desert for Scenario 3 (principal characteristics are given in table 2). These two land-cover types are chosen for two essential reasons: 1) they are present in the current landscape (72% forest, 10% semi-desert), and 2) they are plausible to happen with on-going landscape changes. Forest is increasing under conditions of agriculture abandonment (which is the case in many parts of Europe; this is highly dependent on agricultural and subsidy policies), and semi-desert is naturally increasing under conditions of water stress. Moreover, these two types of land cover have very different behaviours regarding the water cycle and surface energy fluxes.

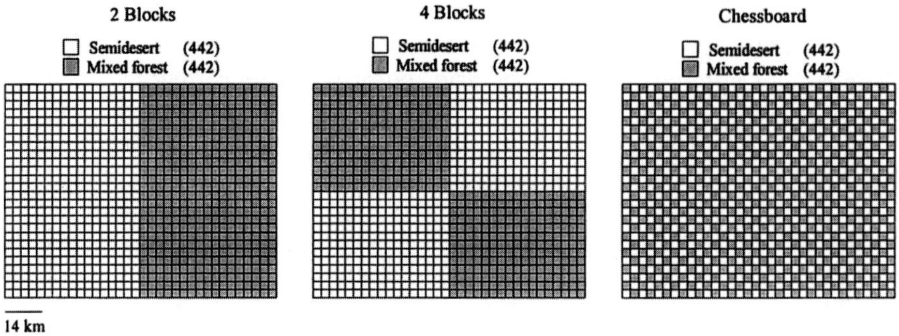

Figure 3. Configuration scenarios: landscape composition is kept constant at 50% mixed forest and 50% semi-desert for three different landscape configurations. Number of cells in each land-cover type is given in brackets.

Table 3. Land cover and configuration of the different scenarios.

scenario	Land cover	configuration
1	USGS current land cover	
2	100% mixed forest	homogeneous
3	100% semi-desert	homogeneous
4	50% mixed forest 50% semi-desert	2 blocks (see Fig. 3)
5	50% mixed forest 50% semi-desert	4 blocks (see Fig. 3)
6	50% mixed forest 50% semi-desert	Chessboard (see Fig. 3)

The next three scenarios represent changes in landscape configuration (Figure 3). They have an unchanged landscape composition of the same two types of land cover, i.e. mixed forest and semi-desert, equally partitioned. The three scenarios are very schematic and have 2 patches completely separated (2 blocks) for Scenario 4, a little more mixed configuration with 4 blocks for Scenario 5, and fully mixed patches in a chessboard-like configuration for Scenario 6. Table 3 gives land cover and composition of the 6 scenarios.

To assess the influence of landscape structure on water cycles and climate in an arid region of Spain, we have implemented a model that simulates changes in screen-height air temperature, energy fluxes, soil moisture, and precipitation, in response to landscape changes. This set of variables is ecologically essential, and precipitation, latent heat flux and soil moisture are water-cycle elements. To avoid boundary influences, the first border cells are not included in our comparisons, thus the total number of analysed cells is 768. Our first comparison is made on the domain average, i.e. averaged value of all the 768 grid cells. This value is often used for assessing sub-grid heterogeneity influences in up-scaling processes. Here, it is used as an indicator of global landscape behaviour, and for understanding impacts that may occur outside the domain.

For analysing behaviours taking place withinthe landscape, it is more relevant to focus on land-cover-type averages, i.e. average value of all the grid cells covered with the same land-cover type. Another average presented here is the one done on the unchanged cells(i.e. all the cells that have the same land cover in two different scenarios). Finally, to the variability between the different changes, the behaviour of particular "unchanged" grid cells are analysed. Actually, landscape ecology typically outlines the importance of the interactions between landscape elements, with a particular focus on landscape structure, for example on patches' neighbours. It is therefore essential to study whether there are particular cases (or particular grid cells), where the impacts are bigger (or smaller) than the corresponding mean values, to see if the impacts are spatially dependent. Averages provide a trend, if there is one, but may also smooth the effects. Even if the standard deviation is an indicator of the dispersion of the response, it does not give any idea where and why there may be a particular grid cell behaviour. All these different averages and values are presented hereafter by differences of values obtained in the various scenarios.

3. RESULTS

3.1 Screen-height air temperature

Table 4 gives the average daily mean air temperature differences between the scenarios, two by two. For the domain average, a difference of almost 1K is observed between Scenario 2 and 3. This result shows that local temperature differs over various land covers, and that semi-desert has a higher mean temperature than mixed forest. For vegetation, the temperature range is also important. In this case, the daily maximum temperature is 6K higher, and daily minimum is 3.5K is cooler, for the case of semi-desert; the total range of temperature is increased by 9.5K. For the case of Scenarios 4 or 6 compared to 2 and 3 (50% forest - 50% semi-desert compared to 100% forest or 100% semi-desert), differences are approximately 0.5K. A 5K increase occurs with Scenario 2, and a decrease of 4.5K occurs with Scenario 3. Differences between the various configurations, i.e. Scenarios 4 to 6, do not have significant domain averages.

Table 4. Mean differences between the scenarios (see numbers in table 3) of averaged daily mean air temperature, in [K]. Standard deviation of the differences is given in brackets.

Average over	2-1	3-1	3-2	4-2	4-3	6-2	6-3	6-4	6-5	5-4
Entire domain	-0.1	0.8	1.0	0.4	-0.5	0.4	-0.6	*-0.1*	*-0.1*	*-0.0*
	(0.1)	(0.4)	(0.4)	(0.2)	(0.2)	(0.2)	(0.3)	*(0.1)*	*(0.1)*	*(0.1)*
Forested areas	*-0.0*			*-0.2*		0.3		0.5	0.2	0.3
	(0.0)			*(0.2)*		(0.1)		(0.2)	(0.1)	(0.2)
Semi-desert		*0.0*			*0.1*		-0.5	-0.6	-0.3	*-0.4*
areas		*(0.2)*			*(0.2)*		(0.2)	(0.2)	(0.1)	*(0.2)*
*Unchanged	-0.1			*0.1*		0.3		0.2	0.1	0.1
forested cells	(0.0)			*(0.1)*		(0.1)		(0.1)	(0.1)	(0.1)
*Unchanged		0.6			-0.3		-0.5	-0.3	-0.2	*-0.2*
semi-desert cells		(0.2)			(0.2)		(0.2)	(0.1)	(0.1)	*(0.1)*

In italic, difference values that have both signs, some days are hotter and some others are cooler.

*Unchanged cells are the cells that have the same land-cover type in the two compared scenarios.

Focussing on the land-cover averages, our results indicate that landscape configuration plays an important role in determining the air temperature. When comparing Scenario 4 (2 blocks configuration) to Scenarios 2 or 3 (complete semi-desert or forest scenario), the differences are insignificant, but comparing the chessboard landscape (Scenario 6) to the same homogeneous scenarios, the differences increase, even if landscape composition (percentages of types of land cover) is the same in these two Scenarios 4 and 6. We also notice that differences are not identical when looking at forest or semi-desert. Semi-desert has greater air temperature sensitivity, and the differences are larger. These configuration influences can also be seen when comparing Scenario 6 to 4, where both have 50% mixed forest and 50% semi-desert.

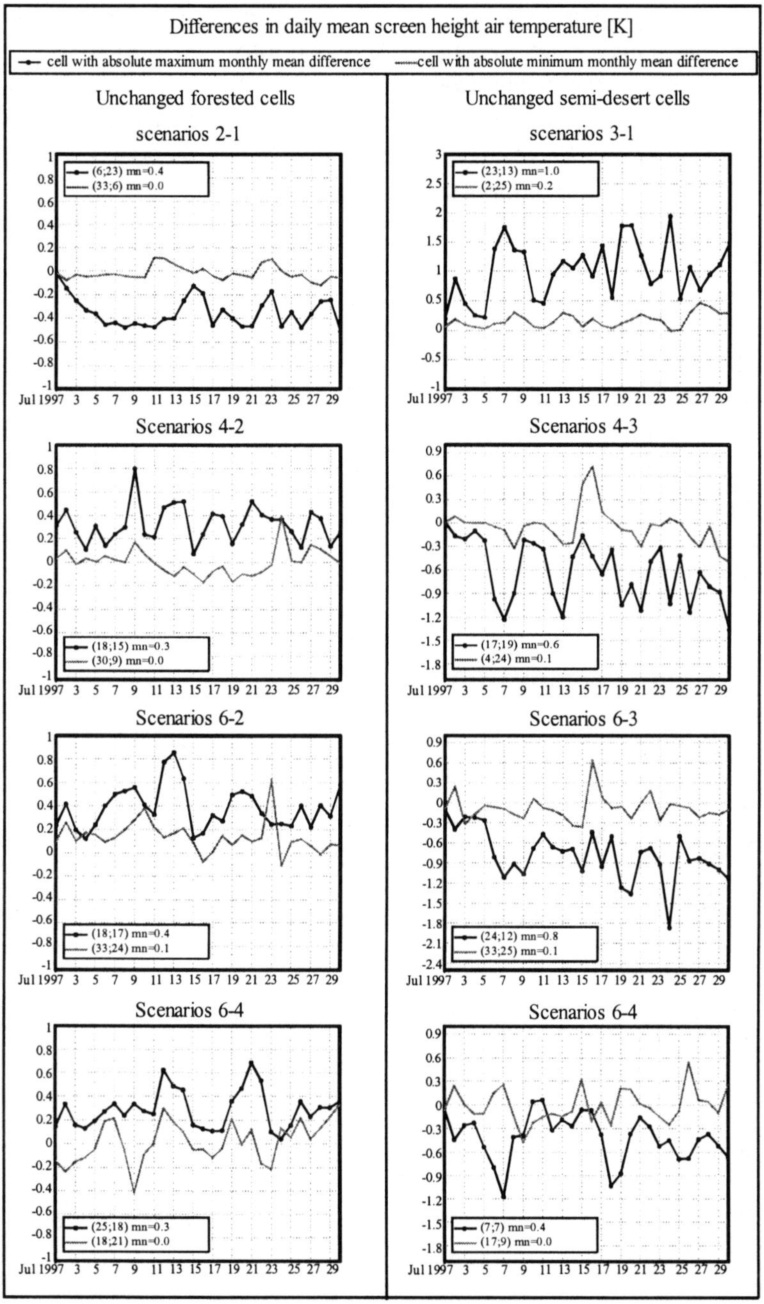

Figure 4. Daily screen-height air temperature differences between cells with the same land cover in two different scenarios [K]. Grid cells (given in brackets) are the one with the minimum or maximum absolute value of difference in monthly mean (mn).

For the case of averaging only the unchanged grid cells, the trends differ from the land-cover type averages. Actually, the "unchanged grid cells" average compares the same number of cells and in the same location between two scenarios, whereas "land-cover type" average compares all the cells with a certain land-cover type. Thus, the differences between these two averages emphasise the importance of grid cell location, and also the limitation of the meaning of average values. Therefore, going further in the analysis of "unchanged" grid cells, we look for grid cells having the minimum or the maximum monthly averaged difference value between two scenarios (Figure 4). The result shows that in every comparison, there are grid cells with a null mean value of change, showing that in every configuration, we find places where the influence of a surrounding landscape change is insignificant. On the other hand, we also always find a semi-desert cell, where the surrounding landscape change effect is significant; the scale effect is related to the scale of the change: Scenario 3 compared to 1 (landscape change of 90% of the land composition) have a 1K change, Scenario 6 compared to 4 (landscape configuration change only) have a 0.4K change. The effects are smoother on air temperature over forested cells.

3.2 Evapotranspiration

Table 5. Mean differences between the scenarios (see numbers in table 3) of averaged daily latent heat flux, in [W/m^2]. Standard deviation of the differences is given in brackets.

Average over	2-1	3-1	3-2	4-2	4-3	6-2	6-3	6-4	6-5	5-4
Entire domain	7.9 (3.4)	-44.4 (18.5)	-52.3 (21.6)	-24.0 (11.0)	28.3 (11.2)	-21.4 (9.2)	30.9 (12.5)	*2.5 (3.2)*	*1.4 (1.6)*	*1.1 (2.8)*
Forested areas	*-3.0 (1.7)*			*5.1 (5.6)*		*10.7 (5.6)*		*5.6 (5.7)*	6.6 (3.4)	*-1.1 (5.9)*
Semi-desert areas		*6.0 (6.6)*			*-0.7 (3.0)*		*-1.2 (3.0)*	*-0.4 (3.2)*	-3.7 (2.4)	*3.3 (3.5)*
*Unchanged forested cells	*-3.1 (1.3)*			*3.4 (2.2)*		*10.7 (5.6)*		*7.7 (5.9)*	*6.2 (3.1)*	*1.3 (2.0)*
*Unchanged semi-desert cells		*0.5 (4.6)*			*-0.1 (2.6)*		*-1.2 (3.0)*	*-1.2 (2.1)*	*-1.3 (2.1)*	*0.5 (1.7)*

In italic, difference values that have both signs.

*Unchanged cells are the cells that have the same land-cover type in the two compared scenarios.

Evapotranspiration is evaluated by analysing the latent heat flux. Table 5 gives the averaged daily mean latent heat flux differences between the scenarios. In the current landscape (Scenario 1), total evapotranspiration averaged over the domain is 111 mm for the entire month of July. Except for differences in the landscape configuration scenarios, all the changes have an important impact. A decrease of 21.4 W/m^2 (i.e., 22 mm of evapotranspiration for July) in Scenario 6 compared to Scenario 2, represent a decrease of approximately 20% of the total monthly evapotranspiration. This is even more dramatic if comparing forest to semi-desert (Scenarios 2 and 3), which has a difference of 54 mm in monthly evapotranspiration. This is similar to the mean monthly amount of evapotranspiration for the case of semi-desert (58 mm). In terms of the water cycle, these quantities are important but are also related to the fact that there is no water stress in the beginning of the simulation. The presence of water stress at the beginning of the simulation would have led to a lower total evapotranspiration amount, but it is difficult to know the magnitude of the relative differences.

Land cover type and unchanged cells averages are also given in Table 5. In every average, there are some days with inverted signs, meaning that domain evapotranspiration may be more important or less important depending on day and on the scenario. Nevertheless, differences of 10.7 W/m^2 between Scenario 6 and 2 on unchanged forested cells are important and roughly 10% of the simulated fluxes. Of even greater importance is the difference of 7.7 W/m^2 in unchanged forested grids cells between Scenario 6 and 4; this confirms the importance of landscape configuration. Note that for evapotranspiration, changes are more important in forested areas than in semi-desert. Figure 5 presents outputs for specific grid cells. Except for Scenarios 6 and 2, in every situation there are cells with mean null values, as was the case for temperature. Forested cells are more sensitive to evapotranspiration changes than semi-desert ones, and reach values of almost 15 W/m^2 for the case of Scenarios 4 or 6 compared to 2. As seen in the averaged value, landscape configuration may induce changes in latent heat flux over a forested cell of 12 W/m^2 in monthly mean, reaching 35 W/m^2 for a single day.

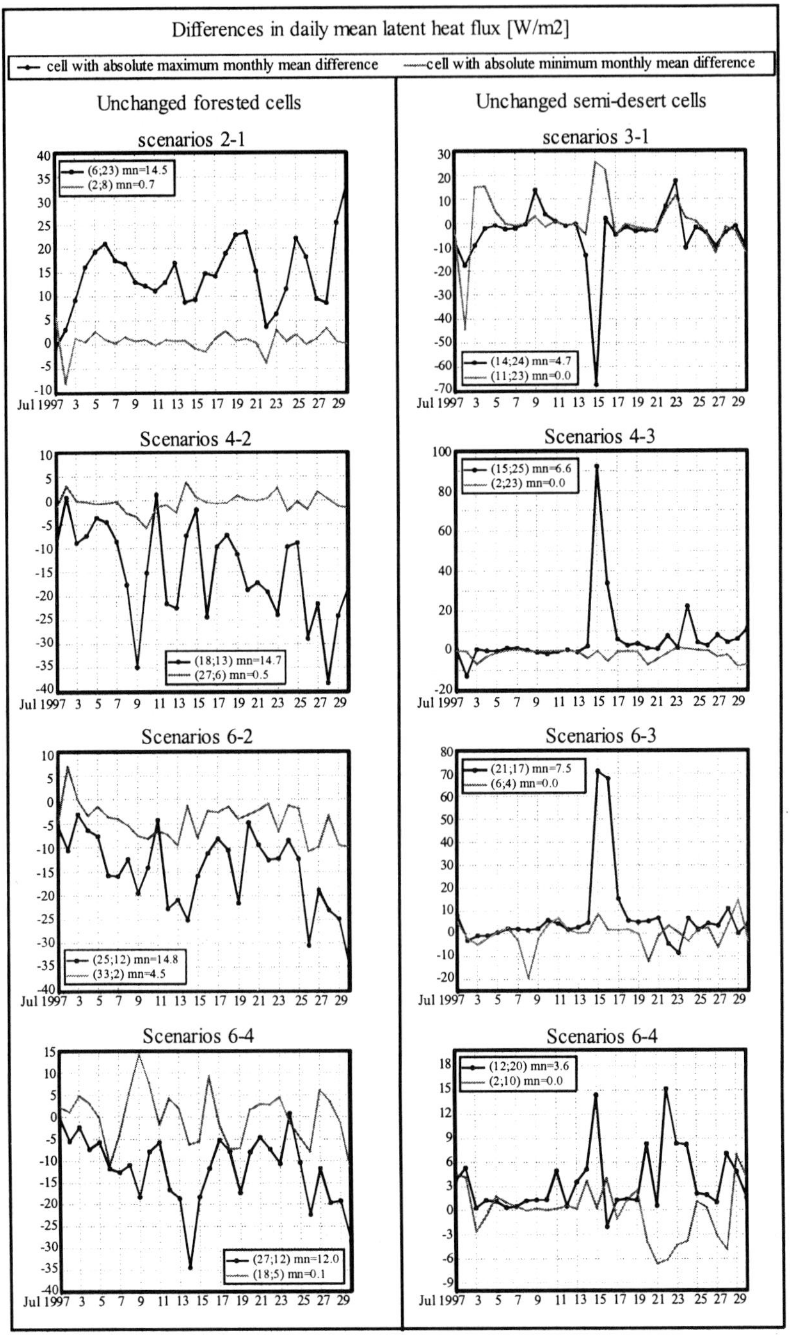

Figure 5. Daily averaged latent heat flux differences between cells with the same land cover in two different scenarios [W/m2]. Grid cells (given in brackets) are the one with the minimum or maximum absolute value of difference in monthly mean (mn).

3.3 Precipitation

Table 6. Differences in averaged monthly amount of precipitation, absolute and relative values.

Value	2-1	3-1	3-2	4-2	4-3	6-2	6-3	6-4	6-5	5-4
Absolute [mm]	1.6	-0.1	-1.7	-0.2	1.5	-1.6	0.1	-1.4	-0.2	-1.2
Relative [%]	4.4	-0.3	-4.8	-0.6	4.2	-4.5	0.3	-4	-0.6	-3.4

Total precipitation for July 1997 was 35 mm. Because of the low precipitation in this region, we expect the absolute differences for the scenarios to be small. Actually, domain averaged values of differences between the scenarios are small (Table 6), but since there is only one particular month of simulation, here we will focus on daily values. Precipitation is related to land-surface processes, but there are less direct links between the land cover of a particular grid cell and precipitation received by that same grid cell. Figure 6 shows the monthly precipitation patterns for the different scenarios, emphasising this lack of direct spatial relation. Nevertheless, these precipitation patterns reveal that there are spatial influences on precipitation, for example higher precipitation in the south-eastern corner of the domain in every scenario, explained by an orographic effect. In Scenarios 4 and 5, we can also guess the transition line between the patches of forest and semi-desert, showing that there may be an influence of large patches of vegetation types. Except for this last slight effect, it seems meaningless to focus on particular grid cells or land-cover type, and comparisons are made on domain averages. Figure 7 gives the averaged daily precipitation and their relative differences, neglecting days with less than 1 mm of rainfall. As expected, absolute rainfall differences are small, but relative values may represent 40% of the rainfall, for a domain average. Since both the timing and magnitude of precipitation events are important for plants, especially under conditions of water stress, these relative differences may be important, even if the absolute values are small. They are particularly relevant if they significantly impact soil moisture content, ultimately enhancing or reducing the plants water stress. In order to determine the impact of landscape changes on precipitation, longer-term model simulations are required.

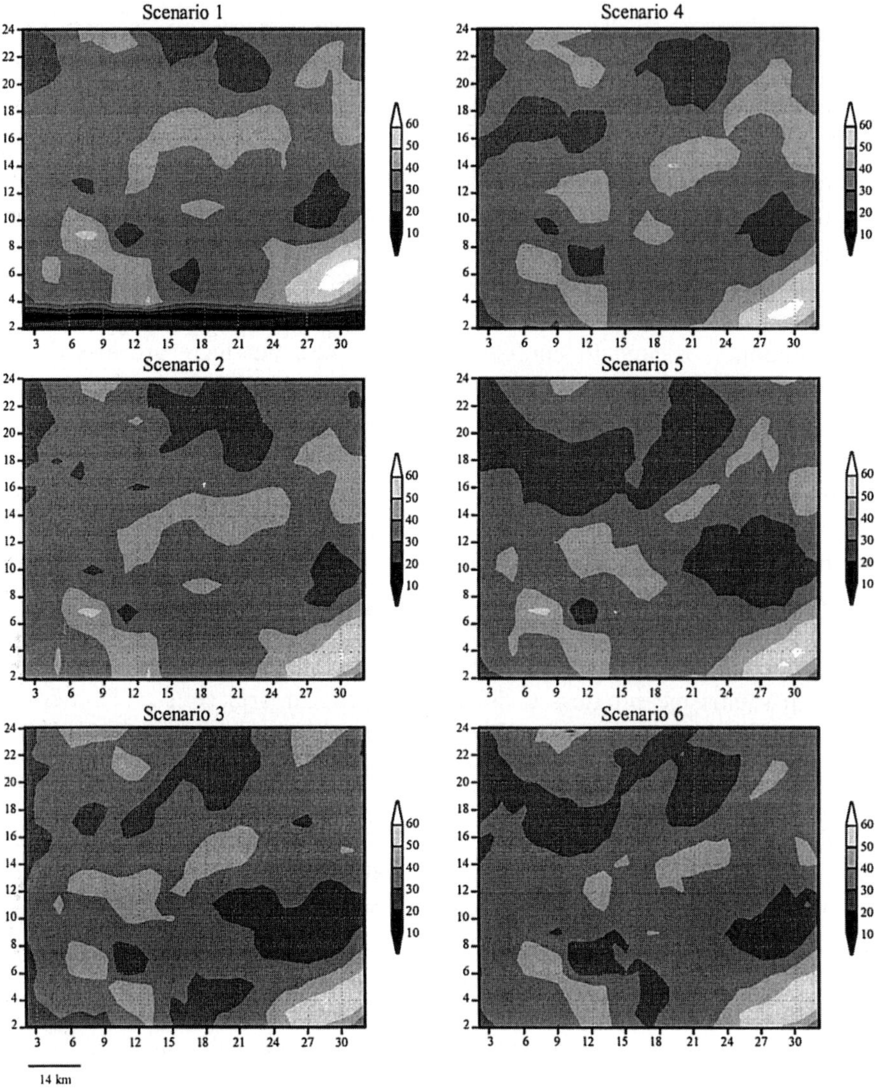

Figure 6. Monthly precipitation patterns in the different scenarios. Precipitation is given in [mm]. The grid cell numbers, or identification, are given on both axes.

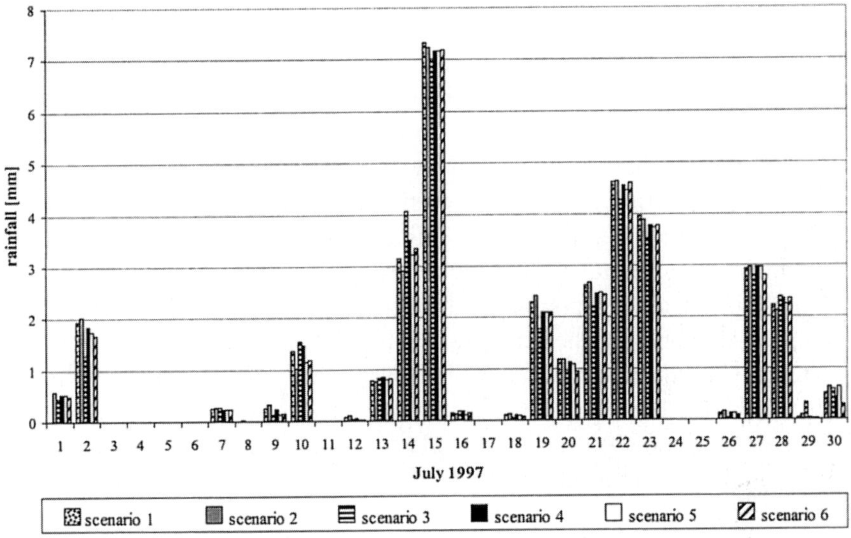

Fig.7a

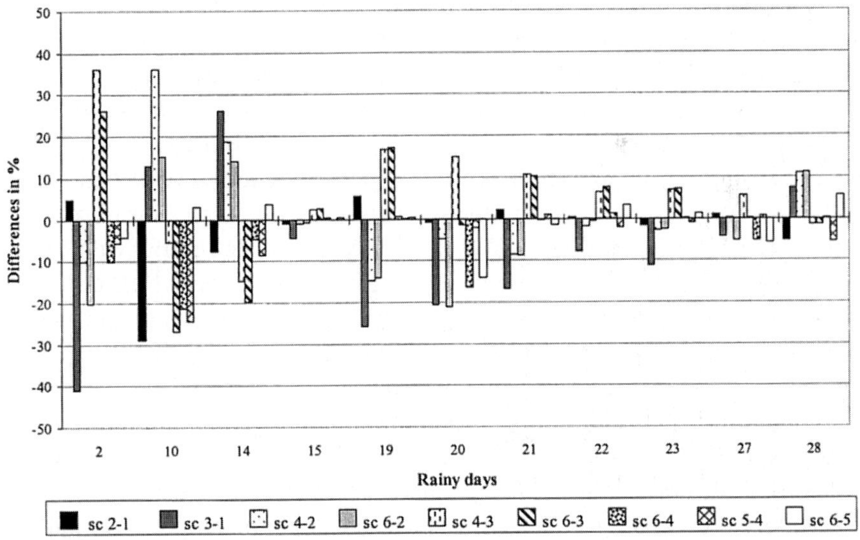

Fig.7b

Figure 7. a. Domain averaged daily precipitation [mm] and *b.* relative differences in % between the scenarios (sc).

3.4 Soil moisture

From ecological perspectives, an important soil moisture variable to assess is the volume of water available for plants. This value depends on soil type, soil structure, and soil texture. Nevertheless, for every soil type we can derive moisture characteristics, in particular the field capacity, and the temporary and permanent wilting points. Field capacity is the amount of water that can be retained by a soil after gravitational water has been removed. Wilting points are the lower limits of water availability for plants, soil water potential being equal or below the minimum osmotic potential of the plant. At or below the temporary wilting point, the plant wilts but can recover. If the permanent wilting point is reached, plants remain wilted and have permanent damages. We focus here on the question of water stress, looking at values of soil moisture under temporary wilting-point conditions. For the domain average, the number of days were this point is reached is 25 to 29 days for the first two soil layers (0-12.5 cm), 5 to 20 days in third layer (12.5-20 cm) and 2 to 5 days in 4th layer (20-30cm). Even though the initial soil moisture value was set above the wilting point, water stress is very quickly reached in the upper soil layers. Soil moisture profiles differ under various types of vegetation, and the absolute water content is usually lower under forested areas but higher in the surface layers. Looking at particular "unchanged cells", we observe that there are differences in the number of water stress days (Table 7). Furthermore, we observe that for the same scenarios (and the same land cover), some grid cells have more days of water stress and some others have less. The differences also depend on the soil layer. The impact of water stress depends on the proportion of roots that are in each particular level. We also notice that the cells with greater differences are not the ones with the maximum latent heat flux difference. This is because water stress is reached when soil moisture is under a certain threshold, which may be reached without much latent heat flux, depending on soil characteristics and rainfall amount.

To compare the scenarios, we also looked at the available water content (amount of water between permanent wilting point and actual soil moisture) in the first 30 cm of soil (usually where most of water extraction by roots takes place). This water content is related to evapotranspiration, rainfall, and capillary rise. Runoff here is negligible or non-existent. For the case of forested unchanged cells, the monthly mean differences are about 1 mm of water content, on a total mean content of approximately 10 mm; thus this difference has an important impact. This can be seen in the changing days of water stress. In a same manner as for latent heat flux, differences are lower for the case of semi-desert grid cells.

Table 7. Maximum difference of number of days under water stress (temporary wilting point) under unchanged land cover, in [d]. Grid cell identification is given in brackets.

	Forested cells				Semi-desert cells			
Soil depth	2-1	4-2	6-2	6-4	3-1	4-3	6-3	6-4
0-5 cm	3 (17;16) -3 (12;17)	3 (18;5) –3 (28;3)	4 (7;6)	4 (21;8)	2 (14;14)	-2 (11;11)	-2 (11;11)	-2 (15;15)
5-12.5 cm	1 (23;24) -1 (9;18)	1 (18;12) –1 (21;7)	1 (2;23) –1 (23;24)	1 (19;20) –1 (18;21)	6 (30;21)	-3 (5;19)	-4 (13;11)	-3 (15;15)
12.5-20 cm	-2 (12;22)	3 (18;14)	3 (10;13)	2 (18;14) –2 (18;5)	3 (23;13)	-2 (17;12)	-3 (33;9)	-2 (15;11)
20-30 cm	-1 (23;12)	1 (18;2)	2 (27;12)	2 (30;15)				

4. DISCUSSION

4.1 Landscape scale

The results presented in Section 3 provide important insights into the influence of landscape structure on local climate and the water cycle in a semiarid area like Castilla-La Mancha, south-eastern Spain. Different land covers have various behaviour, and replacing a forested landscape by a semi-desert one will lead to higher mean temperatures, higher temperature ranges (lower minimums and higher maximums), lower latent heat fluxes (on the order of 50% of the forest latent heat flux values), and lower precipitation (on the order of 5%), at least in July 1997, averaged over the domain of change. These differences may be mostly explained by the fact that mixed forest, compared to semi-desert, has a higher vegetation fraction. Vegetation can extract moisture from the deeper soil layers, and have therefore a lower Bowen ratio (higher latent heat flux) leading to lower temperatures (mean and ranges).

4.2 Land cover type

With this first observation, the next step is to quantify the influence of changes that may occur within the landscape, on different land-cover types. Results obtained from this perspective show that for the case of replacing forest by semi-desert (from totally forested covered to half-half forested and semi-desert covered scenario, in a well mixed, chessboard-like configuration), forested areas have higher average temperatures and higher values of latent heat flux (on the order of 10%). For the case of replacing semi-desert by forest (same kind of scenarios), semi-desert areas showed lower air temperatures and no significant differences in latent heat flux. This set of observations emphasise the fact that response to change depends on land-cover type.

In contrast, these conclusions change when comparing totally forested or semi-desert covered scenarios with the half-half scenario in the 2-block configuration. Here the changes are less important, at least in an average sense. Actually, the changes may be important, but in areas close to the landscape changes. This highlights the importance of landscape configuration.

4.3 Individual grid cells

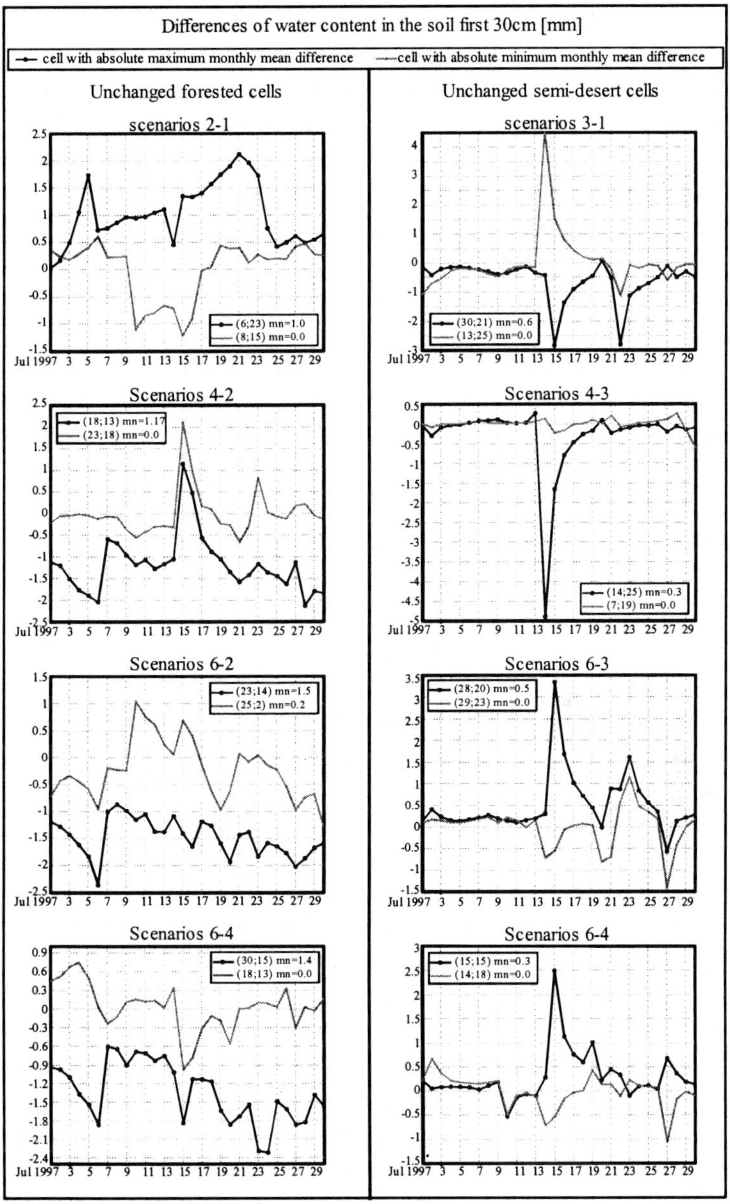

Figure 8. First 30cm of soil water content differences between cells with the same land cover in two different scenarios [mm]. Grid cells (given in brackets) are the one with the minimum or maximum absolute value of difference in monthly mean (mn).

To see in more detail the probable impacts of a landscape change on a surrounding unchanged area, we focused on particular grid cells that have the same land cover in two different scenarios, called "unchanged" grid cells (Figure 8). The average values of air temperature and latent heat flux of these cells show significant impacts on the surrounding landscape in some cases. For example, air temperature is impacted when changing the current landscape to a fully semi-desert land-covered scenario, and latent heat flux is impacted when changing the landscape configuration from a chessboard to the 2-block configuration. In every case, and for all the variables, including soil moisture content or water stress, there are cells where the impact is important and other cells where it is insignificant. This analysis highlights that the cells presenting important responses are the ones spatially close to the landscape changes. For example, for the case of 2 blocks compared to the homogeneous landscapes, influenced cells are those near the transition line between the 2 land-cover types, i.e. x=17 or x=18. This is much more difficult to point out when comparing the chessboard scenario to a homogeneous landscape, since all of the grid cells are close to the changes. In this case, further analysis is needed to identify the influence of factors like elevation, slope, soil type, winds, and other landscape characteristics.

4.4 Ecological consequences

In the previous section, we saw that there are influences of land-cover type on local climate variables, and on water-cycle characteristics. Changes of 1K in monthly mean air temperature, 15 W/m^2 in monthly latent heat flux, 5% in monthly precipitation, some days of water stress and of some millimetres of soil water content are likely to happen in grid cells that have not been changed. It's difficult to identify here the consequence of these changes on the concerned ecosystems, but the magnitude of the changes are in the same range as those foreseen by global climate change studies, and that are claimed to have significant ecosystem impacts (Watson et al., 1996). As well as influencing the rate of plant growth and metabolism, temperature and moisture can influence the timing of development and senescence, or induce changes in species composition and distribution. According to Watson et al. (1996), in Mediterranean mountains, changes in summertime water supply may influence topsoil processes and plant nutrition. When the top layer of the soil desiccates, mineral cycling is blocked. Especially at low altitudes, Mediterranean tree species can replace sub-montane belt species.

4.5 Implications for future works

The study presented here is a first-step sensitivity study and therefore gives an interesting start for further investigations. Thus, there are some particular points that need to be commented on. First, comparing Scenarios 2 and 3 to Scenario 1 is valuable for assessing what could happen to the current landscape while basing the study on the more realistic values of Scenario 1. But the changes we imposed are of different types and it's difficult to identify which of the changes in composition or configuration is of greatest importance. Another interesting aspect is that we have identified changes similar to those obtained by other authors, even though our simulations have typically been at higher spatial resolution. According to Seth and Giorgi (1998), the domain size may impact the response of internal physical processes. Thus, processes like precipitation that may be triggered by more large-scale processes could be influenced by the choice of the domain. On the other hand, precipitation over the studied domain in July is essentially convective, and convective precipitation is known to be spatially heterogeneous and very local. Soil moisture is also a key element in the water cycle, in ecological issues, and in precipitation processes. Our initial soil moisture was set to 50% of the saturation value. Defining the proportion of saturation has the advantage of giving, in a simple manner, different soil moisture contents for the various soil types. It also has the advantage of giving a comparable situation between the scenarios since it is known that atmospheric models can be sensitive to soil moisture conditions. We expect that choosing different initial soil moisture conditions would modify the details of our results. Ideally, soil moisture observations would be used to initialise the model simulations. The description of soil moisture, as a key element in water cycle balance and surface-induced atmospheric circulations, needs to be improved as part of further analyses. Here we have chosen 1997 as an average meteorological year. Recent research results are emphasising the impact of global change on extreme weather events. With this in mind, further research needs to be done on years presenting climatic anomalies, especially drought periods. Here we also have chosen July as a high vegetated period (maximum of leaf area index fixed at mid-July). Longer time integration, like over the whole year or over the entire vegetation period would give a complete view of the vegetation-induced atmospheric behaviour. On another hand, ClimRAMS do not account for vegetation response to atmospheric and moisture forcing (except for stomata response), like changes in growth or in species distribution and composition. It has been neglected partly because of the longer timescale of these processes, but in our topic, it would be interesting to integrate it in the model to see in which way it would change our results.

5. CONCLUSION

The impact of landscape structure on the water cycle and climate is analysed from a model-sensitivity perspective. For that purpose, the ClimRAMS mesoscale model was used to perform simple landscape structure (composition and configuration) scenarios over southeastern Spain. The differences obtained with these different scenarios demonstrate that both configuration and composition have substantial influence on climate and water resources. Changes as large as 1K in daily average air temperature and 9.5K in daily temperature range take place when replacing mixed forest with semi-desert over the whole sub-domain. Mean monthly latent heat flux is 12 W/m2 lower in the (27,12) forested cell case, when the spatial arrangement of the forested and semidesert areas change in the landscape, from a chessboard-like, small scale pattern to a 2 blocks, large scale pattern. Impacts are land-cover type-dependent, for eg. temperature is more influenced in semi-desert areas and forested areas are more moisture sensitive. These preliminary results show that land-use planning, especially in semiarid regions, need to consider vegetation-climate feedback effects for ecosystem and water management. This study also suggests that there is a need to implement a more complete sensitivity study to identifying the magnitude of land-use planning effects on the water cycle and climate.

6. REFERENCES

Anthes, R.A., 1984. Enhancement of convective precipitation by mesoscale variations in vegetative covering in semiarid regions. Journal of Climate and Applied Meteorology, 23: 541-554.

Avissar, R. and Mahrer, Y., 1982. Verification study of a numerical greenhouse microclimate model. Transactions of the American Society of Agricultural Engineering, 25: 1711-1720.

Avissar, R. and Pielke, R.A., 1989. A parameterization of heterogeneous land surfaces for atmospheric numerical models and its impact on regional meteorology. Monthly Weather Review, 117: 2113-2136.

Ben-Gai, T., Bitan, A., Manes, A. and Alpert, P., 1993. Long-Term Change in October Rainfall Patterns in Southern Israel. Theretical and Applied Climatology, 46: 209-217.

Bosch, J.M. and Hewlett, J., D., 1982. A review of catchment experiments to determine the effect of vegetation changes on water yield and evapotranpiration. Journal of Hydrology, 55(1): 3-23.

Calder, I.R., 1993. Hydrologic effects of land-use change. In: D.R. Maidment (Editor), Handbook of Hydrology. McGraw-Hill, New York, pp. 13.1-13.50.

Chase, T.N., Pielke, R.A., Kittel, T.G.F., Nemani, R.R. and Running, S.W., 2000. Simulated impacts of historical land cover changes on global climate in northern winter. Climate Dynamics, 16(2-3): 93-105.

Copeland, J.H., Pielke, R.A. and Kittel, T.G.F., 1996. Potential climatic impacts of vegetation change: a regional modeling study. Journal of Geophysical Research - Atmosphere, 101(D3): 7409-7418.

Cotton, W.R. and Pielke, R.A., 1995. Human impacts on weather and climate. Geophysical Science Series, 2. Cambridge University Press, New York.

de Ridder, K. and Gallée, H., 1998. Land surface-induced regional climate change in southern Israel. Journal of Applied Meteorology, 37(11): 1470-1485.

Dolman, A.J. et al., 1997. The role of the land surface in Sahelian climate: HAPEX-Sahel results and future research needs. Journal of Hydrology, 188-189(1-4): 1067-1079.

Forman, R.T.T., 1995. Land Mosaics. The ecology of landscapes and regions. Cambridge University Press, Cambridge, 632 pp.

Garcia de Pedraza, L. and Rieja Garrido, A., 1994. Tiempo y clima en España, meteorología de las Autonomías. Dossat-2000, Madrid.

Giorgi, F. and Avissar, R., 1997. Representation of Heterogeneity Effects in Earth System Modeling - Experience from Land-Surface Modeling. Reviews of Geophysics, 35(4): 413-437.

Heck, P., 1999. European-Scale Vegetation-Climate Feedbacks Since the time of the Romans, ETH, Zürich, 104 pp.

Heck, P., Lüthi, D., Wernli, H. and Schär, C., 2000. Climate impacts of European-scale anthropogenic vegetation changes: A sensitivity study using a regional climate model., submitted.

Jia, X. and Ye, Z., 1996. A numerical study on impact of crop canopy on mesoscale climate. Science in China (Series C), 39(5): 474-483.

Klaassen, W. and Claussen, M., 1995. Landscape Variability and Surface Flux Parameterization in Climate Models. Agricultural and Forest Meteorology, 73(3-4): 181-188.

Laval, K., 1997. Hydrological Processes in GCMs. In: S. Sorooshian, H.V. Gupta and S.C. Rodda (Editors), Land surface processes in hydrology. Springer-Verlag, Berlin, pp. 45-61.

Liston, G.E. and Pielke, R.A., 2000. A climate version of the regional atmospheric modeling system. Theoretical and Applied Climatology, 66(1-2): 29-47.

Liu, Y. and Avissar, R., 1996. Sensitivity of shallow convective precipitation induced by land surface heterogeneities to dynamical and cloud microphysical parameters. Journal of Geophysical Research, 101(D3): 7477-7497.

Lynn, B.H., Rind, D. and Avissar, R., 1995. The importance of mesoscale circulations generated by subgrid-scale landscape heterogeneities in General Circulation Models. Journal of Climate, 8: 191-205.

Mahfouf, J.F., Richard, E. and Mascart, P., 1987. The influence of soil and vegetation on the development of mesoscale circulations. Journal of Climate and Applied Meteorology, 26: 1483-1495.

Marengo, J.A., Nicholson, S.E., Lare, A.R., Monteny, B.A. and Galle, S., 1996. Application of evapoclimatonomy to monthly surface water balance calculations at the HAPEX-Sahel supersites. Journal of Applied Meteorology, 35(4): 562-573.

McCumber, M.C. and Pielke, R.A., 1981. Simulation of the effects of surface fluxes of heat and moisture in a mesoscale numerical model 1. soil layer. Journal of Geophysical Research, 86(C10): 9920-9938.

McNaughton, K.G., 1989. Regional interactions between canopies and the atmosphere. In: G. Kussel, B. Marshall and P.G. Jarvis (Editors), Plant canopies; their growth form and function. Cambridge University Press, pp. 63-81.

Noilhan, J. and Lacarrère, P., 1995. GCM Grid-Scale Evapotranspiration from mesoscale modeling. Journal of Climate, 8: 206-223.

Nováky, B., Pachner, C., Szesztay, K. and Miller, D., 1985. Water resources. In: R.W. Kates, J.H. Ausubel and M. Berberian (Editors), Climate Impact Assessment. John Wiley & Sons, Chichester, pp. 187-214.

Pan, Z., Takle, E., Segal, M. and Arritt, R., 1999. Simulation of potential impacts of man-made land use changes on U.S. summer climate under various synoptic regimes. Journal of Geophysical Research, 104(D6): 6515-6528.

Pielke, R.A., 2000. Influence of the spatial distribution of vegetation and soils on the prediction of cumulus convective rainfall. submitted to Reviews of Geophysics.

Pielke, R.A. and Avissar, R., 1990. Influence of landscape structure on local and regional climate. Landscape Ecology, 4(2-3): 133-155.

Pielke, R.A. et al., 1999. The Influence of Anthropogenic Landscape Changes on Weather in South Florida. Monthly Weather Review, 127(7): 1663-1673.

Pielke, R.A. et al., 1992. A comprehensive meteorological modeling system-RAMS. Meteorology and Atmospheric Physics, 49: 69-91.

Raupach, M.R. and Finnigan, J.J., 1995. Scale issues in Boundary-layer meteorology: surface energy balances in heterogeneous terrain. Hydrological Processes, 9(5-6): 589-612.

Segal, M., Avissar, R., McCumber, M.C. and Pielke, R.A., 1988. Evaluation of vegetation effects on the generation and modification of mesoscale circulations. Journal of the Atmospheric Sciences, 45(16): 2268-2292.

Seth, A. and Giorgi, F., 1998. The Effects of Domain Choice on Summer Precipitation Simulation and Sensitivity in a Regional Climate Model. Journal of Climate, 11(10): 2698-2712.

Tremback, C.J. and Kessler, R., 1985. A surface temperature and moisture parameterization for use in mesoscale numerical models. In: AMS (Editor), Conference on numerical weather prediction, #7, pp. 355-358.

Wang, J., Bras, R.L. and Eltahir, E.A.B., 2000. The impact of observed deforestation on the mesoscale distribution of rainfall and clouds in Amazonia. Journal of the Atmospheric Sciences, 55: 447-464.

Watson, R.T., Zinyowera, M.C., Moss, R.H. and Dokken, D.J., 1996. Climate change 1995. Impacts, adaptations and mitigation of climate change: scientific-technical analyses, IPCC, Cambridge University Press, Cambridge.

Yang, Z.L., Dai, Y., Dickinson, R.E. and Shuttleworth, W.J., 1999. Sensitivity of Ground Heat-Flux to Vegetation Cover Fraction and Leaf-Area Index. Journal of Geophysical Research, 104(D16): 19505-19514.

Trends in drought in Swiss forested ecosystems

BÄRBEL ZIERL AND HEIKE LISCHKE
Swiss Federal Research Institute for Forest, Snow and Landscape (WSL), Birmensdorf, Switzerland

Abstract: The warming of the atmosphere observed during the last century is supposed to have intensified the hydrological cycle. As a consequence the occurrence of drought is expected to have changed in the last decades. To detect trends regarding drought in forested ecosystems numerous time series of a newly developed ecophysiologically based drought index are analysed. The time series are simulated by the hydrological model WAWAHAMO (WAldWAsserHAushaltsModell, German, i.e. forest water balance model) for the observation period from 1969 to 1998 covering the entire forested area of Switzerland with a horizontal resolution of 1km. Furthermore, time series of related variables are examined for trends, i.e. observations of annual mean temperature, annual and seasonal precipitation totals, and simulations of annual transpiration totals and the length of the vegetation period. Over large parts of the forested area of Switzerland the analysis detected an increase of drought during the observation period. The trend is particularly pronounced in rather dry areas as the inner-alpine valley of Valais or near the northern border towards Germany. Hardly any trend can be detected in the prealpine regions, which are characterised by a huge surplus of water resulting from strong and frequent orographically forced precipitation events. Moreover, there is a definite increase in temperature, transpiration and the length of the vegetation period. No systematic trend could be detected for precipitation, neither for annual nor for seasonal totals.

1. INTRODUCTION

It is now widely accepted that under warmer conditions the hydrological cycle becomes more intense, causing changes in rainfall intensity and seasonality, and changes in evapotranspiration. These changes will affect soil water storage, which in turn might result in longer periods of floods and

M. Beniston (ed.),
Climatic Change: Implications for the Hydrological Cycle and for Water Management, 329–347.

droughts (Beniston et al., 1997, Houghton et al., 1996). Given the enormous social, economic, and ecological consequences of such projections, various studies have been conducted in order to identify any changes that occurred in temperature and precipitation during the 20th century.

Many studies examining temperature trends in the Alps conclude that considerable warming has taken place since the early 1980s. Beniston et al. (1994) examined four observation sites in Switzerland, i.e. Zürich (569 m above mean sea level), Säntis (2500 m), Davos (1590 m) and Lugano (276 m). Their inspection of the time series led them to conclude that minimum temperatures increased at all four stations by about 2 K between 1901 and 1992, maximum temperatures at the low-level stations decreased between 1 and 2 K and maximum temperatures at the high-altitude locations remained unchanged or increased similarly to the minima. In summary, the warming experienced since the early 1980s reaches up close to 1 K for this ensemble average (Beniston et al., 1997). Hasenauer et al. (1999) investigated 20 selected weather stations across Austria and found an increase in mean annual temperature of 0.72 K between 1961 and 1990 for these sites. Auer and Böhm (1993) too detected a strong warming of about 1 K in a spatially averaged temperature time series in Austria since 1890.

In general, the trend in surface temperature observed at various stations in the alpine region resembles the global warming trend with a relatively rapid rise between 1920 and 1950, a slow decrease from 1950 until the mid-1970s, followed by another period of rapid temperature increase. However, the warming experienced in the alpine region by far (app. factor 2) exceeds the global warming trend of about 0.5 K observed in the 20th century (Beniston et al., 1997).

Little evidence can be found for any change in precipitation rates across Switzerland. As annual precipitation totals are highly variable, it is made very difficult to detect a trend. In Switzerland, Beniston et al. (1994) presented time series plots of four stations from 1901 to 1992 and found no trends in the data. Furthermore, Widmann and Schär (1997) performed a more thorough analysis of precipitation trends for the periods from 1901 to 1990 and 1961 to 1990 using 113 stations within Switzerland. For the period from 1901 to 1990 they found a positive trend between 20 and 40 % in wintertime precipitation for most locations and only small trends with both signs in the other seasons. For the period from 1961 to 1990 the trend in wintertime precipitation is again positive almost throughout the entire area of Switzerland, with values up to 15 % per 10 years in the Ticino region, whereas the trend to the north of the Swiss Alps is somewhat smaller.

In Austria, Hasenauer et al. (1999) too could not detect any trend in 20 time series of annual precipitation totals. Moreover, Auer and Böhm (1993) examined state-wide time series of annual precipitation totals within Austria and found no evidence of a systematic trend in precipitation over the last 100 to 150 years.

A study of seasonal precipitation changes by Schönwiese et al. (1993, 1994) covering the entire Alpine region detected some weakly significant trends. In the western parts of the Alps precipitation during the winter months increased by up to 20 %, and decreased by a similar amount during the autumn months considering the period from 1891 to 1990. For the period from 1961 to 1990, Schönwiese et al. observed an Alpine-wide increase of winter precipitation by up to 25 %, an increase of spring precipitation by 30 % in the south-western regions of the Alps and a decrease of spring precipitation by 25 % in the north-eastern regions. Furthermore, they observed an increase (decrease) of precipitation to the north (south) by 25 % during the autumn months. During summertime no significant trends were detected.

Global warming associated with potential changes in precipitation amount and seasonality is expected to increase the likelihood of future droughts and water shortage (Arnell, 1996, Bultot et al., 1992, Gurtz et al., 1997, Herbst and Hörmann, 1998, Mitchell and Warrilow, 1987, Rind et al., 1990, Whetton et al., 1993). However, so far only a few studies have been conducted to investigate, if the occurrence of drought has already significantly changed during the last decades. Balling (1996) conducted a study on variations in United States drought severity during the 20th century using the Palmer Drought Severity Index. He concluded that so far there is only little evidence of a drying trend as predicted by many climate models. In southern Switzerland, an analysis of trends in drought by Rebetez (1999) showed that the frequency and duration of drought episodes (defined solely by precipitation) were significantly increasing towards the end of the 20th century.

So far, there is no study which analyses changes in drought stress a forest stand experiences under dry conditions. There is now an opportunity to examine variations in a newly defined ecophysiologically-based drought index that has been simulated by the hydrological model WAWAHAMO for the entire forested area of Switzerland covering a time period of three decades. In addition, the climate variables annual mean temperatures, annual and seasonal precipitation totals, annual transpiration totals, and the length of the vegetation period are analysed as they are expected to be some of the major factors causing the occurrence of drought.

2. METHODS

2.1 Description of the hydrological model WAWAHAMO

WAWAHAMO (German: WAldWAsserHAushaltsMOdell) is a spatially distributed, physically- and biologically-based water balance model designed to simulate hydrological and ecophysiological processes in forested ecosystems at regional levels. A detailed description of the model is given in Zierl (2000, 2001a). When developing the model, the focus was on the simulation of an ecophysiologically-based drought index indicating the drought stress a plant experiences under the site specific environmental conditions. For this, it was necessary to improve the interaction between the vegetation and water resources, two modelling fields which have historically been one-sided. That is, ecologists tend to minimise hydrological processes and hydrologists tend to minimise vegetation processes.

WAWAHAMO was developed to be applied primarily to the entire forested area of Switzerland. For this reason, the 1km-grid of the Swiss National Forest Inventory (LFI 1, 1983-1985) is used as basic model grid. However, in principle the model can be run on any regional grid, where the necessary input data are available.

The input data required consist of daily meteorological conditions (including air temperature, relative humidity, air pressure, wind speed, cloud cover, absolute or relative sunshine periods, and daily precipitation totals), forest parameters (including ratio of conifers to deciduous trees, predominant tree species, percentage of vegetation cover, and leaf area index), soil physical parameters (including available soil water capacity and permeability), and topographical parameters (including elevation, aspect, slope, and relief shading). The meteorological data were available from approximately 100 weather stations of the Swiss Meteorological Institute and were interpolated to the model grid by altitude dependent, inverse distance weighted interpolation. The topographical data were supplied by or derived from a digital terrain model (DHM25 © Bundesamt für Landestopographie, DV 455.2). The soil physical parameters were derived from a soil map (Bundesamt für Raumplanung 1980) in combination with the mentioned digital terrain model and approximately 700 soil profiles (Waldispühl 1998) using the geostatistical interpolation method external drift kriging (Zierl, 2000). For a detailed description of the preparation of the input data see Zierl (2000, 2001a).

The model system consists of three water storage components: the soil water reservoir, the canopy water reservoir (a reservoir for intercepted precipitation on the canopy) and the snow water reservoir. WAWAHAMO simulates water fluxes, water storage, and the changes of phase of water. Only vertical fluxes are taken into account in the model. No interaction between the different grid points, i.e. no spatial redistribution of the soil water over the landscape due to lateral fluxes is considered. The model employs a simple soil bucket approach and applies the big-leaf Penman-Monteith equation (Monteith, 1975). The basic time step is one day. A scheme of the hydrological model WAWAHAMO, its water reservoirs and the fluxes in between, is given in Figure 1.

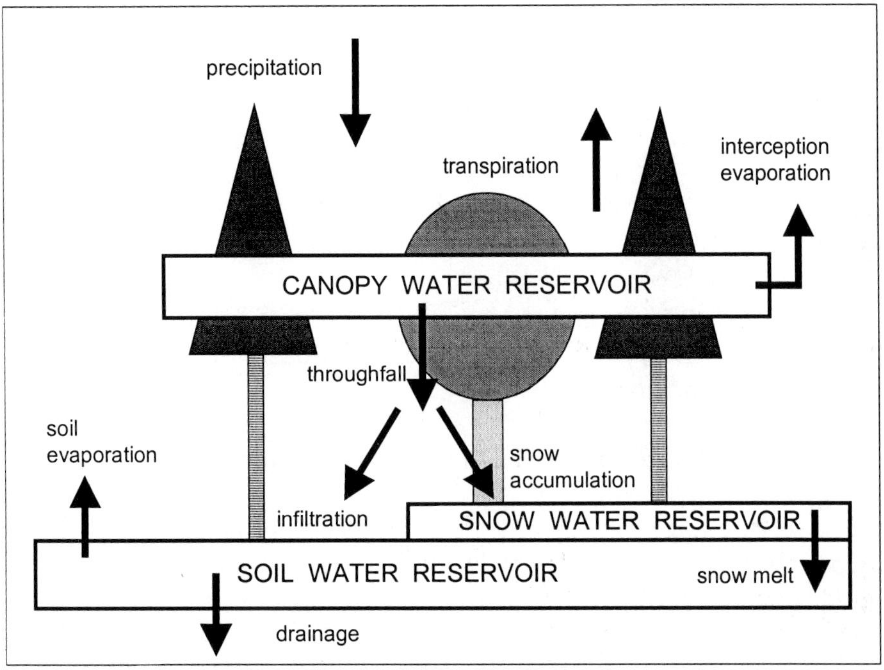

Figure 1. The scheme of the hydrological model WAWAHAMO, its water reservoirs (in capital letters) and the fluxes in between the reservoirs.

Besides the hydrological processes, ecophysiological processes such as bud burst, leaf fall, the seasonal changes in the leaf area index (Kramer, 1996) and the response of stomata to environmental factors (Jarvis, 1976, Lhomme et al., 1998) are simulated by WAWAHAMO.

WAWAHAMO has been validated at various forest sites. To estimate the accuracy, the Nash-Sutcliffe modelling efficiency ME as described by Servat and Dezetter (1991) was calculated from the comparison of daily simulated and observed soil moisture values. The ME values varied between 0.504 and 0.877. Thus, the model performs well in the simulation of the water balance components of forested ecosystems.

2.2 Definition of drought

As an indicator for drought stress the concept of relating actual and potential transpiration is used. The model computes a number of annual drought indexes *ADI* on a yearly or seasonal basis (e.g. bud burst period, vegetation period). However, in this study only the yearly based drought index is considered as any time period defined via ecophysiological processes is highly variable in time depending on the current climate. As a consequence, a comparison of the results is difficult. The yearly based *ADI* is defined by

$$ADI = 1 - \frac{\sum_{t=1}^{365} AT_t}{\sum_{t=1}^{365} PT_t}$$

with AT_t and PT_t indicating daily actual and potential transpiration rates. The values of *ADI* ranges from zero for no drought stress to unity for absolutely dry conditions.

2.2.1 The concept of actual and potential transpiration

The driving force for transpiration is the water potential gradient in the soil-plant-atmosphere continuum which is determined mainly by the atmospheric vapour pressure deficit (Larcher, 1995). However, if the atmospheric demand exceeds the ability of the soil-plant system to supply water from the soil through the plant into the atmosphere, drought stress occurs and plants close their stomata to reduce transpiration rates. This drought stress can be caused by two factors: (1.) high vapour pressure deficits: the atmospheric demand exceeds the transpiration rate the plant is able to provide due to a limited transport capacity of the plant, even if the soil water reservoir is full (Zweifel, 1999); (2.) high soil water deficits: the atmospheric demand exceeds the transpiration rate the soil-plant system is able to provide due to soil water deficits, even if the atmospheric demand is low.

In this study, potential transpiration is consequently defined as the transpiration rate that would occur if the soil-plant system was able to satisfy the atmospheric demand, i.e. if the stomatal regulation was not affected by drought stress. Thus, it is a purely theoretical value. In contrast, actual transpiration is defined as the actually occurring transpiration rate with stomatal opening adjusted to the current drought stress conditions.

In the model, this concept is realised as follows: both transpiration rates are computed using the Penman-Monteith equation (Monteith, 1975) to simulate evaporation and a Jarvis-type model to parameterise the response of stomata to environmental factors (Jarvis, 1976, Lhomme et al., 1998). The latter describes the stomatal resistance r_{st} in the form of a minimal resistance $r_{st,min}$ multiplied by the product of independent stress functions F_i

$$r_{st} = r_{st,\min} \cdot F_1(R_s) \cdot F_2(T) \cdot F_3(\delta e) \cdot F_4(\Psi)$$

where R_s denotes the solar radiation budget, T the air temperature, δe the vapour pressure deficit, and Ψ the soil water potential. Each stress function varies from unity to infinity describing stomatal regulation dependent on environmental factors (Fig. 2).

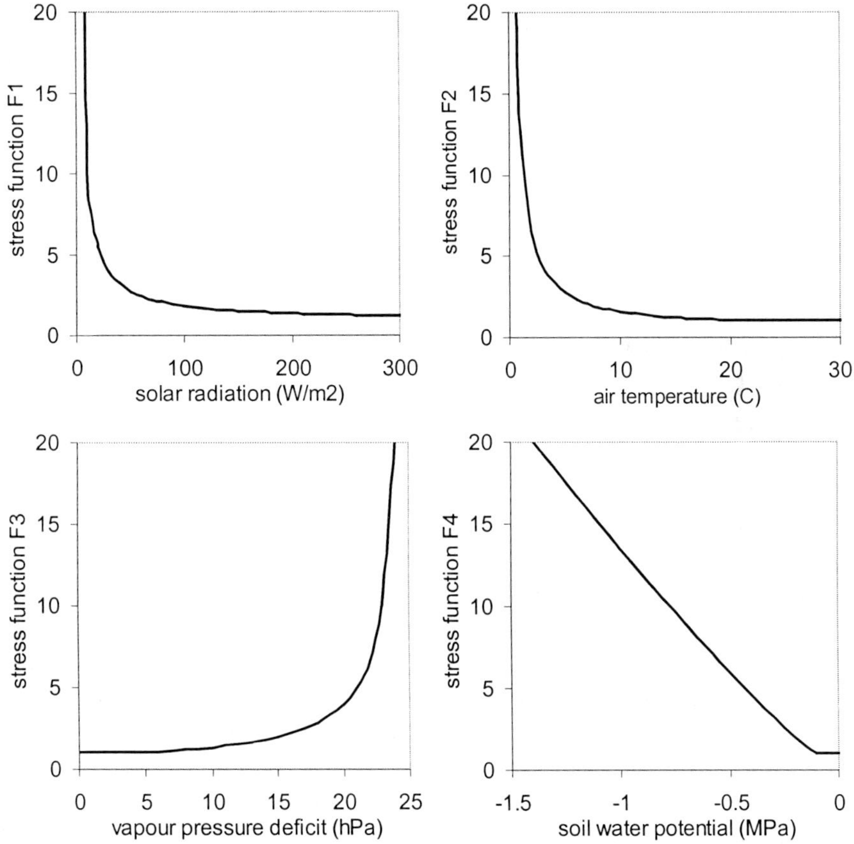

Figure 2. Stress functions for the Jarvis-type model to describe the response of stomata to environmental factors.

However, whereas the full Jarvis-type model is used to compute actual transpiration, potential transpiration is computed by adjusting the stomatal resistance only to solar radiation and temperature, i.e. multiplying the minimum stomatal resistance only with the stress functions F_1 and F_2. Thus, the response of the stomata to drought stress is not included when computing the potential transpiration rate. For a more detailed description of this concept and the used formulas see Zierl (2000, 2001a).

2.3 Time series analysis

The primary objective of this study is to analyse trends in drought and other relevant variables. For this, time series of the annual drought index *ADI*, annual transpiration totals and the length of the vegetation period were computed by WAWAHAMO on a 1 km square grid covering the entire forested area of Switzerland (10605 forest sites). Here, the vegetation period is defined as the period between bud burst and leaf fall. Moreover, values of annual mean temperature and annual or seasonal (May 1st to October 31st) precipitation totals were obtained by averaging or summing up interpolated daily values from the input data set of the hydrological model WAWAHAMO. The observation period for all time series stretches over a time period of 30 years from 1969 to 1998.

The principal trend analysis used here is accomplished by using linear regression with the variable of interest as the dependent variable and t = 1,2,...,n (here n = 30) the time in years as the independent variable. The standard approach is to assume the linear regression model

$$Y_t = a + s \cdot t + E_t$$

where Y_t represents the dependent variable at time t and E_t is the deviation of the data from the straight regression line. The regression slope s is the linear rate of change in the dependent variable with time and, thus, reflects the trend of the considered variable. A test of hypothesis is then proposed for testing if $s = 0$. If this hypothesis is rejected at an appropriate significance level (here 0.05), it is generally accepted that a linear trend is present.

However, the described regression analysis is only appropriate if the residuals E_t of the least squares fit are not autocorrelated, i.e. independent. In the case of autocorrelated residuals, the parameter s can be estimated by assuming that E_t is an AR model (autoregressive model) of order r (Box and Jenkins, 1976).

$$E_t = \alpha_1 E_{t-1} + \ldots + \alpha_r E_{t-r}$$

In this estimation procedure, first the order of the AR model is determined be the Yule-Walker method (function ar.yw in *S-Plus*, Venables and Ripley, 1997). Here, the maximum possible order of the AR model fitted to the residuals is set to 3. As the length of the time series is rather short, a higher order of the AR model is not reasonable. Second, an autoregressive model of order r is fitted to the residuals E_t of the least squares fit, i.e. the parameters α_i with i=1,...,r are estimated. A transformation of the variables is then carried out to remove the autocorrelation of the variables (Cochrane-Orcutt transformation, see Cochrane and Orcutt, 1949, Venables and Ripley, 1997)

$$Y_t^* = Y_t - \hat{\alpha}_1 Y_{t-1} - \ldots - \hat{\alpha}_r Y_{t-r}$$

Finally, a new regression model is fitted to the transformed variables Y_t^* using least squares. In order to perform the Cochrane-Orcutt scheme a *S-Plus* function *autoreg* as programmed by Christian Keller (Seminar für Statistik, Swiss Federal Institute of Technology (ETH), Zürich) is used.

To summarise, for every forest site under investigation a linear regression analysis is carried out. If the corresponding residuals are not autocorrelated (r=0), *s* is taken from this simple regression model. In the case of autocorrelated residuals (r>0), the Cochrane-Orcutt scheme is used to estimate trend *s*.

3. RESULTS

Results of the trend analyses are summarised in Table 1 for the different data records. The corresponding p-values for the hypothesis $s = 0$, it means no trend in the considered time series, are given in Table 2. Throughout the entire forested area of Switzerland there is a positive trend in annual mean temperatures varying between 0.043 and 1.9 K with a mean of 1.2 K in 30 years. As can be seen from the distribution of the p-value this trend is statistically significant on the 5 % level for almost all forest sites under consideration.

Neither for annual nor for seasonal precipitation totals a systematic trend can be detected in the time series. Relative changes in annual totals vary between -37 and +53 % with a mean of +7 % in 30 years, changes in seasonal totals vary between -45 and +58 % with a mean of +10 % in 30 years. Moreover, practically none of these changes in precipitation is statistically significant.

With respect to transpiration, an intensification has occurred during the observation period at most forest sites. The changes vary between –64 and

+224 mm with a mean of +40 mm in 30 years. More than 50 % of these trends is statistically significant. Concerning the length of the vegetation period, a prolongation has been detected for the majority of the considered forest sites. In this case, changes vary between –9 and +36 days with a mean of +15 days. Again, slightly more than 50 % of these trends is statistically significant.

Table 1. Statistics of the computed trends per 30 years for 10605 forest sites. The line ADI stat. sign. includes only statistically significant trends.

Statistics of trends per 30 years	Min	1st Qu	Median	Mean	3rd Qu	Max
Temperature [°C]	0.04	1.0	1.2	1.2	1.4	1.9
Yearly precipitation [%]	-37	2	6	7	11	53
Seasonal precipitation [%]	-42	4	9	10	15	58
Yearly transpiration [mm]	-64	18	38	40	58	224
Length of vegetation period [d]	-5	9	16	15	21	36
ADI	-0.232	0.021	0.039	0.047	0.064	0.265
ADI (stat. sig.)	-0.232	0.041	0.054	0.067	0.081	0.265

Table 2. Statistics of the p-values for the computed trends listed in Table 1.

Statistics of trends per 30 years	Min	1st Qu	Median	Mean	3rd Qu	Max
Temperature	0.0000	0.0003	0.0010	0.0055	0.0030	0.9418
Yearly precipitation	0.000	0.311	0.527	0.525	0.753	1.000
Seasonal precipitation	0.000	0.168	0.410	0.420	0.628	1.000
Yearly transpiration	0.000	0.004	0.034	0.160	0.222	1.000
Length of vegetation period	0.000	0.012	0.048	0.191	0.321	1.000
ADI	0.00	0.027	0.114	0.229	0.370	1.000

Figure 3 displays two examples of time series of ADI for a forest stand in the northern lowland with a rather small interannual variability (upper plot) and for a forest stand in the main valley of Valais with a high interannual variability (lower plot). The computed trend at the forest site in the northern lowland amounts to 0.165 units in 30 year with a p-value of 0.0016. At the forest site in the main valley of Valais the trend is 0.108 units in 30 years with a p-value of 0.41.

Considering the entire forested area of Switzerland, most forest sites in Switzerland experienced an increase in the annual drought index *ADI* from 1969 to 1998 indicating a trend towards drier conditions (Fig. 4). This positive trend is particularly pronounced in the Jura mountains, the northern lowlands, and the very south of Ticino. The alpine area exhibits mostly weak positive trends and in some rare cases negative trends in the drought index (Fig. 4).

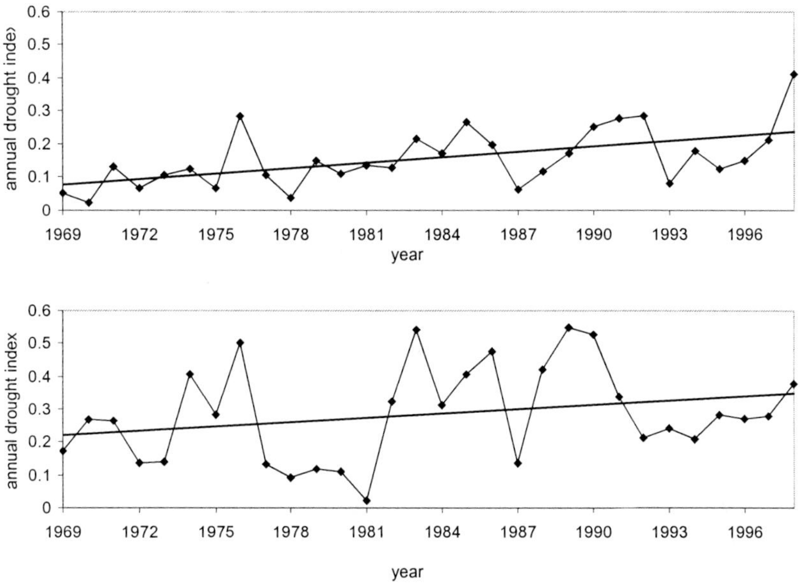

Figure 3. Time series of the annual drought index ADI for a forest stand in the northern lowland with a rather small interannual variability (upper figure) and for a forest stand in the main valley of Valais with a high interannual variability (lower figure). The trends are 0.165 and 0.108 units in 30 years (given by the straight line) with p-values 0.0016 and 0.41.

At large, the trend in the annual drought index *ADI* varies between -0.23 and 0.26 units per 30 years with a mean of 0.046 units per 30 years. The corresponding p-value is less than 5 % for about one third of all forest sites indicating statistically significant trends, and less than 10 % for about one half of all forest sites indicating weak statistical significance. Considering only statistically significant trends, the mean value of the drought trend rises to 0.067 units per 30 years (Fig. 5).

To investigate the influence of changes in temperature, precipitation, transpiration and the length of the vegetation period on the evolution of drought, scatter plots of the statistically significant *ADI* trends against these variables were drawn (Fig. 6). In the case of precipitation and temperature trends no relations can be seen. Regarding transpiration trends, a weak correlation is visible indicating that the drought trend decreases with intensifying transpiration. Finally, the drought index at some forest sites seems to increase with the prolongation of the vegetation period.

Figure 7 illustrates relations between the statistically significant trends in *ADI* and the mean of *ADI* averaged over the observation period. There is a clear positive correlation between these two variables (R^2 =0.6387) indicating that trends in drought are particularly pronounced at the driest forest sites. In contrast to this, the rather wet forest sites experience hardly any change in drought during this period.

Figure 4. **Spatial distribution of the annual drought index in Switzerland.**

Figure 5. Histogram of all trends (gray) and of the statistically significant trends (black) in the

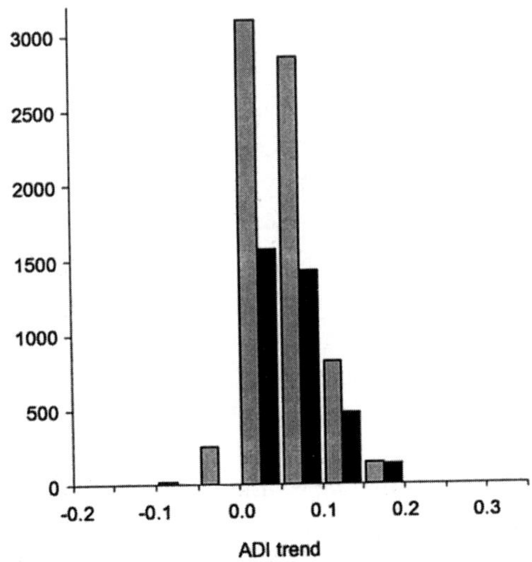

annual drought index ADI.

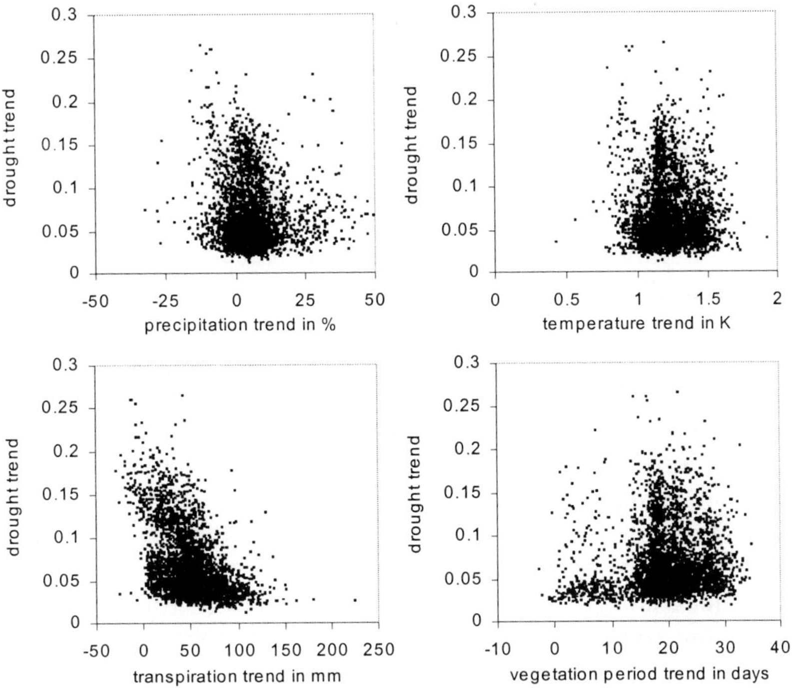

Figure 6. Trend in the annual drought index *ADI* dependent on changes in annual precipitation totals, in annual mean temperature, in annual transpiration totals, and in the length of the vegetation period.

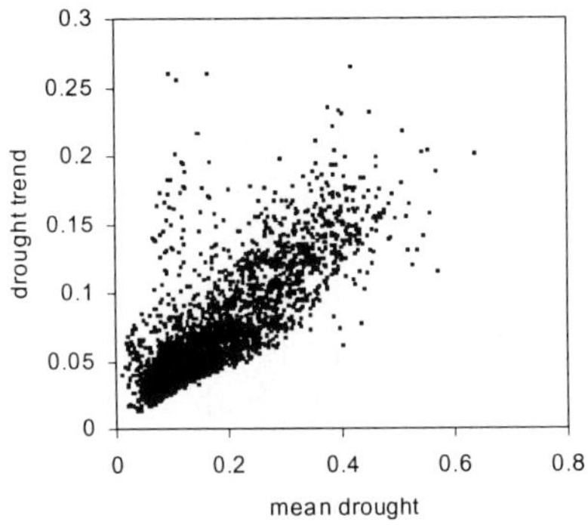

Figure 7. Trend in the annual drought index *ADI* dependent on the 30-years-mean values of *ADI*

4. DISCUSSION

The overall impression of analysing the results is of a Swiss wide warming tendency during the observation period. Furthermore, transpiration intensified, the vegetation period prolonged, and drought increased. No systematic trend could be detected for precipitation, neither for annual nor for seasonal totals.

It is clear from Table 2 that the temperature trend is unquestionable highly statistically significant almost throughout the entire forested area of Switzerland. Regarding annual or seasonal precipitation totals, practically none of the detected positive or negative trends is statistically significant. In the case of transpiration and the length of the vegetation period slightly more than one half of all trends is statistically significant indicating an enhanced water cycle driven by intensified evaporation and a prolonged growing season. Both increases are certainly caused by the strong temperature rise observed during the same time period. The results regarding the vegetation period agree well with findings from Menzel and Fabian (1999).

Concerning drought, some regions can be identified where p-values are uniformly low over a wide area. In these areas as the northern lowland, some prealpine valleys, some areas of the Jura mountains and the low regions in the south of Ticino the computed positive trend can be judged to be really existent. In other areas as in the alpine and prealpine regions and in some parts of the Jura mountains hardly any p-values < 5 % can be found. Therefore, no statement on the evolution of drought in the alpine and in most of the prealpine regions can be made.

Furthermore, it is particularly interesting to point out the fact that even though no significant trends in annual or seasonal precipitation totals could be detected, drought increased almost throughout the entire forested area of Switzerland. Therefore, the contribution of precipitation to the observed drought trend seems to be small, which is also suggested by the upper left plot in Figure 6. However, there is also no definite relation between the change in temperature and the change in drought (upper right plot in Fig. 6). An intensification of drought occurs both in regions with only a slight increase in temperature and in regions with a rather strong increase in temperature. At the same time there are forest stands with a dramatic increase in temperature accompanied by hardly any change in drought.

Some more interesting aspects are the changes in drought related to changes in actual transpiration and the length of the vegetation period (Fig. 6, lower row). Contrary to expectations, annual totals of actual transpiration hardly increased at the forest sites with the strongest drying tendencies, but, in some cases even decreased despite of raised temperatures. This is caused by the strongly limited soil water supply at these forest sites, which leaves no place for the intensification of actual transpiration. In the case of decreasing precipitation, actual transpiration might even decrease. At the same time, however, potential transpiration is strongly enforced by the raised temperatures. This increase of potential transpiration accompanied by more or less unchanged actual transpiration leads to a drastic modification of the annual drought index *ADI* which is defined via the ratio of actual and potential transpiration.

In contrast to this, at forest sites with a huge excess of precipitation the actual transpiration is able to increase similarly to the potential transpiration. The excess of precipitation practically always keeps the soil water reservoir filled. Thus, even if actual transpiration is strongly intensified, the soil water reservoir is not emptied allowing the actual transpiration to increase just as the potential transpiration. This in turn leaves the annual drought index practically unchanged. In summary, however, there is no definite correlation between the change in transpiration and the change in drought as the effect of a change in actual transpiration on drought is strongly dependent on the local drought situation.

Furthermore, in the case of deciduous forest the prolongation of the vegetation period in turn, leads to an earlier onset of transpiration in spring as the date of bud burst shifts towards earlier dates and a later end of transpiration in autumn as the date of leaf fall is postponed towards later dates as a consequence of the rising temperatures. Hence, in the case of limited soil water supply the soil water reservoir is already emptied earlier and refilled later in the year enforcing a more pronounced water shortage during the summer and autumn months. At wet forest sites, however, this prolongation of the vegetation period seems not to have any effect on the occurrence of drought as the huge excess of water is again able to compensate for the intensified consumption.

To summarise, the impact of changing environmental variables seems to be highly variable. The impact of a climate change on the occurrence of drought in forested ecosystems is strongly dependent on the site specific drought conditions. This hypothesis is confirmed by Figure 7 which shows that drought trends are particularly pronounced at rather dry forest sites whereas rather wet forest sites experienced hardly any change in drought during the observation period. That means, the impact of a climatic shift strongly depends on the average drought situation of the forested ecosystems. At forest sites with a limited soil water supply already small changes in environmental variables may have a drastic effect on the water supply of the ecosystem. As a consequence of high water deficits in the soil, actual transpiration is not able to increase similarly to the potential transpiration. Therefore, the ratio of actual to potential transpiration will be effectively reduced resulting in a strongly increased drought index.

In contrast, the soil water supply in most wet forest sites is rather insensitive to the experienced changes in temperature and other associated variables. There, the soil water reservoir remains almost always filled as a consequence of the huge precipitation excess. Because of that, actual transpiration is hardly limited by soil water deficits and, thus, resembles most of the time the potential transpiration rate keeping the ratio of actual to potential transpiration close to unity and, thus, the drought index close to zero.

In other words, the dry forest sites in Switzerland are the most endangered ones with respect to climate change. There, small changes in temperature or precipitation might entail rather serious changes in drought stress. Compared to this, the wet forest sites so far seem hardly to be affected by the warming of the climate.

5. CONCLUSION

The trend analysis performed aimed at detecting whether there were any significant changes in the occurrence of drought and other related environmental variables during the observation period from 1969 to 1998. The results show that temperature increased during this time period throughout the entire forested area of Switzerland on average by 1.2 K, accompanied in some regions by trends towards increasing precipitation and in others towards decreasing precipitation. These results are in agreement with those obtained by various authors (Beniston et al., 1994, Auer and Böhm, 1993, Hasenauer et al., 1999, Schönwiese et al., 1993 and 1994). During the same period, drought increased throughout almost the entire forested area of Switzerland as a consequence of the strong warming. So far, this trend is rather slight for most forest sites. It is only in some areas where it reaches values that indicate a serious increase in drought during the observed 30 years.

The relationship between the rising temperatures and the increasing drought suggests that future climate warming will be accompanied by a prolongation and strengthening of drought periods. Moreover, compared to the past drying trend presented in this study, a much more serious change in the occurrence of drought has to be expected for the future, as the predicted temperature rise of 1.5 to 5.5 K for the alpine region (Gyalistras et al., 1999) by far exceeds the observed temperature rise of 1.2 K. Indeed, studies investigating the impact of future climate change on drought in forested ecosystems (Zierl, 2001b) confirm that the predicted climate change will lead to a fundamental strengthening of the current drought stress situation especially in the dry forested ecosystems. There, the existence of forest stands might be really endangered by the changing climate.

One shortcoming of this trend analysis is that it neglects forest dynamics and adaptation processes caused by the changing climatic and hydrological conditions. Currently, all vegetation parameters are treated as fixed in these model runs, it is the state that was recorded in the years between 1983-1985 (LFI 1, 1983-1985). In other words, no feedback from the changing environment on the forest ecosystems is included in the model. However, this variation in the vegetation parameters certainly has a significant influence on the different hydrological processes. To include the neglected processes, the coupling of WAWAHAMO with a dynamical forest model is necessary.

6. ACKNOWLEDGEMENTS

Many thanks to Heike Lischke, Martin Beniston, and three reviewers whose comments greatly improved the quality of the paper. This research was funded by the Swiss Agency for the Environment, Forest and Landscape, the Swiss Federal Research Institut WSL, DAAD-postgraduate scholarship, and the Carolina-Rüedi-Stiftung postgraduate scholarship.

7. REFERENCES

Arnell, N., 1996. Global warming, river flows and water resources. Wiley cop.

Auer, I., Böhm, R., 1993. Combined Temperature-Precipitation Variations in Austria During the Instrumental Period. Theoretical and Applied Climatology, 49, 161-174.

Balling, R.C., 1996. Century-long variations in United States drought severity. Agricultural and Forest Meteorology, 82, 293-299.

Beniston, M., Rebetez, M., Giorgi, F., Marinucci, M.R., 1994. An Analysis of Regional Climate Change in Switzerland. Theoretical and Applied Climatology, 49, 135-159.

Beniston, M.A., Diaz, H.F., Bradley, R.S., 1997. Climatic Change at high elevation sites: An overview. Climatic Change, 36, 233-251.

Box, E.P., Jenkins, G.M., 1976. Time series analysis, forecasting and control. San Francisco a.o.: Holden-Day, 575 p.

Bultot, F., Gellens, D., Spreasfico, M., Schädler, B., 1992. Repercussions of a CO2 doubling on the water balance - a case study in Switzerland. Journal of Hydrology, 137, 199-208.

Cochrane, D., Orcutt, G.H., 1949. Applications of least squares regression to relationships containing autocorrelated errors. J. Amer. Stat. Assoc. 44, 32-61.

Gurtz, J., A. Baltensweiler,H. Lang, L. Menzel und J. Schulla, 1997. Auswirkungen von klimatischen Variationen auf den Wasserhaushalt und Abfluss im Flussgebiet des Rheins. Projektschlussbericht im Rahmen des nationalen Forschungsprogramms (NFP 31):

Klimaänderungen und Naturkatastrophen. Zürich; vdf, Hochschul-Verlag an der ETH, 147 pp.

Gyalistras, D., Schär, C., Davis, H.C., Wanner, H., 1999. Future Alpine Climate. In: Cebon, P., Dahinden, U., Davies, H., Imboden, D.M., Jäger, C.C. (Eds.): Views from the Alps, The MIT Press Cambridge, 171-223.

Hasenauer, H., Nemani, R.R., Schadauer, K., Running, S.W., 1999. Forest growth response to changing climate between 1961 and 1990 in Austria. Forest Ecology and Management, 122, 209-219.

Herbst, M., Hörmann, G., 1998. Predicting effects of temperature increase on the water balance of beech forest - an application of the 'KAUSHA' model. Climatic Change, 40: 683-698.

Houghton, J.T., Meira Filho, L.G., Callander, B.A., Harris, N., Kattenberg, A., Maskel, K., 1996. IPCC: Climate Change 1995: The Science of Climate Change, Cambridge University Press, Cambridge U.K.

Jarvis, P.G., 1976. The interpretation of the variations in leaf water potential and stomatal conductance found in canopies in the field. Phil. Trans. R. Soc. Lond. 273: 593-610.

Kramer, K., 1996. Phenology and growth of European trees in relation to climate change, Ph.D. Thesis, Landbouw Universiteit, Wageningen.

Larcher, W., 1995. Physiological plant ecology. Springer, Berlin, Heidelberg, New York, 506 pp.

Lhomme, J.-P., Elguero, E., Chehbouni, A., Boulet, G., 1998. Stomatal control of transpiration: Examination of Monteith's formulation of canopy resistance. Water Resources Research, Vol.34, No.9: 2301-2308.

LFI 1, Schweizerisches Landesforstinventar, 1983-85: Datenbankauszug vom 25.6.1997. Eidg. Forschungsanstalt für Wald, Schnee und Landschaft (WSL), Birmensdorf.

Menzel, A., Fabian, P., 1999. Growing season extended in Europe. Nature 397, 659.

Mitchell, J.F.B., Warrilow, D.A., 1987. Summer dryness in northern mid-latitudes due to increased CO2. Nature, 330, 238-241.

Monteith, J.L., 1975. Vegetation and Atmosphere, Vol. 1: Principles. Academic Press.

Rebetez, M., 1999. Twentieth century trends in droughts in southern Switzerland. Geophysical Research Letters, 26, 755-758.

Rind, D., Goldberg, R., Hansen, J., Rosenzweig, C., Ruedy, R., 1990. Potential Evaporation and the Likelihood of Future Drought. Journal of Geophysical Research, 95: 9983-10004.

Schönwiese, C.-D., Rapp, J., Fuchs, T., Denhard, M., 1993. Klimatrend-Atlas Europa 1891-1990. Verlag Zentrum für Umweltforschung, Frankfurt, 218 pp.

Schönwiese, C.-D., Rapp, J., Fuchs, T., Denhard, M., 1994. Observed climate trends in Europe 1891-1990, Meteorol. Z.N.F., 3, 22-28.

Servat, E., Dezetter, A., 1991. Selection of calibration functions in the context of rainfall-runoff modelling in a sudanese savannah area: Journal of Hydrological Sciences, 36: 307-330.

Stahel, W.A., 1995. Statistische Datenanalyse; Eine Einführung für Naturwissenschaftler. Vieweg Lehrbuch, Angewandte Mathematik, 359 pp.

Venables, W.N., Ripley, B.D., 1997. Modern Applied Statistics with S-PLUS. Second edition. Springer-Verlag New York, 548 pp.

Whetton, P.H., Fowler, A.M., Haylock, M.R., Pittock, A.B., 1993. Implications of climate change due to the enhanced greenhouse effect on floods and droughts in Australia. Climatic Change, 25: 289-317.

Widmann, M., Schär, C., 1997. A principal component and long-term trend analysis of daily precipitation data in Switzerland. International Journal of Climatology, Vol. 17, 1333-1356.

Zierl, B., 2000. WAWAHAMO - a hydrological model to simulate drought in forested ecosystems. Ph.D. Thesis No. 1320, University Fribourg, Switzerland.

Zierl, B., 2001a. A water balance model to simulate drought in forested ecosystems and its application to the entire forested area in Switzerland. Journal of Hydrology, 242/1-2, 115-136.

Zierl, B., 2001b. Simulating the impact of climate change on drought in Swiss forest stands. Proceedings of the conference on global change and protected areas, L'Aquila, 1999, published by Kluwer, in press.

Zweifel, R., 1999. The rhythm of trees - water storage dynamics in subalpine Norway spruce. Ph.D. Thesis No. 13391, ETH Zürich, Switzerland.

Feedback between structured vegetation and soil water in a changing climate: A simulation study

HEIKE LISCHKE AND BÄRBEL ZIERL
Swiss Federal Research Institute for Forest, Snow and Landscape (WSL), Birmensdorf, Switzerland

Abstract: Structure and composition of vegetation influence the local water budget by transpiration and interception. On the other hand soil water content crucially affects plant physiological processes such as nutrient transport or photosynthesis. These processes in turn partly determine biomass production, plant growth, survival and competition and, thus, vegetation structure. This vegetation-hydrology feedback has the potential to influence the impacts of a changing climate on vegetation and hydrology.

However, only few simulation studies include this feedback explicitly. In these studies vegetation models are linked to hydrological or climate models. On the local scale, the used vegetation models are complex and slow. On larger scales, the vegetation and its dynamics are usually modelled in a simplified manner. Particularly vegetation structure is either ignored completely, e.g. by using the leaf area index (LAI) averaged over large areas as state variable, or reduced to few canopy layers and plant functional types.

To examine the effect of vegetation structure and the effect of the vegetation-hydrology feedback on drought and vegetation under a changing climate, we coupled the dynamic forest model DisCForM with the soil water budget model WAWAHAMO. DisCForM is a highly efficient model, designed for regional scale applications over decades and centuries. It simulates horizontal, vertical, and species structure in a forest stand.

Simulations of forest development and drought stress occurrence were performed with the coupled and decoupled models at several typical sites in Switzerland. The results indicate that vegetation plays an important stabilising role regarding the impact of a climatic shift on hydrology. By adaptation, i.e. the lowering of LAI and the reduction of cover grade by changed mortality and growth and the exchange of tree species, the vegetation itself is able to counterbalance the influence of climate change on drought. This stabilising effect is particularly pronounced at the currently dry forest sites. At wet forest sites, the positive effect of a warmer climate on biomass production appears to overwhelm the negative effect of the strengthening of drought.

M. Beniston (ed.),
Climatic Change: Implications for the Hydrological Cycle and for Water Management, 349–377.

sites, the positive effect of a warmer climate on biomass production appears to overwhelm the negative effect of the strengthening of drought.
To study the effect of structure, simulations with reduced horizontal and vertical structure and a reduced species set were compared to those with full structure. Reducing structure had minor effects on drought and moderate effects on vegetation in the long-term average. Under medium dry conditions, however, reducing all three kinds of structure simultaneously resulted in lower drought and LAI values. Furthermore, deviations in single years were considerable.
We concluded that neglecting structure in vegetation-hydrology-models can lead to wrong drought values for single years and to a systematic underestimation of drought in the long-term average under dry conditions. In contrast, neglecting the feedback can lead to an overestimation of drought.

1. INTRODUCTION

Large complex models play a central role in the study of future climate and its direct impact on hydrology and its indirect effects on ecological and socio-economic systems. Climate models such as GCMs focus on the climate system by including the atmosphere and the oceans. In most studies, the biosphere is considered constant in time, only varying between regions. However, the biosphere has the potential to influence climate through several processes. Large amounts of carbon are sequestered, stored, or released by the biosphere, in particular by forests and forest soils. The surface roughness of vegetated areas decreases wind speed, and the surface albedo is controlled by the type of vegetation (e.g., deciduous vs. coniferous forest). These interactions depend on several characteristics of the vegetation. These can be aggregated variables such as the overall leaf area index (LAI), stomatal resistance, or biomass, but also more differentiated properties, such as the horizontal and vertical structure of the vegetation or the species composition.

In turn, the biosphere is strongly influenced not only by the carbon dioxide in the air, but especially by the weather. Temperature determines the growth or survival of plants by enhancing or slowing down biochemical reactions or by extreme events such as late frosts or windstorms.

This influence from climate on vegetation and vice versa forms a feedback in the climate-ocean-biosphere system, and therefore has the potential to greatly influence the behaviour of the entire system (Clark, 1993). Additionally, the vegetation is a dynamic system. Once disturbed by its driving forces, it does not reach a new equilibrium instantaneously, but requires a certain time. This transient period can be several decades to centuries in the case of forests. If the impact of climate change is studied on a similar time scale, such transient behaviour can not be neglected.

One important component of the climate system is the hydrological cycle, in particular on the local to regional scale. In the hydrological cycle there is also a feedback between soil water and vegetation (Hutjes *et al.*, 1998). Plants can be considered as an obstacle for water between the atmosphere and the soil in both directions. Plants intercept a part of the rain, so that it is directly evaporated back to the atmosphere. Plants take up water from the soil, transport, store and transpire it again to the atmosphere. Uptake and transpiration depend on current plant characteristics, such as root length and distribution or the number of stomata, and on the plants' current ecophysiological state, such as the supply of water inside the plant or the stomata closure. Internal water supply influences the vigour of a plant; closed stomata reduce the gas-exchange and therefore photosynthesis and synthesis of carbohydrates. Through these processes, the soil water supply eventually influences plant growth and survival.

Vegetation adapts to changes in water supply on different time scales: by closing of stomata (minutes), shedding of leaves (days), changing individual LAI or root length (years), mortality of individuals (years) or exchange of species (decades). All these processes feed back on the water budget of the plants, and through interception and transpiration, to that of the soil. Thus, at least on longer time scales, the feedback of hydrology on the vegetation and the adaptation of the vegetation to the changed water balance potentially influence the overall dynamics of the plant-hydrology-system.

The potential importance of this issue has been recognized by the ecological modelling community (Churkina *et al.*, 1999; Hutjes *et al.*, 1998) in the last decade. On the global scale, the description of the biosphere in climate models has been steadily improved (see reviews in Foley *et al.*, 1998; Pielke *et al.*, 1998). Global vegetation models coupled to climate models cover various approaches. The simplest ones are fixed equilibrium vegetation type distributions (biogeography models, e.g. BIOME, Prentice and Solomon, 1992) with specific climate relevant characteristics. Equilibrium vegetation models determine in each time step the equilibrium vegetation type corresponding to the new climate (e.g. MAPPS, Foley *et al.*, 1996; Neilson and Drapek, 1998). Other biogeography models are iteratively coupled to climate models under constant CO2 and/or climate forcing (Betts *et al.*, 1997). Biogeochemical models describe in detail plant ecophysiological processes attributed to the entire canopy in a simulated grid cell, but do not take into account dynamics, species and spatial heterogeneity of vegetation (Ciencala *et al.*, 1998; Lüdecke *et al.*, 1994; Potter *et al.*, 1993; Running, 1991; Running, 1994; Running and Coughlan, 1988; Woodward *et al.*, 1995). The new DGVMs, i.e. dynamic global vegetation models that are coupled to dynamic climate models (e.g., IBIS, Foley *et al.*, 2000; Foley *et al.*, 1998; Foley *et al.*, 1996; Kucharik *et al.*, 2000) incorporate both ecophysiology and structure. IBIS, for example, includes dynamics of ecophysiological processes as overall carbon fluxes, medium temporal scale canopy dynamics (phenology) and simple, longer temporal scale vegetation dynamics. Growth is controlled by photosynthesis and carbon allocation and competition for light and water by simple rules. Minimal structural aspects are modelled by including two vertical layers and broad plant functional types.

On a medium scale, that is, in a catchment area or region, many hydrological models include vegetation in an oversimplified way, for example, by using standard vegetation parameters for a wide range of different vegetation types or using very simple routines to describe ecophysiological processes (Gurtz *et al.*, 1997; Tiktak and Bouten, 1992). Also more sophisticated hydrological models mostly ignore that vegetation structure can substantially change as a consequence of altered water availability (Band *et al.*, 1993; Zierl, 2001).

On the local scale, that is in a stand, vegetation models can be more detailed because computing time is less limiting and parameter and input values are easier to obtain. However, the plant-soil-water-feedback and structural aspects are rarely taken into account at the same time. For example, stand ecophysiological models (e.g., most models compared by Lexer, 1995; Tiktak and VanGrinsven, 1995) work on a small time scale (season to year), and do not include changes in spatial vegetation structure or species composition. In forest succession models (e.g. Bugmann, 1994; Bugmann, 1996; Lexer *et al.*, 2000; Lischke *et al.*, 1998), hydrology often is oversimplified, and does usually not depend on the vegetation and its dynamics

In other local models the soil-water-vegetation feedback is partly covered. For example in the forest model ForSum (Kräuchi, 1994; Kräuchi and Kienast, 1993), hydrology is rather detailed and a certain feedback is included with the influence of needle tree ratio on the amount of precipitation reaching the soil. However, water uptake is considered as constant and not depending on tree density, LAI, or stomata closure.

The ecophysiological models BIOMASS (McMurtrie and Landsberg, 1992), TREE-BGC (Korol *et al.*, 1995), and HYBRID (Friend *et al.*, 1997) are compromises. BIOMASS includes the feedback from water to ecophysiology and growth, simulates a vertical structure of 3 canopy layers and assumes a random horizontal distribution of the crowns. However, BIOMASS is confined to a medium time scale up to several years, and does not take into account population dynamics such as death (e.g., by disturbance) and birth of trees and the resulting structure. TREE-BGC, the individual based version of of the ecophysiological model FOREST-BGC, combines the structure resulting from an individual based model with the water flux in FOREST-BGC. The ecophysiological model HYBRID is a gap-model and therefore includes species or PFTs, vertical and horizontal structure.

Thus, the degree to which structure (species compositions, spatial vegetation structure) and its dynamics (which are influenced by individual, population and community dynamics) are comprised in models depends on the spatial scale of the models. Very detailed models, for example individual based models, which include explicetly vegetation structure, are confined to the local scale, due to computing time and data constraints. Larger scale models need to be simpler to be manageable.

Spatial structure, however, has been found to affect on the local scale simulated ecosystem variables such as species biomass or LAI (Deutschman *et al.*, 1997; Löffler and Lischke, 2001; Pacala and Deutschman, 1995), which in turn can affect overall hydrology. Probably, fine scale spatial structure and species composition exert an essential influence on the vegetation-hydrology-feedback also on medium and large scales (Hutjes *et al.*, 1998).

In this study we focus on the effects of the local vegetation-hydrology-feedback and of vegetation structure in a vegetation/hydrology model designed for regional scale applications. We ask: What is the influence of the feedback on simulated drought and vegetation characteristics? How important are structural aspects of the simulated forests, such as spatial heterogeneity, vertical layering or species composition for the simulated drought and LAI values? How important is the feedback in climate change studies, where long time periods are simulated with changing environmental conditions? How does the influence of the feedback depend on the current climatic situation, for example inside Switzerland?

To answer these questions, we coupled a forest water balance model with a forest dynamics model, ran simulations under a changing climate at various test sites, and compared the results to simulations without feedback ("decoupled"), i.e. where the drought model used a constant vegetation input. Furthermore, coupled simulations with different extents of vegetation structure were compared.

2. MATERIAL AND METHODS

2.1 Water balance model

WAWAHAMO (WaldWAsserHAushaltsMOdell, German, "forest water balance model" in English) is a physically and biologically based, spatially-distributed water balance model designed to simulate hydrological and some ecophysiological processes in forested ecosystems at a daily time step. Given daily weather variables, the model estimates interception, transpiration, soil evaporation, snow accumulation and melt, soil water storage, stomata regulation, and phenological processes at regional levels. Furthermore, it computes ecophysiology-based drought indexes indicating the drought stress a plant experiences under the site-specific environmental conditions. A detailed description of the model is given in Zierl (2000; 2001).

2.2 Forest dynamics model

To simulate the forest dynamics, we used the distribution based model DisCForM (Lischke *et al.*, 1998; Löffler and Lischke, 2001). This model has been aggregated from the forest gap model ForClim (Bugmann, 1994; Bugmann, 1996). It simulates deterministically the dynamics of trees in discrete height classes. The state variables are the numbers of trees per species and height class. Trees establish, grow and die with empirically derived functions. These functions (taken from the original model ForClim) depend on light, bioclimatic variables calculated from monthly climate variables, and other site conditions. The only competition between trees is by shading. To simulate the spatial variability (which in gap models is given by many replicates of the stochastic processes), in DisCForM we assume that the trees in each height class are randomly distributed in space. From this tree distribution results a certain spatial distribution of the light conditions, and thus of the light dependent process rates. The model calculates species biomasses (t dry weight/ha) and LAI (m^2/m^2) as output variable. From the species biomasses the Shannon-Weaver biodiversity index (Haeupler, 1995; Shannon and Weaver, 1976) is calculated.

2.3 Model Coupling

The models were coupled by interchanging several coupling variables after each simulated year. DisCForM requires as input yearly bioclimatic variables, including an index describing the current plant relevant drought situation. The hydrological model, in turn, requires as input the LAI, the vegetation cover, the main tree species and the ratio of needle to deciduous trees of the current vegetation. These variables had to be derived from the state variables in DisCForM, i.e. the tree numbers per species and height class.

2.3.1 Main tree species, LAI, needle tree ratio

The tree species with the largest total basal area was taken as main species. Needle-tree ratio was defined as the ratio between the sum of the basal area of the coniferous species to the total basal area. LAI was calculated by summing the LAI's of all species in all height classes. To calculate basal area and LAI from tree height, allometric relationships between height and diameter at breast height (*dbh*), and dbh and LAI (Bugmann, 1994) were used.

2.3.2 Percent cover

The percent cover influences the interception of precipitation in WAWAHAMO. It is defined as the portion of the ground in a forest stand, which is below the canopy. It was determined in two steps from the state variables: (a) by deriving empirical species- and height-specific crown projection areas and (b) by calculating the cover grade based on the spatial tree distribution and on the crown projection areas.

To obtain species-specific empirical relationships between the forest model output variables tree height (*h*), *dbh*, or basal area (*ba*), and the crown projection area (*ca*), single tree data from the Swiss National Forest Health Inventory 1997 (Brang, 1998) were evaluated. In the inventory, estimates for the maximum tree crown diameters were available (Dobbertin, 1996). Assuming circular crown projections, their area (*ca*) could be calculated. Also estimated tree heights could be derived from the National Forest Health Inventory data.

Four different linear models (models 1 to 4) were fitted to the crown radius (*cr*) or crown area (*ca*) data. If less than 5 data points were available for a species, the maximum heights (*hmax*) and crown radia (*crmax*) for the species were compiled from Hieke (1989). *hmax* and *crmax* were then transformed to the coefficients of model 5, assuming a crown radius of 0.5 m at height 1.37 m (*dbh*=0 cm).

$$\text{Model 1}: \; ca = b + a_{ba} * ba$$

$$\text{Model 2}: \; ca = b + a_{dbh} * dbh$$

$$\text{Model 3}: \; ca = b + a_{ba} * ba + a_h * h$$

$$\text{Model 4}: \; cr = b + a_{dbh} * dbh$$

$$\text{Model 5}: \; cr = b + a_h * h$$

Table 1: Coefficients of crown area or radius to basal area or dbh relationships and drought threshold parameters.

Species	Model	R^2	b	a_{dbh}	a_{ba}	a_h	Drought threshold
Abies alba	3	0.541	12.833	.	0.013	0.005	0.37
Larix decidua	2	0.369	4.141	0.652	.	.	0.45
Picea excelsa	3	0.513	10.355	.	0.014	0.001	0.41
Pinus cembra	4	0.429	0.682	0.018	.	.	0.43
Pinus montana	3	0.365	5.343	.	0.011	-0.021	0.40
Pinus silvestris	3	0.189	12.686	.	0.010	0.001	0.55
Taxus baccata	3	0.911	-34.869	.	0.266	0.146	0.39
Acer campestre	5	-	0.26		.	0.405	0.46
Acer platanoides	2	0.542	12.669	0.500	.	.	0.42
Acer	3	0.450	14.735	.	0.018	0.018	0.34
Alnus glutinosa	3	0.809	6.911	.	0.019	0.061	0.37
Alnus incana	5	-	0.44		.	0.265	0.34
Alnus viridis	5	-	-0.11		0.000	0.69	0.37
Betul pendula	3	0.396	9.169	.	0.027	-0.027	0.35
Carpinus betulus	3	0.729	27.841	.	-0.009	0.068	0.46
Castanea sativa	3	0.428	14.657	.	0.014	-0.016	0.30
Corylus avellana	5	-	-0.34		.	0.869	0.46
Fagus silvatica	3	0.541	15.570	.	0.031	0.007	0.37
Fraxinus excelsior	3	0.763	9.389	.	0.032	0.034	0.39
Populus nigra	5	-	0.25		.	0.414	0.30
Popul tremula	3	0.398	14.083	.	0.056	0.061	0.41
Quercus petraea	3	0.638	0.204	.	0.048	-0.022	0.44
Quercus pubescens	5	-	0.21		0.000	0.439	0.40
Quercus robur	3	0.259	28.037	.	0.016	0.012	0.44
Salix alba	4	0.3682	0.505	0.0535	.	.	0.39
Sorbus aria	4	0.813	-2.654	0.228	.	.	0.38
Sorbus aucuparia	4	0.735	-3.897	0.341	.	.	0.27
Tilia cordata	3	0.839	10.455	.	0.043	-0.024	0.40
Tilia platyphyllos	3	0.460	23.618	.	0.030	-0.053	0.40
Ulmus scabra	3	0.745	11.721	.	0.014	0.037	0.35

Table 1 shows the coefficients resulting from these fits. For each species the results for the model with the highest coefficient of determination (R^2) are given. The models 1 to 4 yield in many cases high values for the intercepts, i.e. unrealistically large crown areas at dbh=0. This is partly due to the fact that only data for trees with *dbh* > 6 cm were available and the functions were extrapolated to *dbh*=0. Therefore, we used the fitted functions only to *dbh*=6 cm and interpolated linearly between *dbh*=0 and *dbh*=6cm. For this interpolation, we again assumed a crown radius of 0.5 m for trees with *dbh*=0 cm, i.e. of 1.37 m height.

As cover grade (*PC*) we interpreted the probability of any point *x* in the stand to be covered. With *PNC* the probability of the point not to be covered, follows .

If a point is not covered by any tree crown, no tree stands close enough to reach the point with its crown, i.e., no tree stands within the circle around point *x* with the area of the tree's crown area. Because the crown areas are species- and height-class-specific and the height classes are discrete, the crown areas are discrete numbers. Hence, the probability not to be covered can be formulated as a product of probabilities:

$$PNC = \prod_i \prod_s PNC_{i,s} ,$$

with $PNC_{i,s}$ the probability that no tree of height class *i* and species *s* is within the area $ca_{i,s}$ around point *x*.

The $ca_{i,s}$ are determined from the height classes according to model 1 to 5 and table 1. The problem now reduces to determine $PNC_{i,s}$. $PNC_{i,s}$ can be interpreted as the probability that an area of $ca_{i,s}$ does not carry any trees of species *s* and height class *i*. In DisCForM the trees of each height class and species are assumed to be randomly distributed. This results in a Poisson distribution for the tree density per a certain unit area. Here we chose the crown area ($ca_{i,s}$) of these trees as unit area. $PNC_{i,s}$ is the value of the Poisson distribution density for n=0. That value is:

$$f_{Poisson}(n = 0, \lambda_{i,s}) = e^{-\lambda_{i,s}} \text{ with } \lambda_{i,s} = \bar{y}_{i,s} \cdot ca_{i,s} ,$$

where $\bar{y}_{i,s}$ is the population density (in #/m^2).
Thus the cover grade is:

$$PC = 1 - \prod_i \prod_s e^{-\bar{y}_{i,s} \cdot ca_{i,s}} = 1 - e^{\sum_i \sum_s -\bar{y}_{i,s} \cdot ca_{i,s}} .$$

This means the cover grade depends on the overall tree crown projection area. Percent cover is obtained by multiplying *PC* with 100.

2.3.3 Drought index

One of the bioclimatic variables used in DisCForM is a drought index, describing the drought effect on plants on a scale between 0 and 1, where 0 corresponds to no drought stress and 1 to maximum drought stress. The one used up to now is calculated by

$$1-\frac{\sum_{year} Evapotranspiration}{\sum_{year} Demand},$$

where the *Demand* is the demand of the atmosphere to the soil water, i.e. the potential evapotranspiration reduced by the evaporating intercepted water. *Evapotranspiration* is defined by the field capacity and the supply of soil water, which is determined by a simple bucket model (for details see Bugmann, 1996; Bugmann and Cramer, 1997). The trees' growth rate and implicitly also the mortality rate depend with a species-specific function on the drought index, which is determined by one parameter. This parameter, the upper threshold of the drought index, above which no growth occurs and mortality is at maximum (Bugmann, 1996), has been derived from species-specific drought indicator values (Bugmann, 1994; Bugmann, 1996).

In this study, this drought index was replaced by a newly developed drought index calculated by the hydrological model WAWAHAMO, which uses the relation of actual to potential transpiration as an indicator for drought stress. The drought index is calculated separately for each year with a daily time step by

$$1-\frac{\sum_{vegetation\,period} Tr_{act}}{\sum_{vegetation\,period} Tr_{pot}}.$$

Actual transpiration (Tr_{act}) is the actually occurring transpiration rate taking the current drought situation in account. Besides climatic parameters, it is determined by the ecophysiological parameters LAI and stomatal resistance, which in turn is controlled by soil water content, temperature, radiation, and vapour pressure deficit. In contrast to this, potential transpiration is the transpiration rate that would occur if no drought stress would limit transpiration. That means, potential transpiration assumes that neither soil water nor vapour pressure deficit limit transpiration (see Zierl, 2001).

The species parameters of the drought dependence function in DisCForM could not be transferred directly to the new drought index but had to be adapted. To this aim, the drought index was calculated for each sample plot of the first Swiss National Forest Inventory (EAFV, 1988). This yielded a dataset of drought indexes, which could be compared to the occurrence of the species. Then for each tree species its distribution limit concerning the drought index was determined. As distribution limit of a species we defined the 95%-quantile, i.e. the drought value above which were only 5% of all sample plots on which this species occurred. The distribution limits were then used as the threshold parameters in the dependence function. For the species *Pinus silvestris* and *Populus nigra* we shifted these values to a higher (lower) value, to stay consistent with the empirical ranking of drought tolerance derived from other sources, such as reflected in the former parameters of DisCForM and ForClim. The last column in tab. 1 shows the resulting parameter values.

As an additional abiotic drought index, we introduced an index which neglects the influence of the different vegetation at the sites. By this index, the sites could be classified according to their abiotic properties. This standard drought index was calculated with the same spruce forest of LAI=8 m^2/m^2 and percent cover=80% at all sites of the first Swiss National Forest Inventory (see also fig. 7.1 in Zierl, 2000).

2.4 Simulation setup and decoupling

Simulations were run for 26 test sites for 1000 years: 500 years under current climate to reach equilibrium, 50 years of a linear climate change, and 450 years of future conditions.

Our aim was to compare the coupled model to a model that neglects the feedback from the changing vegetation to the hydrology. In such a model, e.g., WAWAHAMO, vegetation is characterised by constant parameters, which are derived from the current vegetation state. Therefore, in the decoupled simulations the coupling variables LAI, percent cover, needle tree ratio, and dominant tree species were set constant. These constant values were taken from the equilibrium under current conditions and served as input to the drought model after the year 500. As no constant values were reached in the equilibrium due to yearly climatic variability, the means over the last 30 years were used The forest model still received the actual drought index values.

2.5 Simulation sites and input data

Simulations were carried out at 26 selected sites in Switzerland, which encompass a wide range of temperature and drought conditions. Fig. 1 shows the location of the test sites in the bioclimatic space, which is spanned by yearly degree day sum and drought. It is combined with the frequencies with which the bioclimatic classes occur on the sample plots of the first Swiss National Forest Inventory (EAFV, 1988).

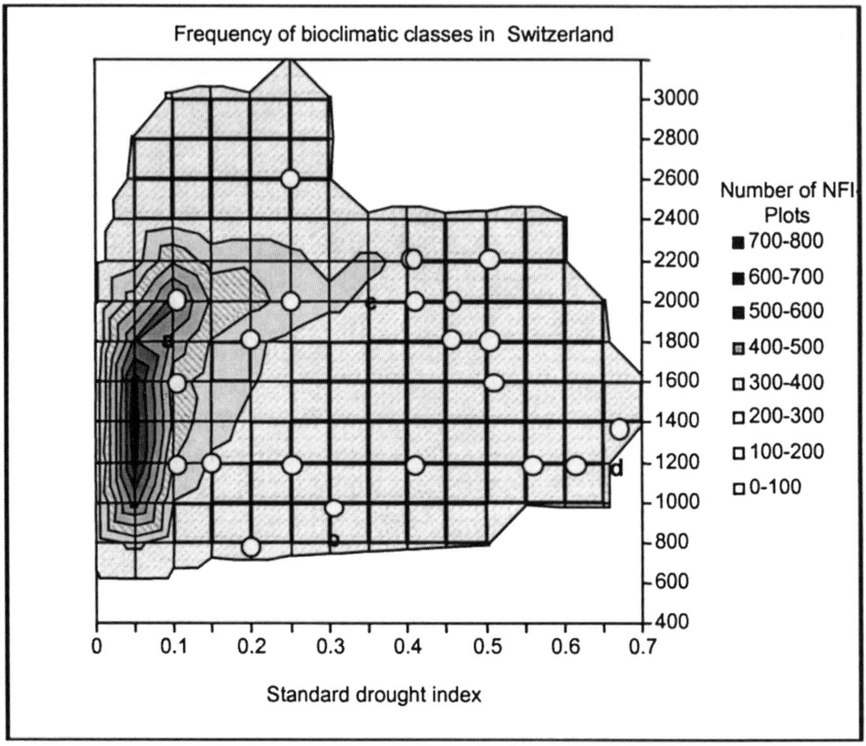

Figure 1: Study sites in the bioclimatic space (circles). Bioclimate is defined by the yearly degree day sum and by the standard drought index, which is determined with a spruce forest of LAI=8 m^2/m^2 and 80% cover. The shading indicates the frequency of the bioclimate classes according to evaluations at the Swiss National Forest Inventory plots. **a**, **b**, **c**, and **d** indicate the sites, for which simulations are shown in fig. 2.

2.5.1 Soil and climate at sites

The required input data regarding soil and climate were available or were derived from the WAWAHAMO-input-datasets that cover the entire forested area of Switzerland with a horizontal resolution of 1km and an observation period of 30 years from 1969 to 1998. In the following, the preparation of this WAWAHAMO-input-dataset is briefly described. For a detailed description see Zierl (2000).

Values of available soil water capacity were estimated on the model grid by external drift kriging using information from a soil map (Bundesamt für Raumplanung, 1980), a digital terrain model (DHM25 ©

Bundesamt für Landestopographie, DV 455.2) and approximately 700 soil profiles (Waldispühl, 1998). Permeability classes (sand, silt and loam) were directly taken from the soil map.

Meteorological data for the observation period were available at approximately 100 weather stations run by the Swiss Meteorological Institute including data on air temperature, air humidity, wind speed, cloud cover, absolute sunshine duration and pressure. Furthermore, daily precipitation totals were available from around 400 stations. These daily observations were interpolated to the model grid using inverse-distance-weighted interpolation. Here, emphasis was put on the correct representation of altitude dependencies, especially the representation of temperature inversions within the winter months.

The 30 years of climate data were repeated to obtain a long term climate series. We chose this approach instead of stochastical sampling from the current climate to detect more easily changes in the regular patterns of output variables.

2.5.2 Climate change scenario

For climate change we applied a scenario (tab. 2) from Beniston *et al.* (1995) and Marinucci *et al.* (1995), which was obtained from a double-nested regional model driven by a global circulation model. This scenario distinguishes between Northern and Southern Alps and between the seasons. We applied climate change linearly over 50 years and kept it constant then. The climate anomalies were added to the current climate data at the study sites.

Table 2: Climate change scenario used in simulations (Beniston *et al.*, 1995; Marinucci *et al.*, 1995). Climate change is applied linearly over 50 years

	ΔT (°C)			ΔP (%)		
	Winter	Spring/Fall	Summer	Winter	Spring/Fall	Summer
North	0.8	2.2	3.5	5	-18	-30
South	1.2	1.9	2.5	15	5	-5

3. RESULTS

3.1 Evaluation of simulation results

The simulated drought indexes, species biomasses, and LAI's were plotted for the simulation period. Additionally, mean and maximum values of the drought index, and the means of the total biomasses, the LAI and the biodiversity index were determined for the equilibrium state before ("current") and after climate change ("future"), and with ("coupled") and without coupling ("decoupled"). To obtain the means or maxima, the last 30 years of the simulation under current and future conditions were used. The absolute values or differences were plotted against the current coupled mean drought values ("current drought"), and simple regression functions were fitted to the data. Only in one plot, the 30-year-maxima of drought were used.

3.2 Sensitivity to modelling of structure

One goal of this study was to assess how strongly structural aspects of the modelled forest influence the simulated drought and LAI values in the coupled model. We ran coupled simulations with model versions where the forest structure and its influence on the dynamics were reduced and compared them to simulations with the full structure. Horizontal structure, i.e. spatial heterogeneity in the standard version of DisCForM is described by assuming that the trees are randomly distributed. In simulations without horizontal structure (simulations H), in contrast, trees were assumed to be uniformly distributed. This corresponds to the spatially homogenous canopy used in biogeochemistry or current DGVMs in entire grid cells. Vertical structure is described in DisCForM by 15 tree height classes. Since tree growth in the model depends crucially on these classes, and discretization errors increase with reducing the number of classes (Lischke *et al.*, 1998), we had to keep the height classes. However, the light conditions differ between the height classes, trees in higher height classes experience more light than those in lower ones. By using in all height classes the light conditions on the forest floor this vertical light structure could be omitted (simulations V). A further structure in DisCForM is given by the 30 different tree species, each of them with its specific fundamental ecological niche, which is characterized by the species parameters. Models to be applied globally, however, must rely on a much smaller number of plant functional types (PFTs). To assess the influence of a reduced species-, or rather PFT-set, we used in one simulation with reduced species diversity (simulations -S) only the species *Abies alba, Fagus sylvatica, Picea abies,* and *Populus tremula*, representing temperate evergreen, temperate deciduous, boreal evergreen, and temperate deciduous species. Furthermore simulations were run, where horizontal and vertical (HV) structure were reduced simultaneously, in another simulation additionally the species diversity (HVS). The latter model setup corresponds to the one in the DGVM IBIS

(Foley *et al.*, 1996), where no intra-grid-cell heterogeneity is simulated and forests consist of one (tree-) layer and (in Switzerland) of four tree PFTs.

4. RESULTS

Figures 2a-d show simulations with the coupled models (left) and decoupled models (right) for four selected sites (**a**, **b**, **c**, and **d** in fig. 1). The periodical oscillations in all variables are the result of repeated 30-year time series of climatic data.

4.1 Influence of climate change

In all simulations, the amplitudes of the oscillations in the biomasses, as well as in the drought stress and the LAI, intensify during and after climate change. In the coupled simulations (fig. 2, right column), climate change impacts depend on the sites, for example the current bioclimatic conditions. At site **a**, a humid and relatively warm forest site which represents a majority of the Swiss forests, the species pattern changes slightly, with a decrease in *Fagus* and an increase in *Castanea*. Biomass and LAI maximum increase. Drought maxima strongly increase. At site **b**, representing a relatively dry subalpine site, the number of species increases after climate change. Biomass is stable. LAI increases slightly, drought maxima strongly. At site **c**, a warm medium dry site, the species are nearly completely exchanged. For example, *Fagus*, which dominates under current climate, is replaced by *Quercus*. This change in species composition is accompanied by a slight decrease in biomass and LAI, and by only a minor increase in the drought maxima. The future drought maxima are considerably lower than at site **a**, although current drought is higher. The future drought means, however, are higher than at site **a**. The drought maxima at site **c** are lower than at site **b**, too. Site **d** is currently one of the driest sites. This is reflected in the high drought maxima already under current conditions. Also here, the species composition changes, however, not as drastically as at site **c**. Biomass and LAI decrease, but drought does not change.

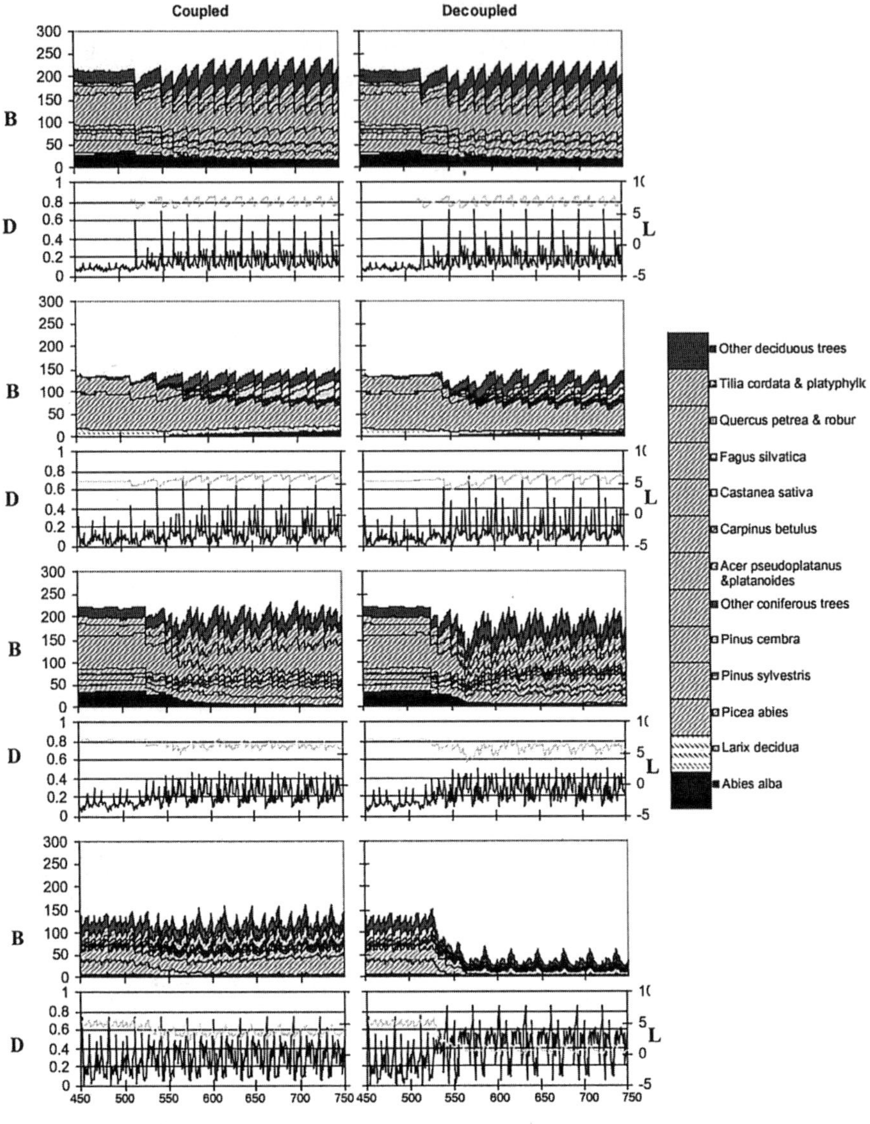

Figure 2: Example simulations. Left: coupled simulations, right: decoupled simulations. a) to d): selected simulation sites. Top pictures: simulated species biomass. Bottom pictures: drought index (black), LAI (grey). Current climate conditions until year 500, then linear climate change from year 500 to 550 according to tab. 2, then future conditions. B= Biomass; D=Drought Index; L=Leaf Area Index (LAI)

Figure 3 shows the differences between future (after climate change) and current (before climate change) values of drought, biomass, biodiversity and LAI vs. the current drought in the coupled simulations for all studied sites. Drought increases in average nearly 0.1 units, independently of current drought, however with a high variability. The extent and direction of the changes of the variables biomass, biodiversity and LAI, however, depend on the current drought condition of the sites. The changes tend to be positive at low current drought values. That means the variables increase with climate change. At high current drought values, the variables decrease with climate change. These trends are clear for all considered variables: biomass (R^2=0.69) and LAI (R^2=0.73), and biodiversity (R^2=0.60).

Δ Drought Index

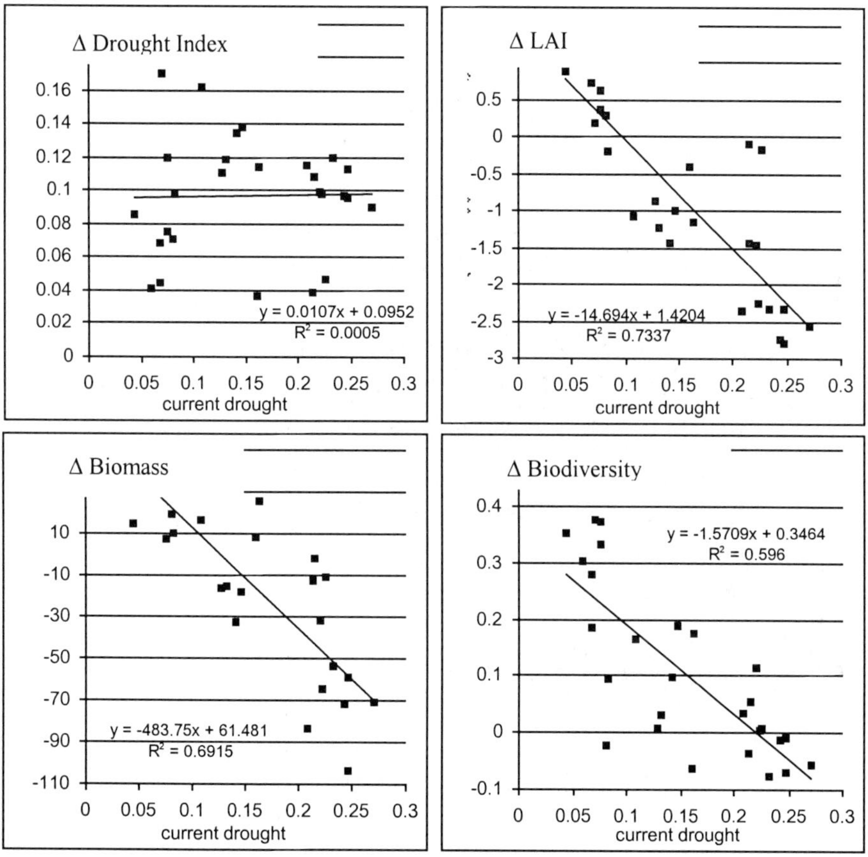

Figure 3: Differences of future (after climate change) to current (before climate change) simulated mean equilibrium values of drought index, LAI, biomass and biodiversity index vs. current mean drought index

4.2 Influence of coupling

The difference between the coupled (fig. 2, left column) and the decoupled simulations (fig. 2, right column) also depends on the current conditions. At sites **a** and **b**, hardly any difference can be detected. At site **c**, the species compositions differ. Biomass and LAI are lower in the decoupled simulations. In this case, however, the changed LAI does not feed back to the determination of drought after climate change, i.e., future drought is determined with the present LAI-value. Since this present LAI is higher than the future LAI, the drought increases more strongly, which in turn decreases biomass and LAI. This effect is even more pronounced at site **d**. Here, in the decoupled simulation biomass decreases drastically. LAI decreases, too. Yet, the drought value, which is already high at current conditions, increases further. In contrast to this, the biomass, LAI and drought decrease less in the coupled simulation.

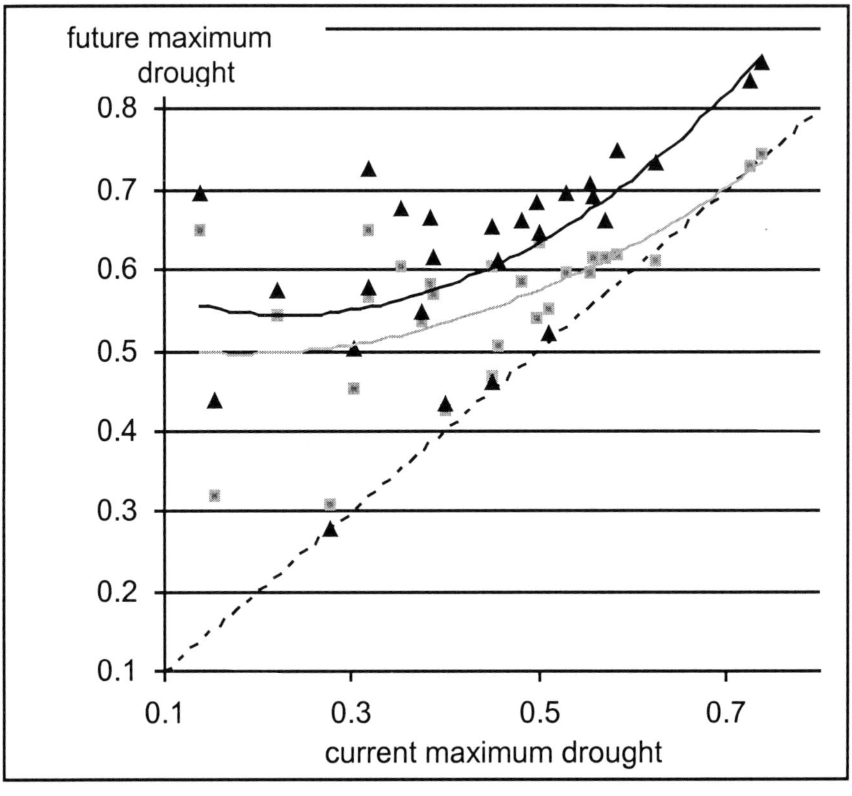

Figure 4: Maximum future drought values for the coupled (squares) and decoupled (triangles) model vs. current maximum drought.

Figure 4 shows that with increasing current maximum drought, the future maximum drought of the coupled simulation (squares) increases less than that of the decoupled simulation (triangles). The maximum future drought values of the coupled simulations are high above the dotted diagonal (current = future drought) for low current values, close to it for high values. In contrast, the future drought maxima of the decoupled simulation increase also at high current drought values.

Figure 5: Difference between coupled and decoupled simulations after climate change in

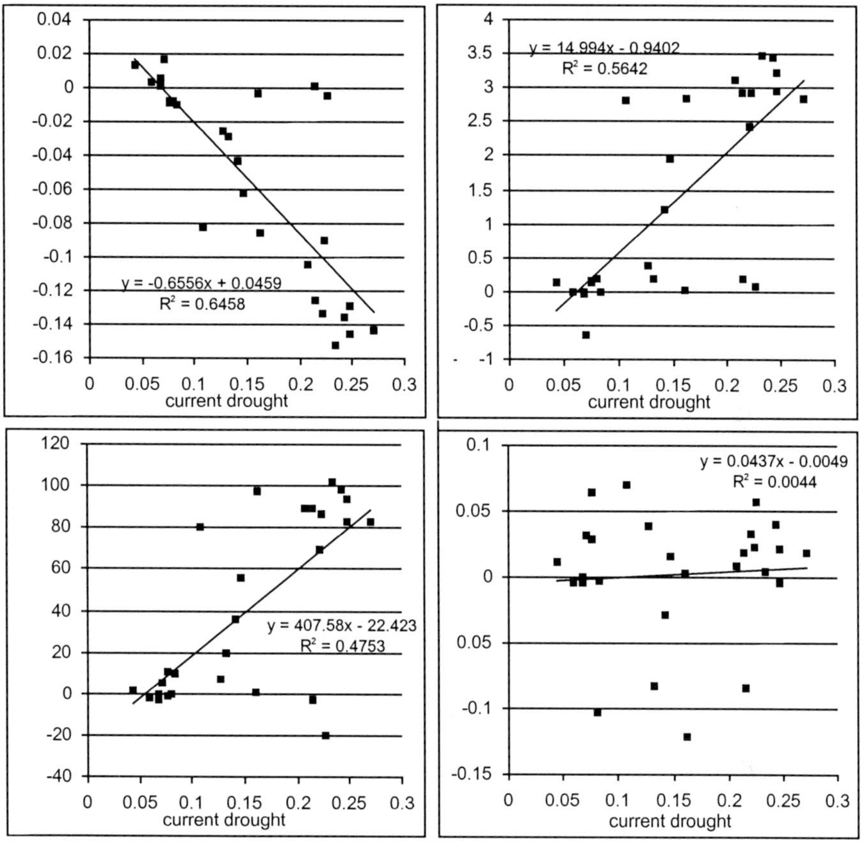

drought (upper right), LAI (upper left), biomass(lower right), and biodiversity (lower left)vs. current mean drought index

Figure 5 shows the difference between coupled and decoupled simulations in mean drought, biomass, LAI, and biodiversity at the future equilibrium vs. the current drought value. For drought a clear negative trend starting from zero can be seen. This means, that the increase in drought in the coupled simulations is less than in the decoupled simulations. This in turn means that in the coupled model the feedback of the vegetation on hydrology tends to keep drought at a lower level than in the decoupled model. The difference in LAI increases with drought. Note that this difference is calculated with the LAI simulated by the forest model in the decoupled simulations, not with the one used to determine drought in the hydrological model.

The biomasses of the coupled and decoupled simulation do not differ at low current drought values, but at high drought values the biomass of the coupled simulation is much higher. In biomass and LAI, the points are aggregated in two dense clusters with some scattered points in between. For biodiversity no clear trend or clustering is detectable.

4.3 Influence of forest structure

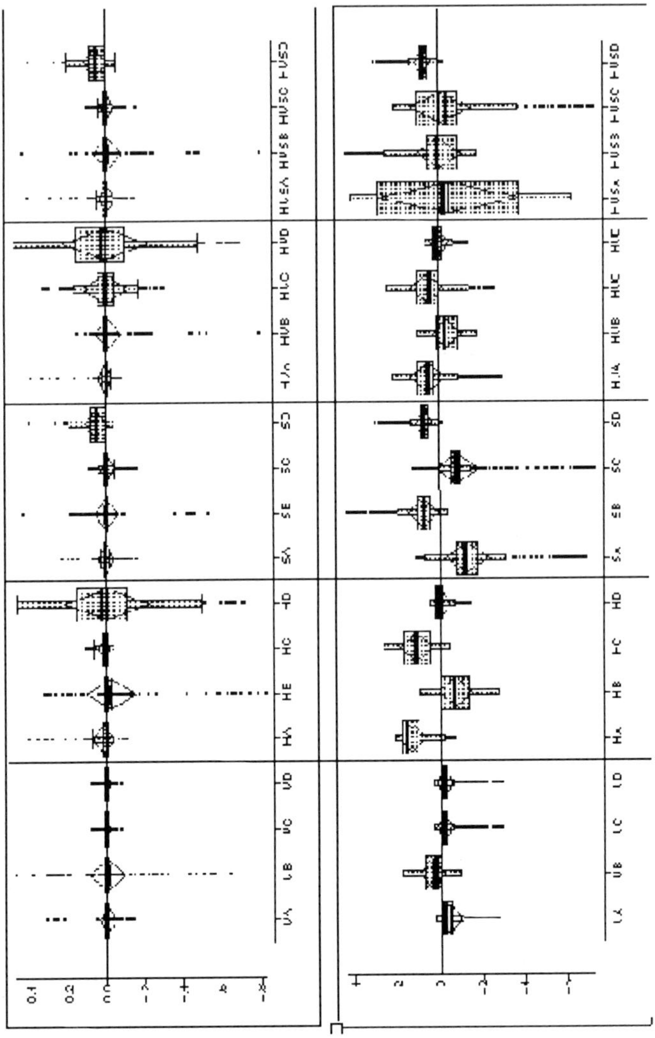

Figure 6: Differences in LAI (m^2/m^2 – lower figure) and drought (upper figure)of coupled simulations (1000 years with climate change) between model with full structure and model with reduced structure H: horizontal heterogeneity by spatial tree distribution, V: Vertical heterogeneity by different light conditions in different tree heights, S: Species diversity. The boxplots show the distributions of the differences. The edges of the boxes indicate the 25% and 75% percentile, the middle line the median. The middle line of the diamonds shows the mean, their upper and lower corners the standard deviation.

Also the influence of forest structure on drought and LAI depends on the site conditions.

In average the sensitivity of drought to reduced structure is small at nearly all sites and for all structures (fig. 6, top). The average LAI is moderately sensitive (fig. 6, bottom). The differences between the simulations with full structure and those with reduced structure range from -1.3 to 1.3 m^2/m^2 and depend on the site. Reduced horizontal structure (H) and species diversity (S) have a larger effect than reduced vertical structure (V) and the combinations (HV, HVS).

In contrast to the averages, the variability of the differences is considerable, with extreme values ranging from about –7.2 to 2.8 m^2/m^2 for LAI, and –0.8 to 0.5 for drought. The variability of the drought differences is high for the vertical and horizontal structure, low for species diversity. In contrast, the variability of the LAI differences is high for the species diversity and low for the horizontal and vertical structure.

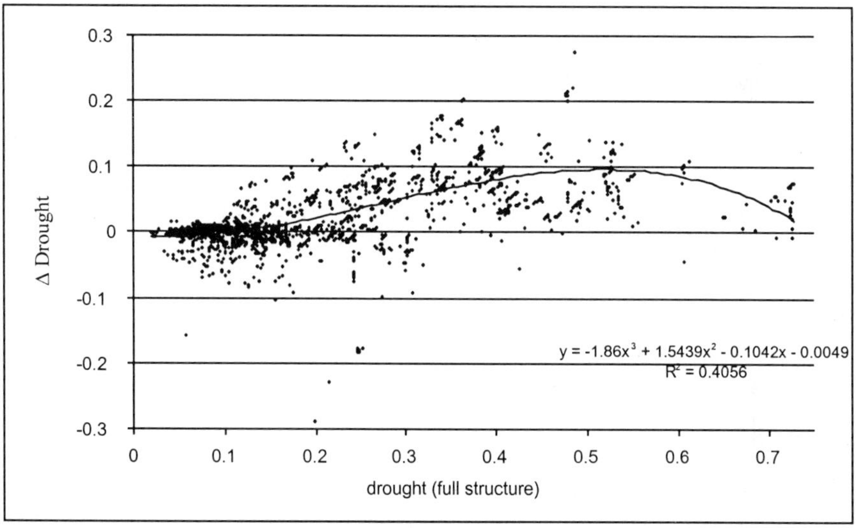

Figure 7: Differences in drought simulated with model with full structure to drought simulated with model with reduced horizontal, vertical and species structure (HVS) vs. drought simulated with model with full structure. All data from sites a, c, and d. Line: Polynominal fit.

Under dry conditions the sensitivity to structure is larger. At the dry site (D), average drought and LAI values are slightly higher in the model with full structure than if vertical, horizontal and species structure were reduced (HVSD). The effect of climate change on drought and LAI was slightly smaller in the reduced model (not shown). The biomass, however, decreased much stronglier after climate change with the reduced model than with the full model. At the temperate to warm sites (A, C, D), the sensitivity of drought to structure depends on the actual drought conditions (fig. 7), increasing to a maximum at about 0.5 and decreasing at higher values. At the cool site (B), however, the sensitivity of drought to structure decreased with increasing drought (not shown).

5. DISCUSSION

The results of this study show that the impact of climate change on hydrology can be significantly modified by the adaptation of vegetation composition and structure to the new environmental conditions.

The coupled simulations show higher future biomasses, LAI's and lower drought values than the decoupled simulations. Thus, the vegetation has a clear stabilising effect on the impact of a climatic shift on hydrological processes, such as the occurrence of drought. Specifically, to a certain extent, the simulated vegetation is able to compensate the influence of rising temperatures and decreasing precipitation on drought stress by adaptation and, thus, is able to keep drought at a lower level than without adaptation.

Modelling studies of other vegetation types have yielded smaller effects of the plant adaptation feedbacks. With simulations of a big-leaf model Wilson *et al.* (1999) studied the effect of feedbacks on the CO2 fertilization on two crop species. They found moderate effects of stomata- and LAI-adaptation. With the big-leaf model FOREST-BGC Kremer *et al.* (1996) simulated the water balance of grass and brush vegetation under climate change. They found a substantial positive influence of stomata closing to soil water content, however, only a minor influence of LAI-change.

The stabilising effect depends on the current environmental conditions. It is particularly strong at currently dry forest sites. There, the vegetation is forced to adapt to the new conditions in order to keep drought at a level where survival is possible. In some cases, the future maximum drought values can even be kept at the current level by the feedback despite of raised temperatures and decreased precipitation. Therefore, the feedback of the biosphere on the hydrological cycle is of great importance at such forest sites. In contrast to the dry sites where the vegetation has to adapt to the climate change to avoid unbearable drought stress, the positive effect of rising temperatures at currently humid sites is apparently stronger than the negative effect of increasing drought. Thus, no drought adaptation of the vegetation is required, but the vegetation is in the position of using the warmer conditions through enhancing biomass production. Such a dependence on environmental conditions was also found by Heck *et al.* (1999). With a big-leaf-/hydrology-model they simulated the spring/summer water balance for different regions of Europe. They found that in spring, in humid regions additionally in summer, an artificial increase of LAI results in an increase of evapotranspiration, since soil water is still sufficient to keep the stomata open. In the dry Mediterranean area during summer, however, the evapotranspiration decreased, although LAI was increased.

The mechanisms of the described stabilisation include the lowering of LAI and cover grade by reduced individual growth and higher mortality. Low LAI and cover grade in turn enlarge throughfall and lessen transpiration. To a certain extent the models achieve this while keeping the biomass at a constant level, through gradually exchanging dominant species with high LAI by light dependent species with lower LAI. Thus, not only the medium-term processes of leaf area reduction and the reduction of the cover grade, but also long-term processes such as the exchange of the dominant tree species are important processes influencing the local water budget.

The found adjustment of the overall LAI to the soil water supply corroborates the theory that an optimal ecosystem LAI maximises NPP with respect to soil water and light (Woodward *et al.*, 1995) which has been used as a rule to determine PFT's in advanced biogeography models (Kergoat, 1998; Neilson and Marks, 1994) or also individual based models. In contrast to the latter models, however, in our study this relationship emerged from the dynamics of competing species with different shadow and drought tolerances, instead of being prescribed.

We focussed on the single feedback from plants to local hydrology and vice versa. However, other interactions between climate and vegetation, such as surface roughness reducing wind speed, evaporation increasing regional precipitation, LAI decreasing albedo, or CO2-fertilization increasing NPP and decreasing stomata closure, can enhance or counterbalance the effects of this feedback. To study the effect of several feedbacks, particularly of CO2 fertilization and LAI adaptation under a double CO2 forcing, three vegetation models were coupled to global climate models. Levis *et al.* (2000) applied the DGVM IBIS (Foley *et al.*, 1996). Betts *et al.* (1997) coupled iteratively the global big-leaf model (Sheffield University Vegetation Model, referred to as "SUVM"). Neilson (1994) used the biogeography model MAPPS, which iteratively adapts LAI and the local soil water conditions.

Under a double CO2 forcing with LAI adaptation, IBIS and SUVM produced an increased, MAPPS a decreased global LAI. In the high northern latitudes, LAI increased with MAPPS and IBIS, same as at the humid sites of our study. In the continental mid latitudes, LAI increased in the IBIS simulations, but decreased in the MAPPs simulations, as well as at the dry sites of our study.

Drought stress was not explicitely calculated by the models. With IBIS, soil water decreased globally, increased at northern latitudes, and remained at the same level at mid latitudes. MAPPS produced a slightly higher global AET, but a much higher PET, which can be interpreted as an increased drought stress, corresponding to the increased drought stress after climate change at all our sites. The influence of LAI adaptation on the water balance was studied with IBIS and SUVM. With IBIS global average soil water was the same with adapting LAI as with fixed LAI. With SUVM temperature decreased and evaporation increased, which can be assumed to result in a lower drought stress. At high (mid) latitudes LAI adaptation resulted in slightly higher (lower) soil water supplies. This is in agreement with the lower (higher) drought values in our coupled simulations at dry (humid) sites.

Thus, the results of our regional study confirm many findings of these global scale studies, although those included additional feedbacks, in particular the CO2-fertilization. However, the effects of LAI-adaptation, i.e. of model coupling, were larger at the dryer sites in our study. This might be because the negative feedback from evaporation over precipitation to drought stress could not be taken into account on the local scale. Increased LAI results in increased evaporation and drought stress. On the large scale the evaporated water forms clouds and increases precipitation, thus eventually decreasing drought stress. We had to rely on a climate scenario, which was not coupled to the vegetation.

The study demonstrates the importance of structure for the modelling of the forest-hydrology-system. The sensitivity test revealed that omitting structure, i.e. spatial heterogeneity, vertical layering of light, or species diversity, has a substantial effect on single year LAI and drought, whereas a smaller impact on long-term averages. However, under dry conditions the reduced structure model yielded also considerably lower average LAI and drought values. In general the sensitivity to the forest structure was highest under strong but not extreme drought conditions, either in time or in space. Thus, simulations with models that neglect vegetation structure might underestimate drought effects, at least under dry conditions.

The example simulations illustrate how for example species diversity affects LAI. By a switch in the species composition to species with lower LAI, drought can be stabilised. However, species depend also on other factors than drought. For example, in the simulation at site **c** more drought-adapted species are available than at site **b**, where even after climate change most deciduous species are prevented by low winter temperatures.

This study revealed also several weak points of the used models, which were not obvious before. WAWAHAMO was developed for the use in vegetated areas. When the LAI reaches low values, its results are no longer reliable, since then in the definition of drought stress both, not only the actual, but also the potential transpiration tends to zero.

DisCForM uses a dependence of the degree day sum with an upper threshold value. This leads to the exclusion of certain species (e.g., of Pine) by already moderate degree day sums, which appears unrealistic. The use of a saturated dependence function, as in newer versions of the model ForClim, should be considered. This is an example for the problems that are associated with using empirical, descriptive process functions. To model climate dependences more reliably, ecophysiological functions and their climate dependences have to be included in more detail, or aggregated systematically from more detailed models (see Lischke, 2001; Lischke *et al.*, 1997a; Lischke *et al.*, 1997b).

This holds particularly for the effects of drought respectively water on the plants. WAWAHAMO focuses on the stomata regulation, but neglects the effects from lack of water in the plant. DisCForM uses a completely empirical drought dependence function, whose threshold parameter was fitted to distribution data. However, since the species distribution is determined by numerous abiotic and also biotic factors, e.g. competition, this approach is rather crude. In further studies, the effects of water in trees and the processes associated with them should be better understood and represented in the coupled model. Their parameters should be derived from those of detailed ecophysiological models.

Our results allow us to roughly assess the average effect of climate change for entire Switzerland. As the overwhelming majority of Switzerland falls into the humid bioclimatic classes (fig. 1), most probably after climate change drought will increase slightly, associated with an increase in LAI, biomass and biodiversity (fig. 3). To obtain more detailed regional assessments, the coupled model shall be applied to the entire country in further studies. As far as possible other feedbacks, particularly CO2-fertilization shall be included in these studies.

For the use on larger, e.g. global scales, vegetation models need to be simple and efficient, and based on a limited set of obtainable parameters. Therefore, the influence of vegetation structure and feedbacks has to be studied in more detail than in this preliminary study. If it turns out to be of major importance, more detailed models, e.g. individual based vegetation models, would have to be aggregated to simpler ones (Hurtt *et al.*, 1998; Huston, 1991), which still reflect the major variabilities contained in the detailed models. To this aim, aggregation approaches as presented by Lischke *et al.* (1998) or Bolker (1997) could be adapted and used.

6. CONCLUSIONS

The found adaptation of vegetation to drought stress implies that in vegetation-hydrology-models that neglect the vegetation-hydrology-feedback the impacts of climate change on drought are probably overestimated. In contrast, the study of the sensitivity of the vegetation-hydrology-model to lack of vegetation structure reveals that neglecting structure in vegetation-hydrology-models can lead to a systematic underestimation of drought in the long-term average under dry conditions. Furthermore, drought values for single years might considerably deviate from those obtained with a full structure model.

7. ACKNOWLEDGEMENTS

We would like to thank Matthias Dobbertin and Andreas Schwyzer for access to and explanation of the WSI-data. We are grateful to Thomas Löffler and Niklaus Zimmermann for the inspiring discussions. Thanks to David Siemens for checking parts of the English.

Thanks to the Swiss Agency for the Environment, Forest and Landscape for funding B. Zierl (Rubrik: 310.3189.016).

8. REFERENCES

Band, L. E., Patterson, P., Nemani, R. and Running, S. W., 1993. Forest ecosystem processes at the watershed scale: incorporating hillslope hydrology. Agricultural and Forest Meteorology, 63 : 63-126.

Beniston, M. A., Ohmura, A., Rotach, M., Tschuk, P., Wild, M. and Marinucci, R. M., 1995. Simulation of Climate Trends over the Alpine Region: Development of a Physically-Base Modeling System for Application to Regional Studies of Current and Future Climate.. 198 pp., Department of Geography, ETH Zürich, Zürich.

Betts, R. A., Cox, P. M., Lee, S. E. and Woodward, F. I., 1997. Contrasting physiological and structural vegetation feedbacks in climate change simulations. Nature (London), 387 : 797-799.

Bolker, B. M., Pacala, S. W. and Levin, S. A., 1997. Moment methods for stochastic processes in continuous space and time. In: U. Dieckmann and J. Metz (Eds.), Low-Dimensional Dynamics of Spatial Ecological Systems, Laxenburg.

Brang, P., 1998. Sanasilva-Bericht 1997. Gesundheit und Gefährdung des Schweizer Waldes - eine Zwischenbilanz nach 15 Jahren Waldschadenforschung, Berichte der Eidg. Forschungsanstalt für Wald, Schnee und Landschaft, Vol. 345, pp. 102. Eidg. Forschungsanstalt für Wald, Schnee und Landschaft, Birmensdorf.

Bugmann, H., 1994. On the ecology of mountainous forests in a changing climate: A simulation study. Diss. ETH, No. 10638, Environmental Sciences, Swiss Federal Institute of Technology Zurich, Zurich.

Bugmann, H., 1996. A simplified forest model to study species composition along climate gradients. Ecology, 77 (7): 2055-2074.

Bugmann, H. and Cramer, W., 1997. Improving the behaviour of forest gap models along drought gradients. No. 24, PIK - Potsdam Institute for Climate Impact Research, Potsdam.

Bundesamt für Raumplanung, 1980. Bodeneignungskarte der Schweiz. Grundlagen für die Raumplanung. , Bundesamt für Raumplanung, Bern,.

Churkina, G., Running, S. W., Schloss, A. L., Bondeau, A., Cramer, W., Colinet, G., Collatz, J., Dedieu, G., W., E., Esser, G., Field, C., Francois, L., Friend, A., Haxeltine, A., Heimann, M., HOffstadt, J., Kaduk, J., Kergoat, L., Kicklighter, D. W., Knorr, W., Kohlmaier, G., Lurin, B., Maisongrande, P., Martin, P., McKEown, R., Meeson, B., Moore III, B., Nemani, R., Nemry, B., Olson, R., Otto, R., Parton, W., Plöchl, M., Prince, S., Randerson, J., Rasool, I., Rizzo, B., Ruimy, A., Running, S., Sahagian, D., Saugier, B., Schloss, A. L., Scurlock, J., Steffen, W., Warnant, P. and Wittenberg, U., 1999. Comparing global models of terrestrial net primary productivity (NPP): the importance of water availability. Global Change Biology, 5 (Suppl. 1): 46-55.

Ciencala, E., Running, S. W., Lindroth, A., Grelle, A. and Ryan, M. G., 1998. Analysis of carbon and water fluxes from the NOPEX boreal forest: comparison of measurements with FOREST-BGC simulations. Journal of Hydrology, 212-213 : 62-78.

Clark, J. S., 1993. Paleoecological perspectives on modeling broad-scale responses to global change. In: P. M. Kareiva, J. G. Kingsolverand R. B. Huey (Eds.), Biotic interactions and global change, pp. 315-332. Sinauer Associates, Sunderland, Massachusetts.

Deutschman, D. H., Levin, S. A. and Pacala, S. W., 1997. Scaling from trees to forest landscapes: The role of fine-scale heterogeneity in light. Ecological Monographs: submitted.

Dobbertin, M., 1996. Relationship between basal area increment, tree crown defoliation, and tree and site variables, IUFRO Conference on Effects of environmental factors on tree and stand growth, pp. 33-44. Technische Universität Dresden, Berggiesshübel near Dresden.

EAFV, 1988. Schweizerisches Landesforstinventar. Ergebnisse der Erstaufnahme 1982-1986, Berichte Eidgenössische. Forschungsanstalt für Wald, Schnee und Landschaft, Vol. 305, pp. 375. Eidgenössische Anstalt für das forstliche Versuchswesen in Zusammenarbeit mit dem Bundesamt für Forstwesen und Landschaftsschutz, Birmensdorf.

Foley, J. A., Levis, S., Costa, M. H., Cramer, W. and Pollard, D., 2000. Incorporating dynamic vegetation cover within global climate models. Ecological Applications, 10 (6): 1620-1632.

Foley, J. A., Levis, S. and Prentice, C., 1998. Coupling dynamic models of climate and vegetation. Global change biology, 4 (5): 561-579.

Foley, J. A., Prentice, C., Ramankutty, N., Levis, S., Pollard, D., Sitch, S. and Haxeltine, A., 1996. An integrated biosphere model of land surface processes, terrestrial carbon balance, and vegetation dynamics. Global Biogeochemical Cycles, 10 : 603-628.

Friend, A. D., Stevens, A. K., Knox, R. G. and Cannell, M. G. R., 1997. A process-based, terrestrial bioshpere model of ecosystem dynamics (Hybrid v3.0). Ecological Modelling, 96 : 249-287.

Gurtz, J., Baltensweiler, A., Lang, H., Menzel, L. and Schulla, J., 1997. Auswirkungen von klimatischen Variationen auf den Wasserhaushalt und Abfluss im Flussgebiet des Rheins. Projektschlussbericht im Rahmen des nationalen Forschungsprogramms (NFP31): Klimaänderungen und Naturkatastrophen. vdf, Hochschul-Verlag an der ETH, Zürich.

Haeupler, H., 1995. Diversität. In: W. Kuttler and K. Steinecke (Eds.), Handbuch zur Ökologie, pp. 99-104. Analytica, Berlin.

Heck, P., Lüthi, D. and Schär, C., 1999. The influence of vegetation on the summertime evolution of European soil moisture. Physics and Chemistry of the Earth. Part B Hydrology, Oceans and Athmosphere, 24 (6): 609-614.

Hieke, K., 1989. Praktische Dendrologie. VEB Deutscher Landwirtschaftsverlag, Berlin.

Hurtt, G. C., Moorcroft, P. R., Pacala, S.W. and Levin, S. A., 1998. Terrestrial models and global change: challenges for the future. Global Change Biology, 4 : 581-590.

Huston, M. A., 1991. Use of individual-based forest succession models to link physiological whole-tree models to landscape-scale ecosystem models. Tree Physiology, 9 : 293-306.

Hutjes, R., Kabat, P., Running, S., Shuttleworth, W., Field, C., Bass, B., Dias, M., Avissar, R., Becker, A., Claussen, M., Dolman, A., Feddes, R., Fosberg, M., Fukushima, Y., Gash, J., Guenni, L., Hoff, H., Jarvis, P., Kayane, I., Krenke, A., Liu, C., Meybeck, M., Nobre, C., Oyebande, L., Pitman, A., Pielke, R., Raupach, M., Saugier, B., Schulze, E., Sellers, P., Tenhunen, J., Valentini, R., Victoria, R. and Vorosmarty, C., 1998. Biospheric aspects of the hydrological cycle - Preface. Journal of Hydrology, 213 (1-4): 1-21.

Kergoat, L., 1998. A model for hydrological equilibrium of leaf area index on a global scale. Journal of hydrology, 212-213 : 268-286.

Korol, R. L., Running, S. W. and Milner, K. S., 1995. Incorporating intertree competition into an ecosystem model. Can. J. For. Res., 25 (3): 413-424.

Kräuchi, N., 1994. Modelling forest succession as influenced by a changing environment. Mitteilungen der Schweizerischen Anstalt für das forstliche Versuchswesen, 69 : 125-271.

Kräuchi, N. and Kienast, F., 1993. Modelling subalpine forest dynamics as influenced by a changing environment. Water, Air, & Soil Pollution, 68 : 185-197.

Kremer, R. G., Hunt, E. R., Running, S. W. and Coughlan, J. C., 1996. Simulating vegetational and hydrologic responses to natural climatic variation and GCM-predicted climate change in a semi-arid ecosystem in Washington, USA. Journal of Aride Environments, 33 (1): 23-38.

Kucharik, C. J., Foley, J. A., Delire, C., Fisher, V. A., Coe, M. T., Lenters, J. D., Young, M., C., Ramankutty, N., Norman, J. M. and Gower, S., 2000. Testing the performance of a Dynamic Global Ecosystem Model: Water balance, carbon balance, and vegetation structure. Global Biogeochemical Cycles, 14 (3): 795-825.

Levis, S., Foley, J. A. and Pollard, D., 2000. Large-scale vegetation feedbacks on a doubled CO2 climate. Journal of Climate, 13 (7): 1313-1325.

Lexer, M., 1995. Anwendung eines "big-leaf"-Modelles zur Simulation des Bodenwasserhaushaltes in Fichtenbeständen (Picea abies(L.) Karst.). Applying a "big-leaf"-model to simulate seasonal sooil water dynamics in Norway spruce stands. Centralblatt für das gesamte Forstwesen, 112 (4): 209-225.

Lexer, M. J., Honninger, K., Scheifinger, H., Matulla, C., Groll, N. and Kromp-Kolb, H., 2000. The sensitivity of central European mountain forests to scenarios of climatic change: Methodological frame for a large-scale risk assessment. Silva Fennica, 34 (2): 113-129.

Lischke, H., 2001. New developments in forest modeling: Convergence between applied and theoretical approaches. Natural Resource Modeling, 14 (1): in press.

Lischke, H., Löffler, T. J. and Fischlin, A., 1997a. Calculating temperature dependence over long time periods: A comparison and study of methods. Agric. For. Meteorol., 86 : 169-181.

Lischke, H., Löffler, T. J. and Fischlin, A., 1997b. Calculating temperature dependence over long time periods: Derivation of methods. Ecol. Model., 98 (2-3): 105-122.

Lischke, H., Löffler, T. J. and Fischlin, A., 1998. Aggregation of individual trees and patches in forest succession models - Capturing variability with height structured random dispersions. Theoretical Population Biology, 54 : 213-226.

Löffler, T. J. and Lischke, H., 2001. Incorporation and Influence of Variability in an Aggregated Forest Model. Natural Resource Modeling, 14 (1): in press.

Lüdecke, M. K. B., Badeck, F.-W., Otto, R. D., Häger, C., Dönges, S., Kindermann, J., Würth, G., Lang, T., Jäkel, U., Klaudius, A., Ramge, P., Habermehl, S. and Kohlmaier, G. H., 1994. The Frankfurt Biosphere Model: a global process-oriented model of seasonal and long-term CO2 exchage between terrestrial ecosystems and the atmosphere. I. Model description and illustrative results for cold deciduous and boreal forests. Clim. Res., 4 : 143-166.

Marinucci, R. M., Giorgi, F., Beniston, M., Wild, M., Tschuck, P. and Bernasconi, A., 1995. High resolution simulations of January and July climate over the western Alpine region with a nested regional modeling system. Theoretical and Applied Climatology, 51 : 119-138.

McMurtrie, R. E. and Landsberg, J. J., 1992. Using a simulation model to evaluate the effects of water and nutrients on the growth and carbon partitioning of Pinus radiata. Forest Ecology and Management, 52 : 243-260.

Neilson, R. P. and Drapek, R. J., 1998. Potentially complex biosphere responses to transient global warming. Global change biology, 4 (5): 505-521.

Neilson, R. P. and Marks, D., 1994. A global perspective of regional vegetation and hydrologic sensitivities from climatic change. Journal of Vegetation Science, 5 : 715-730.

Pacala, S. W. and Deutschman, D. H., 1995. Details that matter: the spatial distribution of individual trees maintains forest ecosystem function. OIKOS, 74 : 357-365.

Pielke, R. A., Avissar, R., Raupach, M., Dolman, A. J., Zeng, X. B. and Denning, A. S., 1998. Interactions between the atmosphere and terrestrial ecosystems: influence on weather and climate. Global change biology, 4 (5): 461-475.

Potter, C. S., Randerson, J. T., Field, C. B., Matson, P. A., Vitousek, P. M., Mooney, H. A. and Klooster, S. A., 1993. Terrestrial ecosystem production: A process model based on global satellite and surface data. Global Biogeochem. Cycles, 7 (4): 811-841.

Prentice, I. C. C., W.Harrison, S.P.Leemans, R.Monserud, R.A. and Solomon, A. M., 1992. A global biome model based on plant physiology and dominance, soil properties and climate. J. Biogeogr., 19 : 117-134.

Running, S. W., Gower, S.T., 1991. FOREST-BGC, A general model of forest ecosystem processes for regional applications. II. Dynamic carbon allocation and nitrogen budgets. Tree Physiol., 9 : 147-160.

Running, S. W., 1994. Testing FOREST-BGC ecosystem process simulations across a climatic gradient in Oregon. Ecological Applications, 4 (238-247).

Running, S. W. and Coughlan, J. C., 1988. A general model of forest ecosystem processes for regional applications. I. Hydrologic balance, canopy gas exchange and primary production processes. Ecol. Model., 42 : 125-154.

Shannon, C. E. and Weaver, W., 1976. Mathematische Grundlagen der Informationstheorie. R. Oldenbourg-Verlag, München, Wien.

Tiktak, A. and Bouten, W., 1992. Modelling soil water dynamics in a forested ecosystem. III: Model description and evaluation of discretization. Hydrological Processes, 6 : 455-465.

Tiktak, A. and VanGrinsven, H. J. M., 1995. Review of sixteen forest-soil-atmosphere models. Ecological Modelling, 83 : 35-53.

Waldispühl, P., . Swiss Institute of forest, snow and landscape research, soil ecology section, 1998. , , Birmensdorf, Switzerland.

Wilson, K. B., Carlson, T. N. and Bunce, J. A., 1999. Feedback significantly influences the simulated effect of CO2 on seasonal evapotranspiration from two agricultural species. Global Change Biology, 5 (8): 903-917.

Woodward, F. I., Smith, T. M. and Emanuel, W. R., 1995. A global land primary productivity and phytogeography model. Global Biogeochemical Cycles, 9 (4): 471-490.

Zierl, B., 2000. WAWAHAMO - a hydrological model to simulate drought in forested ecosystems. PhD thesis no. 1320, Geography, University of Fribourg, Switzerland. Swiss Federal Research Institute WSL, Birmensdorf - Switzerland.

Zierl, B., 2001. A water balance model to simulate drought in forested ecosystems and its application to the entire forested area of Switzerland. Journal of Hydrology, 242 (1-2): 115-136.

Water control and property rights: An analysis of the Middle Eastern situation

URS LUTERBACHER[1] AND ELLEN WIEGANDT[2]
Graduate Institute of International Studies, Geneva, Switzerland
Graduate Institute of International Studies, Geneva, Switzerland and University Institute Kurt Boesch, Sion, Switzerland

Abstract: This paper describes how computer methods can be used to analyze environmental resource problems and evaluate data bases about environmental trends. Resource allocation and use depend on both physical and social processes. Conflicts may erupt over overlapping claims to identical resource pools. But because of the complex interactions determining availability and use, solutions to resource conflicts are difficult to elaborate. This study adopts a formulation that focuses on the crucial linkages between the economic, socio-cultural, political, and demographic parts of a social system on the one hand, and an important resource sector, water, and its interaction with climatic processes on the other. These linkages are expressed within a dynamic simulation model which has been adapted to the particular case of the Middle East. Samples of the data collection that was undertaken as well as model calibration calculations are given in appendices. The goal of our approach is to explore competition, conflict, and possible cooperation between regions and nations in terms of resource use and allocation, particularly water, and movements of people. The argument is made that cooperative strategies and conflict resolution schemes can be effective in achieving sustainable water management in this volatile region. The underlying theoretical framework is that of game theory, which posits that decision makers are engaged in a process in which their preferences, their possible choices and those of other parties, as well as the mutual effects of choices on each participant, all affect its trajectory and outcome. The advantage of the simulation approach is that it permits the investigation of basic conflict and cooperation situations in terms of their game theoretical structures. The report presents several cooperative solutions as that can be elaborated for particular preference functions for given regions.

M. Beniston (ed.),
Climatic Change: Implications for the Hydrological Cycle and for Water Management, 379–409.

1. WATER RESOURCES AND CONFLICT

Satellite photographs of the Jordan River system show green vegetation on the Israeli side, brown desert in the West Bank–dramatically illustrating the inseparability of physical and social components of water resources (our point is illustrated by Figures 7 and 8 in Appendix V). Semi-arid and arid regions are particularly sensitive because low and uneven water availability has direct consequences for well-being. Physical characteristics determining the source of water and its spatial distribution quickly become socio-political issues of control over access and amount available. Some groups can withhold water from others or use it in vastly different quantities, thus precipitating social conflict. Uneven distribution in space, relative rates, and patterns of water use can therefore be as important as strictly quantitative considerations. These quantity and quality issues, moreover, are not just related to the physical element, water, but are embedded in larger questions of climate change, property rights, and economic development. Consequently, numerous regions of the world are particularly at risk, because of potential shortages due either to physical or social causes or because of uncertainties about access generated by rapid and far-reaching political changes. These vulnerabilities can lead to conflict in some cases of the competition resulting from these different developments. Some regions are particularly subject to such resource-based conflicts because of their physical situation and socio-political context.[1]

The Middle East, as a semi-arid region, is particularly vulnerable to changes in water quantity and quality. In addition, it has a long history of regional and international conflicts over political and social issues that have become linked to disputes over water. Two regions in particular highlight the evolution of resource-based conflicts: the Jordan River Basin, from which Syria, Lebanon, Israel, Jordan, and the Palestinian Authority draw their water, and the Euphrates River Basin which originates in Turkey and also supplies Syria and Iraq. In both of these areas, questions of sovereignty, economic development, political autonomy, and property rights are in contradiction with basic physical processes regulating the supply and distribution of fresh and safe water over time. To illustrate the problem we would like to emphasize that the data we have suggest that Israel (with the West Bank and Gaza) uses around 2300 million cubic meters whereas sustainable use should be limited to 1950 cubic meters (Lowi 1993). Jordan

---

[1]For a general discussion on resource conflicts, see Homer-Dixon (1993); for a discussion on water conflicts refer to Gleick (1993).

uses 740 to 750 million cubic meters whereas safe utilization should be only 730 million cubic meters. Such overuse can only exacerbate already existing political disputes.

The Middle East is a particularly striking example of the tensions and open conflicts that can evolve out of competition for vital resources. Geography and history have shaped the region's basic characteristics but complex environment-society interactions through time have exacerbated its problems and lead us to reflect on general physical and social aspects of resource availability and use. In this perspective, resources are the societal definition of environmental goods and thus can become part of explicit political strategies to strengthen allies and weaken opponents. Hence water resources play a crucial strategic role in the political rivalries between Turkey and Syria where they are a factor in relations with domestic opponents as well as in regional territorial disputes.

The work presented here illustrates these processes by focusing on the crucial linkages between the economic, socio-cultural, political, and demographic parts of a social system on the one hand, and an important resource, water, and its interaction with climatic processes on the other. These linkages are expressed within a dynamic simulation model which has been adapted to the particular case of the Middle East. The goal is to explore competition, conflict, and possible cooperation between regions and nations in the context of resource use and allocation, particularly water use. The argument made here is that cooperative strategies and conflict resolution schemes can be effective in achieving sustainable water management in this volatile region. The simulation model explicitly represents demographic and economic trends as well as physical factors affecting water availability, and is able to show that, despite population increase and industrial development leading to increased levels of water use, cooperative strategies can succeed in redistributing resources in a significantly more sustainable way.

2. WATER, ENVIRONMENT AND SOCIETY

The first step in building a model of water use, society, and conflict is to describe the key factors determining water use in order to achieve some predictability about how this use is likely to evolve. Such analysis incorporate the social and economic factors determining demand, including the institutional and regulatory structures that influence distribution. It is also important to better understand the supply side by addressing the geographic and climatic factors that will determine the physical availability of water.

The ultimate goal is to analyze what management, regulatory, and pricing strategies will lead to levels of efficiency and equity that reduce the potential for conflict. Factors, whether climatic, economic, or demographic, that influence the physical supply of water must therefore be seen in the context of social factors that determine how much water is available and where. The conjunction of these two processes will determine the sustainability of water resources, which in turn will influence the future peace and prosperity of regions with water scarcity.

There is a long history of work on the general problem of the use of renewable natural resources and on the particular problems posed by common pool resources such as water, which cannot be unambiguously attributed to a well-defined owner. In these cases, common property rather than private property rules apply and these must accommodate "jointness" and rivalry or crowding. Unregulated use leads to degradation of the resource because, as Dagupta and Heal have shown (1979). Without some control over the number of users or the quantity used, there is an incentive to maximize average return, since each user shares the cost of incremental production with other commons users but retains all of the profits of his share. In the case of diminishing returns, inherent in most forms of production with renewable or exhaustible resources, this means that average return is higher than marginal return. Thus, without any other form of institution or regulation, the result will be dissipation of the resource.

The importance of property rights in determining resource use is further underlined in research about private property arrangements. The seminal 1931 article by the American economist Harold Hotelling showed that efficient solutions to resource allocation require well-defined property rights and that indeed efficiency is a consequence of clear and unambiguous definition of rights and responsibilities. Chichilnisky (1994) further points out the problems of overuse of resources that result from poorly defined rights that can occur in situations where private and collective property regimes coexist in the absence of strong legal regimes.

Achieving efficient and equitable management nevertheless invokes a range of issues in addition to property rights, including preferences over present or future and effects of certain kinds of regulatory or pricing mechanisms. Poor management or misunderstanding of the fundamental causal factors involved can ultimately lead to severe conflict or environmental degradation. The disappearance of significant parts of the Aral sea in the former Soviet Union (now the region of Uzbekistan, Kazakhstan, Tajikistan, and Kyrgyzstan) is an example of the wide-ranging and unintended effects of sectoral resource and financial. Need for hard currency cash to repay the Soviet debt led the government to encourage production for export. As a consequence, most of the waters previously flowing into the sea were diverted to irrigate fields where cotton was produced as an export cash crop. The chain of events, decisions, and their consequences is documentation in Craumer (1992), Fierman (1991) or Klötzli (1994). Many of the same causes explain overuse of ground water resources (Tietenberg, 1992, 177-302).

Social conflicts both within and between societies can arise from uneven distribution of resources in space, from unequal use, or from different natural configurations (such as upstream versus downstream access). Confrontations take on international significance in arid or semi-arid regions of the globe where nations compete for scarce water resources. Current climate change predictions suggest that the risk of such conflicts may be increasing because shortages in some areas could be exacerbated as temperatures rise and the hydrological cycle is modified. An account of such water-related conflicts can be found in Gleick (1993) and Lowi (1993). A discussion of the strategies of conflict connected to water disputes is presented in Güner (1994). The problem of allocating water resources in an efficient and non-conflictual way is discussed in Luterbacher et al. (1996) Luterbacher and Wiegandt (forthcoming) and Kilgour (1995).

The interaction between property rights regimes and population dynamics is also relevant for understanding resource use and overuse. Over time, most societies devise mechanisms that tend toward an equilibrium between resources and population through inheritance rules, marriage patterns, and fertility behavior. Disequilibria are nevertheless evident, and resemble "tragedy of the commons" type situations when household strategies lead to maximization of average rather than marginal rates of returns on production (see Boserup, Lee, Lesthaeghe for discussions of the implications of these dynamics).

The confluence of uneven distribution of resources, climate change, economic growth, institutional change, demographic evolution, and political change increase the probability of regional water shortages. These in turn will aggravate social and economic differences among societies with greater conflict a likely result. Theories and methodologies that allow us to better understand the complexities of these interactions and predict their trajectories have an important role to play in devising effective policy choices.

3. SIMULATION APPROACH TO WATER CONFLICTS.

Our analysis is based on a simulation model. It contributes to a better understanding of the potential outcomes of resource relations among regions or countries because it allows exploration of the interactions among various aspects of a social system, represented in terms of its demographic, political, and economic sectors. The model is thus a tool to explore different decision paths and to identify those that might produce regional cooperation.[2] In our approach, we combine this simulation methodology with the theoretical framework of game theory which posits that decision makers are engaged in a process in which their preferences, their possible choices and those of other parties, as well as the mutual effects of choices on each participant, all affect the trajectory and outcome if the decision process.[3]

Simulation approaches are valuable because they can incorporate game theoretical structures into the models which then permits the investigation of basic conflict and cooperation situations. For example, it becomes possible to evaluate the gains achieved with cooperative solutions as opposed to non-cooperative adaptive behavior (Nash equilibria) for particular preference functions for a given region. In the case of water use, this might mean that it would be reasonable for each region to maximize income while minimizing costs from water consumption. Game theory thus constitutes an approach that can elucidate the decision processes and provide the logic for the dynamic character of the interactions among the actors. In our framework, the relations are expressed in terms of a *differential game*[4], which implies

---

[2]The situation is easier to solve in the case of prisoner's dilemma than in the case of other game structures such as chicken which have multiple Nash equilibria. For a presentation of the problem, see Ward (1993).

[3]Other studies have incorporated strategic decision-making considerations of game theory into their analyses of global change and environmental issues. See, for example Nordhaus and Yang (1996) for a discussion of national climate-change policy strategies or Kilgour and Wolinsky (1997) who engage in a general discussion of the application of game theoretical models to environmental questions.

[4]On the basics of Differential Game Theory refer to Isaacs (1965) for zero-sum games and to Kuhn and Szegö (1971) for more general approaches. The topic is also reviewed by Fudenberg and Tirole (1992). Our choice does not mean that there are no other dynamic game theoretical analyses available such as for instance event oriented sequential analysis or evolutionary game theory (on this Weibull 1995). The latter is more appropriate to the study of whole populations of decision makers. Since we are dealing with resources that can be expressed and controlled in quantitative terms and few decision makers, we thought differential game approaches made the most sense.

that countries will try to control the trajectory toward cooperation and thus have retaliatory instruments in their hands if the other side is perceived as a cheater. Applied to international relations, this perspective argues that in especially difficult international situations in which countries mistrust each other, it is important that they be able to mutually control each others' actions.

These are general considerations about power relations and negotiation that have specific manifestations in different regional and historical situations. Similarly, there are fundamental patterns of opposition that are encountered in resource conflicts which also have specific regional manifestations. We have devised the model presented below by combining simulation and game theoretical analyses based on these general principles and then applied it to the particular aspects of the Middle Eastern case.

4. OUTLINE OF THE SIMULATION MODEL

There are various ways to conceptualize society and the physical processes affecting it. Indeed, models are based on choices researchers make about the variables to include, the nature of their interactions, the time frame etc. The decisions about the content and structure of the model derive from basic theoretical constructs To design a simulation, the researcher elaborates equations that provide a quantitative description of the relationships among selected elements of the social system based on underlying assumptions and substantive and methodological choices.

In elaborating the general model structure, we found it convenient to describe both physical systems and human systems in terms of inputs and outputs in order to establish explicit linkages and interactions between the natural environment and society. In this way, some environmental outputs are inputs to human systems and some human outputs become the inputs to environmental systems. Describing social systems as input and output relationships is quite standard. It is a familiar way of looking at production and consumption, for example, because certain inputs must be consumed for production to take place which, in turn, leads to further consumption of the items produced. Thus most activities can be characterized as successively consumption and production as they move through the process. This conceptualization is consistent with formulations of the economy used by many other macro-models, both general equilibrium and dynamic. Our model can thus be considered as a paradigmatic case of a macro simulation model. The basic inspiration for our simulation model, which emphasizes population and the production and exchange of agricultural goods and on a dual sector type economy, has been the Harris and Todaro migration model

(1970). Some additional sectors have been added and economic and demographic structures have been more fully developed.

Our basic input-output design is used to build sub-models of the general framework in the form of rural and industrial sectors which dominate different but adjacent regions. The two regions interact but they are formulated separately to permit identification of the different effects of environmental changes in each area. The vulnerability of traditional rural society to environmental variables merits our more detailed description of the rural sub-model. It is described by a general structure that includes four main sectors: (1) Population, (2) Economy and Resources, (3) Cultural organization, and (4) Government (cf. Wiegandt 1979).

A demographic structure (1) divides population into age cohorts. Birth, death, and migratory processes drive the evolution of this structure. Births and migration are influenced by economic and resource conditions and by cultural factors.

The economy and resource sector (2) describes a traditional mixed agricultural system that includes both grain and animal production. This type of system is dependent on the establishment of relatively large inventories for survival and, to some extent, on the availability of government subsidies. Land, agricultural capital, the size of herds, manpower, and technology constitute the inputs for potential agricultural production. The link to the population sector is made through the labor force (manpower), whose size depends on the size of the cohorts that are of the age to work (the active population). Actual agricultural production also depends on climatic factors such as moisture, temperature, and solar exposure. These data are derived from climate models.

Cultural organization (3) in this context refers to long-term property and land tenure arrangements which depend significantly upon inheritance rules. These determine the spatial distribution of land and plot size. Cultural factors also play a role in regulating non-community members' access to local resources and in determining the ratio between individual and collective property. Consequences of changes in property rights and management may influence deforestation or reforestation rates. Even if rules themselves do not officially change, their enforcement can vary over time so practice as well as rule must be carefully studied.

The government sector (4) collects taxes, distributes subsidies, and spends resources on some types of capital (such as collective equipment). It is obviously connected with the economic and resource sector.

This marginal region representation has been linked with a model of a more advanced region that also includes population, economic, and governmental sectors. The economic sector of the advanced region contains the descriptions of the main components of the expenditure and productive structure. The human productive process is generally conceptualized as being dependent on certain factor inputs: total production[5] is a function of significant factor inputs such as capital, labor, technology, raw materials, energy, and climate (plus other less salient inputs).[6]

All the above factors of production can themselves be described as outputs of processes specified by other relationships. Capital stock, for example, depends on investments which can be defined as part of consumption, thus linking production with consumption. The labor force obviously derives from demographic processes since the potential labor force is a subset of the active population. Thus labor can be seen as an output of an age cohort-structured demographic model that includes in and out-migration. Similarly, materials and energy are a part of productive processes taking place at an earlier stage. The general model describes the interaction among variables at a given moment. Change is introduced by the independent variables that result from dynamic processes. Technology, for example, is usually defined as a trend; capital stock increases or decreases as investments vary; population changes as marriage, birth, and death rates evolve and therefore the size of the cohort defined as active (working) population will also vary. Water availability depends on investments made for its retention (such as in dams and artificial lakes, for instance) and distribution.

Climate, one of the significant environmental variables considered in our approach can influence various elements of the productive process. Climate is itself determined by a physical climate model partly dependent on parameters generated by human activities. The same can be said for water availability. In addition, water availability and climate are obviously interrelated.

---

[5]Total production can be defined for various levels of analysis. It is commonly calculated for particular countries but it can be aggregated to form a global measure or broken down into regional measures

[6]Here, too, we follow standard descriptions applicable to a whole variety of macro economic simulation models. The list of inputs to production is far from complete: factors like the quality of the transportation infrastructure, the political stability of society, its literacy level, etc. are all relevant. Their inclusion in or exclusion from the model depends on specific research goals and the underlying theoretical framework. Here we chose to limit ourselves to the factors enumerated above.

The definition of production in terms of different factors is very general and describes a wide range of productive processes. More interesting for understanding climate/society linkages, is the division of total production into agricultural and non-agricultural production which together constitute total production. Agricultural production, is dependent on agricultural capital, land, labor, technology, energy, and, ultimately, also the weather or the climate, and water availability. Energy consumption is also an important variable linking society to climate since most greenhouse gas emissions are due to one form or another of energy consumption. Because of differential greenhouse effect of different fuels, it is useful to distinguish among petroleum, natural gas, coal, and electricity. Substitution effects are particularly important because of their impact on greenhouse gas emissions. Specific demand relations and relative price levels of one fuel compared to the other determine the evolution of these consumption levels.

A representation of everything that is purchased and consumed, i.e. total expenditures, in a given national entity highlights the interaction of all these mechanisms. All variables that make up total expenditures measure consumption in a general sense since investments are expenditures necessary to maintain and increase the capital stock; government expenditures, represent consumption and investments by the government; exports, represent expenditures by foreigners of domestic goods and imports are expenditures by nationals of foreign goods. Because this latter item is spent abroad, it appears on the negative side in a national accounting system. This formulation is dynamic because changes in one of the elements will produce a change in the level of total expenditures. Consumption, investments, government expenditures, exports, and imports depend on past levels and will vary according to changes in the categories of demand that they represent. Such variations will produce positive or self-amplifying changes, whereas constraints due to the production structure will introduce negative feedback or self-correcting effects. Indeed, one cannot consume more than is produced or less than is produced without consequence. Differences in population levels and structure will also affect the long term demand evolution of demand. At the same time, population increases or decreases can also be interpreted as resulting in part from investment decisions by parents about whether or not to have children, making these processes also part of the general demand structure[7].

---

[7]The model described above is constructed as a generic structure based on reasonable parameter values, that can be potentially adapted to various empirical situations. Currently, a more elaborate model is being statistically adapted to the Middle East.

5. SIMULATING ENVIRONMENTAL CHANGE AND FRESH WATER RESOURCES

Based on this generalized modeling structure, it is possible to focus explicitly on the *water sector*. Water resources must be considered simultaneously from the point of view of physical availability, distribution, and consumption. Rural and industrial/urban differences in all of these categories must also be taken into account. Moreover, rural consumption can be further distinguished into household and agricultural uses. In the same way, industrial consumption can be separated into use by household and by industry. All water consumption generates waste that must be reabsorbed into the local environment and eventually into the hydrological cycle. Taking into account distribution requirements, physical availability of water resources either through natural or man made reservoirs, as well as regional in and outflows, provides a complete picture of the dynamics of fresh water resources.

All the above considerations have led to the formulation of water sub-models for both rural and urban/industrial sectors. This general model framework, with its emphasis on the water sector, can be used to examine interregional and intersectoral water resource use within specific countries. The model can address various management issues related to water quality and water conservation as well as to equitable distribution among domestic, agricultural, and industrial uses. The examination of various scenarios emanating from different simulation runs can explore impacts of different pricing mechanisms or policies to control quantity and quality. This particular simulation model is designed particularly to study international, transboundary conflicts about water and dispute resolution schemes[8].

As in the general modeling framework, water is conceptualized in terms of supply and demand.[9] Climate change enters into the formulation as a factor that has a potentially destabilizing effect on the relation between supply of and demand for water. High temperatures and periods of drought will increase the demand for water at the same time that these factors diminish the supply through their effects on the hydrological cycle. In rural and industrial regions, various uses of water affect its quantity and quality,

---

[8]The specific mathematical structure is available upon request from the authors. See also Luterbacher and Wiegandt (1995) and (1997).

[9]The basic features of the model are illustrated in a flow diagram.

ultimately influencing the water cycle itself. Water use is thus closely tied to all aspects of global environmental change.

The basic model of the marginal agricultural system incorporates climate change as it manifested through flooding and temperature and moisture changes. Water in the form of irrigation is specified separately from precipitation in the model because irrigation practices interfere much more directly with the hydrological system of a region and influence a variety of ecological variables such as salinity of the water table and the soil. In this context, water availability acts as a type of capital stock (in the form of wells, pipes, and dam systems) that must be maintained by proper investments which depend ultimately on other determinants of capital stock, such as interest rates. It has been designed to apply to such countries with large, relatively unmechanized agricultural sectors and large rural populations as those in the Middle East[10]. This model of socio-economic, climate and water resources has been adapted to the political entities of the Jordan river system: Israel, Jordan and the Palestinian authority.

To construct the simulation model of water use, the following assumptions were made:[11]

- Each Middle Eastern region in the Jordan River system, with the exception of Israel, is represented in terms of an industrial (urban) part and a rural traditional part. Israel is represented as an industrial sector only.
- The agricultural production functions of all the regions specified in the model are influenced directly by water consumption in both the industrial and agricultural regions.
- Maximization over time of utility functions defined as the difference between agricultural production and costs of water consumption (implicit or explicit even if the end consumer is subsidized) is postulated for each region. Each utility function is defined as the difference between agricultural production produced locally minus the total cost of water consumption defined as the per unit price of water in each region multiplied by physical consumption of local water (imports from the other region are subtracted). Sequential maximization of the respective

[10]A first attempt at applying this model to an empirical situation focuses on the Middle East and more specifically, the relations between Israel, the Palestinian Authority and Jordan.

[11]These assumptions also take a mathematical expression presented in Appendix iv.

utility functions of each region can be achieved in this way, thus replicating the basic notions included in the sequential structures discussed above.

- Limits to water consumption are introduced as the result of a dynamic bargaining process. In other words, the regions will progressively agree to maintain their agricultural water consumption within some bounds dependent on the other region's water use.
- Maximizing the utility functions defined above with respect to the constraints defined by the bargaining process specified in point (c), defines a *differential game* similar to a sequential Prisoner's Dilemma game (see figure 1) which leads to a *Nash solution*[12] in which regions will minimize their water uses (a formal proof which shows that this bargaining mechanism works in the case of two regions is presented in Appendix III). Our analysis leads here to a result which is somewhat similar to the one of the *Folk Theorem* (Fudenberg and Tirole 1992:150-160 and 192-197). However, if the regions are too asymmetric in their access to water resources, as in an upstream downstream relationship, the dynamic bargaining process mentioned above will not work.

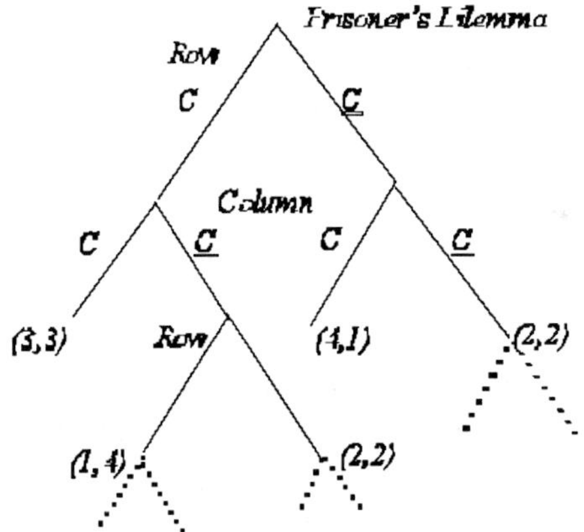

Figure 1. The Sequential Prisoner's Dilemma. The sequential Prisoner's Dilemma game where payoffs are given from 4 (most preferred) to 1 (least preferred) illustrates the notion that retaliatory strategies where both parties will end up using conflict moves (C) resulting in a 2, 2 outcome always exist if there is no foreseeable end to the sequence. The existence of these retaliatory strategies can ensure the existence of a cooperative equilibrium Where both parties make collaborative moves resulting in 3,3 payoffs.

---

[12]A *Nash equilibrium* corresponds to an outcome from which no participant has an unilateral incentive to withdraw. Several *Nash equilibria* might exist for a given interaction.

These results are outlined in Graphs 1 and 2 for a two region case. They depict base runs in water consumption in the industrial and marginal regions (129 is the optimized run whereas 133 is the base run). The graphs show that cooperative limitation of water consumption will slow down water demand in the industrial region. As a result, some reserves will be maintained longer and the decrease in water availability in the rural region will be less steep despite increasing drought.

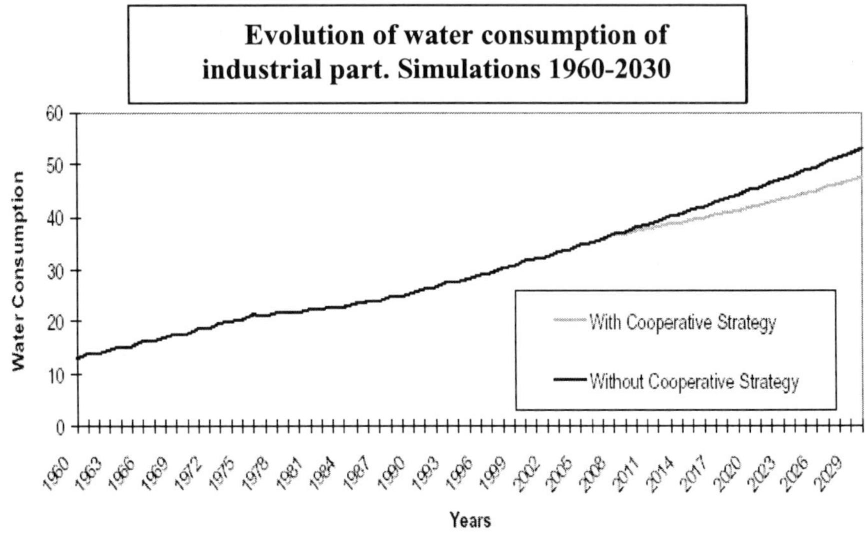

Graph 1.

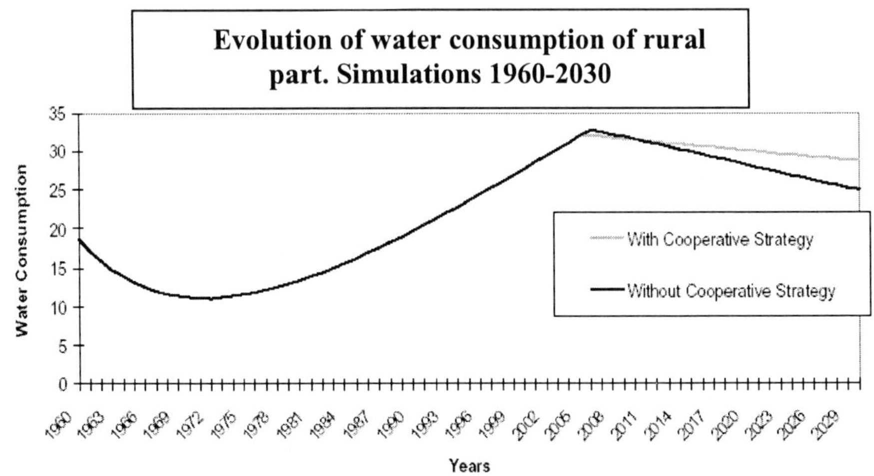

Graph 2

6. THE CASE STUDIES: THE JORDAN RIVER BASIN

As we have seen earlier, the extraction or use of a common resource, such as water, which is slowly exhausted or over- utilized by two or several countries drawing on it together, is a central problem of conflict and cooperation. Our present research focus on the Middle East is a good example of such a common resource problem. It is a complex political environment which merits study in itself. However, given our focus on methodology and strategic decision making analysis in the present chapter, we give here only a very simplified description of the political, geographical, and strategic situation. We nevertheless believe the data are adequate to capture key relationships, construct and run the model, and derive policy-relevant results.[13]

We decided to focus our empirical case on the Jordan river basin. The other major water conflict area in the Middle East, the Euphrates and Tigris river basins are more complicated to deal with and to resolve. It depends to a large extent on wider regional problems such as the respective status of the Syrian and Iraqi regimes and their relations and also on the willingness of the United States to force one of their major allies, Turkey, into a settlement on the water question, an unlikely possibility for the moment. The de facto Turkish Israeli military alliance complicates the issue even more. The Jordan river basin conflict affects directly the peace process between Israel and the Palestinians as well as between Israel and some of its important neighbors (Jordan). Moreover, there are no clear-cut upstream or downstream powers for this part of the Middle East and all entities have some retaliatory powers against each other. Under these conditions, cooperative solutions to the water problem in the Jordan river basin are possible.

The Jordan river basin has already a long history of water management problems with political overtones After the first Arab-Israeli war, Israel decided to promote agriculture and thus proceeded to instal the necessary irrigation facilities. It started the National Water Carrier in 1953 which took water from the Lake of Tiberias in the North to the Negev area in the South.

---

[13]We are fully aware of the fact that the situation we are describing encompasses various levels of complexity. These are currently the object of an ongoing research project but we believe that for the present purpose our outline captures the essential features of the different conflict structures.

The National Water Carrier carries their flow through open canals, tunnels, and water reservoirs. Allegedly the system also channels water for artificial and underground storage which means that mostly during the winter water from the Jordan conduit is recharged into the aquifers in the central region of Israel. This project went on despite heavy protests from Israel's Arab neighbors and almost lead to a military conflict. The occupation of the West bank and the Gaza strip in 1967 gave Israel control over the main aquifers of the region and some additional pumping and irrigation projects were initiated.

These efforts together with a rise in the standard of living lead to huge water consumption increases in Israel. Miriam Lowi (1993:153) has found that Israeli water consumption in the Jordan River Basin in 1991 was between 2100 and 2200 millions of cubic meters, whereas renewable fresh water supply was only 1950 million cubic meters. Agricultural needs account for about 70 percent of total consumption. To these figures it is necessary to add the 125 millions of cubic meters consumed by the Arab population of the West Bank which uses comparatively small amounts but whose needs due to among other factors an increasing population (refer to Graphs 3 and 5) are getting bigger.

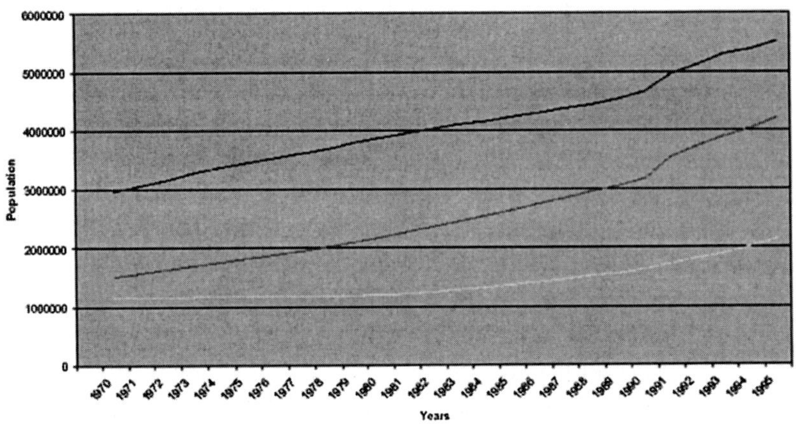

Graph 3. Evolution of Population of Israel, Jordan, and Palestinian Territories (Gaza + West Bank)

All these trends lead to a growing deficit in terms of sustainability of water use. This water deficit has seemingly existed since 1985 and continues to this day. Jordan, according to the same source (Lowi 1993: 159), had a total consumption of 740 million cubic meters, while a safe supply stood at 730 million cubic meters in 1990. It is expected that this deficit will increase even more over time. Various Israeli proposals to increase water supply and to initiate a better management of the major water resources have been put forward. It seems however, that the Israelis want to continue to control most water resources of the region even through the peace process and so far are reluctant to let the West bank and Gaza control and extract more for their growing needs. Thus these Israeli proposals constitute major stumbling blocks in the peace process. More cooperative arrangements for water resources are clearly needed for the peace process to to eventually succeed.

We will now see how analyses conducted with the help of our model can contribute:

- To a better understanding of the basic trends of the region and thus to the specific needs for water resources.
- To elaborate strategies for the initiation of a cooperative process in the sharing of water resources between the main territorial entities of the region.

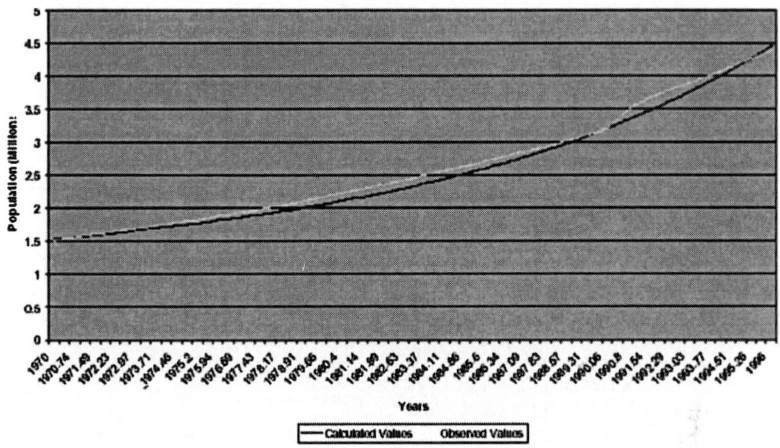

Graph 4. **Population Jordan: Observed and Calculated Values**

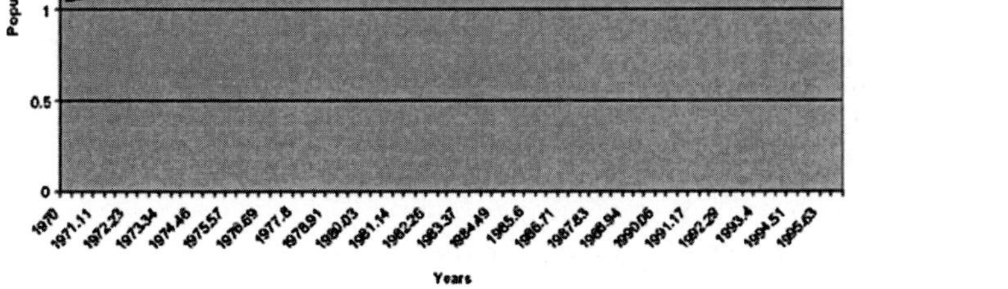

Graph 5. Palestinian Population: Observed and Calculated Values

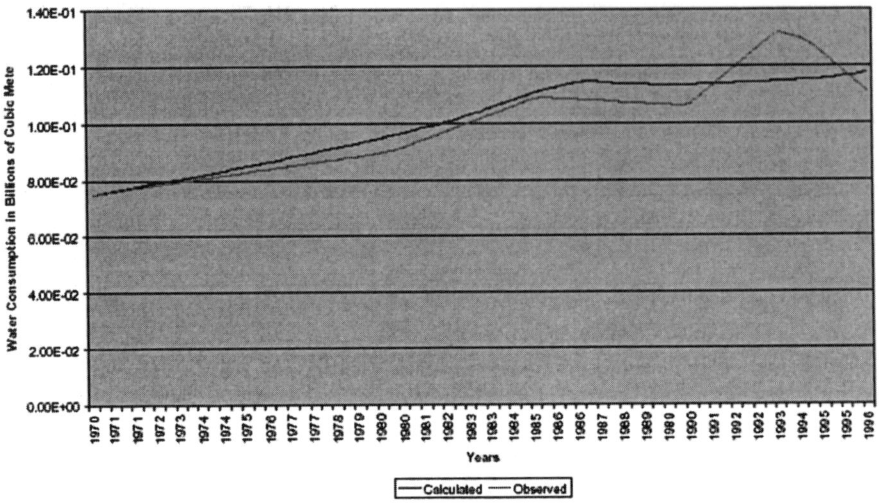

Graph 6. Calculated and Observed Values of Israeli Industrial Water Consumption

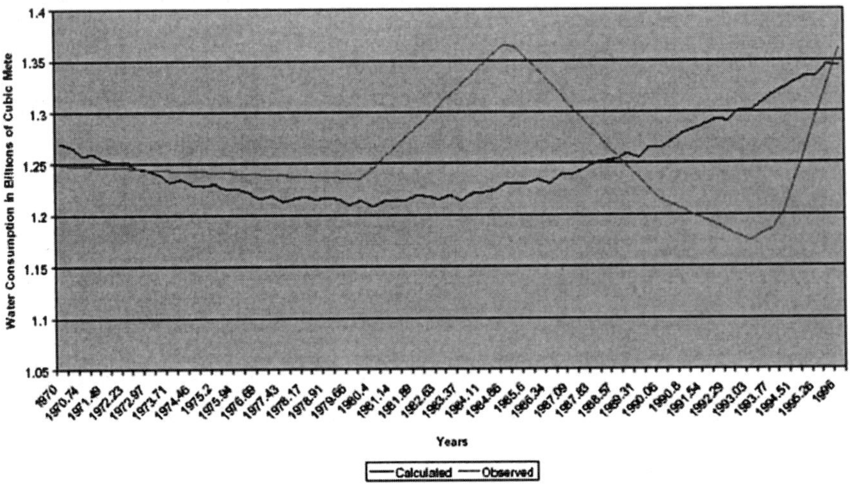

Graph 7. Calculated and Observed Values of Israeli Agricultural Water Consumption

7. SIMULATION RESULTS AND COOPERATIVE SOLUTIONS EMERGING FROM A GAME THEORETICAL PERSPECTIVE

The model that was elaborated for the project is capable of reproducing in satisfactory way trends in indicators. Some results in terms of agreement between observed and calculated values are given for the Jordanian population (Graph 4), the Palestinian population (Graph 5), and the Israeli water consumption for agriculture and industry (Graph 6 and 7). The model results are satisfactory given the data problem for some indicators.

As mentioned earlier, the model allows also for the representation of a bargaining process based on game theoretical notions. Optimal strategies resulting from such a process can be calculated through the steps presented earlier in the section on **Simulating Environmental Change and Fresh Water Resources**. They suggest a sequential response to the water problems in the Middle East and confirm the results presented in Graphs 1 and 2: If Israelis and Palestinians and Jordanians initiate a cooperative approach characterized by voluntary quantitative restriction schemes that could be put into place at a relatively high level by Israel on the one hand and, to a lesser degree, by the Palestinians and Jordanians on the other.[14] Such restrictions could make sense initially because the current conflictual relations between the various parties make it difficult to envisage the establishment of rules that more complex schemes such as the establishment of a water market system would require. We have examined bargaining conditions necessary for the successful negotiation of an agreement where parties would accept voluntary restraints on their water consumption while maintaining a goal of maximizing agricultural production (Luterbacher and Wiegandt 1995, 1997a, 1997b) and present the details of this methodology in the various appendices. It is worth mentioning that the bargaining analysis results presented in Graphs 1 and 2 and imply **that mutual restrictions hold despite trends that could otherwise offset them such as population or agricultural production increases.**

In particular our approach allows us to compare the merits of three proposed solutions to the Jordan river basin problems. These solutions which are not necessarily exclusive of each other are the following:

1) Water could be traded between areas according to market mechanisms. However, standard results concerning market equilibria (Coase 1960, Chichilnisky 1994) show that market mechanisms and trading might lead to inefficient and environmentally damaging outcomes when property rights are ill defined, absent or difficult to establish as in the case of externalities associated with common pool resources (refer to the beginning of our paper and the discussion on commons). Also, since trading concerns a vital common pool resource, a certain degree of trust[15]

---

[14]The Palestinians and Jordanians consume significantly less water.

[15]The importance of this factor should not be exaggerated, however. One must note that the existence of Hong Kong always depended on the supply of water from Communist China.

must prevail between the various participants to such a process, in particular as far as the establishment of clear cut property rights is concerned[16].

2) Technological solutions could enhance water supply. Various schemes have been proposed, ranging from pipe-lines from Turkey to desalination of sea water. All of them are very expensive.
3) The voluntary quantitative restriction schemes presented above which have the advantage of being a) relatively inexpensive and b) not too complex as an initial step.

8. CONCLUSION: WATER USE AS A PROBLEM OF COOPERATION AND NEGOTIATION

The common pool problem that characterizes use of water resources requires a set of rules for their use even in peaceful situations to prevent the over-extraction by some users at the expense of others since, in the absence of regulation, each individual has an incentive to overuse common pool resources.[17] Developing such rules in conflictual situations poses additional problems. Cooperative strategies and conflict resolution schemes can nevertheless be elaborated and can be effective in such instances.

In a multi country situation, common resources are used by several populations or labor forces. If countries try to maximize their surplus independently of each other, they will each end up in a suboptimal position. It should be noted that if the countries act in isolation, they will ignore each other's extraction efforts in their maximization calculations. Such behavior reduces each country's long term surplus since its management occurs without considering the other country's actions. It is quite clear that an agreement between all of them would improve the utility of all parties. Establishing extracting quotas or a pricing scheme that would put countries' surpluses above their purely competitive values, for example, would eliminate each country's negative externality. This situation is a kind of prisoner's dilemma because to extract more than the other who obeys the agreement increases the surplus and, therefore, the utility of the over-

[16]The water trade solution has been advocated by Franklin Fisher (1995) and the Israeli economist, Mordechai Shechter (1997).

[17]This "tragedy of the commons" has been described by Hardin (1968).

extracting country above the agreement level. If, however, as in Prisoner's Dilemma situations retaliatory strategies exist and are in the interest of parties to implement if some other parties cheat, a cooperative scheme can be maintained and improved upon over time even if population or resource trends exist that threaten the existence of the agreement. In the case of the Middle East and the Jordan river basin, we have shown that such strategies leading to the maintenance of a cooperative arrangement can exist under reasonable conditions and can be used.

APPENDIX I: WATER RESOURCE MODEL

Water resources must be considered simultaneously from the point of view of physical availability, distribution, and consumption. Rural and industrial/urban differences in all of these categories must also be taken into account. Moreover, rural consumption can be further distinguished into household and agricultural uses. In the same way, industrial consumption can be separated into use by household and industry. All water consumption generates waste that must be reabsorbed into the local environment and eventually into the hydrological cycle. Taking into account distribution requirements, physical availability of water resources either through natural or man-made reservoirs, as well as regional in and outflows, provides a complete picture of the dynamics of fresh water resources. A pre-existing model linking climate trends and political social and economic factors, CLIMSOC has been supplemented by the inclusion of a complete water sector.

The water sector dynamic models for rural and urban/industrial regions are presented below. In the equations, a dot above a variable name means a derivative with respect to time. Our water sector models are thus conceived, like the rest of CLIMSOC as systems of continuous time dynamic relations (differential equations)[18]. A glossary of terms is presented in Appendix II:

Rural region

CWR=CWRH+CWRA; total water consumption
WASTER=WASTEHR+WASTEA; total waste water rural
CWRH=WPR1*POPM+WPR2*YA-WPR3*PWR-WPR4*CWRH
+WPR5*TEMP; water consumption rural households
WASTEHR=WPR6*CWRH; waste water rural households

[18]The use of continuous time (differential equations) based models is used in various types of socio-economic modeling (cf. for instance Gandolfo 1995). The methodology outlined above is described extensively in Luterbacher et al. 1987.

CWRA=WPR7*YAS-WPR8*PWR-WPR9*CWRA-WPR10*MOIST[19]; water consumption agriculture
WASTEA=WPR11*CWRA; waste water agriculture
PWR=WPR12*(CWR/WR); the price of water
WR=WPR13*WDR*WRESR; water availability link to water cycle model
WDR=IWDR-WPR14*WDR; water distribution rural
IWDR=WPR15*YAS+WPR16*POPM-WPR17*INT; water distribution investments as function of income, existing stock and interest rate
WRESR=WPR18*MOIST+WPR19*FLOWR-WPR20*CWR -WPR21*WASTER+(1-WPR21)*WASTER; total freshwater resource, rural sector
FLOWR=WPR22+OM*SIN(ALPH*0.92845*T)+WPR23*TEMP; net streams flow, rural sector

Industrial Region

CWI=CWIH+CWII; total water consumption industrial
WASTEI=WASTEHI+WASTEII; total waste water industrial sector
CWIH=WPI1*POP+WPI2*Y-WPI3*PWI-WPI4*CWIH+WPI5*TEMP; water consumption households, industrial sector
WASTEHI=WPI6*CWIH; waste water households, industrial sector
CWII=WPI7*Y-WPI8*PWI-WPI9*CWII; water consumption by industry
WASTEII=WPR10*CWII; waste water by industry
PWI=WPI11*(CWI/WI); the price of water link to water cycle model
WI=WPI12*WDI*WRESI; physical freshwater availability
WDI=IWDI-WPI13*WDI; water distribution industrial
IWDI=WPI14*Y+WPI115*POP-WPI16*INT; water distribution investments as function of income, existing stock and interest rate
WRESI=WPI17*MOIST+WPI18*FLOWI-WPI19*CWI-WPI20*WASTEI+(1-WPI20)*WASTEI; total freshwater resource, industrial sector
FLOWI=WPI21+OM*SIN(ALPH*0.92845*T)+WPI22*TEMP; net streams flow, industrial sector

[19] MOIST is an index which is positive for higher than average humidity and negative for lower than average humidity

APPENDIX II: GLOSSARY OF TERMS

Rural Region

CWR	= Total water consumption rural
WASTER	= Total waste water for rural sector
CWR	= Water consumption rural households
WASTEHR	= Waste water rural households
CWRA	= Water consumption agriculture
WASTEA	= Waste water agriculture
PWR	= Price of water rural
WR	= Water availability in physical terms link to water cycle model
WDR	= Water distribution rural
IWDR	= Water distribution investments in the rural sector as functions of income, existing stock and interest rate
WRESR	= Rural total freshwater resource
FLOW	= Net streams flow coming in and out of the rural part

Industrial Region

CWI	= Total water consumption industrial
WASTEI	= Total waste water, industrial sector
CWIH	= Water consumption households, industrial sector
WASTEHI	= Waste water industrial sector households
CWII	= Water consumption by industry
WASTEII	= Waste water by industry
PWI	= Price of water in industrial sector link to water cycle model.
WI	= Physical fresh water availability
WDR	= Water distribution industrial
IWDI	= Water distribution investments in industrial sector as functions of income, existing stock and interest rate
WRESI	= Total freshwater resource, industrial sector
FLOWI	= Net streams flow, industrial sector

APPENDIX III: EXPLORING COOPERATION STRATEGIES ON WATER USE THROUGH TIME: PROOFS.

Problems of cooperation through time arise in terms of the extraction or use of a common resource such as water which is slowly exhausted or over utilized by two or several countries drawing on it together. An example can be found in the case of the Mediterranean sea or landlocked seas or lakes which represent a common resource for many countries in terms of fishing, transportation, etc. The following considerations will show how cooperative strategies and conflict resolution schemes can work in such instances. These theoretical suggestions can very easily be combined with the modeling structure proposed above because they refer essentially to the same concepts:

In a two-country or two-region situation, the utilization of common resources is undertaken by two populations or labor forces. Even if both countries are trying to maximize their surplus independently of each other, they will end up in a suboptimal position. This can easily be shown in the following way. Let's assume that their respective output curve is S shaped and can be represented in terms in terms of some type of production function $Q(R,L)$ where R represents some almost fixed endowment in resources (such as an underground aquifer for instance) and L the labor inputs necessary for the extraction. We could also introduce other inputs such as capital, but this would complicate things uselessly at this point. Since population can be taken as a proxy for labor, we will redefine Q as *Q(WR,P(population))*. Incidentally, P instead of representing population and thus labor could also be thought as representing any type of extractive input. Since two countries are extracting the water resource, each of them will produce a level of output given by respectively $Q_I\ (WR,\ P_I)$ and $Q_R\ (WR,P_R)$ (if the countries are called Industrial and Rural). Obviously we assume here that $\partial Q_I / \partial P_I$ and $\partial Q_R / \partial P_R > 0$ and that eventually $\partial^2 Q_I / \partial P_I$ and $\partial^2 Q_R / \partial P_R < 0$. Each country will also face a cost curve, given here just by population needs (or labor or other extractive costs), mP_I for the Industrial part and nP_R for the Rural part. It seems reasonable to assume that respective utility functions for Industrial and Rural will be given by the respective surpluses which are:

$$U_I = Q_{I'} - mP_I - aQ_R \text{ and } U_R = Q_R - nP_R - bQ_I$$

In the above expressions, partial derivatives are positive with respect to one's own factor input and negative with respect to the other. Thus maximum surpluses can be calculated as:

$$\frac{\partial U_I}{\partial P_I} = \frac{\partial Q_I}{\partial P_I} m = 0$$

and

$$\frac{\partial U_R}{\partial P_R} = \frac{\partial Q_R}{\partial P_R} n = 0$$

Given the assumptions about second partial derivatives, these are obviously maxima.

It should be noted that if the two countries act in isolation, they will ignore each others' extraction efforts in their maximization calculations. Such behavior reduces each country's long term surplus since its management occurs without considering the other country's actions. This can easily be seen if one assumes that Q_I and Q_R are of Cobb-Douglas type. In that case Q_I for instance is $Q_I = WR^{(1-a)}P_I{}^a$. A maximum surplus is thus obtained at $m = aWR^{(1-a)}P_I{}^{(a-1)}$ which gives a value for $P_I = (aWR^{(1-a)}/m)^{1/(1-a)}$. If we now assume now that there is an agreement between the two countries to limit extraction by allowing only a fraction of P_I (such as rP_I) to be used as input, the surplus will be greater since the optimal value of P_I will be lower namely:

$$P_I\,(with\ the\ agreement) = (raWR^{(1-a)}/m)^{1/(1-a)}$$

It is quite clear that an agreement would improve the utility of both parties by, for instance, establishing extracting quota or a pricing scheme that would put countries' surpluses above their purely competitive values, each country's negative externality would be eliminated. In this case, however, we may face the prisoner's dilemma type situation that we mentioned previously, since extracting more than the other who obeys the agreement increases the surplus and therefore a particular country's utility above the agreement level. A mechanism has to be envisaged that will ensure continued cooperation.[20] It is very important to define such a mechanism in terms of a differential game which implies that both countries will try to control the trajectory toward cooperation. In the time domain case, given the fact that cooperation would mean that parties would refrain from overusing a water resource base, the adjustment means that parties would react negatively toward each other's extraction efforts. We would thus have countries maximizing:

---

[20]The situation is easier to solve in the case of prisoner's dilemma than in the case of other game structures such as chicken which have multiple Nash equilibria. For a presentation of the problem see again Ward (1993).

$$\text{Max}(\int(U_I e^{-\theta_1(t-\tau)})d\tau;\ \text{Max}(\int(U_R e^{-\theta_2(t-\tau)})d\tau$$

but subject to the following constraints:

$$\frac{dP_I}{dt} = \varphi P_R ; \frac{dP_R}{dt} = \psi P_I$$

Therefore, the control variables _ and ψ should be at a minimum. We can now demonstrate that this should be the case: that both countries will use maximal restraint in the use of resources if the marginal utility of their productivity is equal to the marginal disutility generated by the other country's productivity of the resource (i.e. _ and ψ are at a minimum). This proof is given in proposition 1.

So we can demonstrate in this case that cooperation can work. We want to prove the following proposition:

Proposition 1: Both countries will use maximal restraint in the use of resources if the marginal utility of their productivity is equal to the marginal disutility generated by the other country's productivity of the resource. The two countries will thus spontaneously over time set _ and ψ at their minimum values.

Again this proposition is valid under the following two conditions:

1. $\partial U_I/\partial P_I = \partial U_I/\partial P_R$,
2. _ = ψ

For both countries the decision made by the other to restrain from the use of the common resource has to be greater or equal than his own which implies _ $\geq \psi$ for the Industrial part and $\psi \geq$ _ for the Rural part which implies _ = ψ. This particular condition suggests a bargaining mechanism that can be envisaged to equalize ϕ and ψ if they are not the same.

Preliminary proposition: If I and R are not too different in terms of their bargaining power[21], their respective ϕ and ψ control variables will converge.

---

[21]An asymmetry in bargaining power would mean that one country could pursue her own policies in terms of the issue under consideration, here water consumption, without being affected by the policies of the other. This is essentially the case if we have a War of attrition type situation. This case is illustrated by Turkey who has much more power than Syria as far as water consumption is concerned because of her upstream position.

The proof for the convergence of ϕ and ψ is the following: If ϕ and ψ are different, then P_I and P_R will be different. If these are different, U_I, respectively U_R will be lower, since a lower use of P_I or respectively P_R will not be compensated by a same decrease in the other variable. Each countries rulers will therefore have an incentive to adjust the values of ϕ and ψ until they are equal. This mechanism suggests also the existence of an equilibrium under conditions of incomplete information.

The function $\alpha_i(t)$ below has to be smaller than 0 for _ and ψ to be at their minimal values.

$$\alpha_i(t) = \sum_{k=1}^{n} \lambda_k b_{kt}(x,t)$$

where the vector $b_{kt}(x,t)$ represents the right hand side of the constraint equations i.e. here

$$-\_P_R \text{ and } -\psi P_I$$

Since these are negative by definition for positive _ and ψ we only need to look at the sign of the sum of the λ solutions of the costate equations making sure that these are positive which will guarantee $\alpha_i(t)<0$.($\partial U_I/\partial P_I = \partial U_I/\partial P_{R,}\_ = \psi$). This is in fact the case because given the assumptions we made about the partial derivatives of the extraction function Q. The following costate equations can be written:

$$\dot{\lambda}_2 = \varphi \lambda_1 + \partial U_I / \partial P_R e^{\theta_1 \tau}$$

for the Industrial Part, and :

$$\dot{\lambda}_{2'} = \varphi \lambda_{1'} + \partial U_R / \partial P_R e^{\theta_1 \tau}$$

for the Rural Part.

The solutions for the λ's can now be computed. Since the λ's and λ's have the same equation structure it is sufficient to look at the solutions for the costate equations of the Industrial Part (i.e. the λ's only). These are:

$$\lambda_1(t) = C_1 e^{\sqrt{\psi\varphi}\, t} + C_2 e^{-\sqrt{\psi\varphi} t} +$$

$$\frac{1}{2}\int [\, e^{\sqrt{\psi\varphi}(t-\tau)} - e^{-\sqrt{\psi\varphi}(t-\tau)}\,]$$

$$[\frac{d/d\tau(\partial U_I / \partial P_I e^{\theta_I \tau})}{\sqrt{\psi\varphi}}$$

$$- \psi\, \partial U_I e^{\theta_I \tau} / \partial P_R] d\tau$$

and

$$\lambda_2(t) = C_3 e^{\sqrt{\psi\varphi}\, t} + C_4 e^{-\sqrt{\psi\varphi} t} +$$

$$\frac{1}{2}\int [\, e^{\sqrt{\psi\varphi}(t-\tau)} + e^{-\sqrt{\psi\varphi}(t-\tau)}\,]$$

$$[\frac{d/d\tau(\partial U_I / \partial P_R e^{\theta_I \tau})}{\sqrt{\psi\varphi}}$$

$$- \varphi\, \partial U_I e^{\theta_I \tau} / \partial P_R] d\tau$$

Adding $\lambda_I(t)$ and $\lambda_2(t)$ together makes all the terms under integral vanish above and leaves only positive terms. This guarantees minimal values for _ and ψ for the Industrial Part. The same can be demonstrated for the Rural part which completes the proof. Minimal values for _ and ψ will also establish a stable or quasi-stable outcome for the dynamic interaction system that is envisaged above. The resulting equilibrium is thus also dynamically stable.

APPENDIX IV: SIMULATION APPROACH FOR A TWO REGION CASE

In order to simulate such interactions within our model, the following assumptions were made:

1. Instead of considering P_N and P_S, the agricultural water consumption of two regions *CWRA* and *CWIA* were considered. The agricultural production functions of two regions are influenced directly by these two components. In the model we have:

YA= f(......,CWRA) i.e the agricultural production function of a rural part is a function of water consumption. And also:

PA = g(.......,CWIA) i.e. The agricultural production function of an industrial part is a function of water consumption for agriculture.

2. Maximization over time of a utility function defined as the difference between agricultural production and costs of water consumption were postulated for each region:

$U_I = PA_I - PWI*CWIA_I - M_i* YA_R$ and $U_R = YA_R - PWR*CWRA_R - M_R* PA_I$ [22]

In other words, each utility function is defined as the difference between agricultural production produced locally minus the total cost of water consumption defined as the price of water in each region multiplied by physical consumption minus imports from the other region.

3. Limits in terms of water consumption were introduced as the result of a dynamic bargaining process. In other words regions I and R will maintain their agricultural water consumption within some boundaries that will depend on the other region's water use so that:

$$\frac{d\,CWIAL}{dt} = \varphi\,CWRAL\,;\,\frac{d\,CWRAL}{dt} = \psi\,CWIAL$$

Maximizing the above utility functions with respect to the constraints defined by the differential equations specified above will lead to a Nash equilibrium where both regions will minimize their water uses. This framework is also applicable to the interactions between more than two regions.

9. REFERENCES

Chichilnisky Graciela (1994) "North -South Trade and the Global Environment" **The American Economic Review**, 84 (4): 851-874.

Coase Ronald (1960) "The Problem of social Cost" **The Journal of Law and Economics**, 3:1-44.

Craumer, Peter R. (1992), Agricultural Change, Labor Supply, and Rural Out-Migration in Soviet Central Asia, in **Geographic Perspectives on Soviet Central Asia**, edited by Robert A. Lewis, London/New York, Routledge, pp. 132-180.

Dasgupta P. S. and G.M. Heal (1979)**Economic Theory and Exhaustible Resources** . Cambridge: Cambridge University Press.

Daoudy, M. (1996) Entre le Tigre et l'Euphrate: Une négociation en eaux troubles. **Les cahiers de l'orient**, 44, 4th Trimestre.

---

[22]Similar definitions can be found in Nordhaus and Yang (1996) who use consumption of goods instead of production.

Fierman William,(1991), The Soviet "Transformation" of Central Asia, in **Soviet Central Asia. The Failed Transformation**, edited by William Fierman, Boulder/San Francisco/Oxford, Westview Press, pp. 11-35.

Fudenberg, D. And J. Tirole (1992) **Game Theory** Cambridge, Mass: MIT Press.

Gandolfo Giancarlo ed. (1993) **Continuous -Time Econometrics: Theory and Applications** London ; New York : Chapman and Hall.

Gleick Peter H. (1991),Water and Conflict **International Security,** 18, 1, 1993: 79-112.

Guner, S. (1997). The Turkish-Syrian War of Attrition: The Water Dispute. **Studies in Conflict and Terrorism,** 20 (1), 105-117.

Homer-Dixon T. F., J.H. Boutwell and G. Rathjens Environmental Change and Violent Conflict (1993) **Scientific American**, 268, 2,: 16-23."

Hardin, Garret, (1968) The Tragedy of the Commons, **Science,** 162:143-48.

Harris, John and Michael Todaro, (1970) Migration, Unemployment and Development: A Two-Sector Analysis, **The American Economic Review**:. 3-27.

Michael A. Heller & Rebecca S. Eisenberg, (1998)"Can Patents Deter Innovation? The Anticommons in Biomedical Research", **Science** 290: 698.

Isaacs, Rufus (1965) **Differential Games**. New York: John Wiley, SIAM Series.

Kilgour, M. & Wolinsky, Y. (1997). Game theory and international environmental policy. Ms.

Klötzli Stephan, (1994) The Water and Soil Crisis in Central Asia- A Source for Future Conflicts? Environment and Conflict Project (ENCOP), **Occasional Paper** No. 11, Center for Security studies and Conflict Research, Swiss Federal Institute of Technology, Zurich.

Kuhn Harold and and G. Szegö (1971) **Differential Games and Related Topics.** Amsterdam: North Holland.

Lowi, M. (1993). **Water and Power.** Cambridge: Cambridge University Press.

Luterbacher Urs, Michael Clarke, Nicolas Kessler and Pierre Allan Simulating the Response of a Small Open Politico-Economic System to International Crises **Management Science,** 33, 2, 1987: 270-287.

Luterbacher, U. & Wiegandt, E. (1995). Social dimensions of resource use: The freshwater case. **Swiss Political Science Review, 1(4),** 131-142

Luterbacher, U. & Wiegandt, E. (1997a). Towards an integrated model of water resource use in different regions. Ms.

Luterbacher, U. & Wiegandt, E. (1997b). Simulation Models, Global Environmental Change, and Policy, Ms.

Moulin Hervé (1982, Second Revised Edition 1986) **Game Theory for the Social Sciences**. New York, N.Y.: New York University Press.

Nordhaus, W. & Yang, Z. (1996). A regional dynamic general-equilibrium model of alternative climate change strategies. **The American Economic Review,** 86 (September). 741-765.

Rosenzweig, C., Parry, M., Fischer, G. & Frohberg, K. (1993). **Climate change and world food supply.** Research Report 3, Environmental Change Unit, Oxford: University of Oxford.

Shechter, M., N. Becker and N. Zeitouni (1997) Reallocating Water Resources in the Middle East through Market Mechanisms **Water Resources Development**, 12 (1):17-31

Weibull, Jörgen (1995) **Evolutionary Game Theory.** Cambridge, MA: MIT Press.

Exploratory economic assessments of climate change impacts in Israel: Agriculture

M. SHECHTER AND N. YEHOSHUA
Natural Resource and Environmental Research Center, University of Haifa, Israel

1. INTRODUCTION

The overall objective of this study is to assess – in monetary terms - the potential damage associated with the impacts of climate change due to the greenhouse effect in Israel. The socioeconomic impacts will affect water resources, agricultural production, biodiversity resources, forestry resources, coastal regions (due to sea level rise), tourism, health levels, and population migration due to desertification and related phenomena. Damage are estimated for a future point in the 21st Century when CO_2 levels in the atmosphere will double from pre-industrial levels. This is a common reference benchmark in the literature, given that there is still a great deal of uncertainty regarding which of the global greenhouse gases (GHG) emission scenarios will actually materialize. There is a widespread tendency in the literature, however, to associate this doubling with the period 2030-2050. The study aims to sum-up damage by sectors, employing the "Bottom-up" approach, assuming present or (when possible) forecasted technological know-how. This is in contrast to the "Top-down" approach that is based on econometric and macroeconomic models.

In this paper we assess the impacts of climate change on Israeli agriculture. During the past fifty years, Israel's highly sophisticated agricultural[1] sector has increased yields 16-fold while introducing a variety

[1] The information presented here is taken from Publication No.2 of Israel's CBS, "Jubilee Publications" series.

M. Beniston (ed.),
Climatic Change: Implications for the Hydrological Cycle and for Water Management, 411–427.

of new crops. In more recent times, as is the case in other developed economies, the comparative advantage of agriculture has declined, and the sector's output has amounted to only 2.4% of GDP (1995). The percentage of labor employed in agriculture was 3.4%, and agricultural exports were 4% of total exports, valued at $740 million. Agriculture's share of export has declined over the years, as well as its composition. Presently, field crops make up of 10% of the total value output, vegetables, potatoes and melons 15%, citrus 8%, other fruits 15%, flowers 8%, poultry 18%, and cattle 16%. The total cultivated area in 1995 was 367,000 hectares, out of which 199,300 are irrigated.

The damage estimates are derived using a simple production function approach, similar to that employed by a number of similar studies, using rather strong assumptions on adaptability potential or the lack thereof. Despite a number of strong assumptions made in this study, we believe it provides a useful order of magnitude estimate of damage with and without some adaptation, which would be of value in policy making regarding proactive adaptation as well as Israel's contribution to the global mitigation efforts under the Kyoto Protocol. We further believe that it is a useful exercise to focus on irrigation water shortage as the major determinant of agriculture's response to climate change in the semi-arid region where the country is located

*

In general, there are three independent categories of costs related to climate change: direct damage (designated as D), adaptation costs (A), and net (of ancillary benefits) mitigation costs (P). Mitigating GHG emissions in turn affects the magnitude of damage, or reduces the need for adaptive measures. Similarly, adaptation reduces exposure to damage. The policy objective is to minimize total costs (T), the sum of the three categories: D+A+P. In this study we focus on the first two cost categories: direct damage and adaptation costs. Clearly, a region the size of Israel or even the whole eastern Mediterranean basin has a negligible impact upon total global GHG emissions and concentration. Therefore, a benefit-cost framework comparing mitigation costs with damage and adaptation cost would make sense only on a global level, due to the common-property nature of the earth's atmosphere with regard to the impact of GHG on global and regional climate.

We employ commonly used assumptions for the economic calculations:

- According to the average emission scenario ("IS92") of the Inter Governmental Panel on Climate Change (IPCC), doubling CO_2 is expected to occur in 2055-2060. (IPCC, 1996a).
- Average temperatures in the eastern Mediterranean are expected to rise by up to 2°C taking into account the cooling effect of aerosols (polluting

particles that "swallow" part of the returned radiation) is considered (IPCC, 1996a).

- Sea level rise by 2060 is expected to range between 10-55 cm, with an average prediction of a rise of 25 cm. (IPCC, 1996a).

Specifically, our estimates are based on a recent research study (Dayan et al., 1999) which has produced climatic forecasts for the coastal region of Israel. The study's forecasts are on a global circulation model (GCM) developed in the University of East Anglia (Palutikof et al., 1996). According to this study, the expected changes in temperature for Israel are:

$$2020:(-2)-(-1)\%$$
$$2050:(-4)-(-2)\%$$
$$2100:(-8)-(-4)\%$$

With corresponding changes in precipitation:

$$2020:0.3-0.4°c$$
$$2050:0.7-0.8°c$$
$$2100:1.6-1.8°c$$

The paper first surveys relevant climate change factors and their potential impact on agriculture in Israel. It then reviews a number of relevant studies and the methodologies employed by them. A brief review of the Israeli agricultural sector is then followed by a description of the methodology and specific assumptions used in this study and their application to the Israeli data. As expected, given the highly sophisticated nature of this sector in Israel, there is a wide variation between estimates assuming no adaptation, compared with those based on optimal crop selection and adaptation to the change in climatic variables, specifically precipitation changes.

2. AGRICULTURE AND CLIMATE CHANGE

Agriculture (and commercial forestry) are the most vulnerable production sectors which would be influenced by climate change, due to its great dependence on climatic variables. The factors that affect agriculture and which are related to climate change can be classified into several subcategories:

1. Climatic factors: temperature, precipitation and soil moisture.

2. Factors that accompany climate change: the effect of CO2 levels on plant development (fertilization effect), and the impact of other gases such as tropospheric ozone and SO2.
3. Factors related to human activity in coping with climate change, such as adaptation measures in various fields.

2.1 Climatic Factors

Temperature largely controls the rate of plant growth, flowering and fruiting responses, seed development, the water vapor flux, plant water status, soil drying, and irrigation practices. On the individual plant level, we can observe that different crops reach their optimum at different ranges of temperature (IPCC, 1996b). For crops in temperate climates the optimal range is between 30-35°C (Parry et al., 1988). Changes in temperature can directly influence livestock, as well as indirectly influence pest distribution (weeds and insects). Another change related to temperature is the length of the growing season. Global warming reduces significantly frost danger to all the Mediterranean region climates. It enables to plant crops earlier in the season and lengthens the harvest season. On the other hand, higher temperatures accelerate development, shorten the growing period and decrease yields if the shortened growing period is not compensated by an enhanced development of the plant (Ellis et al.,1990).

Precipitation is likely to become the most critical component in the structure of stresses that agriculture could face in the future. Any amount of global warming will increase water demand of almost all crops. Changes in annual precipitation affect agriculture directly through soil moisture and non-irrigated crops, and indirectly through refilling water reservoirs. In addition to the total annual rainfall, there is a great importance to the distribution of the precipitation over the year. It should be mentioned, however, that most of the Global Circulation Models (e.g., GISS and UKMO (IPCC, 1996a)) do not provide a uniform picture regarding the predicted changes in precipitation, and forecasts differ from one region to another and from one model to the other.

One of the more significant impacts is the rise in drought incidents. Drought is the most common cause of yield loss, especially in arid to semi-arid regions. Drought is an example of broader phenomena, namely, annual fluctuations in precipitation (and temperature), which influence agriculture.

Other factors are the **moisture** and the amount of **solar radiation** to which the plant is exposed. Moisture affects plant growth rate, fruiting period, evaporation rates and water demands. A constant level of moisture is essential in most phases of crop development. Generally, the hydrological cycle is about to increase with climate change, including precipitation and moisture.

Future agriculture will face a serious challenge due to **Soil degradation**. Among its causes are enhanced erosion, loss of organic matter and accumulation of salts. Climate change is expected to enhance such effects, due to changes in intensities and amounts of precipitation, higher radiation levels and extreme weather events.

- **Pest impacts.** The predicted changes in temperature and CO_2 level will influence the distribution and range of different kinds of pests (weeds, insects and disease-borne organisms). This will probably aggravate the damage to crops. Their influence will be most noticeable in shorter crops and in additional prevention and control costs will become necessary (Rosenzweig et al., 1998).

2.2 Atmospheric changes

Future changes in CO_2 levels. The most important change, which was studied a great deal recently, is the rise in CO_2 concentrations in the atmosphere. This change is predicted to enhance photosynthesis and plant water-use efficiency. The average response of C3 group crops (most of the crops, excluding sugarcanes, maize and several cereals) to doubling CO_2 levels is an increase of 30% in yields, ranging between -10% and +80%. The factors that affect the response include temperature, soil fertility and precipitation (IPCC, 1996b). A recent study, using the FACE method (Free Air Carbon Dioxide Enrichment), found that CO_2 concentration of 550 ppm would cause a 15-16% increase in yields. Other studies point out the effects of changes of CO_2 concentration on the form, shape and compound of crop yields. For example, rice grown under higher CO_2 concentration will contain higher concentration of Amylose, while Fe levels will be lower (Seneweera & Conroy, 1997).

Other atmospheric changes. Another change expected to influence agriculture is the level of plants exposure to tropospheric ozone. Its concentration has doubled in the last century, and has caused an estimated drop in yields in the range of 1-30% (IPCC, 1996b). The depletion of the ozone layer leads to higher exposure to UV-B radiation that has been proved to affect crops. These radiation damage are manifested mainly in high altitudes.

2.2.1 Human induced factors

Crop adaptation is becoming an important component in the process of agriculture's adjustment to climate change. However, research that has examined the independent adjustment capability of crops to changes in growth condition has not yielded encouraging results. Human activity in this field can be divided into two major categories: Choosing an optimal crop mix and biotechnological development. In selecting new genotypes, the factors that should be taken into account are high sensitivity to CO_2 concentration, the maintenance of yield levels even when higher temperatures cause enhanced development, and sustainability to heat waves and water shortage in growth and reproduction phases. Biotechnology is open to research and development of more sustainable crops for a changing environment.

3. A SHORT METHODOLOGICAL SURVEY

To date, numerous studies have dealt with the impact of climate change on agriculture (IPCC, 1996b). The earlier ones forecasted future responses on the basis of statistical regressions on past data. More recent studies using 'dynamic crop models', attempt to model the principal physiological, morphological, and physical processes involving the transfer of energy and mass within the crop and between the crop and its environment. From such relationships, these models derive predictions of crop performance under various conditions (Rosenzweig, 1998). These studies rely on climate scenarios derived from a variety of GCM. Among them CERES, dynamic crop model applied for several crops such as wheat and maize (Ritchie et al., 1989, Godwin et al.,1990) and SOYGRO for soy crops (Jones et al., 1989). A number of works incorporate market responses and long-range adaptation options. Failing to incorporate market reactions probably leads to an overestimation of the impact of climate change and should be classified in a "worst case scenario" category.

More recently, attempts have been made to combine crop responses and economic models in order to evaluate future changes in production and welfare. These works can be divided into two main groups: research based on structural models and evaluations base on spatial models.

Structural models, such as the study by Adams (Adams et al., 1998) specifies the production processes and incorporates it into an economic optimization. The main advantage in this approach is that it can estimate the impact of climate change on market equilibrium. The main omission to date has been overlooking adaptation potential.

Spatial ("Ricardian") models rely on econometric evaluation and are based on historical data. They estimate the relationship between economic data such as land and other asset values and climatic variants. An example is the work by Mendelsohn (Mendelsohn, 1994), which has examined the impact of the warming in a Ricardian approach, in a partial equilibrium framework (agricultural land markets), on the basis of land prices in over 3000 counties in the U.S. Thus, it implicitly reflects (past and present) adaptations to climate-related variables throughout the US. Application of the model to climate change scenarios reveals lower future impacts due to climate change. While including adaptation options, this approach can not represent production processes specifically, and therefore cannot, for example, take explicit account of phenomena such as CO_2 fertilization.

In an attempt to apply this methodology, we examined official documents of the Ministry of Agriculture which provide normative assessments of profitability by crop groups. In theory, we could have used these values as proxies for land prices, assuming land is the only fixed input, and the calculated net profits (after returns to labor) represents computed (i.e., not observable) land rents. Given that most of the agricultural land in Israel is state-owned, and that effectively there is no fully functioning agricultural land market, these computed values could have been used as proxies. However, these values are national averages, and consequently useless in the present context, since regional climatic variability cannot be factored out from the data.

4. OTHER MEDITERRANEAN COUNTRIES STUDIES

Naturally, studies with the highest relevance to this study are those which examined neighboring countries. However, comparisons not easy to make, since one would expect to find differences in methods used (models, the inclusion of adaptation, etc.), the nature of basic climate change assumptions (climate scenarios, CO_2 fertilization), and the type of crops examined.

The Lebanese Ministry of Environment (Ministry of Environment; Republic of Lebanon, 1999), For UNDP and the Global Environmental Facility (GEF) prepared a major study. It examined the vulnerability of Lebanese agriculture to climate change. The study divides Lebanese agriculture into four main crop groups: apple orchards, citrus, olives and sugarcane. Three approaches have been employed by the researchers: (1) "Analogue" – assuming that following climate change, agricultural regions would assume characteristics similar to those of presently lower altitudes; (2) Field studies relying on past data; (3) "Expert judgment". The study's results have been expressed largely in qualitative terms. They predict a drop in yields and a rise in production costs for apples and citrus; inconclusive results for olives, and a negligible impact on sugarcane production. It should be pointed out that the study assumes more severe climate scenarios than ours (a rise of 1.6-4.1°C by 2080).

The Egyptian Environmental Affairs Agency (1999) conducted a similar study, which also examined the influence of climate change on the major crops in Egypt for standard GCM predictions, as well as arbitrary climate change assumptions (+2°, +4°, and 10-20% in precipitation). The study reports a decrease of 18-19% in wheat and maize yields, and an increase of 17% for cotton. Adaptation options were examined using three models: COTTAM, TEAM, DSSAT3. The most important adaptation measures presented by the models are: (1) Improvement of wheat and maize cultivars; (2) Switching from maize to cotton, and replacing wheat with winter crops; (3) Changing agricultural techniques, such as planting dates, water and nitrogen applications and plant density; (4) Removing crops with high water consumption.

Yet another study, reported in the second IPCC report, examined Egyptian agriculture in 2060 (with 2*CO_2) (Yates & Strzepeck, 1998). The study is based on GCM scenarios GFDL, UKMO, and GISS A1. It investigated crop response to climate change with and without adaptation, and included the CO_2 fertilization effect. The forecast (prepared for wheat, rice, other cereals and fruits) indicated a decrease in yields of -5 to –51% for wheat, -5 to –27% for rice and –2 to –21% for other cereals and fruits. The study states that yield damage could decrease by up to 50% if proper adaptation measures are taken, such as changes in crops, fertilizers, and seeding and watering patterns. An earlier study (Eid, 1994) also investigated climate change impact, using the same scenarios. It predicted a more noticeable decrease in yields; for wheat: -18 to –75% and for maize: +6 to – 65%.

5. ASSESSING THE DAMAGE TO ISRAELI AGRICULTURE

5.1 Assumptions

A number of simplifying assumptions were made in this rather preliminary study. A major one limits the impact to one climatic factor, namely – precipitation. That is, climate change would affect the agricultural sector only through the availability of water (including soil moisture) to crop production[2]. The assessment is therefore based on the impact on water supplies, and further assuming that all such shortage, if and when they occur (around 2060) will have to be absorbed by agriculture. However, given the elastic nature of the demand for irrigation water (compared with household demand), this is not a too heroic assumption. The branches included in the calculations are: field crops, vegetables and plants alone[3].

A couple of additional strong assumptions underlie the estimated damage to agriculture, namely, no structural changes will take place due to adaptation, and there relative (real) price levels for agricultural output will remain constant.

A number of important reasons underlie the approach and assumptions adopted for the present analysis.

1. The available climate change data for Israel is scant, at best. Israel's small land area requires climatic forecasts of a very high resolution, which are still lacking. This leads to high uncertainty regarding predicted warming. Recall that according to the forecast used here, only a rise of less than 1°C (Dayan et al., 1999) is predicted for the coastal region of Israel. This makes it hard to estimate impacts using standard models, such as Dynamic Crop Models.
2. The most limiting factor of Israel's agriculture is water. This will be aggravated in the future given the expected rise in domestic and industrial water demand.
3. Because of the nature of land ownership in the Israeli agricultural sector (most of the land is owned by the state), there is no developed market for agricultural land, and therefore there is no possibility to apply the spatial model approach.
4. Only partial consideration is given to adaptation. Adaptation options that have not been analyzed are: (1) crop mix changes (including cultivars development); (2) development of new agro-technologies in agriculture and water management; (3) changes in crop location.
5. The study does not incorporate the CO2 fertilization effect and other atmospheric changes that might alter crop yields significantly.

[2] According to the Israel National Report on climate Change (Ministry of Environment, 2000) the decreased annual rainfall and lengthened intervals between rains, coupled with increased temperatures and evaporation, will reduce plant productivity, with a concomitant decrease in soil organic matter. This will reduce water holding capacity and soil permeabilty. An increase in runoff velocity and intensity will erode the most fertile topsoil, further reducing productivity. Soil salinity will increase due to higher evapotranspiration and lower leaching effect of the reduced rains.

[3] Flowers and livestock were not included in the calculation due to lack of relevant data. In any case, we assume that these branches will not incur any water cutbacks.

5.2 Damage Calculations

A number of factors will affect water supplies in connection with climate change: a decrease in precipitation, enhanced evaporation from water reservoirs (as a result of the rise in temperature), aquifer salinization due to over pumping and sea water intrusion. Since it is not easy to evaluate the impact of the two last factors in quantitative terms, we use only the predicted precipitation shortfall (Dayan et al., 1999), employing the upper limit of 4% as a "worst-case" scenario. Given that the average annual supply is 2000 million cm, the supply shortfall assumed in the calculations is 80 millions cm.

As mentioned earlier, the approach used in this assessment is the production function approach. i.e., using an estimated production function for water (other inputs held fixed) for a number of key crops, we estimated the drop in yields due to cuts in irrigated water and natural precipitation. We used recently estimated production functions by Vered (Vered, 2000), which examined crop yield responses to water of different quality (fresh, recycled, and brackish).

The production function with water as single variant, and all other inputs constant (evaluated at their means and added to the intercept):

$$Y_i = a + b(W_i) + c(W_i)^2$$

Where:
Y_i = yield per dunam (0.1 hectare) for crop i
W_i = irrigation water input per dunam of crop i.
a = constant (incorporates yield with only natural rainfall).
b, c = coefficients.

The estimated functions in Vered (2000) incorporate additional variables, which influence output beside water quantity and quality. These are: the geographical region (there are significant differences in climatic and soil characteristics between regions) and the type of irrigation system (dripping, etc.). However, for our purpose (crop yields as a function of water input) we assumed them to remain constant, and incorporated them into the constant term.

Table 1: Crop Groups

Group	Selected crop
Citrus	Oranges
Fruits (non citrus)	Avocado, apples
Cereals and oils	Wheat
Fibers	Cotton
Vegetables	Tomato, Watermelon, Potato[4]

5.2.1 Scenario I (the "Naive" scenario)

The underlying assumption in this scenario is that production cutbacks will be undertaken in an arbitrary fashion due to the water shortfall. There will be a proportionate cutback in water use by each crop group, relative to its present water consumption. Total damage is given therefore by summing the value of yield cutbacks for all groups, namely,

$$(1) TD = \sum_{i=1}^{n} \Delta Yi \times Pi$$

Where:
TD= annual damage in monetary terms.
i = crop group.
Pi = Average price per ton for crop group i[5].
ΔYi = Change in yield for crop group i.

Yield change is a function of a change in water allocation to the respective crop group and the change in the value of the constant coefficient (due to decline in rainfall), given by (2):

$$(2) \Delta Yi = f(\Delta Wi) + \Delta a$$

where
Wi= Amount of water consumed by crop group i.
Δ Wi = The change in the amount of water consumed by crop group i.

The change in the water allocated to each crop group is proportionate to its present consumption,

[4] Because of the large size of the vegetables group, three representing crops, equally weighted, were taken for it.

[5] This price is calculated according to data from the "agriculture Statistics quarterly (CBS, 1997) containing total products (tones) and total monetary return in average prices for 1995.

$$(3)\Delta Wi = \Delta W \times \frac{Wi}{TW}$$

where
TW= total amount of water consumed by all groups
Δ W= total change in water consumption.

$$\sum \Delta Wi = \Delta W$$

Out of the assumption that production will be affected not only by irrigation cutbacks but also due to decline in rainfall, one must incorporate this impact as well. The intercept in the response function is supposed to capture this effect, as well as other factors not explicitly represented in the model specification. We estimated its weight, roughly to be around 66% of the value of the intercept.[6]. This part of the intercept would therefore need to be re-calculated along with the change in the irrigation water input[7], as given in equation (4)

$$(4)\Delta a = 0.66 \times 0.04 \times a$$

Table 2 summarizes the loss in production value in the naïve scenario. The total annual loss is about $208 million in present prices.

Table 2: Total annual damage by crop groups, Scenario 1.

Sub group	**Total damage (mil. $)**
Cereals and oils	5.5
Fibers	51.7
Citrus	86.5
Vegetables, Potatoes and Melons Vegetables (representing crop-tomato) Melons (representing crop- water melon) Potatoes Total vegetables sub group	 21 16 11 48
Fruits (no citrus)	16.5
Total	208

5.2.2 Scenario II: Partial Adaptation

It is most likely, even certain, that in reality costs will be significantly lower, given the agricultural sector's ability to carry out adaptive measures over time. Here we assume that (partial) adaptation will be in the form of

[6] Based on discussions with Vered.

[7] Greenhouse crops (e.g., tomatoes) were excluded because there is no significance to precipitation in their growth.

economic adjustment, by which we mean adjusting crop areas according to the water-use efficiency of the different crops. The naive scenario ignores adaptation altogether, and therefore overlooks significant savings in production costs due to a response in the form of crop adjustment (among many other, of course).

In Scenario II we consider two components of damage {given in equation 5}:

1. Decrease in precipitation as an element having an impact on all crop groups. {Equation 6}
2. Cutbacks in water allocated to crops whose water use efficiency is relatively low, based on the marginal value product (MVP) of water (= the partial derivative of the response function of the representative crops with respect to the water input, evaluated at the average water input per irrigated dunam of the given crop, multiplied by the average price per ton).

$$(5) TD = \sum_{i=1}^{n}(\Delta Yi \times Pi) + \Delta Yj \times Pj$$

$$(6) \Delta Yi = f(\Delta a)$$

where
j = the crop group for which MVP is the lowest.

$$\frac{\partial Yj}{\partial W} \times Pj \leq \frac{\partial Yi}{\partial W} \times Pi$$

for all i.

However, we should also consider the possibility that whenever a cutback in irrigation reduces a given crop acreage (and not just yield per dunam), there will be a corresponding reduction in production costs. Consequently, *net* damage costs will be correspondingly lower. Production costs for the different crops were taken from the annual agricultural survey (CBS, 1998).

The calculation of net damage for the second scenario is given by:

$$(7) Dj = f(\Delta Wj) - \lambda TC(\Delta Yj)$$

$$(8) \lambda = \frac{\Delta Yj(w)}{Yj(w)}$$

$$(9) \Delta Wj = \Delta W \qquad 0 \leq \lambda \leq 1$$

where:
Dj = total damage value to group j.
$\Delta Yj =$ yield change, group j.

TC= total production costs for group j.
ΔWj = change in water consumption of group j.

Table 3 gives the MVP for the different crop groups.

Table 3: Marginal Revenue by Crop Groups

Sub group	Marginal revenue($ / cm)
Cereals & oils	0.195
Vegetables	2.3
Citrus	2
Fruits	17.75
Fibers	0.4

The crop group with the lowest MVP is Cereals and Oil. It turns out that if it absorbed the entire needed cutback to meet the projected drop in precipitation, production will drop by about 66%, with a corresponding annual decline in output valued at $105 million. However, net damage, after taking into account the corresponding decrease in production inputs (Equation 5} is $40 million.

In addition to the cutback achieved through the reduction in crop acreage of crop group j, we should take into account the impact of precipitation decrease on all crops, i = 1,...n, represented by ΔY_i in equations 5 and 6 above. Table 4 gives the monetary value of this impact on crops. Thus, the total annual damage due to a projected decrease in mean precipitation levels amounts to $62 mil. Combining the two components of the forecasted impact (62 + 40), total annual damage under this scenario, adds up to $ 102 mil.

Table 4: Damage resulting precipitation decrease:

Sub group	Damage (mil. $)
Vegetables	
Potatoes	4
Melons	3
Cereals and oils	1.5
Fibers	43
Fruits	2
Citrus	9
Total	61

The outcome is described in Figure 1, which depicts the demand (= MVP) for and supply of irrigation water in agriculture. The expected impact of climate change through the decline in precipitation is depicted by the leftward shift in the supply schedule. The decline in economic welfare (i.e., social damage costs) is given by the triangle formed by the intersection of the demand and the two supply curves.

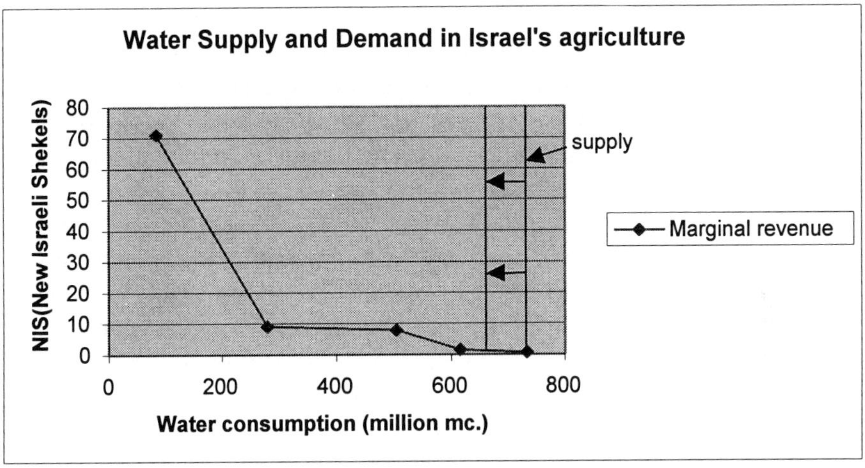

Figure 1: Water supply and demand for Israeli agriculture

5.2.3 Scenario III

The third scenario considered in this study examined the possibility of augmenting domestic freshwater water supplies by desalinized water at current production costs. In this case damage is represented by the replacement costs of supply shortage due to climate change will be the sum of the damage resulting from precipitation decrease (Table 4) and the costs of desalinizing 80 million cm. (the assumed water shortage). This is described by equation 10:

$$(10) TD = \sum_{i=1}^{n} (\Delta Yi \times Pi) + TCs(\Delta W)$$

where:
TCs = total desalination costs

The current cost of desalination stands at about 80 cents per cm. This yields an annual bill for this scenario of about $ 126 million.

6. CONCLUDING REMARKS

Our aim in this exploratory study has been to provide a range of rough preliminary estimates of expected future damage to agriculture in Israel resulting from climate change. In addition to the inherent uncertainty in forecasting climate change impacts, there is the uncertainty regarding what adaptation options will be available to farmers several decades from now, the nature of demand for water, and the role of agriculture in the national economy. There are also several additional climatic factors which we were not able to incorporate in the analysis due to a lack of usable data for Israel: change in temperature, climatic fluctuations (temperature and precipitation), the role of CO_2 fertilization, and more.

However, even this modest exercise, tells us a great deal about the importance of adaptation and correct proactive planning in counteracting the adverse effects of climate change. In addition to providing a range of quantitative economic estimates of costs, this is probably the major lesson of the study.

Table 5: Total damage by scenario (mil. US $)

Scenario	**I**	**II**	**III**
Total Damage	208	101.5	125.5

7. REFERENCES

Adams, R.M, McCARL, B.A., Segerson, K, Rosenzweig, C, Briant, K.J, Dixon, B.L, Conner, R., Evenson, R.E, and Ojima, D., 1999. *The Impact of Climate Change on the U.S Economy* ,Cambridge: Cambridge University Press.

CBS (Central Bureau of Statistics),1997. *Agricultural Quarterly,* Volume 28 Jerusalem: CBS.

Dayan, U. and Koch, .J, 1999. *Implications of Climate Change on the coastal region of Israel,* Mediterranean Action Plan. Athens: UNEP.

Egyptian Environmental Affairs Agency, 1999. *The Arab Republic of Egypt: Initial National Communication on Climate Change.* Prepared for the UNFCCC.

Eid, H.M.,1994: Impactof climate change on simulated wheat and Maize yields in Egypt. In: *Implications of climate change for international Agriculture:* U.S Environmental Protection Agency , Egypt Chapter, Washington , DC, pp.1-14

Ellis, R.H., Headley ,P, Roberts, E.H., and Summerfield, R.J., 1990: Quantitative relations between temperature and crop development and growth. In : Belhaven, *Climate change and Plant genetic resources,* London, UK, pp.85-115.

Godwin, D.C et al. 1990, A User's Guide to CERES-Rice-V2.10 International Fertilizer Development Center. Muscle Shoals ,Al.

Israel, Ministry of Environment. 2000, *Israel National Report on Climate Change,* Gabbay S., Jerusalem, Ministry of Environment's Publications unit.

IPCC 1996. *Climate Change 1995: The science of climate change,* Houghton, J.T., Jenkins, G.J and Ephraums, J.J. (eds). Cambridge: Cambridge University Press.

IPCC 1996 :Agriculture in a Changing Climate: Impacts and Adaptation. In: Watson, R.T., Zinyowera, M.C., Moss, R.H., Dokken, D.J. *Impacts, Adaptation and Mitigation of Climate Change: Scientific- Technical Analysis*., New York., Cambridge University Press.

Jewish Agency to Israel, Agricultural Planning and Development Authority, 1975-1996. *Agricultural Branches.* Tel-Aviv.

Jones,J.W. et al. 1989. SOYGROv5.42 soybean crop growth simulation Model: Users Guide. Florida Agriculture Experimental Station Journal, University of Florida, Gainesville.

Kol, N.,1998. Central Bureau of Statistics, *Agriculture, publication no.2* from Jubilee Publications", Jerusalem:.

Mendelsohn, R., Nordhaus ,W.D., and Daigee, S., 1994 the impact of Global Warming on agriculture: A Ricardian Analysis. The American Economic Review. 84 754-771

Palutikof, J.P.,X. Gou, T.M.L., and Wigley, 1996. "Developing climate change scenarios for Mediterranean Region" in: L. Jeftic, S.Keckes, and J.C Pernetta, eds., Climate Change and the Mediterranean, Vol. 2, pp.27-56. London: Arnold.

Parry, M.L., Carter, J.H., Konijn, N.T., 1988. The Impact of Climate Variation on U.S Agriculture. Vol.1. *Assessment in semi arid regions.* Kluwer Academic Publishers, Dordercht, Netherlands, 764 pp.

Pinter et al. 1997.,

Republic of Lebanon: Ministry of Environment 1999., Climate change Assessment Agriculture. In: *Technical Annex to Lebanon's First National Comunication. Final Report.* Republic of Lebanon; UNDP; Global Environment Facility.

Ritchie , J.T., Baer, B.D., Chou T.Y., 1989., Effect of global Climate Change on the Agriculture in the great Lakes region. In Smith, J & Tirpak ,D. *The potential effect of global climate change on the United states.* Office of Policy , Planning and Evaluation U.S. Environmental Protection Agency. Washington D.C appendix C. pp. 1-1-1-21

Rosenzweig, C., Hillel, D., 1998. *Climate Change and the global harvest.* New York. Oxford University Press.

Seneweera & Conroy, 1997.

Vered A., 2000. *Yields function according to water quality* (M.Sc. Thesis), Hebrew University, Faculty of Agriculture, Rehovot, Israel.

Wittwer, S.H., 1995 *Food, Climate, and Carbon dioxide*. Boca Raton, Florida. CRC Press.

Yates & Strezepeck .1998.

Changing climate and increasing costs – Implications for liability and insurance

LAURENS M. BOUWER AND PIER VELLINGA
Institute for Environmental Studies, Vrije Universiteit, Amsterdam, The Netherlands

Abstract: This paper discusses the probability of changing frequencies and intensities of extreme weather events and their financial aspects. The main concerns are with precipitation related events such as rainfall, flooding and drought. Particular attention is given to damage costs and the role of insurance. The recent increase of economic and insured losses may well be an early indication of what climate change has in store. Attribution analyses of damage costs can show what part of the costs is related to increased exposure of people and capital and what part is related to climate change. In Europe, a large part of flood related damage is taken up by governments or state-owned companies and individuals. Under a changing climate regime, the organisation, the geographic scale of covering risk, and the roles of private and public insurance will have to be reconsidered. We argue that new ways of covering the cost of risk mitigation and insurance should be explored. Elaboration of the polluter pays principle and liability approaches in an international context would help to identify ways of dealing with the risk and damage cost related to climate change.

1. INTRODUCTION

Losses resulting from weather related natural catastrophes, measured in monetary terms, have been increasing drastically over recent decades. Economic losses from weather related disasters have increased from approximately 4 billion US dollar average per year in the 1950s to approximately 40 billion US dollar average per year in the 1990s (see Figure 1). In the same period, insured losses rose from a negligible amount to 10 billion US dollar per year. While probably a major part of this increase can be attributed to changes in society and economy, changes in the

M. Beniston (ed.),
Climatic Change: Implications for the Hydrological Cycle and for Water Management, 429–444.

occurrence and intensity of extreme weather events are likely to have contributed to the increase of costs (Vellinga *et al.*, 2001), and may become even more important in the future.

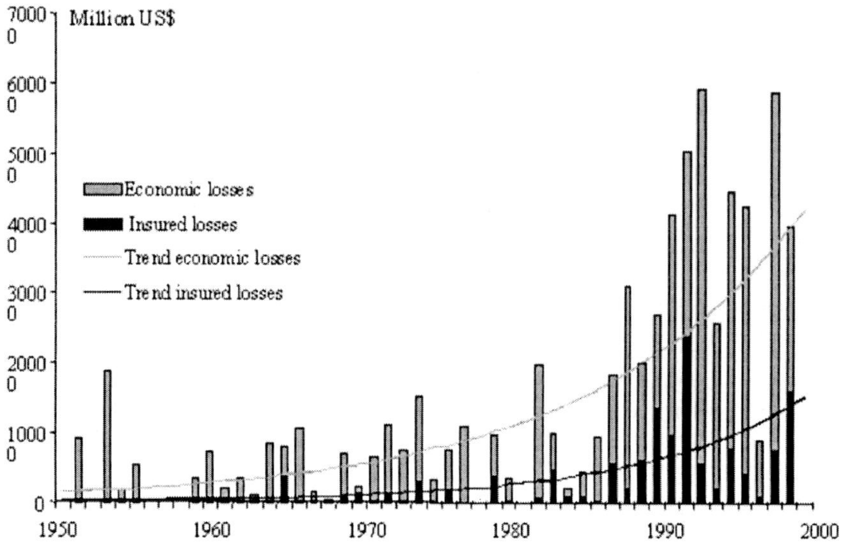

Figure 1. Economic and insured losses from great weather-related natural disasters 1950-1999 in 1999 US$. Great disasters are defined as events during which the ability of the region to help itself is distinctly overtaxed, making interregional or international assistance necessary. This is usually the case when thousands of people are killed, hundreds of thousands made homeless, or when a country suffers substantial economic losses, depending on the economic circumstances generally prevailing in that country (Munich Re, 1999). If the cost of small, non-catastrophic weather-related events were included, the figures would be a factor two larger. (After Munich Re, 2000: page 18)

Of relevance here are extreme weather phenomena, such as extreme high or low temperatures, extreme amounts of – or lack of – precipitation, and high-speed winds (storms). Here we discuss changes in precipitation related extreme weather events and the causes and consequences of their costs. Much of what will be presented may also be of relevance for sea-level rise and resultant storm surges and flooding in coastal and other lowland areas. In a global warming world, events that were considered extreme and rare over the last decades and centuries can become more frequent when the mean state of the weather and/or the variance of the distribution of events change (Meehl *et al.*, 2000). It is important to stress the geographical aspect of changes of extremes. Areas that are under a given climate condition, are adjusted to the extremes of that regime. Shifts in the frequency, and the intensity of the extremes due to a change in the climate, cause stresses to an area that is not able to cope with these new circumstances, because it is historically not adjusted. The weather is ultimately causing damage to natural and human systems. This, as a result of spatial shifts in patterns of for instance storm tracks or the appearance of totally new phenomenon (flash floods, hail, or ice storms) in places where they never occurred before.

Extreme weather events are by definition rare, and records of change over the last century do not show clear signals for every region of the world. For the future, direction and magnitude of change of many extreme weather events are not known, but changes in temperature and precipitation are among the best understood. As temperature increases, potential evapotranspiration will increase, and thus more water can be taken up and be released by the atmosphere (IPCC, 2001). Therefore, the amount of precipitation is most likely to increase in many parts of the world, due to this enhanced hydrological cycle, in particular in the mid and high latitudes in the northern hemisphere.

Table 1 lists precipitation-related weather events that are important in the context of extreme weather impacts and monetary losses. Directions of change may not be the same for every region of the world as some areas could face less extreme forms of certain weather events or see a decrease in (mean) precipitation. The sectors of society and their activities that are affected are listed in the last column.

Findings on the relation between global warming and precipitation processes and patterns include for instance increases in heavy precipitation (Karl and Knight, 1998; Easterling *et al.*, 2000), and increases in the number of hailstorms (Dessens, 1995). Ice storms will occur when precipitation falls as rain, but freezes when it reaches a solid surface. Temperatures just around or below freezing are ideal circumstances for ice storms to occur. These storms can cause a tremendous amount of damage to property, as for instance shown by a recent ice storm in Canada and north-eastern United States, of which insured losses amounted to 1.5 billion US dollars (Lecomte *et al.*, 1998). In a warming climate, areas experiencing temperatures well below the freezing point may be confronted with more of these events.

The relationship between climate change, precipitation, and extreme flooding is not as obvious as it seems. Basin characteristics, such as land use change, changes in vegetation cover, and river training all influence the response of catchments to changing meteorological phenomena. For instance, in the United States, observed river discharges have generally increased, but mainly in the small and medium classes (Lins and Slack,

1999). However, there has been some discussion around the validity of these findings (Pielke and Downton, 1999), as the frequency of extreme precipitation has increased (e.g. Karl and Knight, 1998). It may well be that an increase in extreme precipitation does not directly reflect itself into more extreme streamflow.

Table 1. Precipitation related (extreme) phenomena, their time scales, expected changes due to climate change, and their effects on society. Levels of confidence for present observations and future directions of change are indicated. (After Vellinga *et al.*, 2001: table 8-1)

Type of extreme and impact	**Relevant timescale**	**Direction of change in climate extreme**	**Recently observed changes**	**Future changes**	**Sectors and activities in society at risk**
			Level of confidence		
Flash flood	Hourly-daily maximum	More frequent in mid/high latitude in northern hemisphere	High	Very high	Human settlements
Flood, inundation, mudslide	Weekly-monthly maximum	More frequent in mid/high latitude in northern hemisphere	High	High	Agriculture, forests, transport, water quality, human settlements, tourism
Summer drought, land subsidence, wildfire	Monthly-seasonal minimum	More frequent in many interior continental locations	Low	High	Forests (tree health), natural resources, agriculture, water resources, (hydro) energy supply, human settlements
Snowstorm, ice storm, avalanche	Hourly-weekly	More frequent in mid/high latitude continental interiors	Medium	Unknown	Forests, agriculture, energy distribution and reliability, human settlements, tourism, mortality
Hailstorm	Hourly	Increased frequency	Medium	Medium	Agriculture, vehicles, buildings

In Europe, the relationship between changes in circulation patterns, due to the North Atlantic Oscillation (NAO), winter precipitation, and more frequent extreme high water levels in rivers has been extensively studied. Werner *et al.* (2000) report that the period 1981-1990 is the first statistical outlier in terms of the residence time of western circulation, which forms the major source of precipitation in western and central Europe. According to Bárdossy and Caspari (1990) the frequency of the zonal circulation since 1973 forms a break in the time series in the period 1881-1989. Prolonged precipitation periods lead to water saturation of the top soil of catchments, in turn giving rise to more surface runoff and possible flooding. Further, the intense floods of 1990, 1993 and 1995 in Germany (and The Netherlands) were related to the intense and prolonged western circulation and associated precipitation (Caspari and Bárdossy, 1995; Schönwiese, 1999).

Drought is not an extreme weather event in the sense that it does its harm within a short timeframe. But the aggregated loss of crop revenues induced by droughts can add up considerably, not to mention the impact droughts have on water availability and food security in many regions of the world. Numbers on drought losses are generally not included in annual loss figures published by insurance companies (Munich Re, 1999). A consequence of drought that is also often not included in loss figures is the gradual process of land subsidence, even though it amounts to a large proportion of property damage world-wide. Because of persistent droughts, buildings slowly subside and become damaged. In a global warming scenario, not only with increasing, but also decreasing precipitation amounts in many areas in the world, there is a great potential for this peril to become even more important. Probably the major problem with subsidence is that it is not covered by insurance in many countries (Radevsky, 1999), so that the cost becomes the burden of individual property owners.

2. EXTREME WEATHER AND LOSS TRENDS

The occurrence of natural catastrophes and damage costs, is determined by the interplay between the natural and human systems. Thus, in assuming that extreme weather has an increasing impact on society, one must also determine what factors with consequences for vulnerability have changed in society. Influences on the impact of natural disasters that are not part of the natural system, the socio-economic factors, consist of a large number of differently operating factors. Some studies have considered an increasing vulnerability of society as the main driver of the increase in losses from natural disasters over the last decades (e.g. Kunkel *et al.*, 1999; Changnon *et al.*, 2000). Notably, a growing world population in particularly vulnerable areas such as coastal areas and floodplains, and an increasing world gross domestic product are often mentioned as the most important components.

Table 2 lists a number of the socio-economic factors, together with some important natural factors, that influence the cost of natural disasters.

Changes in these factors can have either positive or negative influences. When considering the sum of insured losses, changes in insurance penetration, insurance cover, height of deductibles and changes in approaches to claims also have to be taken into account.

Earthquake losses have not grown as rapidly as losses caused by other (mainly weather related) natural disasters (Bruce *et al.*, 1999), which suggests that aggravating weather may be of importance. One global analysis shows that if global disaster losses are corrected for increased wealth (by means of global GDP) and population growth, these losses still show a factor two increase over the period 1970-1999, caused by other factors, such as a change in the climate (Munich Re, 1999: page 80).

Table 2. Some major factors that contribute to changes in the costs of natural disasters. The ordering is random, as each factor may be of different importance for different weather events or different areas in the world.

Natural factors	Socio-economic factors
- Frequency of extreme event - Intensity of extreme event - Duration of weather events	- Increasing population - Habitation of particularly vulnerable areas (coastal zones, flood plain areas) - Increasing wealth (more expensive construction of buildings, larger amount of increasingly expensive goods, larger stocks, larger number of increasingly expensive vehicles) - Increasing urbanisation - Change in management of rivers and coastal zones - Land use change (environmental deterioration, changed agricultural techniques, vegetation cover) - Changing or negligence of building codes and construction methods - Improving forecast, warning, and evacuation schemes - Increasing awareness of population

A clear, direct relationship between natural phenomena, socio-economic factors, and disaster losses is in most cases not yet established. Some studies have attempted to quantify the proportions of each contributing factor. For hurricane damage in the United States, the trend of increasing economic losses has shown to be mainly the product of socio-economic factors; after correction for wealth, population, and inflation, the trend in costs is actually downward (Pielke and Landsea, 1998). A study of floods in the United States has shown that changes in the climate (i.e. frequency and intensity of the floods) play an important role, besides socio-economic factors (Pielke and Downton, 2000). However, the two mentioned studies do not take into account certain measures, such as forecasting and early warning systems, improved flood protection, and risen awareness amongst the population, that may have reduced the amount of loss of life and damages to property. For instance, the number of lives lost in Bangladesh caused by flooding has decreased considerably, as a result of early warning and the building of flood shelters. Additionally, the study by Pielke and Landsea (1998) does not consider the influence of strong El-Niño events in the 1990s, which were damping the formation of tropical storms.

In Europe, as well as in the rest of the world outside the USA, relatively little work has been carried out on the analysis of individual causes of the increasing natural disaster costs. In the case of flooding, population migration to river valleys and coastal areas is not so likely to be an important factor in Europe, compared to the situation in developing countries. But higher winter temperatures in Europe are leading to precipitation that more often falls as rain, instead of snow. Additionally, long persistent precipitation periods have been observed. These aspects of the weather have caused major occurrences of flash-floods and landslides, most recently in Italy and Switzerland in fall 2000.

Exact quantification of the factors contributing to damages is difficult, as each event is more or less unique in terms of occurrence, duration, and affected area. Additionally, good quantitative, homogenous, and long data series for both the natural meteorological event and the socio-economic factors are lacking in many cases, and data from most areas in the world are simply not available. In summary, it can be stated that too little is known yet to state any general valid conclusion on the relative contribution. But the observed increasing trend of losses, despite significant and ever increasing efforts at fortifying infrastructure and enhancing disaster preparedness, is one that is to be expected under a changing climate regime.

3. DISASTER LOSSES: WHO PAYS?

Disasters can have an important impact on the economy and economic development of countries. For instance, in 1998, the massive flooding in China produced losses in the order of 5–7% of Chinese GDP. Hunduras experienced losses up to as much as half of its GDP after hurricane Fifi struck in 1974 (Hooke, 2000). Sudden-impact disasters damage and destroy capital stocks and infrastructure, conceivably boosting levels of investments as lost assets are replaced. However, subsequent reconstruction and rehabilitation efforts may not increase the net capital stock except to the extent that they entail upgrading. Meanwhile, disasters can act as an disincentive to new investment – or even to the replacement of existing investment. In assessing investment performance in post-disaster circumstances, it is therefore important to distinguish between investment, which replaces or repairs destroyed and damaged assets, and new investments, which may be depressed (Benson, 1997). Arriens and Benson (1999) have estimated that in the period 1988-1999 on average 5.6% of the loans of the Asian Development Bank were used for disaster rehabilitation. Top year was 1992 with 20% of the total of loans. The World Bank has recognised the importance of integrating disaster prevention and mitigation measures in its development activities, to protect investments and foster future development activities (Kreimer and Arnold, 2000).

The insurance sector has recognised climate change as a threat to their business (Changnon *et al.*, 1997; Munich Re, 1999; Swiss Re, 1999a; Vellinga *et al.*, 2001). Most likely, insurance companies will not immediately run a too great risk from natural disasters, also because the markets keep globalising, companies are inventing new tools to transfer risks, and are merging with other than risk-markets. Moreover, adjusting premiums, increasing deductibles, and withdrawing from markets seems reasonable and justifiable, in order to cope with increasing claim amount (Tucker, 1997; Tol, 1998). But others warn that the insurance industry will have to incorporate preventive planning, assess liabilities, and adjust catastrophe losses properly to deal with the ever increasing losses (Berz, 1999). Additionally, the sector can make a contribution to protective measures by incentives to prevent losses and by raising awareness amongst governments and individuals.

From the total economic losses in the world caused by weather related natural disasters, about 45% results from flooding (Munich Re, 1999). In a review of European flood insurance systems, Van Schoubroeck (1997) concludes that the types of insurance vary widely amongst countries, and that in many commercial insurance does not cover the needs. In these cases the cost is born by the victims, supplemented with ad hoc aid and governmental relief. Large corporations have schemes of self insurance against large natural catastrophes. The insurability of some natural disasters has indeed been questioned, due to possible changes in their nature (Kunreuther, 1998). In Europe, major parts of natural catastrophe losses are covered by governments, to compensate for damages for which no conventional insurance can be bought from the private sector. Such as in The Netherlands, where the government has implemented a law to reserve money for taking up part of the costs due to flooding caused by precipitation. But recently, investigations have been made to shift parts of the flooding risk from individuals (and the government) to private insurers (HKV, 1999). Other governments provide collective insurance through partially or wholly state-owned (re)insurance companies for perils not covered by commercial insurers, on the basis of solidarity among citizens. For instance in France, where payment is made after an event is declared "a disaster" by the State (CCR, 1999).

In developing countries, insurance is largely absent as a tool to mitigate costs, as shown by the market share of premiums collected world-wide. In 1998, 1.3% of world insurance premiums was collected in Africa, and 5.5% in Asia, excluding Japan (Swiss Re, 1999b). As in many countries adequate insurance is lacking, a large part of the rebuilding costs will be taken up by federal or local governments, and especially individuals. It is also important to note that in most areas in the world one cannot insure oneself against flood and drought-related disasters. With the flood hazard, victims are too easily identified, and the concept of insurance, covering the cost of a few by the premiums of many, does not work.

In the climate change discussion over the last decade, a number of proposals for covering the costs of disaster damage due to climate change have emerged. A number have considered insurance as an important tool, however, not all of them may be considered as insurance in the strictest sense (Tol, 1998). A scheme proposed by the Association of Small Island States (AOSIS) for inclusion in the United Nations Framework Convention on Climate Change (UNFCCC), consisted of a pool for funding the cost of adapting to climate change (Wilford, 1993). These funds must then finance the adaptation to sea-level rise. Rather than an insurance pool, this is a liability claim from small island states to the industrialised nations, as the latter would have to provide the "premiums". This mechanism was however not included in the final draft of the Convention. In the Kyoto Protocol, the Clean Development Mechanism (CDM) provides a basis for funding adaptation, as it states that a portion of the proceeds from the CDM projects is to be used to meet the needs of "particularly vulnerable" Parties for adaptation (United Nations, 1992; UNFCCC, 1997). However, there are still many issues to be resolved, as different interpretations of the documents remain (Werksman, 1998; Yamin, 1998).

Wilford (1993) discusses issues of liability, such as oil pollution and nuclear damage, for which there exist conventions on liability and the payment of costs. Modern environmental liability insurance is an effective measure for companies to pay for pollution, accidents and unforeseen harm to society or the environment. An illustration of modern liability is the current pressure on the tobacco industry, where victims of their products (cigarettes) demand compensation. Skogh (1998) indicates that as time proceeds, new catastrophes and industrial hazards place new demands on states and governments to compensate, as states have often followed the idea of the welfare state. Also, natural disasters are mentioned as potential new hazards, together with examples such as dam breaks, the mad cow disease, and chemical accidents. In these cases, conventions have to be agreed upon in an international context, to provide a basis for industrial liability.

A recent proposal to meet the increased risks of countries where funds for adaptation are lacking, and insurance is not a tool as commonly available as in industrialised countries, is suggested by Linnerooth-Bayer and Amendola (2000). They argue that subsidised risk transfer can be an efficient and equitable way for industrialised countries to assume partial responsibility for the increasing disaster losses in developing countries. This subsidy could consist of the purchasing of alternative risk transfer mechanisms from developing countries at low interest rates, for instance catastrophe bonds, by banks such as the World Bank and the Asian Development Bank. These banks can then issue these to the private investors. Not only on the basis of liability, but also on the basis of solidarity among nations this might be a way to proceed.

4. AN ALTERNATIVE MECHANISM

Conventional insurance consists of the transfer of risk from an individual to a group, whereby premium is paid by many, to cover the costs of a few. However, as some regions of the world are likely to suffer increasingly more severe damage due to weather-related disasters than other regions, and insurance is not yet available everywhere, this system might need a world-wide expansion. The global private insurance market is already ever expanding, merging, and offering new products to effectively transfer risk. However, the present situation shows that there are perils in certain regions that cannot be insured due to their high loss potentials in comparison with the number of potential buyers, or due to the fact that the local population simply cannot afford to buy the insurance. To subsidise insurance, one can imagine a pool in which part of the premiums is collected. The money for this pool should be provided by countries on the basis of their relative contribution to greenhouse gas emissions.

Figure 2 illustrates such a mechanism. It attempts to reduce the cost-impact of natural disasters by funding risk transfer, tries to finance physically adaptation to the changing climate and weather patterns, and also seeks to prevent increases of additional losses by urging to mitigate greenhouse gas emissions.

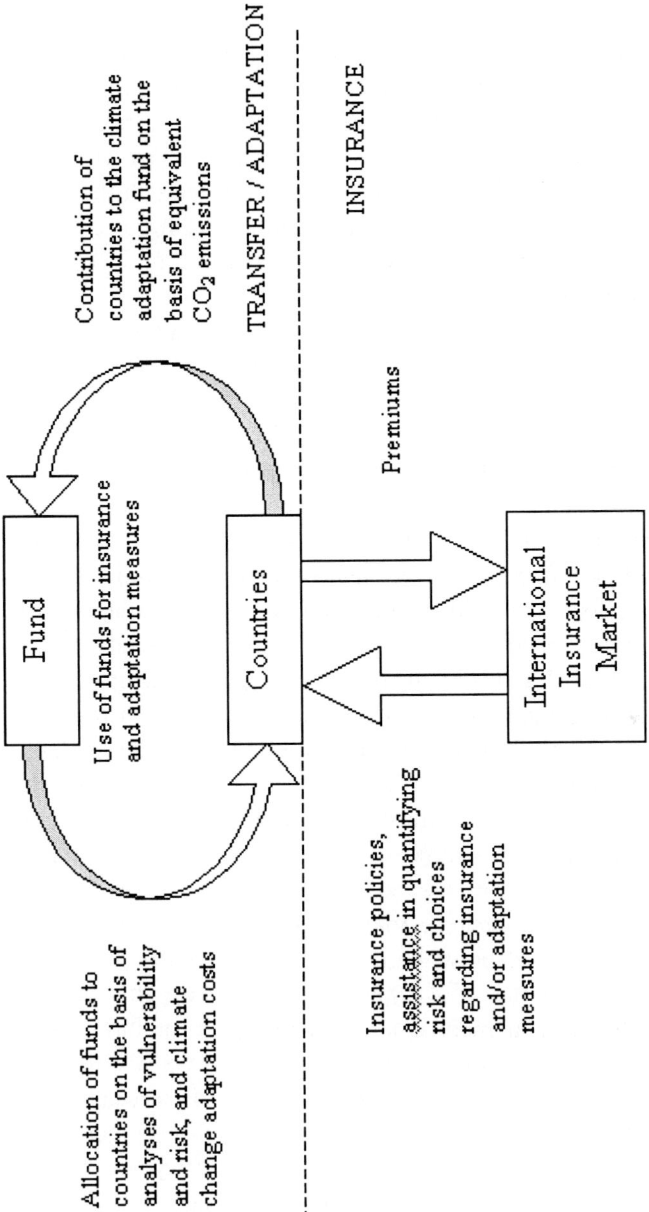

Figure 2. Schematic illustration of a mechanism to fund insurance and adaptation measures.

A panel reviews the emissions of each country, and requires a sum of money according to equivalent tons of CO_2 emissions (the transfer part). The fund is then used to subsidise insurance via the world insurance market (the insurance part), and to finance adaptation to natural disasters to reduce damage costs. The money is not transferred directly to the insurers, but first to individual countries. The fund can finance part of the premium, which is the cost of the "new" risk arising from global climate change, in addition to the "old" risk. This fund could for instance be managed by the Global Environment Facility (GEF), which already executes projects in the context of climate change adaptation. As the sums are paid according to equivalent emissions, this is an incentive to reduce emissions; it could be incorporated in the compliance mechanisms of the Kyoto Protocol.

The funds are to be used for insurance and adaptation measures and they are distributed between these two on the basis of own risk assessments by the countries. Most likely, many Annex I countries will have to finance the fund, and many Annex II countries will receive financial assistance. However, as a number of Annex I countries are also faced with problems regarding the financing of risk transfer or adaptive measures, for instance in the event of sea-level rise, floods or storms, they also receive assistance from the fund. In the end, a balance between the amount of pollution and the amount of adverse effects of severe weather (and cost of adaptation) is reached for each country. The world insurance market plays an important role in the scheme, both as a player in the insurance market, in setting up new businesses, and in acting as an advisor for adaptation and/or initiator of adaptation measures. Also, the market can assist in assessing risks and advising on the most efficient ways to deal with the new risk, whether it is insurance or adaptation. Indeed, the most cost-effective options have to be identified, before deciding upon which action to take. Adaptation measures can be very costly, and inventive schemes where part of the risk is shared between the insurer and the insured, may be very effective for insuring against natural catastrophes, as shown by Loubergé and Schlesinger (1999).

An example of how the risk of damage due to climate change can be analysed and quantified, and what options for coping with the increased risk is given below. Risk in the form of extreme weather varies due to climate change. This variation in risk can be quantified in terms of damage cost. For example, a certain river flood event in Country A, which previously had a return period of 100 years, now has a return period of 50 years under the changed climatic circumstances. Assuming that the damage in Country A for a one in a hundred year flood is 1 billion US$, the cost per year for this flood event has changed from 10 to 20 million US$ per year. If the country wishes to insure this risk, the premium would have to be in the order of 20 million US$ per year. Alternatively, the country could adapt to the changed risk by improving its flood protection, which may amount to only 30% of the cost of damage per year, i.e. 6 million US$. A risk assessment, performed by the country itself, assisted by insurance and risk-experts, would be necessary to identify the most efficient way to adapt, probably by a combination of increasing flood protection and partially insuring the risk of damage due to flooding. Similar schemes can be envisaged for storms and seawater flooding damage. To calculate the necessary funds, one adds up all costs world-wide due to climate related damaging events. These costs then can be divided by the greenhouse gas emissions, to reflect a "price for emitting". The amount of funds will not be enormous. Ideally, the fund would cover 100% of the cost, but the proposed scheme tries to subsidise just a part of the risk.

This scheme is by no means a complete answer to increasing natural catastrophe risks. Many questions remain pending, such as; what countries are relatively the "victims" of climate change, and which ones are the "polluters"? How is the ability (or inability) of countries to buy insurance measured? How much money should be involved? Vulnerability, adaptive capacity, and emissions and sinks of greenhouse gasses have to be investigated and accurately quantified on a national basis, and internationally communicated and agreed upon. Especially the theoretical basis for deciding whether to invest in adaptation measures or insurance schemes needs to be further explored.

5. CONCLUSIONS

The present knowledge of climate change indicates that society will see more frequent and more intense extreme weather in the future, as global warming continues. However, as argued above, the level of confidence is not yet well known. But even in the absence of a permanent change, or of clearly a more extreme nature of weather events, society has already developed itself towards more sensitivity and vulnerability. In this light, both adaptation to, and mitigation of climate change (avoidance), seem justified. Estimates of the price of adaptation vary, but a quick look at present and future measures of protection against flooding and sea-level rise show that the costs are likely to be enormous. Funding the adaptation would be most equitably carried by the industrialised countries in the present distribution of wealth and contribution to greenhouse gas emissions. Meanwhile, as time goes by, a liability claim to the industrialised world from countries benefiting negatively from global warming, seems likely. An international scheme, which both offers adaptation possibilities to victim countries, and gives incentives for mitigating the causes of the disasters, is an effective and powerful tool to cope with the risk of climate change. The proposed levy on greenhouse gas emissions may be implemented to give an incentive for reducing greenhouse gas emission and to compensate for increasing natural disaster damage.

Further analysis of the exact present and possible future impacts of a changing climate on weather extremes is needed. Changes in vulnerability, which obviously play an important role in the increase in costs – if not in some cases a determining one – need to be studied. Additionally, effective measures comprising both insurance and adaptation measures need to be identified. These analyses are required to motivate agreements for measures to compensate countries that suffer most from increases in natural disasters.

6. REFERENCES

Arriens, W.J.T., and C. Benson, 1999: Post disaster rehabilitation: the experience of the Asian Development Bank. Paper for IDNDR-ESCAP regional meeting for Asia: risk reduction and society in the 21st century, Bangkok, 23-26 February 1999. Asian Development Bank, Manila, Phillipines, 17 pp.

Bárdossy, A., and H.J. Caspari, 1990: Detection of climate change in Europe by analyzing European atmospheric circulation patterns from 1881 to 1989. *Theoretical and Applied Climatology,* 42(3), 155-167.

Benson, C., 1997: The economic impacts of natural disasters in Fiji. Working Paper 97, Overseas Development Institute, London, UK, 108 pp.

Berz, G.A., 1999: Catastrophes and climate change: concerns and possible countermeasures for the insurance industry. *Mitigation and Adaptation Strategies for Global Change,* 4(3/4), 283-293.

Bruce, J.P., I. Burton, and M. Egener, 1999: *Disaster Mitigation and Preparedness in a Changing Climate.* Research Paper No. 2, Institute for Catastrophic Loss Reduction (ICLR), Ottawa, Canada, 36 pp.

CCR, 1999: *Natural Disasters in France.* Caisse Centrale de Réassurance, Paris, France, 22 pp. (in French and English)

Caspari, H.J., and A. Bárdossy, 1995: Markieren die Winterhochwasser 1990 und 1993 das Ende der Stationarität in der Hochwasserhydrology infolge von Klimaänderungen? *Boden und Wasser,* 47(3), 18-24. (Do the winterfloods of 1990 and 1993 mark the end of stationary high-water hydrology as a consequence of climate change?) (in German)

Changnon, S.A., D. Changnon, E.R. Fosse, D.C. Hoganson, R.J. Roth Sr., and J.M. Totsch, 1997: Effects of recent weather extremes on the insurance industry: major implications for the atmospheric sciences. *Bulletin of the American Meteorological Society,* 78(3), 425-431.

Changnon, S.A., R.A. Pielke, Jr., D. Changnon, R.T. Sylves, and R. Pulwarty, 2000: Human factors explain the increased losses from weather and climate extremes. *Bulletin of the American Meteorological Society,* 81(3), 437-442.

Dessens, J., 1995: Severe convective weather in the context of a global nighttime warming. *Geophysical Research Letters,* 22(10), 1241-1244.

Easterling, D.R., G.A. Meehl, C. Parmesan, S.A. Changnon, T.R. Karl, and L.O. Mearns, 2000: Climate extremes; observations, modeling, and impacts. *Science,* 289, 2068-2074.

HKV, 1999: *Onderzoek van waterschade ten gevolge van neerslag.* HKV – lijn in water, Lelystad, The Netherlands, 66 pp. (Research of damage caused by precipitation) (in Dutch)

Hooke, W.H., 2000: U.S. participation in International Decade for Natural Disaster Reduction. *Natural Hazards Review,* 1(1), 2-9.

IPCC, 2001: *Climate Change 2001 – The Scientific Basis. Contribution of Working Group I to the Third Assessment Report of The Intergovernmental Panel on Climate Change.* Cambridge University Press, Cambridge, UK. (in press)

Karl, T.R., and R.W. Knight, 1998: Secular trends of precipitation amount, frequency, and intensity in the United States. *Bulletin of the American Meteorological Society,* 79(2), 231-241.

Kreimer, A., and M. Arnold, 2000: World Bank's role in reducing impacts of disasters. *Natural Hazards Review,* 1(1), 37-42.

Kunkel, K.E., R.A. Pielke Jr., and S.A. Changnon, 1999: Temporal fluctuations in weather and climate extremes that cause economic and human health impacts: a review. *Bulletin of the American Meteorological Society,* 80(6), 1077-1098.

Kunreuther, H., 1998: Insurability conditions and the supply of coverage. In: *Paying the Price, The Status and Role of Insurance in the United States.* Eds. Kunreuther, H., and R. Roth. Joseph Henry Press, Washington D.C., USA, p. 17-50.

Lecomte, E., A.W. Pang, and J.W. Russell, 1998: *Ice Storm '98.* Institute for Catastrophic Loss Reduction, Toronto, Canada, 39 pp.

Linnerooth-Bayer, J., and A. Amendola, 2000: Global change, natural disasters and loss-sharing: issues of efficiency and equity. *The Geneva Papers on Risk and Insurance – Issues and Practice,* 25(2), 203-219.

Lins, H.F., and J.R. Slack, 1999: Streamflow trends in the United States. *Geophysical Research Letters,* 26(2), 227-230.

Loubergé, H., and H. Schlesinger, 1999: Optimal catastrophe insurance with uncorrelated catastrophes. Working Paper 99.03, Department of Economics, University of Geneva, Geneva, Switzerland, 16 pp.

Meehl, G.A., T. Karl, D.R. Easterling, S. Changnon, R. Pielke Jr., D. Changnon, J. Evans, P.Ya. Groisman, T.R. Knutson, K.E. Kunkel, L.O. Mearns, C. Parmesan, R. Pulwarty, T. Root, R.T Syves, P. Whetton, and F. Zwiers, 2000: An introduction to trends in extreme weather and climate events: observations, socioeconomic impacts, terrestrial ecological impacts, and model projections. *Bulletin of the American Meteorological Society,* 81(3), 413-416.

Munich Re, 1999: *Topics 2000, Natural Catastrophes – The Current Position.* Munich Reinsurance Company, Munich, Germany, 126 pp.

Munich Re, 2000: *Topics, Annual Review of Natural Catastrophes 1999.* Munich Reinsurance Company, Munich, Germany, 46 pp.

Pielke, Jr., R.A., and C. Landsea, 1998: Normalized hurricane damages in the United States 1925-95. *Weather and Forecasting,* 13(3), 621-631.

Pielke, Jr., R.A., and M. Downton, 1999: U.S. trends in streamflow and precipitation: using societal impacts data to address an apparent paradox. *Bulletin of the American Meteorological Society,* 80(7), 1435-1436.

Pielke, Jr., R.A., and M.W. Downton, 2000: Precipitation and damaging floods: trends in the United States, 1932-97. *Journal of Climate,* 13(20), 3625-3637.

Radevsky, R., 1999: *Subsidence, a Global Perspective.* General Insurance Research Report No. 1, Association of British Insurers, London, UK, 44 pp.

Schönwiese, C.D., 1999: Trends mittlerer und extremer Niederslage als Ursache von Abflußtrends? In: *Extreme Naturereignisse und Wasserwirtschaft – Niederslag und Abfluß*

– *Internationales Symposium.* Bayerisches Landesamt für Wasserwirtschaft, Munich Germany, p. 105-114. (Trends of middle and extreme precipitation, a cause of streamflow trends?) (in German)

Skogh, G., 1998: Development risks, strict liability, and the insurability of industrial hazards. *The Geneva Papers on Risk and Insurance – Issues and Practice,* 23(87), 247-264.

Swiss Re, 1999a: *Climate Research Does Not Remove the Uncertainty – Coping With the Risks of Climate Change.* Swiss Reinsurance Company, Zurich, Switzerland, 8 pp.

Swiss Re, 1999b: *World Insurance in 1998.* Sigma Report 7, Swiss Reinsurance Company, Zurich, Switzerland, 29 pp.

Tol, R.S.J., 1998: Climate change and insurance: a critical appraisal. *Energy Policy,* 26(3), 257-262.

Tucker, M., 1997: Climate change and the insurance industry: the cost of increased risk and the impetus for action. *Ecological Economics,* 22(2), 85-96.

UNFCCC, 1997: *Kyoto Protocol.* United Nations Framework Convention on Climate Change (UNFCCC), Bonn, Germany, 23 pp.

United Nations, 1992: *United Nations Framework Convention on Climate Change.* United Nations, New York, USA, 32 pp.

Van Schoubroeck, C., 1997: Legislation and practice concerning natural disasters and insurance in a number of European countries. *The Geneva Papers on Risk and Insurance – Issues and Practice,* 22(83), 238-267.

Vellinga, P., E. Mills, G. Berz, L. Bouwer, S. Huq, L.A. Kozak, J. Palutikof, B. Schanzenbächer, G. Soler, C. Benson, J. Bruce, G. Frerks, P. Huyck, P. Kovacs, X. Olsthoorn, A. Peara, S. Shida, and A. Dlugolecki, 2001: Insurance and other financial services. In: *Climate Change 2001 – Impacts, Adaptation and Vulnerability. Contribution of Working Group 2 to the Third Assessment Report of the Intergovernmental Panel on Climate Change.* Cambridge University Press, Cambridge, UK. (in press)

Werksman, J., 1998: Compliance issues under the Kyoto protocol's clean development mechanism. In: *The clean development mechanism.* Draft working papers WRI/FIELD/CSDA, October. Foundation for International and Environmental Law and Development (FIELD), London, UK, p. 32-39.

Werner, P.C., F.W. Gerstengarbe, K. Frädrich, and H. Oesterle, 2000: Recent climate change in the North Atlantic/European sector. *International Journal of Climatology,* 20(5), 463-471.

Wilford, M., 1993: Insuring against sea level rise. In: *The Global Greenhouse Regime – Who Pays?* Eds. Hayes, P., and K. Smith. Earthscan, London, UK, p. 169-187.

Yamin, F., 1998: Adaptation and the clean development mechanism. In: *The clean development mechanism.* Draft working papers WRI/FIELD/CSDA, October. Foundation for International and Environmental Law and Development (FIELD), London, UK, p. 41-52.

Scenario analyses in global change assessment for water management in the lower Rhine delta

H. MIDDELKOOP[1], J.C.J. KWADIJK[2], W.P.A. VAN DEURSEN[3], AND M.B.A VAN ASSELT[4]

[1]*Department of Physical Geography, Utrecht University, Utrecht, The Netherlands*
[2]*WL|Delft Hydraulics, Delft, The Netherlands*
[3]*Carthago Consultancy, Rotterdam, The Netherlands;*
[4]*ICIS, Maastricht, The Netherlands*

Abstract: The river Rhine fulfils important socio-economic and environmental functions for the riparian states in its 185,000 km^2 large basin. Even under present-day conditions, the demands for flood protection, inland navigation, ecology, hydropower, water availability for agriculture, industry and drinking water are high. Climate change is expected to enhance discrepancies between water supply and demand in the Rhine River. This issue has been investigated in a sequence of climate-impact studies in the Rhine basin. These studies firstly focused at hydrological changes and at the implications for various water-related sectors in the Netherlands. These results indicated that the water systems and user functions are sensitive to climate change. In addition, the study demonstrated that there are major uncertainties, not only in the climate scenario, but also in the socio-economic developments, which together determine the vulnerability of the water systems to global change. In an ongoing project we will establish a set of integrated scenarios for the Rhine and Meuse basins, representing different world-views, based on the so-called Perspectives method. Using these scenarios we attempt to define water management strategies that are robust under uncertainty.

1. INTRODUCTION

The river Rhine fulfils important socio-economic and environmental functions for its riparian states and, in particular, for the Netherlands. Safety against flooding and the water demands of the user functions, such as ecology, inland navigation, hydropower, agriculture, industry and drinking

M. Beniston (ed.),
Climatic Change: Implications for the Hydrological Cycle and for Water Management, 445–463.

water are under pressure already under present-day conditions. Because of global change discrepancies between water supply and demand in the Rhine River may be enhanced. Therefore, several projects have been carried out over the past years to assess the possible implications of climate change in the Rhine and Meuse basins for flood protection, water availability and water management (Grabs, 1997; Middelkoop et al., 2000). One of these projects, carried out by Utrecht University, RIZA, WL|Delft Hydraulics and Carthago Consultancy within the framework of the Dutch National Research Programme on Global Air Pollution and Climate Change (NRP), addressed the hydrological changes in the Rhine basin in response to climate change, as well as the sensitivity of water-related functions and water management in the Netherlands to the assumed changes (Middelkoop, 2000). These analyses have shown that water demand and supply may considerable change. The uncertainties around the estimates of future impacts, however, are large. Climate change and the hydrologic response are major causes of this uncertainty, as they may affect water *availability*. The uncertainties also depend on various unknown socio-economic and agro-economic developments (such as population growth, industrial expansion, land use changes, and use of different crop types), that will affect water *demand*. The question that rises is then: which is, given the uncertain future, the best water management strategy?

The objective of this project is to analyse the sensitivity of the water-bound functions in the Netherlands to global change and, subsequently, to provide an integrated framework for decision making in water management in the lower Rhine basin to identify policy options that are robust under uncertain future conditions.

2. METHODS

As a first step, a sensitivity test of the hydrological regime in the lower Rhine basin (figure 1) to changes in climate and land use was carried out. This was done using a suite of inter-linked models that firstly determine the hydrological changes. Subsequently, for the water systems in the Netherlands, a more detailed impact assessment regarding the demands and vulnerability of the user functions of the water systems was carried out. Here, a distinction is made between the sub-systems 'Rivers', 'Lake IJsselmeer', and 'Terrestrial areas'.

In the second step, existing socio-economic, demographic, land use and climate change scenarios as well as existing water management strategies were structured using the Perspectives method into a set of consistent integrated scenarios. These represent future changes in demand and supply for the Rhine and the Meuse, as well as the according water management policies. This step in the project has recently been finished. In the forthcoming stage of the project, the hydrological and socio-economic consequences for the water systems in the Netherlands will be determined using the models for different combinations of scenarios against policy strategies. The outcome will be the basis for the evaluation of the water management strategies.

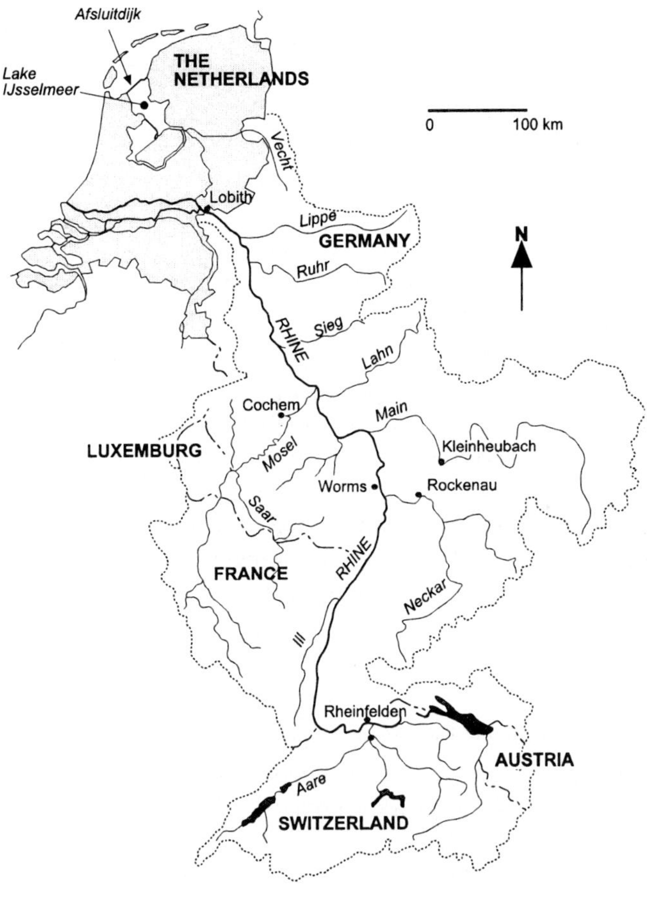

Figure 1:The Rhine basin.

2.1 Sensitivity analyses

2.1.1 Climate change scenarios

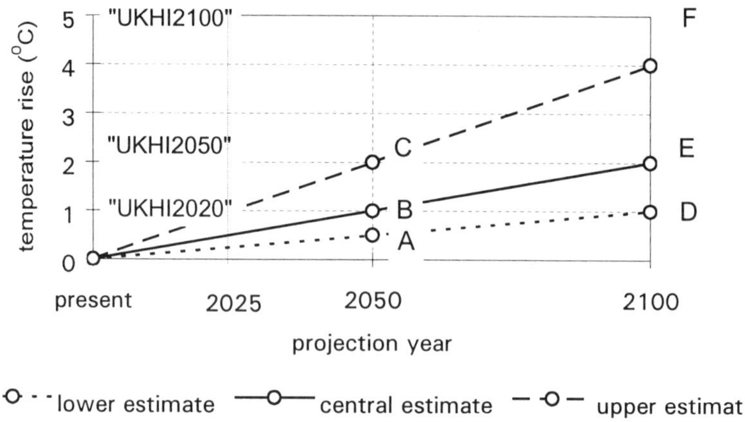

*Figure 2:*Climate change scenarios used in this study: lower, central and upper estimates for different projection years, and relation with the UKHI GCM experiment.
A = lower estimate for 2050; B = central estimate for 2050; C = upper estimate for 2050; D = lower estimate for 2100, same climate change as B; E = central estimate for 2100, same climate change as C; F = upper estimate for 2100

For the first sensitivity analyses, three climate change scenarios were considered, corresponding to a central estimate of climate change for the Rhine basin with a temperature rise equal to 2°C in the year 2100, with a lower estimate of +1°C and an upper estimate equal to +4°C (IPCC, 1996; Können & Fransen, 1996; Können et al., 1997). The changes projected to the year 2050 were assumed to be half of those in 2100 (figure 2). The construction of climate scenarios was based on the Hadley centre's high-resolution atmospheric General Circulation Model (UKHI) (Hulme et al., 1994; Grabs et al., 1997), which has been processed by the Climatic Research Unit, University of East Anglia. For each scenario, anomalies of mean monthly temperature, precipitation, and potential evaporation have been determined. These have been scaled in accordance with the assumed temperature rises for the lower, central and upper estimates for the projections years 2050 and 2100. Under conditions of the central estimate in 2050 precipitation increases in winter between 4% in the Alps and 8% in the northern part of the Rhine basin. In summer, precipitation decreases by about 2% in the Alps, while in the northern part it increases by about 2%. Using these anomalies, the baseline climate series (1960-1990) for the hydrological models was adapted. For all scenarios the corresponding sea level rise (SLR) was established in accordance to NW4 (1997). SLR varies between 20 cm for the lower estimate in 2050 to 110 cm for the upper estimate in 2100.

2.1.2 Land use scenarios

The sensitivity analyses also considered a scenario for changes in land use in the Rhine basin (Veeneklaas et al., 1994) for the year 2050, reflecting changes caused by socio-economic developments and changed climate. The scenario envisages an increase in crop production due to increased CO_2 concentrations and a higher temperature, and a reduction in de demand for agricultural products due to an assumed low population growth. This results in a reduction of the area of agricultural land between 30% and 60%, depending on the area and crop type. For the Netherlands, a scenario with autonomous changes in land use was considered, in which the area of agriculture land reduces by about 20% and is replaced by nature (Haasnoot et al., 1999).

2.1.3 Other developments

In addition to climate and land use, many other, climate-independent, boundary conditions may change over the years to come. Examples that are important for the Netherlands include soil subsidence of the western part of the Netherlands, nature development plans, changes in the amounts of goods to be transported by inland navigation, or changes in water demand for agriculture, industrial and domestic use. In the sensitivity analyses, all these changes together have been considered as a scenario, projected to the year 2050, describing the (hypothetical) situation without climate change, which serves as a reference situation (REF2050) for the scenario analyses that consider both climate change and these climate-independent boundary conditions.

2.2 Modelling tools

The effects on the river regime, including runoff, sediment production, transport and deposition, have been assessed using mathematical models embedded in a Geographical Information System. Hydrological changes in response to climate and land use changes in the Rhine and Meuse basins are calculated on a 10-daily basis using the GIS-based water balance model RHINEFLOW-2 (Kwadijk, 1993; Van Deursen, 1999). For each of the water systems 'Rivers', 'Lake IJsselmeer', and 'Terrestrial areas' in the Netherlands separate models have been developed to simulate the hydrological changes. Estimates of changes in peak flows for the lower

Rhine River are based on statistical downscaling of the results of the coarse-scale water balance models. The impacts on the IJsselmeer lake area is studied using a water balance model BEKKEN (Buiteveld & Lorenz, 1999), and the effects on terrestrial areas is assessed using the nation-wide groundwater models MOZART-NAGROM (Haasnoot et al., 1999). Subsequently, tools have been developed to evaluate the consequences for the water-related user functions considered (safety, nature, navigation, agriculture, industry, and drinking water production) in each sub-system. Examples are the Landscape Planning DSS for rivers, to evaluate scenarios of landscape planning in relation to changes in design discharges, the SHIPS@RISK tool that allows interactive evaluation of the effects of changing discharges on inland navigation, and empirical functions to estimate water logging and drought damage from ground water tables calculated by the MOZART model (Middelkoop, 2000). The suite of models is schematically shown in figure 3.

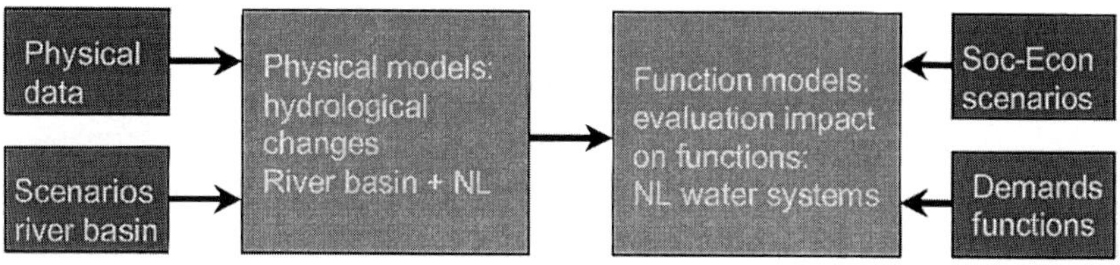

Figure 3: Outline of the suite of models (light grey) and inputs (dark grey). The scenarios include climate changes and land us changes.

3. RESULTS OF THE SENSITIVITY ANALYSES

3.1 Hydrological changes

3.1.1 River discharge

Due to climate change, the Rhine will change from a combined rainfed–snowfed river to a predominantly rainfed river. Winter precipitation will increase, while higher temperatures cause that in the Alpes a smaller proportion of precipitation is stored in the form of snow, but directly comes to runoff. The reduced snow accumulation causes a reduction of meltwater contribution in late spring. Together with a net precipitation shortage in the entire basin, summer discharge will decrease. The overall result is an increase in the intra-annual variation of Rhine discharge (figure 4).

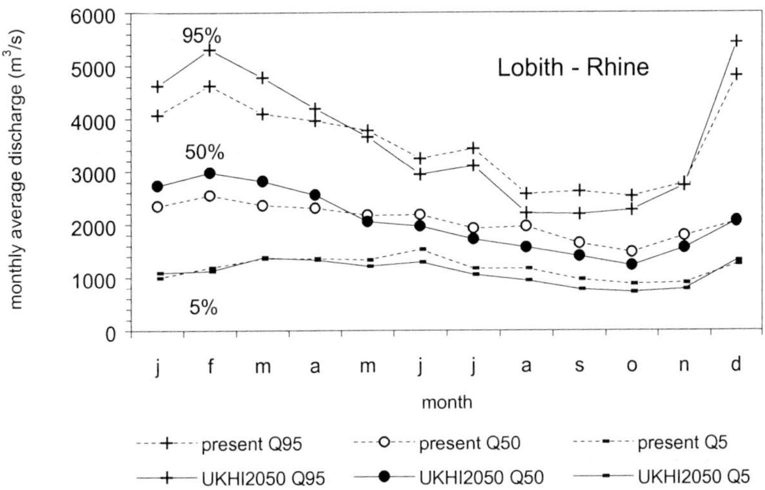

Figure 4: Monthly average discharge of the River Rhine at Lobith according to the UKHI2050 scenario. The changes correspond to the upper estimate for the year 2050 and central estimate for 2100.

3.1.2 Water systems in the Netherlands

The changes in river discharge will affect the upstream boundary of the water systems in the Netherlands. At the downstream border of the Netherlands sea level may rise by several dm. Finally, direct climate parameters will change: winters will be warmer and wetter, while water deficit will be larger during summer, although heavy rainfall events become more intensive. Under the present-day management, the current target lake levels in the IJsselmeer will be exceeded most of the time in winter due to sea level rise, while a reduction of inflow of Rhine water during summer will cause water levels falling below the target level. If no measures are taken, peak water levels in the IJsselmeer lake will increase almost proportionally to sea level rise. Changes in wind speed and direction may also affect extreme water levels. However, there are presently no scenarios that provide plausible estimates for changes in extreme wind fields.

Geographically, the ongoing land subsidence and climate change will enhance the present-day differences between the higher Pleistocene areas in the SE part of the Netherlands where infiltration is accelerated on the one hand, and the low lying polder areas in the W part where upward seepage flows intensify on the other hand. This results in a lowering of groundwater levels in infiltration areas and higher groundwater levels in seepage areas. Since the designs of waterways, flood protection measures, infrastructure and regulations in water management are based on discharge extremes, it may be expected a-priori that these are sensitive to a climate induced amplification of extremes. This was explored using the user-function models.

3.2 Implications for water-related functions in the Netherlands

The implications for water-related functions in the Netherlands are summarised in table 1. Climate change will firstly affect the current safety levels for flood protection. Along with increased winter discharges, the design discharge (i.e. the discharge with a 1250-year recurrence time) for safety standards along the rivers will increase. Under conditions of the central estimate of climate change the increase for the Rhine may be in the order of 5% by the year 2050 and 10% by the year 2100. This demands enlargement of the discharge and storage capacity of the river channels, leaving less physical space for other functions such as nature, agriculture or industry. In case of the upper estimate of climate change, far-reaching flood reduction measures within the high-water bed are demanded in combination with water retention basins along the rivers. Extreme water levels along the IJsselmeer lakes rise almost proportionally with sea level, reducing the discharge possibilities of excess water from the surrounding polders in winter.

Inland navigation will be mainly affected by changing water levels on the Rhine between Rotterdam and Basle. An increasing peak flow frequency in winter will stop the navigation traffic more often, while a reduced summer flow reduces the navigation depth and hence limits the loading capacity of the ships more frequently and over longer periods. In case the navigation sector would not be able to adapt, the hydrological changes would gradually reduce the transport reliability, and increase transport costs by inland shipping by hundreds of millions DFL per year.

Agriculture will experience an intensified water deficit during summer and accelerated intrusion of salt seepage water in the coastal area. At the same time, there will be less water available from the Rhine and Meuse rivers. Wetter conditions during winter and spring will cause an increase of the water logging damage. The potential effective damage, however, is spatially highly variable, depending on the different local hydrological conditions. The largest damage may be expected in the northern and eastern parts of the Netherlands. Finally, the total damage greatly depends on changes in land use and adaptation by using new crop variants or different crop types over the forthcoming century.

Wet terrestrial ecosystems that depend on shallow groundwater tables will profit from intensified upward seepage in the lower regions. Moreover, larger areas will be suitable for wet ecotopes. Floodplain ecology will experience more frequent flooding in winter and accelerated overbank sedimentation. Low flows during summer, however, result in low river water levels, which in turn leads to lower groundwater levels for wet floodplain ecotopes. Flood reduction measures within the floodplain will drastically change the abiotic conditions of ecotopes. Higher winter levels and low summer levels in the IJsselmeer area are supposed favourable for ecology. Higher water temperatures of the lakes, however, area may increase the risk of algae bloom.

Table 1: Qualitative summary of the implications for water related functions in the Netherlands.

Function	Impact (central estimate, projected to the year 2050)	Socio-economic factors and water management measures that affect the impacts
Safety		
Rivers	Increase in design discharge (recurrence time = 1250 yr) by 5% (Rhine) to 10% (Meuse).	Implementation of flood retention areas; landscaping priorities along the rivers.
Rhine-Meuse estuary	Increase of design water levels (recurrence time = 10,000 yr) by 20-25 cm.	Management rules of sluices in the Haringvliet barrier dam.
IJsselmeer	Increase in maximum levels by about 20 cm. Changes in wind are important but uncertain.	Agreed target lake levels; design of sluice gates in the Afsluitdijk; design of shallow shores.
Inland navigation	Longer periods of low flow in summer when ships cannot be fully loaded. Increase in annual transportation cost by the order of 50-100 Mfl.	Economic development – goods supply; ship types; stock prizes; channel improvements.
Agriculture	Increase of biomass due to higher CO_2 concentrations; increased water deficit in summer; larger damage due to intensified rainfall; salt intrusion in the coastal area.	EC-regulations; introduction of new crop varieties.
Ecology	Rising water tables and	Political choices on the

	intensified upward seepage beneficial for wet ecosystems; opportunities due to win-win strategies in flood mitigating measures; reduced water quality in summer.	development of nature areas; landscape planning strategies along the rivers; implementation of shallow shores in the IJsselmeer.

3.3 Prospects and research issues

The sensitivity analyses have demonstrated that the hydrological system is sensitive to climate changes, which is expected to affect, directly of indirectly, several water-related functions. Several aspects of climate-impact on the water systems were not fully addressed in the present study, and therefore can be put on the agenda for ongoing research. Among these are:

- Impacts for the Meuse River. This river has a considerable different nature when compared to the Rhine. The catchment is much smaller and it is an entirely rainfed river. Consequently, discharge peaks are steeper, and flow can be almost zero during dry summers. These characteristics are very relevant with respect to the effects of climate change on peak flows, but also on reduced water availability shortage during summer.
- The impacts on low flows and water shortage for various users deserves to be considered as a separate issue, particularly for the Meuse but also for the Rhine. While a lot of attention has been paid to peak flows and mitigating measures, less is known about the mechanisms behind low flows (even in our temperate region) and possible measures for preventing or mitigating this problem.
- Water quality has been a major issue in addition to water quantity in the densely populated and heavily industrialised Rhine and Meuse basins. Climate change may affect water quality in different ways. Firstly, biological activity, oxygen content and algae bloom directly depend on water temperature. Secondly, climate induced changes in river flow, resulting in more intense periods of low river flow may lead to increased concentrations of contaminants in the river. Also, sea level rise will intensify the intrusion of salt water into delta areas. Finally, measures to mitigate hydrological impacts of climate change may indirectly affect water quality. For example, stagnant water in flood retention basins along the Rhine produces toxins would prevent any further uptake of infiltration water from the river banks for drinking water production.
- Drinking water production is sensitive to climate-induced changes both in water quantity and water quality. The implications for this sector in the

densely populated Rhine-Meuse delta should therefore be explicitly addressed.

- Analyses of climate impacts on water systems and water users demand a cross-sectorial integration of climate, socio-economic developments, and mitigation measures. Impact analyses therefore should be based on scenarios that combine climate change and socio-economic developments in a coherent and consistent way.
- The results of climate impact studies are still surrounded by major uncertainties, in spite of many efforts to reduce these. Since the uncertainties have many sources, and part of the total uncertainty is structural (i.e. cannot be eliminated, for example by using 'better' models), combined hydrological and socio-economic research should be undertaken to develop water management strategies that are robust under uncertainty.

The present study focuses on the development of integrated scenarios and the development of water management strategies under uncertainty, mentioned in the last two points. This is the subject of ongoing research, which is introduced below.

4. ANALYSIS OF PERSPECTIVE-BASED SCENARIOS

4.1 Introduction

The analyses carried out so far have demonstrated the sensitivity of the water systems and water-related functions to changes in water supply. The analyses also demonstrated that the vulnerability of the system not only depends on climate but also on socio-economic developments (such as changes in land use, water demand, and acceptance of flood risk), and on mitigation measures that might be implemented. Examples are given in the right-hand column of table 1. It is also important to note that most developments are inter-related and thus cannot be considered independently from each other. Although it is clear that demand and supply will change, these developments are surrounded with considerable uncertainties. Together, they will provide the uncertainty ranges for the boundary conditions of water management in future. The question is then how societies can develop water management strategies and how societies can adapt to climate change, despite the inherent uncertainties. Because of the large uncertainties around future conditions, different possible futures can be envisaged, depending on the perspectives people may have. Different perspectives in turn will lead to the adoption of different water management strategies. Of course, water management may adapt to the changing conditions over the years, but in several cases long-term strategies (e.g. large scale widening of the river channels) have to be initiated already before climate changes become apparent, or cannot be adapted at low cost. The next step in the analyses, therefore, was to explore different water management strategies and their robustness under various future conditions.

4.2 Concepts

Uncertainties originate from different sources and result from a number of unknown or partly known social, economic and environmental processes at different scale levels. Scenarios can be helpful tools in developing, analysing and evaluating future strategies. Many different scenarios already exist in the fields of economy, demography, lifestyle, agriculture, physical environment and climate change. All these scenarios, however, differ in terms of underlying assumptions and perspectives. For integrated assessment studies in the Rhine and Meuse basins, therefore, methods are required to structure and inter-relate various categories of future developments into integrated and consistent sets of scenarios.

The basic concept for the next step in the project was to produce a limited set of integrated scenarios that include climate and socio-economic developments in a coherent and consistent way. This is done using the Perspectives method, as developed by the TARGETS research group at RIVM in the Netherlands (Rotmans & De Vries, 1997). A perspective is a consistent description of the perceptual screen through which people interpret the world, and which guides them in acting. A perspective comprises both a worldview (i.e. how people interpret the world) and a management style (i.e. how they act upon it), focusing either on Control (Controllist), Environment (Egalitarian) or Economy (Individualist) (figure 5). If the worldview coincides with the management style (which means that at last, the future will have turned out to be in accordance with the a-priori expectations), we speak of a utopia; otherwise it is called a dystopia. By making different combinations of worldviews and management styles, a matrix of perspective-based scenarios can be developed or existing scenarios can be interpreted and tested.

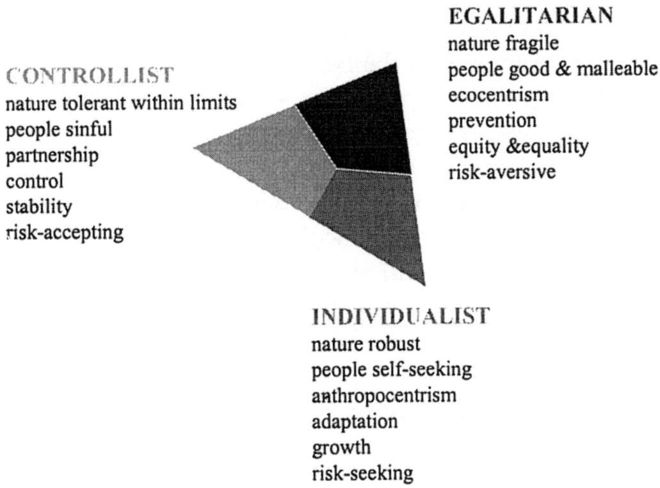

Figure 5: Perspectives and their view on the world.

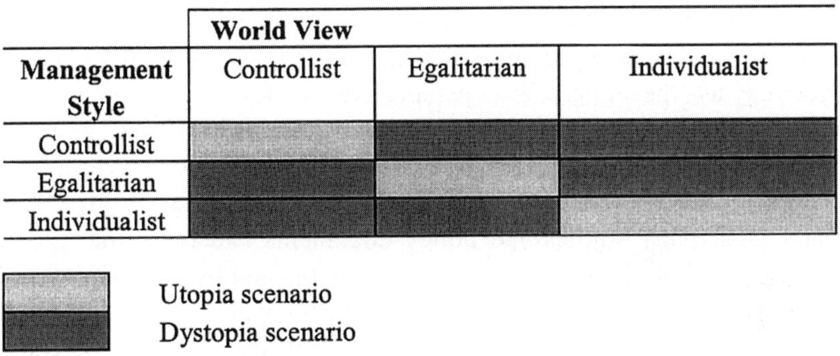

Figure 6: Matrix of Utopias and Dystopias.

4.3 Approach for the Rhine study

In an inventory analysis, a large number of existing plan studies, policy papers, scenario studies and other documents from various different water-related sectors were analysed, and the visions and management strategies of these were categorised according to the three perspectives. Also, expert meetings and interviews were held to convert all the existing scenarios into three coherent and consistent sets of world (water) view and management style, according to the Egalitarian, Controllist and Individualist Perspectives. From these results the utopia-dystopia matrix could be filled-in.

For the modelling analyses, all nine combinations of world views and management styles might be evaluated using the models. However, some combinations may represent unrealistic or less interesting situations. To determine which scenarios from the matrix should at least be considered, a workshop with stakeholders from various water-related sectors was held. In this workshop, various futures and events were put forward by the participants. From these a set of storylines of future developments in the Rhine basin are established, in each of which different combinations of worldviews and management styles occur. These storylines thus primarily determine which scenarios will be evaluated. Secondly, they allow implementing the adaptation of a water management style in response to new events and insights that may become apparent in the course of time.

4.4 Definition of scenarios

4.4.1 Implementation of perspectives

With regard to the Rhine study, the perspectives can be characterised as in table 2. When the documents analysed in the inventory are characterised of in terms of the three perspectives, it appears that most of these can be categorised as the Controllist perspective, a few of them match the Egalitarian perspective and none can be considered as having the Individualist perspective. This indicates that the Dutch style of water management largely matches the Controllist management style. Also, it appeared that many studies and policy documents had the same 'roots', as they were based on the same scenarios, or established by related institutions. The absence of the Individualist perspective in water-related studies may indicate that market oriented sectors do not treat water management explicitly as a separate issue. Apparently, these sectors consider water as one of their (economic) boundary conditions, and water may be a less relevant issue to them when compared to other socio-economic factors. The definition of the Individualist perspective for the present study, therefore, was for a large part based on expert meetings.

Table 2: World view and management style according to the Perspective archetypes, applied to water management in the Rhine basin.

	Egalitarian	Controllist	Individualist
World view	– Climate is very sensitive, large changes must be expected; – Nature is vulnerable; – Society will adapt to environment, not vice versa – Water functions will radically change; – Public awareness of nature has increased; – Economy and water use will decrease;	– Climate is sensitive, changes must be expected, additional research is required; – Nature is vulnerable, but can be managed by defining adequate standards; – Safety has priority; – Uncertainty demands research	– Climate is insensitive, some changes must be anticipated; – Economy is more important than Nature; – The environment can be exploited for economic use; – Water is considered as economic goods; – Economy and water use will grow; – Risk can be calculated and offers challenges
Management style	– Natural processes are guiding in water management; – Water guides landscape planning, – Pro-active, sustainable water management with long time horizon (>50 yr); – Prevention and adaptation instead of mitigation	– Sustainable water management, combined with win-win solutions and negotiation between different interests functions; – Water is integral part of landscape planning; – Water management should be based on standards; – Strong (top-down) control by government (including EC); – River basin approach, also institutional; – Time horizon in planning varies, depending on the function (10-50 yr)	– Cost-benefit analyses and economic risk assessment guide water management; – Short time horizon in planning (10 yr); – Governmental control is weak, role of supervisor; – EC is monetary union, no further control; – Privatisation of water and water-related sectors

To carry out the modelling analyses, each perspective was implemented in the modelling in terms of boundary conditions (e.g. climate scenario, land use change), model parameters (e.g. water holding capacity of upstream areas, discharge capacity of sluice gates), and rules for water demand (navigation, drinking water, industry) and management style (e.g. flood reduction measures). Examples for various models and water sectors are given in table 3.

4.4.2 Storylines

The stakeholder meeting resulted in a large number of events, socio-economic responses and consequences for water management strategies. Examples of such events can be natural floods or periods of drought (both resulting in considerable damage, and possible due to climate change), calamities (e.g. pollution accident) or socio-economic events (e.g. malfunctioning EC regulation, economic recession). Such events may lead to the utopia or dystopia situations, and eventually may lead to the adaptation of the water management strategy. After overlying all acquired pieces of information three storylines were deduced, representing plausible sequences of water management policies, events and responses. Each storyline includes one or more utopias and dystopias from the matrix, but also gives an implementation of adaptation of water management style in response to changing perspective due to natural or socio-economic events.

Table 3: Implementation examples of perspectives in scenarios and model boundary conditions.

Boundary condition or management rules	Egalitarian	Controllist	Individualist
Climate scenario	Climate is very sensitive, upper estimate for temperature change, both wet or dry variants are possible	Central estimate of changes	No change (because of short planning time horizon), or lower estimate for change
Rhine			
Land use changes	More nature, agriculture more adapted to local conditions	Current land use, ER-scenario, minor expansion urban area	Expansion of urban area, land use develops guided by economic processes in EC
Flood retention measures	Maximal renaturalisation of small rivers, increase of water holding capacity of upstream areas, maximum capacity and assumed efficiency of flood retention areas along the rivers	Minor increase in water holding capacity of upstream areas, some renaturalisation, few retention basins, moderate assumed efficiency	No increase in water holding capacity of upstream areas, few retention basins, moderate assumed efficiency, cheapest efficient option
Netherlands			

water systems			
Rivers			
Landscape planning and flood protection	Make way for the river, aiming at maximum resilience of rivers, maximum nature development, no dike raising, retention polders within the Netherlands	Always tries to find win-win solutions for different interests, slow implementation process of flood reduction measures	Cheapest solution for landscape planning, dike raising where necessary
Water distribution	Removal of weirs from Nederrijn, more natural water distribution between river branches	No changes	If necessary, additional weirs for inland navigation
Inland navigation	Small vessels, adapted to shallow rivers, minor growth of goods stream	Unchanged fleet composition, minor growth in goods stream	Large vessels (push-barges), strong growth of good transport, river channels improved by technical means
IJsselmeer area			
Lake level management	No target levels, natural fluctuations	Adapted target levels, standards in accordance to inevitable changes	Current target level maintained
Safety measures	Larger sluice gates in the Afsluitdijk barrier, shallow lake shores, no dike rising	Larger sluice gates in the Afsluitdijk barrier, shallow lake shores, local dike rising	Larger sluice gates in the Afsluitdijk barrier, installation of pumps if necessary, dike rising
Terrestrial areas			
Soil subsidence	Soil subsidence reduced in some polder areas, little subsidence due to reduced natural gas exploration	Soil subsidence as extrapolated from past decennia	Soil subsidence as extrapolated from past decennia
Land use	Main Ecological Network established, no growth of urban area, agriculture adapted to environment	Minor expansion of nature areas, minor expansion of urban area	Agriculture based on economic cost/benefit, major growth of urban area, nature in residual areas
Water management	Natural ground water level fluctuations, inundation of some polders	Present management	Drainage conditions optimised for agriculture

4.5 Scenario analyses and evaluation

Each utopia and dystopia considered relevant from the storylines will be explored using the suite of physical and 'user-function' modelling tools, together with detailed models developed for sub-catchments in the Rhine and Meuse basins. Every utopia or dystopia will be evaluated in terms of cost benefit analyses. The costs include 2 components: the first is the implementation cost of the water management measures, such as retention basins, landscaping measures, dike raising, construction of discharge pumps and do forth. The second component is the cost that will arise after implementation of measures, in terms of the functioning of water-related sectors (e.g. areas of agriculture, nature, preservation of valuable cultural landscape, annual navigation cost) and the cost that may arise due to extreme events (dry periods, floods). The perception of cost and benefits, however, is not the same for each perspective. For example, an egalitarian will consider high implementation cost of reserving large areas for water not as unnecessary wasted money if no flood has occurred, since these measures are an inevitable consequence of the egalitarian's world view. An individualist will accept the loss of nature areas or cultural-historic sites due to dike enforcement if this provides an economically efficient solution for flood protection.

On the basis of the intercomparison between the utopias and dystopias of the different perspectives, the robustness of different management styles will be evaluated. A management style in which not only the utopias lead to benefits, but also for which even the dystopias do not cause high cost, may be considered robust under different futures. In this way, the results may help decision-makers in operationalising robust strategies for water management against a dynamic, uncertain background of anticipated global change.

5. CONCLUSIONS

Even under present-day conditions, the demands for flood protection, inland navigation, ecology, hydropower, water availability for agriculture, industry and drinking water in the lower Rhine basin are high. Climate change is expected to enhance discrepancies between water supply and demand in the Rhine and Meuse rivers. The sensitivity analyses have shown that climate change may have considerable impacts on flood protection, landscaping of the floodplain, inland navigation, water availability, agriculture and nature. The vulnerability of the water systems, however, also strongly depends on climate-independent socio-economic developments, as well as the implementation of water management measures. Both the climate scenarios and the socio-economic developments are surrounded by major uncertainties. Research should of course aim at reducing these uncertainties. However, some of these uncertainties are structural in the sense that these cannot be reduced. This particularly holds for socio-economic developments. Therefore, research also should focus at developing water management strategies that are able to cope with uncertainty. Scenarios are valuable tools for exploring different futures and analysing the consequences of different water management strategies. The Perspectives method provides a useful framework for establishing a limited set of coherent integrated scenarios for water management in the Rhine and Meuse basins. By analysing and comparing the results of these scenarios a sound basis may be obtained for identifying robust strategies for water management under uncertainty.

6. REFERENCES

Buiteveld, H. & N.N. Lorenz (1999), *The impact of climate change on the IJsselmeer Area. Report of the NRP project 952210.* RIZA report 99.066. Lelystad/Delft: RIZA/Delft|hydraulics.

Grabs, W., K. Daamen, D. Gellens, J.C.J. Kwadijk, H. Lang, H. Middelkoop, B.W.A.H. Parmet, B. Schädler, J. Schulla and K. Wilke (1997), *Impact of climate change on hydrological regimes and water resources management in the Rhine basin.* CHR-report I-16. Lelystad: CHR.

Haasnoot, M., J.A.P.H. Vermulst & H. Middelkoop (1999), *Impacts of climate change and land subsidence on the water systems in the Netherlands. Terrestrial areas. Report of the NRP project 952210.* RIZA report 99.049. Lelystad: RIZA.

Hulme, M., D. Conway, O. Brown & E. Barrow (1994), *A 1961-1990 baseline climatology for future Climate Change Scenarios for Great Britain and Europe. Part III: Climate Change Scenarios for Great Britain and Europe.* Norwich: Climate Research Unit, University of East Anglia.

Können, G.P., W. Fransen (1996), *De toestand van het klimaat in Nederland.* De Bilt: KNMI.

Können, G.P., W. Fransen & R. Mureau (1997), *Meteorologie ten behoeve van de Vierde Nota Water-huishouding.* De Bilt: KNMI.

Kwadijk, J.C.J. (1993), *The impact of climate change on the discharge of the River Rhine.* Thesis. Netherlands Geographical Studies 171. Utrecht, The Netherlands.

Middelkoop, H. (red.) (2000), *The impact of climate change on the river Rhine and the implications for water management in the Netherlands. Summary report of the NRP project 952210.* RIZA report 2000.010 / ICG-report 00/04. Utrecht: Dept. of Physical Geography, Utrecht University.

Middelkoop, H., K. Daamen, D. Gellens, W. Grabs, J.C.J. Kwadijk, H. Lang, B.W.A.H. Parmet, B. Schädler, J. Schulla & K. Wilke (2001), Impact of climate change on hydrological regimes and water resources management in the Rhine basin. *Climatic Change* 49, 105-128.

NW4 (1997), *Klimaatverandering en bodemdaling: gevolgen voor de waterhuishouding van Nederland. Resultaten van een onderzoek in het kader van de voorbereidingen van de vierde Nota waterhuishouding.* 's-Gravenhage: RIKZ.

Rotmans, J. and H.J.M. de Vries (1997), *Perspectives on Global Change: The TARGETS approach.* Cambridge University Press, Cambridge, UK.

Van Deursen, W.P.A. (1999), *RHINEFLOW-2. Development, calibration and application.* Rotterdam: Carthago Consultancy.

Veeneklaas, F.R., L.M. Van den Berg, D. Slothouwer & G.F.P. IJkelenstam (1994): *Rhine basin study: land use projections based on biophysical and socio-economic analysis. Volume 4, Land use: past, present and future.* Wageningen, DLO Winand Staring Centre, Report 85.4.

Climate change and water resources in an arid continent: Managing uncertainty and risk in Australia

R. N. JONES AND A. BARRIE PITTOCK
Department of CSIRO Division of Atmospheric Research, Aspendale, Victoria, Australia

1. INTRODUCTION

Most of the Australian environment is water-limited. Generally low rainfall, low runoff, and large seasonal and ENSO-related variations, result in a significantly different hydrology compared to most continents and regions (Finlayson and McMahon, 1988; Chiew et al., 1998). Australian rainfall and runoff is two to four times more variable than that in north-western Europe and North America for the same climatic zones. Potential evaporation is generally higher than in equivalent Northern Hemisphere latitudes. Droughts are frequent, extensive and prolonged, and are interspersed with sequences of above-average rainfall. Most of Australia's major rivers have ceased to flow at least once during the last 100 to 200 years, and groundwater levels show remarkable variations through time.

Despite large water surpluses in northern Australia during the summer monsoon season, and in Tasmania, water resources are stressed in many areas, with intense competition for water supply. Competing demands come from agriculture, power generation, urban areas and environmental flows. Australia is well adapted to interannual rainfall variability having one of the highest volumes of water storage capacity per head of population in the world (Smith, 1999). Irrigation has been a national success story, building significant agricultural wealth. However, a climate history where episodic and prolonged water shortages have concentrated salts in groundwater, followed by more recent vegetation clearing and over-use of irrigation water has resulted in a significant and widespread salinity problem (MDBMC, 1999). This has entrained a process of land degradation that will play out

M. Beniston (ed.),
Climatic Change: Implications for the Hydrological Cycle and for Water Management, 465–501.

over the coming century, during which time climate change is expected to significantly affect Australia's hydrology. Increased evaporation and possible decreases of rainfall in many areas would adversely affect water supply, agriculture, and the survival of aquatic and flood plain ecosystems.

Sustainable management and industry reform is being encouraged through a user pays policy and tradeable water rights with caps on irrigation supplies. Through previous research and experience of climate variability, it is widely acknowledged that Australia's water resources are sensitive to climate change. However, significant uncertainties attached to climate change scenarios and to the outcomes of impact studies, have meant that Australian water managers have been reluctant to consider adaptation measures beyond those already captured within current policy framework. This chapter describes recent research contributing to the *Projections of Climate Change for Australia* (CSIRO, 2001) and the application of risk assessment methods to climate change in the Macquarie River catchment in the eastern Murray-Darling Basin. This research, and an earlier project on the same catchment (Hassall and Associates, 1998) has contributed to a change in attitude. Adaptation is increasingly being seen as a necessary step in managing the risk of climate change, and as a contributing rather than competing factor in regard to other land and water management issues.

2. HISTORIC CHANGES IN AUSTRALIAN RAINFALL

Australian rainfall records date from the mid 19th century and show large variations in annual mean rainfall over the multi-decadal time scale. Mean rainfall and rainfall intensity have increased slightly over the 20th century (Hennessy et al., 1999; Collins and Della-Marta, 1999), with statistically significant increases in the frequency of heavy rainfalls in some parts of Australia (Hennessy et al., 1999). However, records from the late 19th century indicate that changes in mean rainfall during the 20th century could be part of long-term climate variability. The largest and most statistically significant change has been a decline in rainfall in the winter-dominated region of the far south-west of Western Australia, where in the period 1910-1995 winter (JJA) rainfall declined by some 25% since the mid 1970s (e.g. Allan and Haylock, 1993; Yu and Neil, 1993). This decrease has been attributed to atmospheric circulation changes, presumably due to natural variability (Smith et al., 2000). Lake level changes in south-eastern Australia indicate the regional climate of the 19th and 20th century is the driest for some 9,000 years (Jones et al., 1998, 2001).

Marked interdecadal variations over northern and eastern Australia in summer half-year rainfall are dominated by ENSO-induced variations (Power et al., 1999a). Interannual and decadal variations in central and eastern Australian rainfall are associated with Indian and Pacific Ocean sea surface temperatures (Power et al., 1999b). The strength of the relationship between eastern Australian climate and ENSO has been observed to vary over the last century which seems to be linked to longer-term climate oscillations such as the North Pacific Decadal Oscillation (e.g. Power et al., 1999). Storm frequency and tropical cyclones also appear to be affected by long term variations in the Australian region (Hopkins and Holland, 1997; Leighton et al., 1997; Radford et al., 1996; Nicholls et al., 1996a and 1998).

3. CLIMATE CHANGE PROJECTIONS

Climate change projections for Australia showing ranges of change for selected variables, with information on possible changes to climate variability and extremes, were recently released by CSIRO (2001). The ranges of change are based on the output from nine climate models sourced from CSIRO and the Intergovernmental Panel on Climate Change Data Distribution Centre (IPCC DDC) and the global warming projections from IPCC (2001a). Projections for temperature, rainfall and potential evaporation were created from patterns of change for each variable calculated monthly and seasonally. The method used extracts the local climate signal for each variable in each grid as a function of mean global warming from a transient simulation forced by a greenhouse gas scenario. This procedure eliminates the individual climate sensitivities of different GCMs, allowing the comparison of regional patterns from those models and the subsequent construction of projected ranges of regional change utilising the full range of projected global warming. Change for each grid square as a time series from the entire run is regressed against mean global warming over the same period to give a change per degree of global warming (Whetton, 2001). Compared to the more common method of averaging 30-year periods (see IPCC TGCIA, 1999; Mearns and Hulme, 2001), this method removes most of the simulated decadal variability (Jones et al., 1999). It is a robust method except in very dry regions where several extreme rainfall events over a century can influence the outcome. Such patterns can be scaled assuming different greenhouse gas emission scenarios and climate sensitivities to produce a wide range of possible outcomes (Mearns and Hulme, 2001). The scenario construction techniques used here are described in detail in IPCC-TGCIA (1999), and Mearns and Hulme (2001).

The major finding for water resources is that the models simulate decreases in winter–spring (June–November) rainfall over the southern half of the continent while potential evaporation increases in all simulations. Changes in summer–autumn (December–May) rainfall are biased towards increases, especially to the north of the continent. Mean atmospheric moisture demand, measured as rainfall minus potential evaporation, becomes more negative in most models in most regions (CSIRO, 2001).

Seasonal changes for rainfall and potential evaporation changes for five regions of Australia are shown in Figure 1. The units are percentage change per degree of global warming. They show the mean of the eight models used in CSIRO (2001) for which potential evaporation could be calculated, shown with one standard deviation. North-eastern and north-western Australia show very similar changes biased towards rainfall increases in the first half of the year, although directional uncertainty remains high. This would indicate a longer and perhaps more intense wet season. Potential evaporation changes are modest in percentage terms, although current values are so high, that in absolute terms increases are generally larger than projected for southern Australia. South-eastern and south-western Australia show a winter–spring (June–November) decrease in rainfall with correspondingly greater increases in potential evaporation. Rainfall decreases are greater in the south-west. The direction of summer–autumn rainfall remains uncertain, particularly to the west of the continent. Tasmania shows a different pattern to the rest of the country, with decreases in rainfall in the Austral summer and increases in the winter. This is interpreted as an intensification of the westerly rain belt which moves south in response to warming, reducing its influence over the mainland. This is supported by increases in pressure over southern Australia and directly to the south, with decreases at higher southern latitudes (not shown).

These changes show that possible increases in rainfall in the summer–autumn period may lead to a longer and more intense Australian Monsoon, although a weaker monsoon cannot be discounted. Increased seasonal rainfall would lead to greater water surpluses in areas of northern Australia where the current level of human use is low. In southern Australia, where winter rainfall becomes progressively more important and competition for resources between both natural and human uses is already high, reductions in water supply appear to be much more likely. Tasmania is an exception, but is another area of largely existing surplus, although the state has a large hydroelectric industry. In the remainder of this chapter we concentrate on the Macquarie River catchment, which is situated in the south-east of Australia within the eastern central part of the Murray-Darling Basin, using newly developed techniques for risk assessment to explore the full range of uncertainty for climate change as summarised by CSIRO (2001).

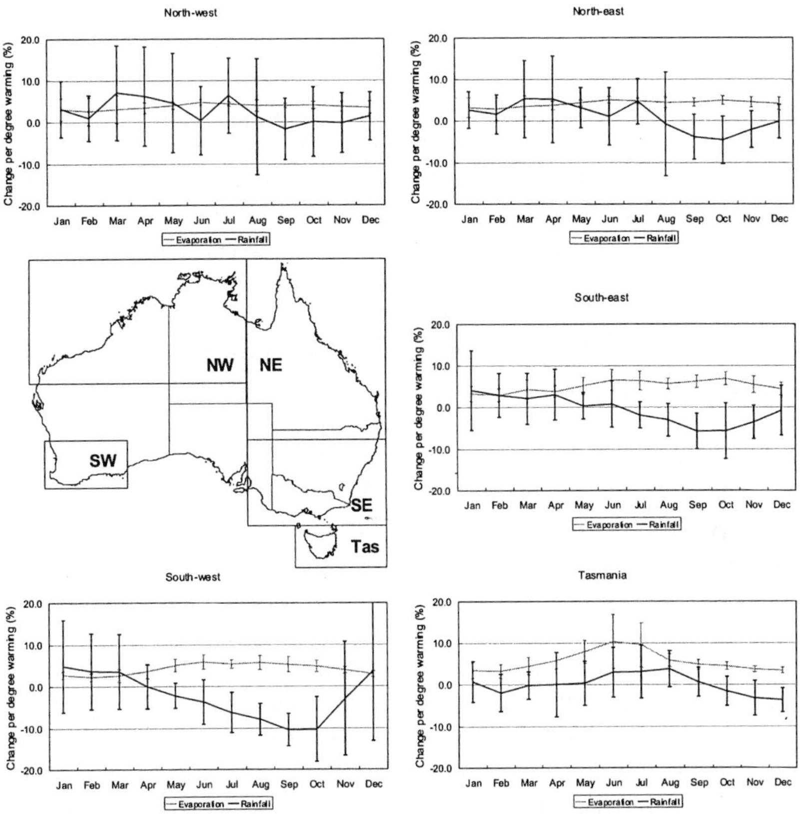

Figure 1: Change per degree of global warming for rainfall and potential evaporation for five regions of Australia (percentage change). Projected ranges of change for a particular date can be obtained by multiplying these data by ranges of global warming from IPCC (2001). Rainfall is negligible in northern Australia (except on the east coast) during the Austral winter, and low on the southern mainland during the Austral summer, especially in the south-west.

4. THE MACQUARIE CATCHMENT

The Macquarie River Catchment is situated in the central eastern part of the Murray-Darling Basin in New South Wales and covers about 75,000 km^2. It rises on the western slopes of the Great Dividing Range 100 km west of Sydney and flows WNW into the Darling River. Rainfall (P) ranges from about 1200 mm pa in the upper catchment to <400 mm pa in the lower catchment. Potential evapotranspiration (Ep) follows an inverse spatial pattern to rainfall, ranging from about 1,200 mm pa in the upper catchment to 2,200 mm pa in the lower catchment. The bulk of runoff is generated in the headwater catchments and flows into the Burrendong Dam, the main storage in the catchment with a capacity of 1,189 GL (NSW EPA, 1995). Downstream, there is a net loss of streamflow due to evaporation, recharge of shallow aquifers and abstractions for irrigation. Competition for water is high, with supply to towns, irrigation, wetlands and streamflow. The catchment is affected by both dryland and irrigation salinity and a fifteen-year salinity management plan is currently being devised and implemented.

We focussed on three major aspects of water supply within the catchment: supply into Burrendong Dam, irrigation allocations met at the end of the irrigation season, and flows into the Macquarie Marshes.Water allocations are classed as either high security or general security. High security water is a fixed allocation amount guaranteed in all but the lowest flow years. Town water supply is a high security water allocation. General security water is allocated according to water availability and most irrigation water is of this type. Every irrigator is licensed for a fixed volume of water, but the proportion of their entitlement that they actually receive in any year is calculated after dead storage and high security water allocations have been allocated. Irrigators are more vulnerable to reductions in catchment runoff than recipients of high security water, and they often receive less than 100% of their licensed allocation in a water year. Beef cattle, sheep, wool, wheat and cotton are the major contributors to the agricultural economy. Wheat dominates the dryland cropping sector; cotton production comprises the majority of revenue from irrigated summer cropping, and accounts for approximately 60% of consumptive water use in the Macquarie (Foreman et al., 1998). The area planted to cotton more than tripled between 1985 and 1995, and has continued to expand since.

The Macquarie Marshes, located in the lower reaches of the catchment, are a large and diverse system of wetlands. They have particular significance as a refuge and breeding area for waterbirds (Kingsford and Johnson, 1998). The Macquarie Marshes Nature Reserve is listed under the RAMSAR Convention of Wetlands of International Importance. Under the Water Management Plan for the Macquarie Marshes 1996 (DLWC and NWPS, 1996), 50 GL of high security water and 75 GL of general security water have been allocated to the Marshes to ensure their ecological sustainability. The Marshes are very sensitive to changes in flow regime, having shrunk by more than 40% since the construction of Burrendong Dam in 1967 (Kingsford and Thomas, 1995).

4.1 Model structure

Changes in flow and the impacts of climate uncertainties were investigated by coupling the Australian climate scenario generator, OzClim, to the Integrated Quality Quantity Model (IQQM), developed by the NSW Department of Land and Water Conservation (DLWC). OzClim, is an Australian climate scenario generator constructed by CSIRO Atmospheric Research in collaboration with the International Global Change Institute at the University of Waikato (Jones and Page, 2001). OzClim contains regional climate patterns for precipitation (P) and potential evapotranspiration (Ep) that are combined with global warming scenarios to create regional scenarios of P and Ep. P is calculated directly by the models, but Ep was calculated using the Complementary Relationship Areal Evaporation (CRAE) Method (Morton 1983), to provide estimates of changes in A-Class pan equivalent Ep for input into IQQM. IQQM consists of the Sacramento model to simulate runoff (Burnash et al., 1984), river routing routines and extensive management rules, including those for water supply, allocations, crop demand, crop area and wetland management (DLWC, 1995a and b). The model uses the 1996 management rules developed by DLWC to manage the Macquarie catchment more sustainably (DLWC and NWPS, 1996). Output consists of streamflow at a large number of sites, with irrigation and urban/industrial allocations, and effluent flows. Input data consists of daily P from 23 stations and Ep from 17 stations throughout the catchment dating from 1890 to 1996. Gaps in P have been infilled using regression from nearby stations. Three of the stations, Mudgee, Rylstone and Dunildra are members of the extended high quality historical rainfall data set for Australia (Lavery et al., 1997) and are homogenous, high quality records. The historical rainfall series based on average anomalies from these three stations is shown in Figure 3. Ep is Class A pan data dating from the 1960s. Before this, Ep has been constructed using regressions based on daily temperature readings. This is a valid assumption in a stationary climate but will not hold if climate becomes non-stationary due to climate change.

Output is a simulated record based on how historical climate may have affected the Macquarie under current development and management rules, and does not simulate historical water supply. OzClim "drives" IQQM by applying climate change scenarios to these historical climate series, which are then used to simulate rainfall-runoff, streamflow and allocations. Both input and output are available as daily data.

Australia's high rainfall variability and its dominance over water supply, requires close attention to be paid to the management of uncertainty, particularly with regard to how variability may change with global warming. This takes on several aspects:

1. Terminology – climate variability can mean anything from variability on timescales of less than a day to variability on decadal to century timescales. Careless use of terminology has led to a situation where the capacity of Australian water managers to cope with some forms of variability, has led to the idea that this adaptive capacity covers all climate variability.
2. Predictability – some modes of change are more difficult to predict than others. CSIRO (2001) contains projections of change for mean annual and seasonal rainfall that appear to be more robust than past estimates (e.g. CSIRO, 1996). Another robust outcome is that daily rainfall also appears likely to become more intense, unless accompanied by significant decreases in mean. However, changes to interannual and decadal variability are highly uncertain.
3. Criticality – criticality within a system can be linked to a level of change that is associated with an unacceptable level of harm. This is a function of both sensitivity and existing adaptive capacity. Although there is a level of harm associated with extreme events such as floods and individual droughts, these events can usually be managed. Criticality is a function of accumulated stress caused largely by sequences of events, such as repeated droughts.

Three aspects of the baseline precipitation climate were investigated with regard to their impact on water supply and how they may change over the coming decades to century. They are mean rainfall, decadal-scale variability and interannual variability. Mean climate is the historical mean over the history of record (in this case, 1890–1996). Decadal rainfall variability in this case is rainfall averaged over a period of decades. A break between a drought-dominated and flood-dominated period is shown in Figure 2.

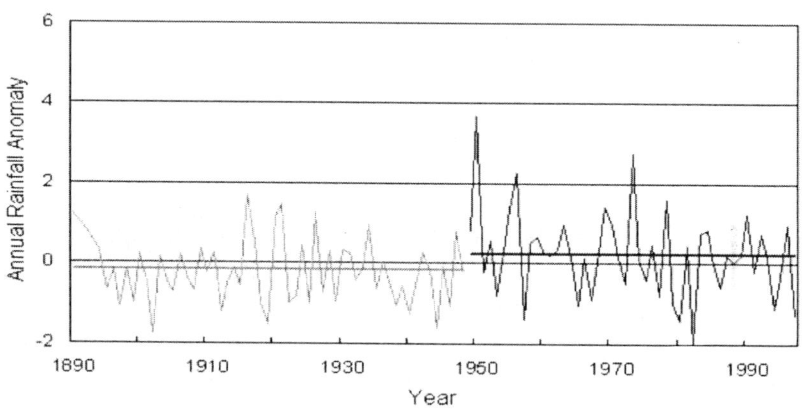

Figure 2: Historical rainfall anomalies averaged for three high quality rainfall stations in the Macquarie Catchment.

Interannual variability is the variation in total annual rainfall, which is highly influenced by the El Niño-Southern Oscillation (ENSO) phenomenon. The Macquarie River has one of the most variable records of annual streamflow in Australia (NSW EPA, 1995).

The homogeneity of the input rainfall records was investigated using the bivariate test (Potter, 1981) showing an inhomogeneity in 1948 for almost half of the input rainfall stations. This change is not linked with any systematic change of rain gauges or measurement methods. A similar regional change in western Victoria was interpreted as a differential change in aerodynamic loss by rain gauges in exposed and unexposed locations brought on by a rapid increase in rainfall and rainfall intensity (Jones, 1995). Similar inhomogeneities also occur in NSW and Queensland. Therefore, the historical rainfall record for eastern Australia can be divided into two periods: a drought-dominated and a flood-dominated regime dating from 1890–1948 and 1949–1998 respectively. Rainfall anomalies for 1890–1998 averaged for the three high quality stations mentioned above are shown in Figure 2.

Warner (1987) described this phenomenon based on river morphology and rainfall in central eastern NSW citing a rapid shift in rainfall. Pittock (1975) also noted this increase, suggesting a possible relationship with ENSO. Earlier phase shifts of this type have also been noted for around 1893–4 (Kraus, 1955; Whetton et al., 1990) affecting much of the tropics and extra-tropics. More recent shifts involve decadal-scale changes in rainfall in the Sahel and southwestern Western Australia (Flohn, 1986; Allan and Haylock, 1993; Hunt, 2000). This is distinguished from other forms of decadal variability, such as the Interdecadal Pacific Oscillation and the Indian Ocean Dipole, that appear to affect interannual variability through the modulation of phenomena such as ENSO (Power et al., 1999).

Streamflow also reflects this change in rainfall regime. Table 1 shows the summary statistics for P and mean annual storage in the Burrendong Dam under current climate. Simulated streamflow from IQQM is a product of the infrastructure and management rules as set in 1996 whereas historical streamflow has changed in response to catchment modification as well as climate. Simulated streamflow from IQQM increases in the second half of the 20th century, whereas historical streamflow has decreased over time due to increasing extractions of water for irrigation and other purposes (Kingsford and Thomas, 1995). Table 2 shows that both rainfall and stream flow are significantly affected by decadal-scale variations in rainfall, deviating by >20% from the long-term historical mean. The flood-dominated regime contains both the wettest and lowest records for the entire series.

Table 1: Average anomalies for P and mean annual storage volume the Burrendong Dam for drought-dominated and flood-dominated periods under current climate simulated using the 1996 flow management rules.

Drought-dominated period	1890–1948
Flood-dominated period	1949–1996
Annual total rainfall coefficient of variability	30%
Mean annual flow (MAF) coefficient of variability	100%
Rainfall anomaly – drought-dominated regime	–20%
Rainfall anomaly – flood-dominated regime	+23%
Flow anomaly – drought-dominated regime	–23%
Flow anomaly – flood-dominated regime	+27%

Australia's high rainfall variability and substantial evaporative deficit has led to major developments in irrigated agriculture. An understanding of how such systems may adapt to change requires an understanding of the climatic conditions under which they were constructed and operated. Most of the irrigation infrastructure development in the Macquarie River catchment followed the building of the Burrendong Dam in 1967, so is the product of a flood-dominated regime. The behaviour of today's irrigators will be influenced by their personal experiences within that climate. Does the coping range for irrigation within the Macquarie catchment encompass the whole of the historical climate, or just the second half? If the operation of the catchment, especially security of streamflow and extraction allocations, is dependent on rainfall and water supply available during a flood-dominated period, a repeat of conditions characteristic of a drought-dominated regime may alter this security substantially. This suggests that although over 100 years of climate data is available, the true baseline should reflect the period reflecting conditions to which a system is adapted, in this case, the post-1948 flood-dominate period.

Control water supply and bulk allocation of irrigation was simulated using historical climate from 1890–1996 applied to allocation rules and infrastructure of 1996. Figure 3 shows Burrendong Dam and Macquarie Marshes inflows along with the percentage of irrigation allocations met within a water year (July 1 to June 30) for 1890/91–1995/96. The effect of decadal variability is obvious. Flows and irrigations allocations are much lower for the first half of the 20th century than they are for the second half. During the period 1947/48 to 1978/79, irrigation allocations only fall below half on one occasion, and 100% allocations were supplied almost 80% of the time. Between 1947 and 1996, three extreme shortfalls (<25%) in allocations are simulated, whereas in the first half of the century, simulated irrigation allocations fall below 25% on fifteen occasions.

Water supply for the catchment is now fully allocated to over-allocated when all competing needs, such as irrigation, environmental flows and water quality, are taken into account (see NSW EPA, 2000). Although recent droughts and the imposition of irrigation caps have limited supply, it is likely that sustained periods of low rainfall typical of the first half of the 20th century would cause severe hardship if they were to occur again. Severe years of drought, including several years in succession such as occurred in 1979 to 1983 can be managed within the current coping range. However, could a longer sequence of dry years reminiscent of 1900–1910 or 1935–1945 cross the threshold into vulnerability? In future, different modes of climate variability occurring on daily to decadal timescales may continue independently of climate change, or may be modulated by it. Changes in mean climate may occur gradually or may be abrupt. These phenomena will need to be managed in combination and is why climate scenarios must explicitly reference climate variability <u>and</u> mean climate change.

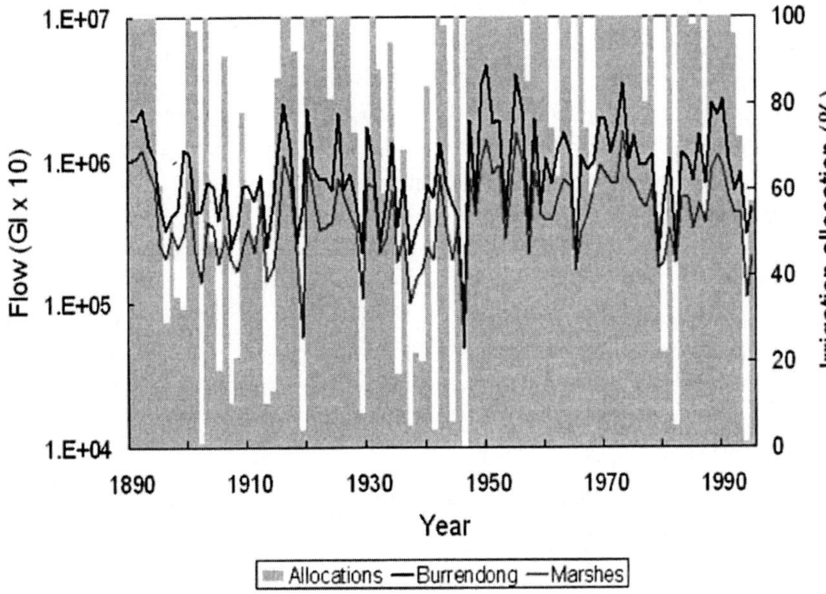

Figure 3: Simulated flows into Burrendong Dam and the Macquarie Marshes in Gl and bulk irrigation allocations in percentage for the Macquarie River based on 1890/91 to 1995/96 climate and 1996 management rules.

5. CLIMATE CHANGE

Output from nine climate models was used to explore the fullest possible range of regional change in P and Ep. Three CSIRO models, the Mark2 GCM and the DARLAM 60km and 125 km regional climate model simulations and six international models, the Hadley Centre (UK) HADCM3, the Deutsches Klimarechenzentrum (DKRZ Germany) ECHAM3 and ECHAM4 models, the Canadian Centre for Climate Modelling and Analysis (CCCma) CGCM1 model, the Geophysical Fluid Dynamic Laboratories (GFDL, US) R-15a model and the National Center for Atmospheric Research (NCAR) DOE model were used (Table 2). All models are transient greenhouse gas only simulations based on historical emission to 1990 and either 1% compounding CO_2 or IS92a greenhouse gas scenarios from 1990 (these are similar). These models are the same as used in CSIRO (2001) save that HADCM2 is omitted and the CSIRO DARLAM 60km model is added.

Table 2: Model runs used to produce regional projections of climate change.

Centre	Model	Emission Scenario	Features	Years
CSIRO, Australia[1]	Mk2	IS92a	No sulphates, GM ocean	1881–2100
CSIRO, Australia[2]	DARLAM 125 km	IS92a	Nested in CSIRO Mk2	1961–2100
CSIRO, Australia[2]	DARLAM 60 km	IS92a	Nested in CSIRO Mk2	1961–2100
DKRZ, Germany[3]	ECHAM4/OPYC3	IS92a	No sulphates	1860–2099
DKRZ, Germany[3]	ECHAM3/LSG	IS92a	No sulphates	1880–2085
Hadley Centre, UK[4]	HADCM2	1% CO_2 pa	No sulphates	1861–2100
Canadian CCMA[5]	CGCM1	1% CO_2 pa	No sulphates	1900–2100
GFDL USA[6]	GFDL-R15a	1% CO_2 pa	No sulphates	1958-2057
NCAR USA[7]	DOE-PCM	IS92a	No sulphates	1960-2099

[1]Gordon and O'Farrell (1997)
[2]McGregor and Katzfey (1997)
[3]DKRZ-Model (1992), Oberhuber (1992), Maier-Reimer and Mikolajewicz (1991)
[4]Johns et al. (1997)
[5]Flato et al. (2000)
[6]Haywood et al. (2000)
[7]Washington et al. (2000)

Precipitation patterns were linearly scaled to match the OzClim grid for the Murray-Darling Basin. There was no further downscaling. Morton's (1983) Complementary Relationship Areal Evaporation (CRAE) model was used to calculate Ep, requiring modelled average temperature, dewpoint temperature (or an equivalent moisture variable) and downward shortwave radiation. The CRAE model was used because:

- An evaporation climatology for 1961-1990 has been prepared for Australia (Wang et al., 2001) using the same method, allowing comparisons to be made between model output and Ep derived from observations.
- The input variables required from the climate model are not overly dependent on the performance of model land surface schemes. The atmospheric input variables were also readily available from the IPCC Data Distribution Centre.
- The CRAE method compares favourably with other methods for calculating Ep for rainfall-runoff modelling (Chiew et al., 1993), and has been used in modelling the impacts of past (Jones et al., 1998, 2001) and future climate change (Chiew et al., 1995).
- Aerodynamic input is not needed, reducing the margin of error associated with wind speed data taken from both observational and simulated records (Morton, 1983).

The methods for calculating Ep from GCM output using the CRAE model are described and assessed using error analysis in Walsh et al. (2000). Having access to an unprecedented set of estimates of P and Ep, we were able to examine the relationship between P and Ep under different aspects of climate (mean climate, inter-annual variability and decadal-scale variability) and as simulated by the climate models for climate change.

Within OzClim, each model pattern was multiplied by the projected global warming for a particular date (e.g. 2030) to produce individual estimates of δP and δEp in percent, which were then used to scale the input data in the IQQM model. This produces a scenario with average monthly changes in P and Ep based on an individual model pattern multiplied by global warming based on a given emission scenario and level of climate sensitivity superimposed on historical climate variability (1890–1996). Climate model output was linearly downscaled for use in OzClim rather than using more realistic methods of downscaling for the following reason. Faced with the choice of using a large number of GCMs to represent a wide selection of regional change, or using more resource and time intensive methods to produce more precise regional results from fewer GCMs, we have opted for the former. The need for better regional downscaling is not as acute in much of Australia as it is in other regions (e.g. the European Alps, New Zealand) because the relationship between topography and climate is not overly complex.

5.1 Precipitation and potential evaporation relationships under climate change

The relationship between the P and Ep under climate change was investigated to determine whether they showed dependence in the same way that changes to temperature and precipitation for many regions appear to be related (e.g. Carter et al., 2000). When climate is stationary, the relationship between P and Ep follows the complementary relationship of Bouchet (1963), where annual values of P and Ep show an inverse relationship in environments that are moisture limited. This relationship can also be investigated at the decadal scale. Figure 4 shows P and Ep from the Chifley climate station in the upper Macquarie catchment from 1890–1998 with 29-year running means over that period obeying a linear relationship. Plotted with this is a time series from the DKRZ4/OPYC3 climate model grid cell for the same location plotted in a 29-year running mean from 1860–2100. The model P and Ep anomaly reproduces the historical relationship until about the year 2000 (covering the period c. 1985–2015), then departs along a vector where δP and δEp are roughly equal. All the climate models used produce similar behaviour but shift along different vectors of P and Ep change, some getting wetter, some getting drier, but all with increasing Ep. Decadal variability of the type seen in the historical climate, accompanies this shift, although generally it is much smaller than observed variability.

Average δP and δEp for the Macquarie catchment from all nine models were obtained for the summer, autumn, winter and spring quarters. Figure 5 shows the resulting relationship, which is consistent with the complementary relationship. Increases in P are associated with small increases in Ep, whereas decreases in P are associated with large increases in Ep. This is consistent with increasing cloud cover (decreasing solar radiation) and humidity accompanying rainfall increases, limiting the affects of increasing temperature. The correlation is -0.77 and the R^2 shown in Figure 7 is 0.58. This relationship has been confirmed for all of eastern Australia and shows that P and Ep are co-dependent under a changing climate (Walsh et al., 2001). This allows P and Ep changes from individual models to be re-scaled using a regression based on the linear relationship shown in Figure 5.

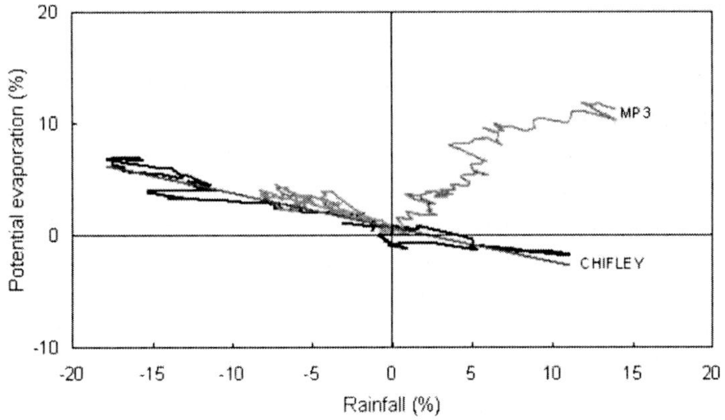

Figure 4: Relationship of δP and δEp as percent anomalies from 1961–1990 for the upper Macquarie catchment Chifley station for 1890–1998 and from the DKRZ4/OPYC3 (MP3) climate model for the same location plotted in a 29-year running mean from 1860–2100.

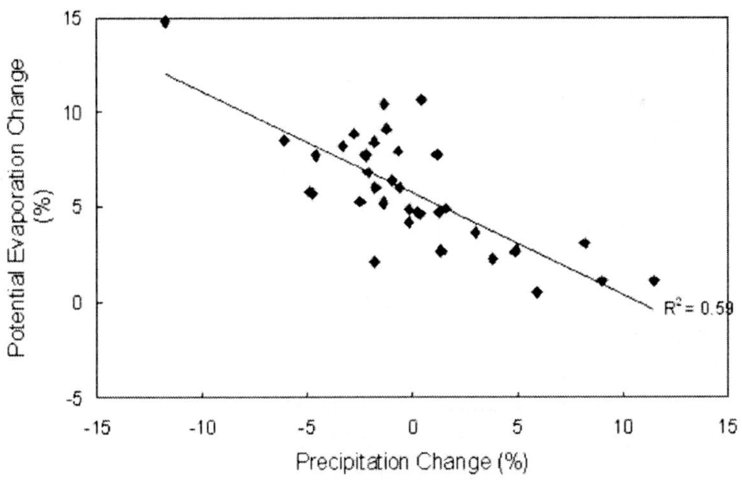

Figure 5: Relationship between δP and δEp per degree of global warming from nine models on a quarterly basis for the Macquarie River catchment.

Annual average change in P per degree of global warming for all nine climate models lies within ±4% of zero. Ep increases by between 3 and 11% showing that most models lean towards drier conditions. The P–Ep deficit increase (i.e. becomes drier) for all models on an annual basis with all models showing a fairly similar pattern on a seasonal basis. As described earlier for south-eastern Australia, rainfall increases dominate over the summer–autumn period and decreases dominate over winter–spring. Figure 6 shows the mean pattern and ±1 standard deviation for all nine models. Only the GFDL and NCAR models do not fully conform to this pattern.

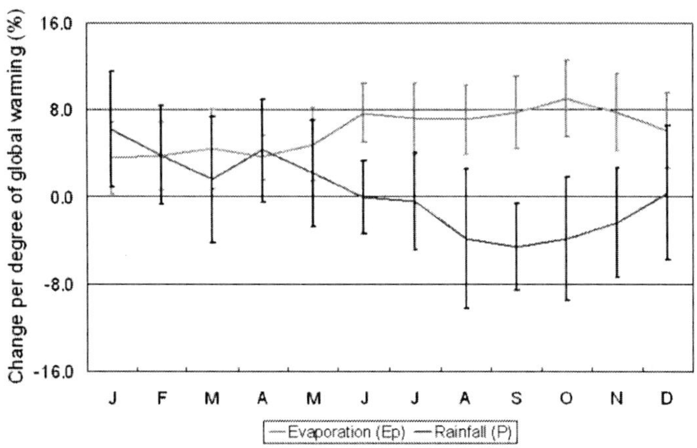

Figure 6: Average monthly changes in P and Ep with ±1 standard deviation for nine climate models for the Macquarie River catchment. Change is percent per degree of global warming.

5.2 Climate change results

The IQQM model was coupled to OzClim, and three measures of output were selected for analysis:

1. mean storage in Burrendong Dam, the major water storage,
2. environmental flows into the Macquarie Marshes, and
3. allocations of seasonal irrigation water rights.

These outputs encompass most of the large-issues concerning water supply in the Macquarie Catchment. High security flows to urban centres is a fourth important factor but will be threatened only if irrigation allocations are zero in a season.

Two sets of results were obtained for 2030, based on the model changes per degree of global warming, and warmings IPCC (1996) and (2001). Firstly, the IS92a, c and e scenarios at low, mid and high climate sensitivities producing global warmings 0.36, 0.55 and 0.79°C were applied to each of the nine climate models, producing a series of 27 results. The results in terms of percentage change in Burrendong Dam storage are shown in Table 3. The range of changes in P is very small at ±3%, Ep ranges from +1% to +8% and changes in Mean Annual Flow (MAF) range from +1% to –22%. Flow changes into the Macquarie Marshes are similar, although flows appear to be slightly more sensitive than for Burrendong Dam. Irrigation allocations fall within a slightly smaller range and show a similar pattern to the other two outputs.

Table 3: Change in P, Ep, Burrendong Dam storage and Macquarie Marshes flows and Irrigation allocations for the year for nine climate models for 2030. The range of global warming applied is 0.36–0.79°C (IPCC, 1996). Input is scaled daily P and Ep from 1890–1996.

Model	**Change in 2030**				
	P	Ep	Burr	Marsh	Irrigation
Mark2, CSIRO Australia	–0.5 to –1.0	1.8 to 4.0	–3 to –10	–5 to –10	–3 to –9
DARLAM 125, CSIRO Australia	0.3 to 0.6	1.8 to 3.9	–2 to –6	–4 to –7	–3 to –5
DARLAM 60, CSIRO Australia	0.3 to 0.6	1.7 to 3.7	–2 to –7	–4 to –8	–3 to –6
ECHAM4, DKRZ Germany	1.2 to 2.6	1.1 to 2.4	0 to 1	0 to 1	0 to –1
ECHAM3, DKRZ Germany	1.3 to 2.9	1.2 to 2.6	0 to 1	–1 to 1	–1 to –2
HADCM3, Hadley Centre UK	–1.5 to –3.2	3.8 to 8.3	–10 to –22	–10 to –22	–8 to –20
CGCM1, CCMA Canada	–0.4 to –0.8	1.8 to 3.9	–3 to –8	–4 to –9	–3 to –6
GFDL-R15-a, GFDL USA	0.4 to 1.0	2.1 to 4.6	–2 to –5	–4 to –6	–2 to –4
DOE-PCM, NCAR USA	–0.9 to –1.9	2.4 to 5.4	–6 to –13	–8 to –14	–5 to –12
Combined range			1 to –22	1 to –22	0 to –20

The two DKRZ models are the wettest and maintain water supply at current levels, while all the other models produce decreases in supply. The Hadley Centre HADCM3 produces the driest outcomes. A global warming of 0.36°C produces a range of change in supply of +1 to –10%, 0.55°C produces a range of +1 to –16% and 0.79°C produces a range of +1 to –22%.

Climate-impact response relationships were needed to rescale the 2030 results to take account of the recently released IPCC SRES emission scenarios (Nakicenovic and Swart, 2000) and IPCC Third Assessment

Report findings (IPCC, 2001a), and to construct risk-response relationships for the Macquarie catchment. The IPCC Third Assessment Report findings became available in January 2001, but the data for the global warming projections have not yet become available. We currently have access to provisional estimates simulated using the same simple climate model used to estimate the global warming projections in IPCC (1996) and have scaled these appropriately. They are accurate to within a standard error of ±0.1°C. By placing these into OzClim, we produced the means for calculating average δP (change in P) and δEp (change in Ep) for the Macquarie catchment.

For 2030, the 9 × 3 simulation runs provides a total of 27 samples by which to measure future climate. A linear regression was performed for Burrendong Dam storage, flows into the Macquarie Marshes and for change in irrigation allocations in the form:

$$\delta MAF = a\delta P + b\delta Ep$$

where a and b are constants. The results are shown in Table 4.

Table 4: Regression relationships for percent change for Burrendong Dam storage, Macquarie Marshes inflows and for percent change in irrigation allocations derived from the results shown in Table 3. SE is standard error.

Output	**a**	**b**	**R^2**	**SE**
Burrendong inflows (δMAF in %)	2.06	-1.74	0.98	0.86
Macquarie Marshes inflows (δMAF in %)	1.88	-2.04	0.99	0.67
Irrigation allocations (change in %)	1.32	-1.68	0.95	1.00

All three relationships show a high correlation and relatively low standard error so are suitable for scaling across the range of P and Ep projected for 2030. The regressions confirm an almost linear response of the IQQM model to changes in climate within the limits of the projected ranges for P and Ep in 2030. Other simulations for 2070 covering a wider range of uncertainty were non-linear, but can be approximated by a linear relationship at low values.

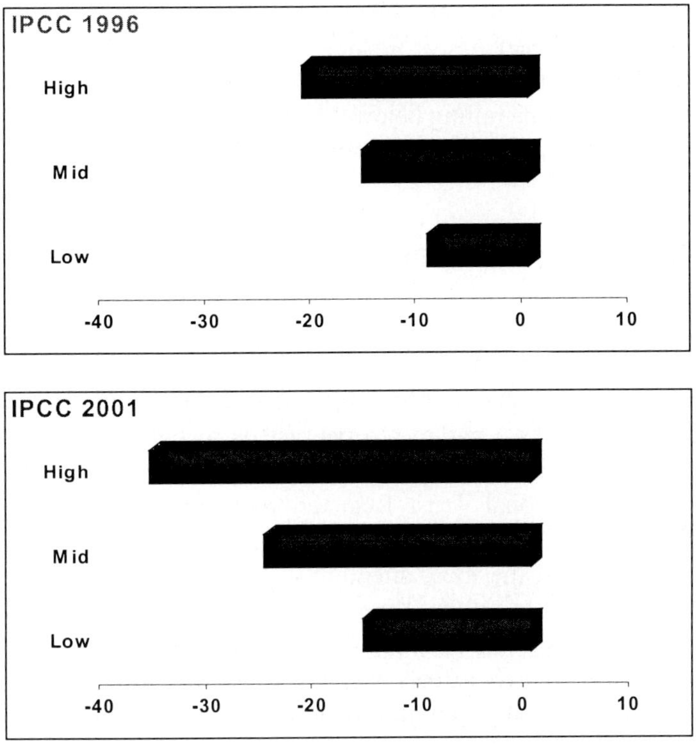

Figure 7: Comparison of low, median and high projections of global warming (IPCC, 1996 and 2001) combined with δP and δEp from 9 climate models on mean annual storage in the Burrendong Dam.

The global warming projections of IPCC (2001) in 2030 are substantially higher than those of IPCC (1996), ranging from 0.55°C to 1.27°C, up from 0.36°C to 0.79°C. When Equation 2 is used to rescale the results from Table 3 using these new warming projections, the range of possible changes to MAF in Burrendong Dam increases from +1% to –22% to +1% to –34%. These comparisons are shown in Figure 7. The upper limit has not changed, because δP and δEp in the two DKRZ models approximately balance themselves out. However, the projections for all other models become drier.

6. CRITICAL THRESHOLDS

A critical threshold marks the point at which an activity or system faces an unacceptable level of harm. It delineates the coping ranges and area of vulnerability, and provides a criterion by which to measure risk. If a critical threshold can be expressed in climatic terms and quantified under climate change, it then becomes possible to estimate risk if a probability distribution for the climate inputs can be determined. In this project, two critical thresholds were examined:

1. Bird breeding events in the Macquarie Marshes
2. Irrigation allocations falling below a nominated level

Both thresholds are a measure of accumulated stress rather than a single extreme event.

6.1 The Macquarie Marshes

The Macquarie Marshes lie on a complex system of braided streams and anabranches of the lower Macquarie River. The core area of the Marshes is 40,000 ha with another 220,000 ha subject to occasional inundation. 18,000 ha has been gazetted as a nature reserve, and is recognised by the Ramsar Convention of Wetlands of International Importance, the National Heritage Register and the National Trust Register. Although both vegetation and geomorphology are highly significant, waterbirds, including migratory species, have received the most attention in terms of conservation biology (Kingsford and Thomas, 1995; Kingsford and Johnson, 1998; Kingsford, 1999a). Because the Marsh wetlands have evolved in a system of high climate variability, which affects inflows and water levels, they require environmental flows that continue a pattern of floods and droughts. Semi-regular flooding is required by colonially-nesting waterbirds, who build nests on emergent vegetation. The flooding needs to be of a size and frequency to sustain bird populations (Johnson, 1998; Kingsford and Johnson, 1998). Abstractions from the Macquarie catchment over the longer term have had a significant impact on bird numbers in the Macquarie Marshes (Kingsford and Thomas, 1995; Kingsford and Johnson, 1998). Vegetation has also declined due to grazing pressures, changed flow conditions and due to the establishment of irrigated cotton farming on the floodplain adjacent to high conservation areas (Kingsford, 1999b; Lemly et al., 2000).

In 1996, the NSW government released the Macquarie Marshes Water Management Plan (DLWC and NPWS, 1996). This plan prescribes the water allocation rules within the version of IQQM that we use. Lemly et al. (2000) conclude that these rules effectively halt the continuing degradation in the Macquarie Marshes caused by uncontrolled access to off-allocation (not covered by a formal water right) irrigation water. Kingsford and Johnson (1998) nominate three thresholds based on historical data: below 200 GL total annual flow at the Oxley gauge in the Macquarie River, the birds do not breed; between 200 and 500 GL breeding is usually below 20,000 nests and above 500 GL breeding ranges up to 90,000 nests. The relationship between climate change and significant flow events was estimated using bird-breeding thresholds as a performance measure. Figures are not yet available to translate the historical flows into thresholds for flows into the Macquarie Marshes but Johnson (1998) estimated that 350 GL could be used as an indicator of significant bird breeding.

The effective breeding life of key bird species is estimated to be less than 10 years Johnson, 1998). The control case has 45% of years below this nesting threshold with one period of 8 years and one period of 7 years both of which are approaching a critical threshold for local breeding (Table 5). With decreasing flows into the Macquarie Marshes of >10%, the longest periods extend to 11 and 7 years and with decreases of >15%, they extend to 11 and 10 years. The next increase in the number of critical thresholds is exceeded with decreases in MAF of >30% for periods of 15, 14, 13, 12 and 10 years. Although colonially nesting birds will migrate to different wetlands to breed, a threshold of 10 years' flow below the nesting threshold is considered critical both locally and regionally because the drying shown by a number of climate models is widespread and other wetlands are likely to be similarly affected.

Table 5: Sequences of years below 350 GL simulated flow into Macquarie Marshes, linked to change in mean annual flow from the control.

Sequences below threshold (years)	Change in MAF						
	+5%	0	-10%	-15%	-30%	-40%	-50%
16							1
15					1	1	2
14					1	1	2
13					1	2	3
12					1	2	3
11			1	1			
10				1	1	1	1
9							
8		1					
7	1	1	1				1
6						2	1
5	1	1	1	1	2	1	
4	3	2	2	3	4	2	3
3	2	1	3	4	3	3	1
2	4	7	4	2	2		1
1	4	3	7	5	3	3	
Percent of total years below threshold	40	45	52	56	63	71	79

6.2 Irrigation allocations

Irrigation allocations within the Macquarie catchment are permitted to 100% of the water right as an entitlement, depending on the amount of water available. Irrigation allocations in IQQM are registered at October 1, January 1 and June 30. The October allocation has the greatest impact on cropping decisions and the January update allows some flexibility. Here, we investigate the end of June allocation, which represented the total allocation over a season. This measures the performance of the system in terms of allocations but not of irrigation performance, that relies on the management of irrigators in planning for years when 100% of allocations may not be available.

Irrigation allocations were investigated in the same way as for flow thresholds into the Macquarie Marshes. In general, farmers are well adapted to individual dry years, such as those associated with El Niño events. They are less well adapted to sustained periods of drought or floods when reserves are taxed to the point of collapse. In this case, we look at sequences of below 50% allocations in irrigation water. Under current climate, a 50% allocation is unmet in about ¼ of years, producing 4 sequences of 2 years and 1 sequence of 3 years (Table 6). Climate scenarios from different climate models produce quite variable outcomes due to the seasonal expression of changes, so the results in Table 10 do not always appear consistent. With a 10% reduction, a typical outcome has 2 sequences of 4 years. This would cause severe hardship amongst irrigators. Above a 30% reduction in average irrigation allocations, there is a rapid deterioration in performance, and 50% of allocations are not met in over half of all years.

We have defined a critical threshold as a sequence of 5 years of irrigation below the 50% allocation but would expect severe hardship in a sequence of 3 or 4 years. In individual scenario runs, this threshold is exceeded when mean annual allocations reduce by between–7% and –12%. It is exceeded twice when mean allocations reduce by > 15%. Therefore, the outlook for irrigation becomes critical when allocations decrease by 10% to 15% slightly less sensitive than for the Macquarie Marshes. Because both critical thresholds are measured as a sequence of years below a given level of performance they are a product of both decadal and interannual rainfall variability. The previous section on estimating risk deals with the single event probability of mean climate change imposed on the historical climate of 1990–1996, and its affect on mean flows. Estimating the exceedance of critical thresholds based on unique combinations of climate variability requires estimating both single-event (estimating the magnitude of climate change) and frequentist probabilities (the likelihood of climate variability exceeding a simple threshold *n* times).

Table 6: Sequences of years below 50% water allocations, linked to change in mean annual water allocation.

Sequences below threshold (years)	Change in mean average allocation: +5%	0	-10%	-15%	-30%	-40%	-45%
16							
15							1
14							1
13							1
12							1
11						1	0
10						1	1
9							
8							
7							1
6					1	1	
5				2	2		1
4			2		2	4	5
3	1	1	1	2	4	1	1
2	5	4	6	6	5	6	2
1	10	13	11	12	7	4	4
Percent of total years below threshold	22	23	34	38	50	58	64

7. ESTIMATING RISK

This section presents the results of risk analysis techniques developed by Jones (1999, 2000a), applied to Burrendong Dam storage, Macquarie Marshes inflows and to irrigation allocations. Although the techniques used in the risk assessment are simple, the concepts are fairly complex. Two large ranges of uncertainties are sampled here: global warming based on the projections of IPCC (2001) and local ranges of δP and δEp.

Each of the scenarios run in the previous section is a sample within a population of possible outcomes. Here we investigate two such populations: the simulated changes for regional P and Ep sampled separately, and global warming for 2030 from IPCC (2001). These ranges are both plausible but their probability, and the probability distributions of the populations they comprise are unknown. By displaying the outer limits of the combined ranges from Table 3, we are quantifying the fullest quantified range of outcomes from the published data available from IPCC (2001) and the IPCC-DDC. Jones (1999a, 2000a and b), Hulme and Carter (1999) and New and Hulme (2000) have shown that when these component ranges of uncertainties are combined, the results are not uniform but are a function of assumed prior probabilities of the component ranges.

The results in Figure 7 show that the range of change produced using individual model results shows that Burrendong Dam storage changes by approximately 0% to –34%. How likely are these changes to occur? If three ranges of uncertainty, global warming and regional δP and δEp are combined, the probability of three extremes combining would be less than any other result (e.g. 2 extreme, 1 moderate; 1 extreme 2 moderate or 3 moderate samples). This produces a peaked distribution. In estimating probabilities of change to water supply, we used Monte Carlo methods (repeated random sampling) to sample the IPCC (2001a) range of globally warming for 2030, used these to scale a range of change per degree of global warming on a quarterly basis for P, then sampled Ep for that quarter based on the relationship shown in Figure 5. The quarterly changes for P and Ep were then totalled for annual δP and δEp. The transfer function in Equation 1 and Table 4 was then used to estimate the likelihood of various outcomes for Burrendong Dam storage, Macquarie Marshes inflows and to irrigation allocations.

The assumptions used in constructing the ranges of change were as follows:

1. The range of global warming in 2030 is 0.55–1.27°C and has a uniform distribution.
2. Changes in P are taken from the full range of change for each quarter from the sample of nine climate models (Table 5).
3. Changes in P for each quarter are independent of each other (seasonally dependent changes were tested in each of the climate models but no lag correlation between seasons could be found).
4. The difference between samples in any consecutive quarter could not exceed the largest difference observed in the sample of nine models (Table 5).
5. Ep is partially dependent on P ($\delta Ep = 5.75 - 0.53\delta P$ with a standard error of 2.00 randomly sampled using a Gaussian distribution).

Table 7: Rainfall in change per degree (%) for the Macquarie catchment from nine climate models, showing the resultant range and maximum difference from the previous quarter.

Model	Summer	Autumn	Winter	Spring
CSIRO Mk2	1.3	-0.2	-1.8	-4.9
CGCM1	0.2	3.8	-1.8	-6.1
HADCM3	-1.8	-1.3	-1.3	-11.8
ECHAM4	11.5	1.4	-0.2	-1.4
DARLAM125	4.8	0.4	-1.0	-2.1
DARLAM 60	4.9	3.0	-2.2	-3.3
ECHAM3	9.0	8.2	-4.6	1.6
GFDL-R15a	-0.7	5.9	1.2	-0.6
NCAR DOE	-2.5	-2.8	0.4	-4.8
Range	11.5 to – 2.5	8.2 to – 2.8	1.2 to – 4.6	1.6 to – 11.8

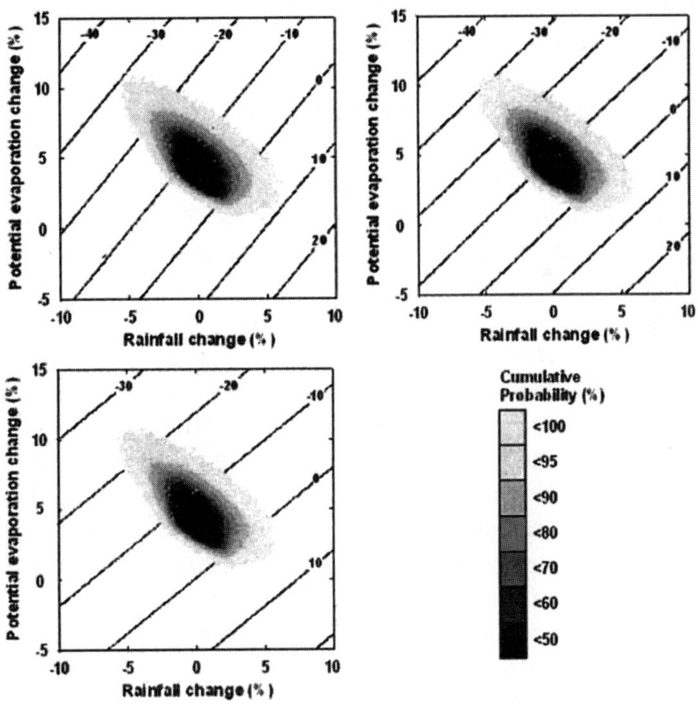

Figure 8: Risk response surfaces for Burrendong storage, Macquarie Marsh inflows and irrigation allocations for 2030 based on Monte Carlo sampling of δP and δEp for the Macquarie catchment within the IPCC (2001) ranges for temperature using the relationships described in Table 6.

One million samples were run for 2030, but the filter described in Assumption 4 cut this down to just under 700,000 samples. Figure 8 shows risk-response surfaces for the three major flow variables. These are tallied from the perspective of examining the most likely climate to the least likely, and the response is measured in cumulative percent. The dark core registers the most likely 50% of climates, while all the shades through to the lightest encompass the most likely 95% of all climates. The surfaces are Burrendong storage on the top left, Macquarie Marsh inflows on the top left and irrigation allocations on the lower left. The most likely climates tend to fall in the range of a small increase in flows to –20%. Note that the upper limits of this distribution (declines in flow) appear to be less than those in the lower panel of Figure 7, while the lower limit (flow increases) appears to give wetter flows than shown in Figure 8. The critical thresholds introduced in the previous section correspond to the –10% contours in a drought-dominated climate, –20% in a normal climate and –30% in a flood-dominated climate. The –10% contour intersects the most likely climate outcomes, whereas the –20% and –30% outcomes appear less likely. We investigate this question further in the next section.

These results are influenced by the sampling methodology acting in combination with the ranges of δP and δEp. The driest three of four quarters is produced by one model, HADCM3, but the wettest quarters are shared by a number of models (Table 5). By sampling quarterly, rather than six-monthly or annually, the likelihood of encountering the very dry outcome represented by the HADCM3 model is reduced. Wetter outcomes than represented by any single model, are more likely because three models share the wettest quarters. This sampling strategy selects against combinations of extreme outcomes from independent ranges of uncertainty. If in the real world these seasonal changes as simulated by the models are independent from one another, then the range of change in water supply is more likely to favour a moderate outcome rather than a very wet or very dry outcome.

How sensitive are these results to the input assumptions? Here global warming is assumed to be uniform across its range, but is also a combination of several ranges of uncertainty so that the extremes may be less likely than the more central estimates (Jones, 2000b; Schneider, 2001). Therefore, uniformity accentuates the extremes more than a peaked distribution such as that produced by Webster et al. (2001) or New and Hulme (2000). However, if the "real" distribution is biased towards low or high warming, then this may affect the results also. A distribution biased towards low global warming, not properly represented, could lead to the risk being overstated.

7.1 Climate variability and risk

As mentioned in the section on baseline climate, it is useful to distinguish between the different types of climate variability and their impacts on water supply to better understand how they may behave under climate change. For example, Figures 2 and 3 show the importance of decadal rainfall variability on water supply in the Macquarie Catchment, where drought-dominated and flood-dominated regimes can alter water supply by as much as ±20% from the long-term historical mean. Figure 3 shows that water supply during the two regimes is very different and can be partitioned according to decadal rainfall regime. Irrigation allocations are particularly noticeable, with only 4 years falling below the 50% allocation level during the flood-dominated period and 20 years during the drought-dominated period. In this section, we draw together our understanding of climate variability and change, as it is likely to affect water resources. The examples used are all from the Macquarie catchment but the implications are more widely relevant to many Australian catchments where water supply is highly allocated as a proportion of total streamflow (e.g. NSW EPA, 2000).

The vulnerability of water resources to climate is related to sequences of bad years where stress is cumulative, rather than to individual very wet or very dry years. Both the critical thresholds for irrigation demand and water supply into the Macquarie Marshes are based on sequences of years below a given water supply threshold. Current adaptation strategies have aimed to overcome inter-annual variability in water supply, and the success of irrigation agriculture is testament to this. However, the capping of water supply for the Murray-Darling Basin (Ref.) means that beyond the cap, demand cannot be met and in the years that demand is highest, supply is likely to be least. Furthermore, most of the development and allocation planning of water resources has taken place since 1948, during the most favourable rainfall regime post 1890.

The results above have been presented as a simulated change from the long-term mean from 1890–1996, but are quite different if partitioned into drought-dominated and flood-dominated periods. When the two critical thresholds relating to irrigation allocations and bird-breeding events are investigated, they both occur predominantly during the drought-dominated portion of the input climate data. For instance, the critical threshold for irrigation allocations is not reached in the flood-dominated period until the mean reductions exceed –30% to –35%. For bird breeding events in the Macquarie Marshes, the range is similar.

Both critical thresholds are a function of long-term low rainfall and a sequence of dry years. Therefore, the vulnerability of water supply to climate change in the Macquarie catchment is a combination of climate change, decadal climate variability and interannual climate variability. Climate change, expressed though mean changes in P and Ep, looks at best to maintain current water balance, and at worst to reduce water balance leading to over 30% reductions in water supply by 2030 if "normal" climate is used as a baseline. How decadal climate variability combines with these changes is critical. If rainfall is in a drought-dominated mode, then critical thresholds may be exceeded with changes in mean supply of –10%. If rainfall is close to the historical average, critical thresholds will be exceeded at about –20% in mean supply, and if in a flood-dominated mode, irrigation and the Macquarie Marshes, could possible adapt with reductions of up to –30% in mean supply. However, as both these thresholds have been chosen fairly conservatively, considerable hardship would be experienced first.

Figure 9 shows how the probability distributions in Figure 8 can be tallied in terms of "wettest" to "driest", i.e. from the lower left of Figure 8 to the upper right. According to the assumptions used in the Monte Carlo sampling, large increases in flow are possible, but not very likely. About 95% of the single event probabilities show that flow decreases are more likely. The most likely outcomes, where the probability distribution function is steepest, shows that the 10% to 90% outcomes for Burrendong Dam inflows range from a –3% to –20% change in flow. The Macquarie Marshes are slightly more sensitive than Burrendong storage. Irrigation allocations are more sensitive to increasing flows but less sensitive to decreasing flows.

Figure 9 combines this information the critical thresholds to estimate their exceedance under the assumptions applied in the risk analysis. In a drought-dominated climate the probability of the two thresholds being exceeded are 45% for the Macquarie Marshes and 30% for irrigation allocations. That is, under the assumptions we have applied to the risk assessment, if a drought-dominated climate is present in 2030, there is a marked probability chance of a climate change exceeding both critical thresholds. This would threaten the operation of the whole system in terms of both agricultural productivity and environment, but significant hardship would be felt even if those thresholds were approached. In a normal climate, mean water supply needs to reduce by 20% which appears likely in only 5% of climates. Exceeding the critical thresholds is possible in a flood-dominated climate, but very unlikely.

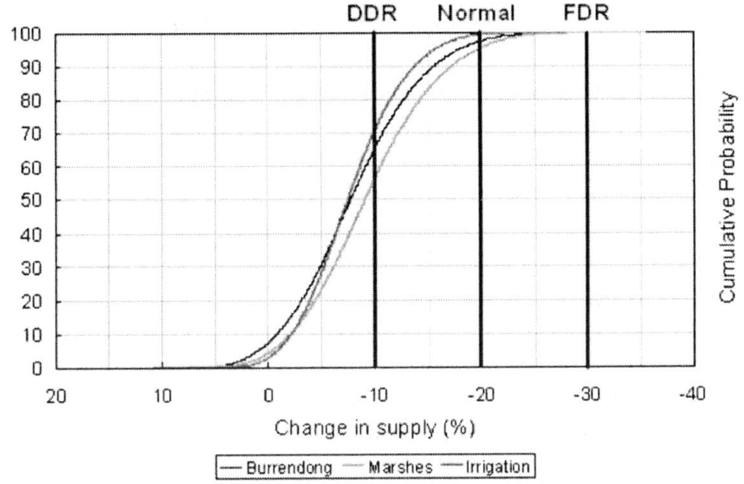

Figure 9: Cumulative probability diagram for Burrendong storage, Macquarie Marshes inflow and irrigation allocations in 2030 shown with changes in mean water supply needed to exceed critical thresholds under three modes of decadal rainfall variability. DDR is a drought-dominated regime and FDR, a flood dominated regime.

8. DISCUSSION

8.1

These results have brought together a number of recommendations from recent summaries or exploratory studies on water resources under climate change into a single project undertaking supply and demand relationships for a single catchment. Jones (1999, 2000a) explored the use of Monte Carlo methods for undertaking risk analysis on irrigation demand but could not couple this to supply, Hulme and Carter (1999) developed this theme but again did not have access to a working catchment model, and New and Hulme (2000) and Jones (2000b) showed how altering the input assumptions allowed Bayesian analysis of such input assumptions. Arnell (2000) recommends the use of important system thresholds for undertaking such analyses. This project is the first to bring together all these methods using an operational catchment management model.

The next steps are to present the results to stakeholders to determine whether the risks outlined here warrant further assessment of adaptation measures, conducting Bayesian analysis to test the robustness of the results under different inputs of uncertainty, to further examine the role of decadal rainfall variability and its influence on water resources, and to repeat this exercise for other catchments. Results combining the impacts of climate change with reductions in streamflow due to reforestation of the upper Macquarie catchment have already been presented to the Murray-Darling Basin Commission for consideration as part of its dryland salinity strategy. The outcome of that project has been to show that unplanned tree-planting in the upper reaches of the catchment could reduce streamflow, producing flow reductions under all combinations of reforestation and climate change (Herron et al., submitted). This indicates that the joint impacts of climate change, land-use change and natural resource use must be managed carefully and in an integrated manner.

Broader implications for water supply in the MDB, elsewhere in south-eastern and south-western Australia can be imagined by comparing δP and δEp in Figure 7 with the changes Australia-wide shown in Figure 1. In both south-eastern and south-western Australia these changes are likely to lead to reductions in water supply, especially as the decreases in P are larger in the southern halves of those two areas (except perhaps on the south coast of Victoria). These are consistent with studies based on single model outputs. Using the NCAR CCMO Kothavala (1999) found that the Palmer Drought Index showed longer and more severe drought in north-and south-eastern Australia. Based on the HadCM2 and HadCM3 models Arnell (1999) found marked decreases in runoff over most of mainland Australia, but some increases over Tasmania. For the Murray-Darling Basin he found decreases in mean flow by the 2050s ranging from about 12 to 35%, with decreases in the magnitude of both 10-year maximum and minimum monthly runoff.

This project was unable to uncover much extra information about the relationship between climate change and interannual variability as to how it may affect water resources. Interannual variability is subject to the influences of decadal variability, to the modulation of atmosphere-ocean decadal oscillations and to the behaviour of ENSO and other shorter lived phenomena. ENSO is very likely to be present in a greenhouse climate, and many climate models indicate a more ENSO-like mean state of the Pacific Ocean (IPCC, 2001a). It is not certain whether this ENSO-like pattern observed in these models is responsible for the simulated rainfall decrease. Given that many similar reductions are simulated elsewhere in the sub-tropics and mid-latitudes of continents, latitudinal changes in weather patterns are highly likely to be playing a part. Periods in the historical climate when several El Niños occurred closely together in a drought-dominated climate, e.g. 1914–1916 and 1939–1941, feature strongly in the exceedance of critical thresholds when further reduced by climate change. These sequences are currently unpredictable, so it is impossible to accurately quantify the risk of exceedance of critical threshold over the longer term. However, as ENSO seem likely to continue, if long-term decadal rainfall variability remains at or below the historical mean, such exceedances are quite probable. In regions where the capacity to deal with individual events is high, as it is in Australia, then current moves to implement medium range forecasting for predicting extreme events is a prudent form of risk management.

Given that current methods for diagnosing decadal modes of rainfall only work in hindsight, requiring extended series of historical data, it is very difficult to diagnose the current mode of decadal rainfall variability. South-eastern Australia has recently experienced several years of reduced rainfall which has stressed a number of systems. South-west Western Australia is currently in a thirty-year drought dominated climate (Allan and Haylock, 1993), which has caused a significant degree of water stress, and is projected to become drier under climate change (CSIRO, 2001). The results of this project show that the mode of decadal variability is extremely important for assessing the risk of exceedance of critical thresholds. In conclusion, knowledge of what the current mode of long-term variability is, and what it may be in the future, is extremely important for understanding the likely risks of climate change. While research into short-term rainfall variability is fairly well advanced, we know very little about the long-term.

9. CONCLUSIONS

This project provides results of an impact modelling and risk assessment exercise for the Macquarie River, a catchment within the eastern portion of the Murray-Darling Basin in Australia. By investigating changes to P and Ep that are consistent with IPCC (2001) global warming projections and recently released regional projections for Australia (CSIRO, 2001), we show that there is a significant risk to water resources in the catchment. Two critical thresholds tied to irrigation allocations and environmental flows have greater than a 1 in 3 chance of being exceeded by 2030, if a drought-dominated rainfall regime coincides with a climate change occuring within the range expected. If baseline rainfall is normal, the likelihood that critical thresholds will be exceeded by 2030 is less than approximately 5%. The most likely range of change in mean water supply in 2030 (between a probability of 10% and 90%) is 0 to –15%. Even if the climate regime resembled that of the early 20th century with little change due to greenhouse, we would expect irrigation allocations of <50% of the water right in up to 4 out of every 10 years.

This project shows that long-term variations in rainfall are very important for water resource planning, switching between modes in ways that are poorly understood. This is counter to the current paradigm of investigating secular trends in historical rainfall, and some of the phenomena attributed to trends in the past may in fact be due to these phase changes. Recent reductions in winter rainfall in south-western Australia have led to AU$275 million being spent on drought-proofing for Perth (Boer, 2000). Therefore, understanding the dynamics of decadal rainfall variability and the relationship between decadal variability and secular trends are both valuable short to medium research goals.

Based on climate projections, any reductions in water supply experienced by the Macquarie catchment are likely to be more widespread, threatening water supply in south-eastern and south-western Australia. Climate projections show that reductions in winter–spring rainfall combined with increased evaporation will lead to reduced flows, unless summer rainfall increases can compensate. In the Macquarie, this compensation occurs in only two of the models we examined, both from the same laboratory. Therefore, planned adaptations to implement market reform, ameliorate salinity, make agriculture more productive, rehabilitate degraded lands and to protect natural habitat should factor in management for reduced rainfall and water supply and increased water demand. Our assessment also shows that system failure, measured by two critical thresholds for irrigation allocations and for environmental flows is possible by 2030. Even if we face a drying trend that may not exceed these critical thresholds by 2030, it is possible they could be exceeded at a later date. Further research is needed to undertake Bayesian analysis to test the robustness of these results, to expand the analysis to other catchments, to better understand decadal rainfall variability, to define critical thresholds more accurately along with communication of the risks to the community so that they can commence adaptation.

10. REFERENCES

Allan, R.J. (2000) ENSO and climate variability in the last 150 years, in Diaz, H. F. and Markgraf, V. (eds), *El Niño and the Southern Oscillation: multiscale variability and global regional impacts*, Cambridge University Press, Cambridge, 3–55.

Allan, R.J., and Haylock, M.R. (1993) Circulation features associated with the winter rainfall decrease in southwestern Australia, *Journal of Climate*, **6**, 1356–1367.

Arnell, N.W. (1999) Climate change and global water resources, *Global Environmental Change*, **9**, S31–S46.

Arnell, N.W. (2000) Thresholds and response to climate change forcing: The water sector, *Climatic Change*, **46**, 305–316.

Arnell, N.W. and Liu, C. (2001) Hydrology and water resources, in McCarthy, J.J., Canziani, O.F., Leary, N.A., Dokken, D.J. and White, K.S. (eds.) *Climate Change 2001: Impacts, Adaptation, and Vulnerability*, Contribution of Working Group II to the Third Assessment Report of the Intergovernmental Panel on Climate Change, Cambridge University Press, Cambridge, 191–233.

Bouchet, R.J. (1963) Evapotranspiration reelle et potentielle, signification climatique, *Int. Assoc. Sci. Hydrol. Pub.*, **62**, 134–142.

Burnash, R.J.E., Ferral, R.L. and McGuire, R.A. (1984) *A Generalised Streamflow Simulation System*, Joint Federal-State River Forecast Centre, Sacramento, California.

Cai, W., Baines, P.G. and Gordon, H.B. (1999) Southern mid-to-high latitude variability, a zonal wavenumber 3 pattern, and the Antarctic Circumpolar Wave in the CSIRO coupled model, *J.Climate*, **12**, 3087–3104.

Chiew, F.H.S., Stewardson, M.J. and McMahon, T.A. (1993) Comparison of six rainfall–runoff modelling approaches, *J. Hydrol.*, **147**, 1–36.

Chiew, F.H.S., Whetton, P.H., McMahon, T.A. and Pittock, A.B. (1995) Simulation of the impacts of climate change on runoff and soil moisture in Australian catchments. *J. Hydrol.*, **167**, 121-147.

Chiew, F.H.S., Piechota, T.C., Dracup, J.A. and McMahon, T.A. (1998) El Niño Southern Oscillation and Australian rainfall, streamflow and drought – links and potential for forecasting, *Journal of Hydrology*, **204**, 138–149.

Collins, D.A. and Della-Marta, P.M. (1999) Annual climate summary 1998: Australia's warmest year on record, *Australian Meteorological Magazine*, **48**, 273-383.

CSIRO (1996) *Climate Change Scenarios for the Australian Region*, Climate Impact Group, CSIRO Division of Atmospheric Research, Melbourne, 8 pp.

CSIRO (2001) *Climate Change Projections for Australia*, Climate Impact Group, CSIRO Division of Atmospheric Research, Melbourne, 8 pp.

DLWC (1995a) *Integrated Quantity-Quality Model (IQQM) Reference Manual*, NSW DLWC, Sydney.

DLWC (1995b) *IQQM – Macquarie River System Calibration Report,* TS94.041, NSW DLWC, Sydney.

DLWC and NPWS (1996) *The Macquarie Marshes Water Management Plan*, DLWC and NPWS, Sydney, 40pp.

DKRZ-Model User Support Group (eds) (1992) ECHAM3 – Atmospheric General Circulation Model. Report No. 6, Deutsches Klimarechenzentrum, Hamburg. http://www.dkrz.de/forschung/reports/ReportNo.6.ps

NSW EPA (1995) New South Wales State of the Environment 1995, NSW Environment Protection Authority, Chatswood, http://www.epa.nsw.gov.au/soe/95/index.asp

NSW EPA (2000) New South Wales State of the Environment 2000, NSW Environment Protection Authority, Chatswood, http://www.epa.nsw.gov.au/soe2000

Finlayson, B. and McMahon, T.A. (1988) Australia v. the World: A comparative analysis of streamflow characteristics, in Warner, R.F. (Ed.) *Fluvial Geomorphology of Australia*, Academic Press Australia, Sydney, 17–40.

Flato, G.M., Boer, G.J., Lee, W.G., McFarlane, N.A., Ramsden, D., Reader, M.C. and Weaver, A.J. (2000) The Canadian Centre for Climate Modelling and Analysis global coupled model and its climate, *Climate Dynamics*, **16**, 451-467.

Flohn, H. (1986) Singular events and catastrophes now and in climatic history, *Naturwissenschaften*, **73**, 136–149.

Fu, C.B. and Wen, G. (1999) Variation of ecosystems over East Asia in association with seasonal, interannual and decadal monsoon climate variability, *Climatic Change*, **43**, 477–494.

Foreman, I., Fritsch, S. and Woldring, H. (1998) Consequences for the agricultural economy, In Hassall and Associates, *Climate Change Scenarios and Managing the Scarce Water Resources of the Macquarie River*, Report prepared for the Australian Greenhouse Office, Canberra, 46–60.

Gordon, H.B. and O'Farrell, S.P. (1997) Transient climate change in the CSIRO coupled model with dynamic sea ice, *Mon. Weather. Rev.*, **125**, 875—907.

Hassall and Associates (eds.) (1998) *Climate Change and Managing the Scarce Water Resources of the Macquarie River*, Australian Greenhouse Office, Canberra, 113 pp.

Haywood, J.M., Stouffer, R.J., Wetherald, R.T., Manabe, S. and Ramaswamy, V. (1997) Transient response of a coupled model to estimated changes in greenhouse gas and sulfate concentrations, *Geophys. Res. Lett.*, **24**, 1335.

Hennessy, K.J., Suppiah, R., & Page, C.M. (1999) Australian rainfall changes, 1910-1995, *Aust. Met. Mag.*, **48**, 1–13.

Herron, N., Davis, R. and Jones, R.N. (2001) The effects of large-scale afforestation and climate change on water allocation in the Macquarie River Catchment, NSW, Australia, Journal of Environmental Management (submitted).

Hulme, M. and Carter, T. R. (1999) Representing uncertainty in climate change scenarios and impact studies, in Carter, T. R., Hulme, M. and Viner, D. (eds.), *Representing Uncertainty in Climate Change Scenarios and Impact Studies*, ECLAT-2 Workshop Report Number 1, Climatic Research Unit, Norwich, UK, 11–37.

Hunt, B.G. (2000). Natural climatic variability and Sahelian rainfall trends, *Global and Planetary Change*, **24**, 107–131.

IPCC (1996) Technical Summary, in Houghton, J.T., Meira Filho, L.G., Callander, B.A., Harris, N., Kattenberg, A. and Maskell, K. (eds.), *Climate Change 1995: The Science of Climate Change*, Contribution of Working Group I to the second assessment report of the IPCC, Cambridge University Press, Cambridge, 9–49.

IPCC (2001a) Summary for Policymakers, in Ding, Y. et al. (eds.) *Climate Change 2001: The Scientific Basis*, Contribution of Working Group I to the Third Assessment Report of the Intergovernmental Panel on Climate Change, Cambridge University Press, Cambridge, 944 pp.

IPCC (2001b) Summary for Policymakers, in McCarthy, J.J., Canziani, O.F., Leary, N.A., Dokken, D.J. and White, K.S. (eds.) *Climate Change 2001: Impacts, Adaptation, and Vulnerability*, Contribution of Working Group II to the Third Assessment Report of the Intergovernmental Panel on Climate Change, Cambridge University Press, Cambridge.

IPCC-TGCIA (1999) *Guidelines on the Use of Scenario Data for Climate Impact and Adaptation Assessment.* Version 1. Prepared by Carter, T. R., Hulme, M. and Lal, M., IPCC Task Group on Scenarios for Climate Impact Assessment, 69pp.

Johns, T. C., Carnell, R. E., Crossley, J. F., Gregory, J. M., Mitchell, J. F. B., Senior, C. A., Tett, S. F. B. and Wood, R. A. (1997) The second Hadley Centre coupled ocean-atmosphere GCM: model description, spinup and validation, *Climate Dynamics*, **13**, 103–134.

Johnson, W. (1998) Consequences for the Macquarie Marshes, In Hassall and Associates, *Climate Change Scenarios and Managing the Scarce Water Resources of the Macquarie River*, Report prepared for the Australian Greenhouse Office, Canberra, 61–68.

Jolly, I. D., Williamson, D. R., Gilfedder, M., Walker, G. R., Robinson, G., Jones, H., Zhang, L., Dowling, T. I., Dyce, P., Nathan, R. J., Nandakumar, N., Clarke, R. and McNeill, V. (2001) Historical stream salinity trends and catchment salt balances in the Murray-Darling Basin, Australia, *Mar. Freshwater Res.*, **52**, 53–63

Jones, R.N. (1995) *Modelling the Hydrologic and Climatic Controls of Closed lakes, Western Victoria*. Ph. D. Thesis, University of Melbourne (unpublished), 283 pp.

Jones, R.N. (1999) Climate change impact, thresholds and risk, in Howden, S.M and Gorman, J.T. (eds.) Impacts of Global Change on Australian Temperate Forests, Working Paper Series 99/08, Resource Futures Program, CSIRO Wildlife and Ecology Canberra, 40–52.

Jones, R.N. (2000a) Analysing the risk of climate change using an irrigation demand model, *Climate Research*, **14**, 89–100.

Jones, R.N. (2000b) Managing uncertainty in climate change projections – issues for impact assessment, *Climatic Change*, **45**, 403–419.

Jones, R.N. (2001) An environmental risk assessment/management framework for climate change impact assessments, *Nat. Hazards*, **23**, 197–230.

Jones, R.N., Bowler, J.M. and MacMahon, T.A. (1998) A high resolution Holocene record of P/E ratio from closed lakes in Western Victoria. *Palaeoclimates*, **3**, 51–82.

Jones, R.N. and Page, C.M. (2001) OzClim: a climate scenario generator for Australia, CSIRO Atmospheric Research, Melbourne, CD-RoM.

Jones, R.N., McMahon, T.A. and Bowler, J.M. (2001) Modelling historical lake levels and recent climate change at three closed lakes, Western Victoria, Australia (c.1840–1990), *J. Hydrol.*, **246**, 158–179.

Kingsford, R.T. (1999) Aerial survey of waterbirds on wetland as a measure of river and flood plain health, *Freshwater Biology*, **41**, 425–438.

Kingsford, R.T. and Thomas, R.F. (1995) The Macquarie Marshes in arid Australia and their waterbirds: a 50 year history of decline. *Environmental Management*, **19**, 867–878.

Kingsford, R.T. and Johnson, W. (1998) Impact of water diversions on colonially-nesting waterbirds in the Macquarie Marshes of arid Australia, *Colonial Waterbirds*, **21**, 159–170.

Kraus, E.B. (1955) Secular changes of east-coast rainfall regimes, *Quarterly Journal of the Royal Meteorological Society*, **81**, 430–439.

Kothavala, Z. (1999) The duration and severity of drought over eastern Australia simulated by a coupled ocean-atmosphere GCM with a transient increase in CO_2, *Environmental Modelling & Software*, **14**, 243–252.

Lavery, B., Joung G. and Nicholls, N. (1998) An extended high-quality historical rainfall dataset for Australia. *Aust. Meteorol. Mag.*, **46**, 27–38.

Leighton R.M., Keay, K., and Simmonds, I. (1997) Variation in annual cyclonicity across the Australian region for the 29-year period 1965–1993 and the effect on annual all-Australia rainfall, In *Proceedings Workshop on Climate Prediction for Agriculture and Resource Management*, Bureau of Resource Sciences, Canberra.

Lemly, A.D, Kingsford, R.T. and Thompson, H.R. (2000) Irrigated agriculture and wildlife conservation: conflict on a global scale, *Environmental Management*, **25**, 485–512.

Maier-Reimer, E. and Mikolajewicz, U. (1991) The Hamburg Large Scale Geostrophic Ocean General Circulation Model (Cycle 1); Deutsches Klimarechenzentrum (DKRZ), http://www.dkrz.de/forschung/reports/ReportNo.2.ps

McGregor, J. L. and Katzfey, J. J. (1997) Climate modelling for the Australian region using DARLAM, *Proceedings of the International Workshop on Regional Modeling of the "General Monsoon System" in Asia*, Beijing, China (TEACOM Report, 4), START Regional Centre for Temperate East Asia, 13–18.

MDBMC: 1999, *The Salinity Audit of the Murray-Darling Basin*. Murray-Darling Basin Commission, Canberra, 39 pp. http://www.mdbc.gov.au/Download/SalinityAudit/Final_Salt_Audit.pdf

Mearns, L.O. and Hulme, M. (2001) Climate scenario development, in Ding, Y. et al. (eds.) Climate Change 2001: The Scientific Basis, Contribution of Working Group I to the Third Assessment Report of the Intergovernmental Panel on Climate Change, Cambridge University Press, Cambridge

Morton, F.I. (1983) Operational estimates of areal evapotranspiration and their significance to the science and practice of hydrology, *J. Hydrol.*, **66**, 1–76.

Nicholls, N., Gruza, G.V., Jouzel, J., Karl, T.R., Ogallo, L.A. and Parker, D.E. (1996) Observed Climate Variability and Change, In J.T. Houghton, L.G. Meira Filho, B.A. Callander, N. Harris, A. Kattenberg and K. Maskell (eds.) *Climate Change 1995: The Science of Climate Change, Contribution of Working Group 1 to the Second Scientific Assessment Report of the Intergovernmental Panel on Climate Change*, Cambridge University Press, Cambridge, 137–192.

Nicholls, N., Landsea, C. and Gill, J., 1998: Recent trends in Australian region Tropical Cyclone activity, *Meteorol. Atmos. Phys.*, **65**, 197–205.

NSW EPA (1995) *New South Wales State of the Environment 1995*, Environmental Protection Authority, Chatswood NSW.

NSW EPA (2000) *New South Wales State of the Environment 2000*, Environmental Protection Authority, Chatswood NSW.

Oberhuber, J. M. (1992) The OPYC Ocean General Circulation Model, Report No. 7, Deutsches Klimarechenzentrum, Hamburg. http://www.dkrz.de/forschung/reports/ReportNo.7.ps

Pittock, A.B. (1975) Climate change and the pattern of variation in Australian rainfall, *Search*, **6**, 498–503.

Potter, K.W. (1981) Illustration of a new test for detecting a shift in mean in precipitation series. *Mon. Wea. Rev.*, **109**, 2040–2045.

Power, S., Casey, T., Folland, C., Colman, A. and Mehta, V. (1999a) Decadal modulation of the impact of ENSO on Australia, *Climate Dynamics*, **15,** 319–324.

Power, S., Tseitkin, F., Mehta, V., Lavery, B., Torok, S., and Holbrook, N. (1999b) Decadal climate variability in Australia during the 20th century, *International J. Climatology*, **19**, 169–184.

Radford, D., Blong, R., d'Aubert, A.M., Kuhnel, I. and Nunn, P. (1996) *Occurrence of Tropical Cyclones in the Southwest Pacific Region 1920-1994*, Greenpeace International, Amsterdam, Netherlands, 35 pp.

Schneider, S.H. (2001) What is 'dangerous' climate change? *Nature*, **411**, 17–19.

Smith, D.I. (1999) *Water in Australia: Resources and Management*, Oxford University Press, Melbourne, 384 pp.

Smith, I.N., McIntosh, P., Ansell, T.J., Reason, C.J.C. and McInnes, K.L. (2001) South-west Western Australia rainfall and its association with Indian Ocean climate variability, *International Journal of Climatology,* in press.

Walsh, K.J.E., Hennessy, K.J., Jones, R.N., Pittock, A.B., Rotstayn, L., Suppiah, R. and Whetton, P.H. (2000) *Climate change in Queensland under enhanced greenhouse conditions: second annual report, 1998–1999* , CSIRO Atmospheric Research, Melbourne, 130 pp.

Walsh, K.J.E., Hennessy, K.J., Jones, R.N., McInnes, K.L., Page, C.M., Pittock, A.B., Suppiah, R. and Whetton, P.H. (2001) *Climate change in Queensland under enhanced greenhouse conditions: third annual report, 1999–2000*, CSIRO Atmospheric Research, Melbourne, 108 pp.

Wang, Q.J., McConachy, F.L.N., Chiew, F.H.S., James, R., de Hoedt, G.C. and Wright, W.J. (2001). Maps of Evaporation. In: *Climatic Atlas of Australia*, Bureau of Meteorology, Melbourne.

Warner, R.F. (1987) The impacts of alternating flood- and drought-dominated regimes on channel morphology at Penrith, New South Wales, Australia, *IAHS Publ.*, **68**, 327–338.

Washington, W.M., Weatherly, J.W., Meehl, G.A., Semtner Jr., A.J., Bettge, T.W., Craig, A.P., Strand Jr., W.G., Arblaster, J.M., Wayland, V.B., James, R., and Zhang, Y. (2000) Parallel climate model (PCM) control and transient simulations, *Climate Dynamics*, **16**, 755–774.

Webster, M.C., Forest, C.E., Reilly, J.M., Sokolov, A.P., Stone, O.H., Jacoby, H.D. and Prinn, R.G. (2001) Uncertainty analysis of global climate change projections, MIT Joint Program on the Science and Policy of Global Change Report No. 73, Massachussets Institute of Technology, Cambridge, 3pp.

Whetton, P.H., Adamson, D. and Williams, M.A.J. (1990) Rainfall and river flow variability in Africa, Australia, and East Asia linked to El Niño – Southern Oscillation events, *Geological Society of Australia Symposium Proceedings*, **1**, 71–82.

Yu, B. and Neil, T.D. (1993) Long-term variations in regional rainfall in the south-west of Western Australia and the difference between average and high-intensity rainfall, *Internat. J. of Climatology*, **13**, 77–88.